AF386613

Indexed by Scopus

The series Springer Proceedings in Physics, founded in 1984, is devoted to timely reports of state-of-the-art developments in physics and related sciences. Typically based on material presented at conferences, workshops and similar scientific meetings, volumes published in this series will constitute a comprehensive up to date source of reference on a field or subfield of relevance in contemporary physics. Proposals must include the following:

- Name, place and date of the scientific meeting
- A link to the committees (local organization, international advisors etc.)
- Scientific description of the meeting
- List of invited/plenary speakers
- An estimate of the planned proceedings book parameters (number of pages/articles, requested number of bulk copies, submission deadline).

Please contact:

For Americas and Europe: Dr. Zachary Evenson; zachary.evenson@springer.com
For Asia, Australia and New Zealand: Dr. Loyola DSilva; loyola.dsilva@springer.com

Aaron E. Craft · Hassina Z. Bilheux
Editors

Proceedings of the 12th World Conference on Neutron Radiography

WCNR-12, June 2–7, 2024, Idaho Falls, Idaho, USA

Editors
Aaron E. Craft
Idaho National Laboratory
Idaho Falls, ID, USA

Hassina Z. Bilheux
Spallation Neutron Source
Oak Ridge National Laboratory
Oak Ridge, TN, USA

ISSN 0930-8989   ISSN 1867-4941   (electronic)
Springer Proceedings in Physics
ISBN 978-3-032-15002-8   ISBN 978-3-032-15003-5   (eBook)
https://doi.org/10.1007/978-3-032-15003-5

This Springer imprint is published by the registered company Springer Nature Switzerland AG
The registered company address is: Gewerbestrasse 11, 6330 Cham, Switzerland

If disposing of this product, please recycle the paper.

Preface

We are pleased to present the Proceedings for the 12th World Conference on Neutron Radiography (WCNR-12), which was held in Idaho Falls, Idaho, USA, June 2–7, 2024. Hosted by Idaho National Laboratory (INL) and co-organized with Oak Ridge National Laboratory, WCNR-12 continues the tradition of a quadrennial conference series that began in San Diego, California, in December 1981, alternating among the United States, Japan, South Africa, and Europe, with the most recent one (WCNR-11) held in Sydney, Australia, in 2018. Originally scheduled in 2022, WCNR-12 was postponed by two years due to the COVID-19 pandemic.

In keeping with tradition, WCNR-12 convened distinguished scientists, researchers, and practitioners from around the world to share their expertise, latest innovations and advancements in the field of neutron imaging. The global winds brought together participants from large-scale user facilities such as reactors and spallation neutron sources, national laboratory and university reactor facilities, domain scientists from academia, and researchers from national laboratories and industry.

Neutron imaging is a powerful non-destructive and non-invasive technique which has advanced significantly in a variety of scientific and industrial domains. At continuous sources, the method provides detailed insights into objects relating to materials science, in particular energy materials, nuclear materials, advanced manufacturing, engineering, archaeometry, biology, plant physiology, etc. Pulsed sources provide techniques such as energy-resolved imaging (e.g., Bragg edge and resonance imaging), high temporal resolution for dynamic processes, and contrast tuning using wavelength/energy selection. By leveraging the time-of-flight nature of pulsed neutron beams, imaging facilities provide spatially resolved information about crystallographic structure, strain, and elemental composition. This makes pulsed neutron imaging particularly powerful for studying complex materials such as those made by additive manufacturing, nuclear energy materials, and engineering materials, to name a few.

We extend our sincere gratitude to all authors, reviewers, and session chairs for their immense contributions to the scientific breadth and quality of this conference. We also thank the Scientific Advisory Board, Local Organizing Committee, and our sponsors for their tireless efforts in making WCNR-12 a success.

These proceedings, published by Springer Nature in the *Springer Proceedings in Physics* series, reflect the diversity and depth of research presented at WCNR-12. We hope they will serve as a valuable resource for the neutron imaging community and newcomers alike and inspire continued innovation and collaboration.

We look forward to seeing you at WCNR-13 and hope the winds of change bring more groundbreaking research!

Aaron E. Craft
Chairman, WCNR-12

Hassina Z. Bilheux
Co-chairwoman, WCNR-12

WCNR-12

12th World Conference for Neutron Radiography

On June 2–7, 2024, around 130 providers and users of neutron imaging converged from 30 countries to meet in Idaho Falls, Idaho, in the United States for WCNR-12.

WCNR-12 included geographically diverse colleagues from Algeria, Australia, Austria, Bangladesh, Brazil, Burundi, Canada, China, Czechia, Denmark, France, Germany, India, Indonesia, Iran, Israel, Italy, Japan, Netherlands, Pakistan, Philippines, Poland, South Korea, Spain, Sweden, Switzerland, Taiwan, Thailand, the United Kingdom, and the United States.

The conference was hosted by Idaho National Laboratory (INL), which was established in 1949 originally as the National Reactor Testing Station to serve as the United States' proving ground for first-of-a-kind nuclear reactors for peaceful uses of nuclear energy. In INL's storied history, 52 nuclear reactors have been built at the INL site west of Idaho Falls. Four reactors continue operating at INL, including the Advanced Test Reactor (ATR), the Advanced Test Reactor Critical (ATRC) facility, the Transient Reactor Test (TREAT) facility, and the Neutron Radiography Reactor (NRAD). Neutron imaging has been used at INL for over 60 years, and neutron imaging efforts continue using NRAD and TREAT reactors, primarily for post-irradiated examination of nuclear fuels and materials.

The conference opened with a keynote presentation from Dr. Joseph Bevitt titled *Neutron Tomography and the Virtual World of Palaeontology*. Technical presentations spanned the full range of the neutron imaging field, including sessions on nondestructive testing and standards, facility overviews, advanced techniques, material science applications, engineering applications, facility upgrades, software and machine learning, energy and environmental applications, new facilities, nuclear engineering applications, cultural heritage, detectors, neutron grating interferometry and dark field imaging, and neutron optics. A total of 65 oral presentations were given in a single auditorium between Monday and Thursday. A total of 72 poster presentations were shared in two lively sessions following lunch on Monday and Tuesday.

The American Society for Testing and Materials International (ASTM International) Subcommittee E07.05 that develops standards for neutron radiography held the semi-annual meeting co-located at WCNR-12. Notably, the subcommittee is pursuing new standards focused on digital neutron imaging techniques, and the meeting at WCNR-12 enabled broader engagement from the community than typically available.

The conference also included a special plenary session focused on the exciting new VENUS instrument at the Spallation Neutron Source at Oak Ridge National Laboratory, which included presentations on state-of-the-art technologies and approaches, representing in many ways a culmination of the technical developments and lessons learned in the nearly 70-year history of neutron imaging.

The conference dinner was held Thursday evening at the Experimental Breeder Reactor 1 (EBR-I), a national historical landmark known as the world's first nuclear power plant. Attendees had the unique opportunity to tour EBR-I, gaining insights into its historical significance and groundbreaking contributions to nuclear energy. However, the evening took an unexpected turn when a fierce windstorm swept through, causing quite a gust of excitement. Despite the blustery conditions, spirits remained high as participants weathered the storm, making the dinner event a whirlwind of memorable moments. Conversations were filled with a breath of fresh air, as the wind of change seemed to blow through the discussions. The unruly weather couldn't dampen the enthusiasm of the attendees, who breezed through the evening with remarkable resilience. In the end, the storm provided an atmospheric backdrop to an otherwise illuminating and breezy gathering at EBR-I.

Forty-four attendees toured Idaho National Laboratory on Friday, June 7, 2024. Three groups rotated between the TREAT facility, the Hot Fuel Examination Facility where the Neutron Radiography Reactor is located, and the Irradiated Materials Characterization Laboratory.

Local Organizing Committee

Aaron Craft, Co-chairman
Hassina Bilheux, Co-chairwoman
Tandy Bales, Co-organizer
Lisa Wells, Co-organizer

Scientific Advisory Board

Jean Bilheux, USA
Thomas Bücherl, Germany
Ulf Garbe, Australia
Dan Hussey, USA
Winfried Kockelmann, UK
Yasushi Saito, Japan
Floriana Salvemini, Australia

Javier Santisteban, Argentina
Burkhard Schillinger, Germany
Takenao Shinohara, Japan
Markus Strobl, Switzerland
Alessandro Tengattini, France
Anton Tremsin, USA
Pavel Trtik, Switzerland

Exhibitors

Amsterdam Scientific Instruments
Phoenix Neutron Imaging
Radiation Monitoring Devices (RMD)
McClellan Nuclear Research Center (MNRC)/University of California – Davis
Nray Services
Photonis
RC Tritec

ISNR

International Society for Neutron Radiography

The International Society for Neutron Radiography (ISNR) was established to foster and actively promote the field of neutron imaging, primarily by facilitating the free exchange of neutron imaging developments from the worldwide community of providers and users of neutron imaging. The ISNR accomplishes its mission primarily through organizing two conference series: The World Conference for Neutron Radiography (WCNR) and the International Topical Meeting for Neutron Radiography (ITMNR). The history, organization, elected members, and other information can be found at the ISNR webpage (*currently* www.isnr.de).

ISNR officers and board members are democratically elected by the WCNR attendees. Accordingly, elections were held at WCNR-12, and the new members are listed below. The previous President and host of the current WCNR, Dr. Aaron Craft in this case, serves as a de-facto member of the ISNR Board through the next WCNR.

Most notably, the President, Dr. Michael Schulz, will be hosting the next World Conference, WCNR-13, in Munich, Germany in 2028.

President
Michael Schulz, Germany
Vice-president
Markus Strobl, Switzerland

Secretary
Tobias Neuwirth, Germany

Board Members
Hassina Bilheux, USA
Aaron Craft, USA
Daniel Hussey, USA
Anders Kaestner, Switzerland
Winfried Kockelmann, UK
Burkhard Schillinger, Germany
Takenao Shinohara, Japan
Alessandro Tengattini, France
Anton Tremsin, USA
Pavel Trtik, Switzerland

The next ISNR conference is ITMNR-10, to be hosted by Dr. Alessandro Tengattini (Institut Laue–Langevin, ILL) in Grenoble, France, in 2026.

Fig. 1. Attendees of the 12th World Conference on Neutron Radiography standing on the west bank of the Snake River in Idaho Falls

Fig. 2. Conference attendees enjoyed an excursion to Yellowstone National Park on Sunday, June 2, 2024. This group photo was taken along the Madison River at the west entrance

Fig. 3. Collage of conference attendees

Fig. 4. Dr. Michael Schulz fielding questions from the audience

Fig. 5. Dr. Tobias Neuwirth was awarded Best Poster for his work on "Study of failure mechanisms in additively manufactured steel samples using grating interferometry and Bragg edge
imaging"

Fig. 6. Dr. Mohammad Samin Nur Chowdhury was awarded with the Best Presentation for his work on "Fast hyperspectral reconstruction for neutron tomography using subspace extraction." His efforts reduced the number of projections for a tomography data set from 1,200 radiographs to just nine while maintaining the same image quality

Fig. 7. (Left) The first Karel and Illona Trtikovi Memorial Award for Best Presentation, which comes with a monetary award from Dr. Pavel Trtik in honor of his parents' memory, and (Right) Dr. Samin Chowdhury presented with his award on a bus after retreating from the conference dinner at EBR-I

Fig. 8. ISNR Honorary Membership, the highest award given by the Society, was given to two deserving members: Prof. Nobuyuki Takenaka, Emeritus Professor of Kobe University, and Dr. Thomas Bücherl from the Technical University Munich. (Left) ISNR Honorary Member Eberhard Lehmann presenting the Honorary Membership Award to Prof. Nobuyuki Takenaka (unable to attend in person), accepted by Professor Hitoshi Asano. (Right) ISNR President Aaron Craft presenting the Honorary Membership Award to Thomas Bücherl

Fig. 9. Attendees visiting the Hot Fuel Examination Facility at INL's Materials & Fuels Complex, the world's largest inert-atmosphere hot cell, which is the starting point for post-irradiation examination of nuclear fuels. The NRAD reactor beneath the hot cell is used for neutron imaging of irradiated nuclear fuels

Fig. 10. Participants enjoying the TREAT facility, a pulsing reactor at INL's Materials & Fuels Complex where experimental nuclear fuels are tested at powers up to 19,000 MW to determine how they perform in over-powered and under-cooled conditions, ultimately informing the safe operating envelope for new nuclear fuels

Contents

Time-Series Neutron Tomography of Rhizoboxes 1
A. P. Kaestner, S. Di Bert, R. Jia, P. Benard, and A. Carminati

Bimodal Neutron and X-ray Imaging and Fusion 10
*M. Shakoorioskooie, E. Lehmann, M. Strobl, Q. Zhan, P. Trtik,
M. Krieg, D. Mannes, and A. Kaestner*

Limits in the Neutron Imaging Quantification of Hydrogenous Materials 19
Eberhard Lehmann and Mahdieh Shakoorioskooie

How to Perform Cultural Heritage Research by Using Neutron Imaging?
Considerations Based on PSI's Experiences 25
Eberhard Lehmann and David Mannes

Cr-coated Zircaloy-4 Performance During Ramping Conditions 32
*Aaron W. Colldeweih, David W. Kamerman, Malachi Nelson,
Aaron Craft, Nathan Capps, Caleb Massey, Mackenzie Ridley,
Pavel Trtik, Michael Meyer, and Okan Yetik*

Shapes of Mobile Interfaces in Tubes Revealed by Neutron Imaging:
Sensitivity Analysis for Three-Phase High-Pressure Systems
from Methane, *p*-xylene, and Water 40
*Ondřej Vopička, Tereza-Markéta Durďáková, Petr Číhal, Pierre Boillat,
Jongmin Lee, Martin Melčák, Jonatan Šercl, Štěpán Tvrdý, Jan Heyda,
and Pavel Trtik*

Measuring the Influence of Pulsed Injection on the Liquid Fraction
Distribution in Foam Using Neutron Radiography 47
*Friedrich Walzel, Muhammad Ziauddin, Leon Knuepfer,
Tobias Lappan, Pavel Trtik, and Sascha Heitkam*

Application of Monte Carlo-Based Surrogate Model for Laser-Driven
Neutron Radiography Source Optimization 55
*D. P. Broughton, O. F. Erdem, C.-K. Huang, S. H. Batha,
C.-S. Wong, R. E. Reinovsky, T. R. Schmidt, Z. Wang, B. T. Wolfe,
M. Alvarado Alvarez, and A. Junghans*

Lithium-Containing Semiconductors for Neutron Imaging 64
  M. A. Benkechkache, E. Hoegberg, J. Gallagher, R. Golduber,
  A. Kargar, H. Hong, J. Christian, M. Kanatzidis, E. Qian, K. Saheb,
  S. Imam, and E. Lukosi

Noise Reduction in Neutron Tomography 76
  Qianru Zhan, Mahdieh Shakoorioskooie, David Mannes,
  and Anders Kaestner

Spatially Resolved Evaluation of Texture Variations in Cold-Rolled
Zircaloy-4 Cladding by Bragg Edge Imaging 84
  Florencia Malamud, Miguel A. Vicente Alvarez, Valentina Stella,
  Javier R. Santisteban, and Markus Strobl

Monte Carlo Simulation of Prompt-Gamma Ghost Imaging: Proof
of Concept 93
  Cheng-i Chiang, Alaleh Aminzadeh, Wilfred K. Fullagar, Ulf Garbe,
  Filomena Salvemini, Joseph J. Bevitt, Jeremy M. C. Brown,
  David M. Paganin, and Andrew M. Kingston

TRIXIE – Development and Construction of a New Neutron Imaging
Instrument in the Czech Republic 101
  Jana Matouskova, Nikolay Kardjilov, Burkhard Schillinger,
  and Lubomir Sklenka

Neutron Generators: From Logical Assumptions to totally different
behavior 109
  Burkhard Schillinger, Tomas Bily, and Jana Matoušková

How to Avoid Going Wrong in Neutron Imaging Facility Design 119
  Burkhard Schillinger, Aaron Craft, and Nikolay Kardjilov

A Vertical Neutron Beam Device Used to Examine Liquids 129
  Burkhard Schillinger, Stephan Sponar, Alessandro Tengattini,
  and Clemens Trunner

Future from the Past: Long-Term Corrosion of Iron Analyzed Using
Bimodal Neutron and X-ray CT 140
  Ocson Cocen, Mahdieh Shakoorioskooie, Elodie Granget,
  David Mannes, Anders Kaestner, Jean-Marie Drezet,
  and Laura Brambilla

2D and 3D Visualization of Localized Plastic Deformation 149
  Khanh Van Tran, Thawatchart Chulapakorn, Nikolay Kardjilov,
  Henning Markötter, Stephen A. Hall, Ingo Manke, and Robin Woracek

In-Situ Neutron Radiography with Hydrogenated Tensile Samples
in the INCHAMEL Facility 158
    Sarah Weick, Mirco Grosse, Conrado Roessger, Martin Steinbrueck,
    and Anders Kaestner

Boron-Based Neutron Scintillator Screen Characterization with X-rays
and Neutrons 168
    William Chuirazzi, Burkhard Schillinger, Steven Cool, Aaron Craft,
    Zoltan Kis, and Laszlo Szentmiklosi

Visualization of Liquid Water Behavior in a Polymer Electrolyte Fuel Cell
Under High-Temperature Operation Using a Neutron Radiography 180
    Yuto Obayashi, Hideki Murakawa, Katsufumi Nakagawa,
    Katsumi Sugimoto, Hitoshi Asano, Yuto Shirase, Christopher Schreiber,
    Solomon Wakolo, Eric Dzramado, Junji Inukai, Daisuke Ito,
    Naoya Odaira, and Yasushi Saito

The Performance of Neutron Sensitive Scintillators for Imaging Purposes 188
    E. Lehmann and B. Walfort

Review on Digital Neutron Imaging 200
    Eberhard H. Lehmann and Burkhard Schillinger

Fast Hyperspectral Reconstruction for Neutron Computed Tomography
Using Subspace Extraction 206
    Mohammad Samin Nur Chowdhury, Diyu Yang, Shimin Tang,
    Singanallur V. Venkatakrishnan, Andrew W. Needham,
    Hassina Z. Bilheux, Gregery T. Buzzard, and Charles A. Bouman

Preliminary Design Study on Neutron Imaging Facilities in a New
Research Reactor in Japan 216
    Daisuke Ito, Naoya Odaira, Yasushi Saito, Takenao Shinohara,
    and Yoshiaki Kiyanagi

Experimental MeV Neutron Imaging at RADEN with Monte Carlo
Simulations 222
    Naoya Odaira, Yusuke Tsuchikawa, Daisuke Ito, Yasushi Saito,
    Takenao Shinohara, Yoshiaki Kiyanagi, Saerom Kwon, Kentaro Ochiai,
    and Satoshi Sato

Condensation of the Vapor in Sessile Water Droplets and Open-Ended
Water-Filled Capillaries Revealed by Neutron Radiography 235
    Hyeonjun An, Youngtak Koo, Jiyong Cheon, Pavel Trtik,
    and Joonwoo Jeong

Introduction of Camera System for CNRF at JRR-3 244
  Keisuke Kurita, Hiroshi Iikura, Isao Harayama, Yusuke Tsuchikawa,
  Satoshi Morooka, Tetsuya Kai, Takenao Shinohara, Naoya Odaira,
  Daisuke Ito, Yasushi Saito, and Masahito Matsubayashi

Crystallographic Phase Transformations and Corresponding Temperature
Distributions During GTAW of Supermartensitic Stainless Steel Visualized
by Neutron Bragg-Edge Imaging 252
  Axel Griesche, Tobias Mente, Alessandro Tengattini, Stefano Dal Pont,
  and Nikolay Kardjilov

Neutron Tomography and the Virtual World of Palaeontology 259
  Joseph J. Bevitt

Enhancement of Spatial Resolution in High-speed Neutron Imaging Using
a Multi-slit Collimator 268
  Yasushi Saito, Daisuke Ito, Naoya Odaira, Isao Harayama,
  and Keisuke Kurita

Characterization of Dual-Mode Scintillators for Fast Neutron and X-ray
Radiography 276
  Nicholas Anastasi, Stuart Miller, Pijush Bhattacharya, Tawan Jamdee,
  Edgar van Loef, Guillermo Velasco, Charles Sosa, Paul Rose,
  Bipin Singh, Lakshmi Soundara Pandian, and Vivek V. Nagarkar

Overview of MARS: The Multimodal Advanced Radiography Station
at the High-Flux Isotope Reactor 290
  James Torres, Hassina Z. Bilheux, Jean-Christophe Bilheux,
  Chen Zhang, Lowell Crow, Erik Stringfellow, Ian Turnbull,
  Violet Freudenberg, Roger Hobbs, Mike Harrington, Thomas Huegle,
  Erik Iverson, Rohit Srivastava, and Yuxuan Zhang

Friction Stir Additive Manufacturing: Neutron Interferometry and Bragg
Edge Imaging 299
  Saber Nemati, Huan Ding, Kyungmin Ham, Shengmin Guo,
  James Torres, Anton Tremsin, and Leslie G Butler

Neutron Imaging of Hydrogenous Gases in Porous-Carbon Materials
Around Phase Transitions 308
  Frederik Ossler, Charles E. A. Finney, Louis J. Santodonato,
  Jean-Christophe Bilheux, Hassina Z. Bilheux, Douglas P. Armitage,
  Rebecca A. Mills, Yuxuan Zhang, and Luke L. Daemen

Commercial Neutron Imaging with an Accelerator-Based Neutron Source 317
  Martin Wissink

iBeatles: Bragg Edge Imaging Software for the VENUS Beamline
at the Spallation Neutron Source 326
Jean C. Bilheux and Hassina Z. Bilheux

Enhancing Reconstruction of Time-of-Flight Neutron Computed
Tomography Using Artificial Intelligence 337
*Shimin Tang, Mohammad Samin Nur Chowdhury, Diyu Yang,
Singanallur V. Venkatakrishnan, Kyle D. Anderson, Ryan D. Ross,
Gregery T. Buzzard, Charles A. Bouman, Yuxuan Zhang,
and Hassina Z. Bilheux*

Early Commissioning Activities at the Spallation Neutron Source VENUS
Imaging Beamline 347
*Hassina Z. Bilheux, Harley Skorpenske, Kevin Yahne, Greg Guyotte,
Matthew Pearson, Matthew J. Frost, Jean C. Bilheux, Shimin Tang,
Chen Zhang, Yuxuan Zhang, James Torres, and Matthew Balafas*

Data Workflow and Software Tools at VENUS 355
*Jean C. Bilheux, Hassina Z. Bilheux, Anton Khaplanov,
Matthew J. Bedynek, Su-Ann Chong, Cornelius Donahue Jr,
Greg Guyotte, Kazimierz J. Gofron, Rob Knudson IV,
Jamie Molaison, Harley Skorpenske, Shimin Tang, Zach Thurman,
Singanallur Venkatakrishnan, and Chen Zhang*

Neutron Grating Interferometry at the High Flux Isotope Reactor 364
*Yuxuan Zhang, Erik Stringfellow, Jean-Christophe Bilheux,
Hassina Z. Bilheux, James Torres, Ian Turnbull, Roger Hobbs,
Leslie G. Butler, and Kyungmin Ham*

Conceptual Optics Design of the CUPI^{2}D Beamline at the Spallation
Neutron Source Second Target Station 372
Hassina Z. Bilheux, James Torres, Adrian Brügger, and Jiao Lin

Stroboscopic Bragg Edge Radiography of Twin Formation in a Magnetic
Shape Memory Alloy 380
S. Kabra, W. Kockelmann, G. Burgess, M. J. Gutmann, and A. S. Tremsin

Preliminary Campaign of Bragg Edge Neutron Imaging Round Robin 389
  Ranggi S. Ramadhan, Winfried Kockelmann, Florencia Malamud,
  Matteo Busi, Saurabh Kabra, Takenao Shinohara, Thilo Pirling,
  and Anton Tremsin

Author Index 399

Time-Series Neutron Tomography
of Rhizoboxes

A. P. Kaestner[1]($\boxtimes$), S. Di Bert[2], R. Jia[2], P. Benard[2], and A. Carminati[2]

[1] PSI Center for Neutron and Muon Sciences, Paul Scherrer Institute, Villigen,
Switzerland
anders.kaestner@psi.ch
[2] Department of Environmental Systems Science, ETH Zurich, Zurich, Switzerland

Abstract. This study investigates the feasibility of tilt-series neutron tomography for analyzing rhizoboxes used in root-soil interaction studies. Traditional neutron imaging methods are limited by constrained root growth volumes and poor penetration in moist soil. Using a vertical acquisition axis, the tilt-series approach avoids constraints typical of laminography. This method allows simultaneous radiographic and tomographic data acquisition, enhancing time resolution and providing detailed insights into root networks and soil water content. Experiments involved scanning six-week-old maize plants in rhizoboxes filled with sand. Results show that tilt-series tomography can effectively reconstruct root networks despite some artifacts from the missing wedge. While the tilt-series tomographic data qualitatively reveal water distribution changes, radiographic data remain essential for quantitative analysis. This approach demonstrates the potential for dynamic root-soil interaction studies, offering a valuable tool for agricultural and environmental research by providing comprehensive insights into the rhizosphere.

Keywords: Neutron imaging · Tilt-series · Time-series · Rhizobox

1 Introduction

Processes in the rhizosphere are well-studied using neutron imaging. This task is challenging as it requires compromises in sample design and imaging conditions. Studying roots requires a certain unconstrained volume for the roots to grow unhindered on one side, while the neutrons limit the rhizosphere volume due to the short penetration depth in moist soil. This resulted in two main rhizobox designs for neutron imaging experiments. For young plants, it is possible to use round soil samples with dimensions suitable for tomography. This geometry can only be used until the root network grows to be bounded by the container walls and is unnaturally bent to become clustered at the walls. Older plants have much larger root networks and would be too constrained by the wall of the small cylinders. Instead, slabs are used for these plants. These samples are often designed for radiography time series acquisition as the thin slab can be

© The Author(s) 2026
A. E. Craft and H. Z. Bilheux (Eds.): WCNR 2024, SPPHY 348, pp. 1–9, 2026.
https://doi.org/10.1007/978-3-032-15003-5_1

considered an approximately two-dimensional representation of the distribution of the material. The slab is also limited as the roots can only grow unconstrained in two directions, while the remaining direction is narrowed down to a minimum. This approximation initially works well, but at some point, with refined models, it becomes increasingly important to obtain information about the location of sample features in the beam direction. This information can only be obtained using computed tomography. The slab is far from the ideal shape for tomography because of the unbalanced aspect ratio, making it impossible for the neutron beam to penetrate the sample in the direction parallel to the plane.

Two approaches usually obtain tomographic images from slabs: a full turn with a tilted acquisition axis or incomplete sample rotation with a vertical acquisition axis [1,2]. Scans with a tilted axis are often referred to as laminography. The incomplete scan lacks a wedge of the sample rotation and has different names depending on the application: tomosynthesis (medicine)[3] and tilt series (microscopy). Both cases have to compromise because the Radon space is incompletely filled, which results in more or less severe artifacts in the reconstructed images. The tilt series were identified as less valuable due to the nature of the artifacts caused by the missing wedge in the Radon space [4], which are more severe and also asymmetric compared to laminography. We argue, however, that this method is still relevant for specific use cases where the sample is unsuited for the rotation around a tilted acquisition axis. The sample may contain liquids or loosely packed grains that could rearrange or the sample may be wired in a manner that constrains its ability to perform a complete rotation. The use of laminography for rhizoboxes was demonstrated by Rudolph-Mohr et al. [5]. This approach requires further compromises for the scan; the content of the rhizobox must be fixed to avoid reordering the slab content, and it is not possible to perform a complete turn scan if the shoot is intact. Furthermore, combining radiography and tomography scans is impossible as the sample needs two different mountings. These compromises are obsolete with tilt-series as the sample will be scanned using a vertical acquisition axis.

The aim of this study is to investigate the feasibility of using a tilt series for rhizoboxes. We will explore the method's limitations and compare the results to a full-turn scan. We will also investigate the possibility of using a dynamic scan sequence to obtain radiographs and tomographic data in the same scan. This method will provide a time series of radiographs and tomographic data with flexible time resolution. The method will be demonstrated on a maize plant in a rhizobox. The results will be used to quantify the root network morphology and water content in the rhizobox. The results will be compared to a full-turn scan. The study will provide a method for obtaining quantitative data from rhizoboxes using a tilt series. The method will be useful for studying root-soil interactions in the rhizosphere. Due to the reconstruction artifacts, the missing wedge tomography can not be used to quantify the water content in the rhizobox. Radiography time series are needed for this. At first sight, this looks like two separate scans. However, we will show that the two can be interleaved in the same scan sequence. This will provide a time series of radiographs and tomographic data with flexible time resolution. The tomography time series will be achieved using the golden ratio to determine the scan angle sequence [6,7].

2   Methods and Samples

2.1   Sample Design

The rhizoboxes are designed to study root-soil interactions, Fig. 1a. The rhizoboxes used in the following study are made of aluminum to minimize attenuation from the box. The box has the outer dimensions $140 \times 140 \times 16\,\mathrm{mm}^3$ and the inner dimensions for the soil slab $125 \times 125 \times 10\,\mathrm{mm}^3$. The mounting is solved using a base mount with three pins in-line attached to the turn table, and each rhizobox has matching holes in the bottom plate, Fig. 1b. This mounting allows for precise repositioning of the sample in the beam. This is important as the experiments often involve several boxes that will be scanned at different times of the day or before and after treatment.

The rhizoboxes are filled with sand material with a porosity of 42% and planted with maize seeds. The maize plants were grown for six weeks to produce a root network filling out the rhizobox.

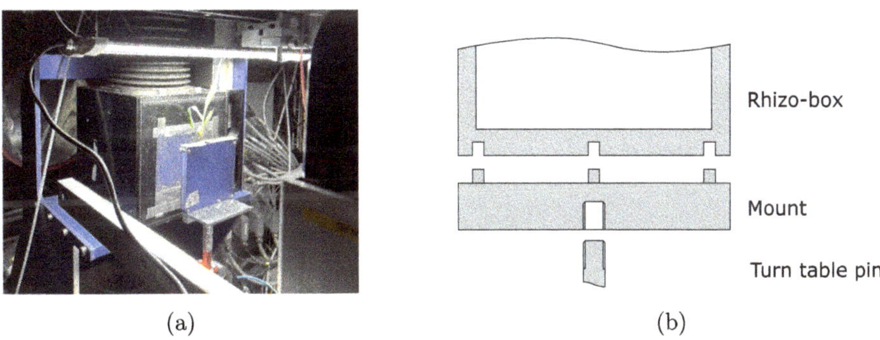

(a)   (b)

Fig. 1. A rhizo-box with a maize plant installed for scanning on the experiment position at the neutron imaging instrument ICON, Paul Scherrer Institute (a). The mounting consists of two parts, one mounted on the turn table and the rhizo-box with matching holes in the bottom plate (b).

2.2   Determining the Tilt Range

The choice of the tilt angle for a neutron tilt series is balanced between how well sample features need to be represented on one side and the ability to penetrate the sample and blurring on the other side. The fidelity of the reconstructed features to shapes in the sample requires a full tomography scan. In contrast, the maximum scan range is limited by transmission, and the penumbra blurring is caused by the required distance between the sample and detector for a given scan wedge.

The transmission is primarily determined by the porosity of the soil material, the slab thickness, and water content. Beer-Lambert's law gives the transmission at oblique beam incidence to the slab plane. In the case of a rhizobox, this equation reduces to a sum of three materials: water, sand, and the wall of the sample container. The container is made of aluminum. Thus, we can write the transmission as

$$T = \frac{I}{I_0} = e^{-(\mu_{Al} \cdot d_{Al} + \mu_{Sand} \cdot d_{Sand} + \mu_{H2O} \cdot d_{H2O})}, \tag{1}$$

The wall and sand thicknesses are constant, while the water length depends on the sample water saturation, Θ. The length through the sample at different view angles increases with the tilt angle, α. Equation 1 can be revised into

$$T = \frac{I}{I_0} = e^{-(\mu_{Al} \cdot d_{Al} + \mu_{Sand} \cdot d_\alpha \cdot (1 - \phi_{Sand}) + \mu_{H2O} \cdot d_\alpha \cdot \Theta \cdot \phi_{Sand})}, \tag{2}$$

when we take the thickness $d_\alpha = d_0 / \cos\alpha$, porosity ϕ_{Sand}, and saturation Θ of the pores into account. The thickness and saturation are variable during the experiment. We aim for a transmission greater than 10% to calculate the maximum tilt angle. Lower transmission is not advisable as noise and scattering amplitudes are in this order of magnitude. In Fig. 2b, we show the greatest allowed tilt angle for a sand material with $\phi_{Sand}=42\%$ for the rhizobox described above in Sect. 2.1. The greatest tilt angle depends on the degree of water saturation and slab thickness. Here, we used $d_0=10$ mm. The red band to the left indicates tilt ranges that produce too strong missing wedge artifacts. The tilt series scan will be in the interval $\alpha_{scan} = [-\alpha, \alpha]$. The tilt angle further impacts the distance between the sample and detector, l in Fig. 2a. This distance causes a penumbra blurring given by the collimation ratio of the neutron beam. The blurring represented by the vertical axis to the right in Fig. 2b is computed for the rotation axis. In reality, l varies between just a few mm to $2l$ during the scan. In this plot, we used L/D=340, corresponding to the ICON's nominal value when the 20 mm neutron aperture is used [8].

The distance between the sample and the detector determines the width of the penumbra blurring, which is the major source of unsharpness in the setup used. A lower bound on the tilt angle is given by the severity of missing wedge artifacts. These include fidelity to the original shape of objects in the sample and the amount of cross-talk in the beam direction. The artifacts are more severe for narrow scan ranges.

2.3 Dynamic Scan

Typical rhizobox experiments are performed as a time series of radiographs. A method that has proven to be useful for the study of root-soil interactions. The time series is used to quantify the dynamics of water movement in the

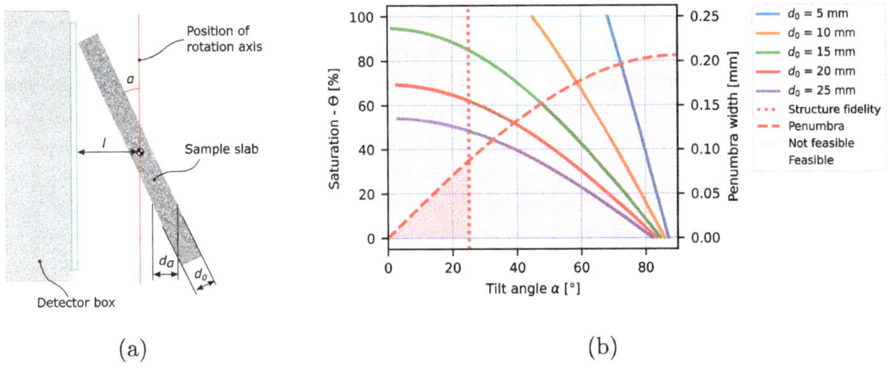

Fig. 2. The setup geometry (a) limits the tilt angle, θ. The greatest tilt angles for different sample thicknesses and saturation levels in a 42% porosity material.

soil and root network. The radiographs have a limitation in that they are two-dimensional and do not provide information about the location of features in the beam direction. This information can only be obtained using tomography. The dynamics of the sample require the use of golden ratio sequences [6], which have two important features: 1, motion artifacts are minimized, and 2, any sequence of consecutive projections in time covers the full range of tilt angles. The second feature is important as it allows the scan to be broken down into a series of sub-scans, each covering the full scan range with an almost uniform distribution of the acquisition angles. Craig *et al.* [7] showed that the golden ratio sequence works equally well for tilt series as for the complete scan. Here, we modify the original golden ratio sequence to only cover a limited scan range of $\pm\alpha_{max}$. The acquisition angles are, thus, given by

$$Golden(i) = (i \cdot 2\alpha_{max} \cdot \phi \quad \mathrm{mod} \ 2\alpha_{max}) - \alpha_{max}, \tag{3}$$

where $\phi = (1 + \sqrt{5})/2$ is the golden ratio, and α_0 is the greatest tilt angle from the view where the slab is parallel to the detector. This sequence can be used for $i \rightarrow \infty$ or until the experiment is terminated without repeating any angles, thanks to the irrational nature of the golden ratio.

In our experiment, we are interested in obtaining both radiographs for their capacity to quantify the local water content and tomographic data for the root network topology. Normally, the radiography time series and tomography would be acquired separately after each other. This would, however, mean that the tomography can only capture the observed process's initial and final state. Thus, we propose to interleave the radiographs with the tomographic data in the same scan sequence such that,

$$\begin{cases} i = 2n & \rightarrow \alpha = Golden(i/2) \\ i = 2n + 1 & \rightarrow \alpha = 0°. \end{cases} \tag{4}$$

This scan provides a time series of radiographs and tomographic data with flexible time resolution, observing the same process. The tomographic temporal resolution is naturally lower due to the number of projections needed for the reconstruction.

3   Experiments and Analysis

3.1   Tomography Scans

The tomography data was acquired at the ICON neutron imaging beamline at PSI [8] using the so-called midi setup on experiment position 2. The setup was configured with an Andor iKon L camera, an Otus 55mm lens and a 30 μm thick GadOx scintillator. This combination resulted in a FOV of 150×150 mm^2 and the pixel size 90.9 μm.

Interleaved Time-Series Scan. The rhizoboxes were scanned in a day-night-day cycle, and plant lights were installed above the plants during the scan. The plants are expected to consume more water during the day than at night, which will show in soil water distribution.

The demonstrated scan was obtained during the daytime and consisted of 320 radiographs. From which every second is a radiograph at 0°, and the other is obtained from the golden ratio sequence in the interval $\alpha = \pm 60°$.

Full-Turn Scan. At the end of the time series, a full 360° tomography scan was performed to provide a reference. The slab is assumed to be sufficiently dry to allow reconstruction. The slab was moved further from the detector to allow this scan without colliding with the equipment.

3.2   Processing Projection Data

The processing has two tracks, one for the radiography data and the other for the tomography. Both data sets were normalized using the scattering correction method described in [9,10] to reduce artifacts originating from scattered neutrons. The data is further also filtered to remove outlier spots and ring artifacts.

The radiography data was processed using Python scripts based on NumPy and SciKit-Image packages.

The tomography data was reconstructed using the filtered back projection algorithm provided in the tomography reconstruction tool MuhRec [11,12] includes support for golden ratio projection data and weighting needed to handle the missing wedge. All pre-processing was done in MuhRec.

4   Results

4.1   Tilt Range

The full tomography of a dry slab allowed us to reconstruct the entire root network. This scan was reduced to several sub-scans with wedge sizes between 20°

and 80°. The reconstructed slices are shown in Fig. 3. The reconstructions clearly show the degradation of sample features when the covered wedge size decreases. The choice of using $\alpha = \pm 60°$ is confirmed. The missing wedge artifacts are present, but the main shape has still been well reconstructed.

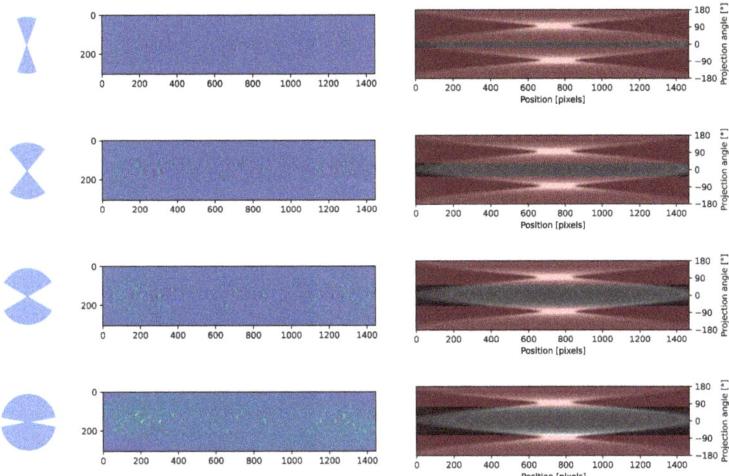

Fig. 3. Reconstructions from the full scan with covered wedges from 20° and 80°. The sinograms to the right show each slice's used fraction of the full sinogram.

4.2   Time-Series

The time series scan consists of 160 golden ratio projections, which we reconstructed into three time frame volumes. Each volume does not show much change

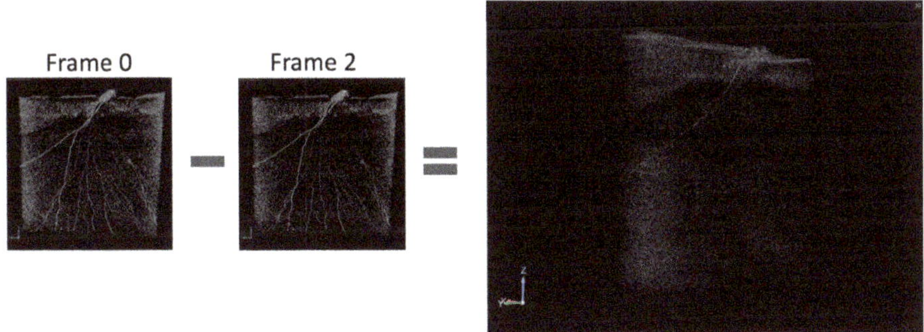

Fig. 4. The difference between the last and first volume time frame in the scan sequence.

at first sight, but subtracting frame 0 from frame 2 shows the change in water contents, Fig. 4. Due to bias and reconstruction artifacts, the missing wedge volumes cannot be used for quantitative water content analysis. It is, however, still possible to qualitatively identify the location of the water from the tomography.

5   Conclusions

These preliminary results using missing wedge tomography combined with radiography for time-series acquisition of water movement in rhizoboxes show a promising path towards improving model accuracy for soil-root interactions. Upcoming experiments will focus on obtaining longer time series and systematic studies to improve image quality. Iterative reconstruction techniques are also expected to improve the image quality.

Acknowledgement. This work is based on experiments performed at the Swiss spallation neutron source SINQ, Paul Scherrer Institute, Villigen, Switzerland.

References

1. Helfen, L., et al.: Nucl. Inst. Methods Phys. Res. A **651**(1), 135 (2011). https://doi.org/10.1016/j.nima.2011.01.114
2. Osterloh, K., et al.: Limited view tomography of wood with fast and thermal neutrons, vol. 128 (Deutsche Gesellschaft für Zerstörungsfreie Prüfung (DGZfP) (2011)
3. Three-Dimensional Digital Tomosynthesis. AFMLRME, Springer, Wiesbaden (2014). https://doi.org/10.1007/978-3-658-05697-1
4. Xu, F., Helfen, L., Baumbach, T., Suhonen, H.: **20**(2), 794. https://doi.org/10.1364/oe.20.000794
5. Rudolph-Mohr, N., Bereswill, S., Tötzke, C., Kardjilov, N., Oswald, S.E.: Neutron computed laminography yields 3D root system architecture and complements investigations of spatiotemporal rhizosphere patterns. Plant Soil, 489–501 (2021). https://doi.org/10.1007/s11104-021-05120-7
6. Kaestner, A., Münch, B., Trtik, P., Butler, L.: Opt. Eng. **50**(12), 123201 (2011). https://doi.org/10.1117/1.3660298
7. Craig, T.M., Kadu, A.A., Batenburg, K.J., Bals, S.: **15**(11), 5391. https://doi.org/10.1039/d2nr07198c
8. Kaestner, A., Hartmann, S., Kühne, G., Frei, G., Gruüzweig, C., Josic, L., Schmid, F., Lehmann, E.: Nucl. Inst. Methods Phys. Res. A **659**(1), 387 (2011). https://doi.org/10.1016/j.nima.2011.08.022
9. Boillat, P., et al.: Opt. Express **26**(12), 15769 (2018). https://doi.org/10.1364/oe.26.015769
10. Carminati, C., et al.: PLoS ONE **14**(1), e0210300 (2019). https://doi.org/10.1371/journal.pone.0210300
11. Kaestner, A.: Nucl. Inst. Methods Phys. Res. A **651**(1), 156 (2011). https://doi.org/10.1016/j.nima.2011.01.129
12. Kaestner, A., Carminati, C., Tasev, D., Potrzebowski. W.: Muhrec ver. 4.2.2 (2023). https://doi.org/10.5281/zenodo.10321810

Bimodal Neutron and X-ray Imaging and Fusion

M. Shakoorioskooie[1]([✉]), E. Lehmann[1], M. Strobl[1], Q. Zhan[1], P. Trtik[1], M. Krieg[2],
D. Mannes[1], and A. Kaestner[1]

[1] PSI Center for Neutron and Muon Sciences, 5232 Villigen PSI, Switzerland
`mahdieh.shakoorioskooie@psi.ch`
[2] Site et Musée romains d'Avenches (SMRA), Avenches, Switzerland

Abstract. Bimodal neutron and X-ray imaging provides a robust approach for non-destructive analysis of complex materials, particularly in engineering, archaeological, and life-science specimens. Over a decade ago, our group pioneered this technique, enabling a wide range of applications and inspiring similar implementations across other beamlines. This study presents examples of possible data fusions relevant to our activities. We employed three primary fusion methods: direct bivariate histogram-based fusion, Laplacian Pyramid image fusion, and PCA-based feature-level fusion. The bivariate histogram method effectively highlighted correlations and divergences between neutron and X-ray datasets, aiding material differentiation. The Laplacian Pyramid method enhanced spatial resolution and contrast, providing detailed visual and analytical representations of material phases. The PCA-based method integrated intensity and structural information, maximizing variance representation for comprehensive material characterization. Future work will focus on integrating state-of-the-art fusion techniques, such as DeepFuse-inspired methods, to further improve the quality and applicability of bimodal imaging in engineering and archaeological contexts.

Keywords: neutron imaging · X-ray Tomography · Image Fusion · Laplacian Pyramid · Principal Component Analysis (PCA)

1 Background and Objectives

In the dynamic field of materials science, the integration of X-ray tomography with neutron imaging stands as a landmark advancement, revolutionizing the non-destructive characterization of materials [1]. This innovative approach harnesses the distinct yet complementary capabilities of each modality. X-rays, with their adeptness at mapping electron density distributions, provide contrast based on atomic number variations, whereas neutrons offer a unique window onto the involved nuclei, revealing highly complementary insights, particularly with respect to hydrogen content. This synergy enables a more comprehensive understanding of material properties and structures. More than a decade ago, our group at PSI led the way in bimodal imaging, introducing instrumental innovations that enabled this powerful technique [2, 3]. The establishment of bimodal imaging at two distinct beamlines at the spallation neutron source SINQ, namely ICON

A. E. Craft and H. Z. Bilheux (Eds.): WCNR 2024, SPPHY 348, pp. 10–18, 2026.
https://doi.org/10.1007/978-3-032-15003-5_2

(simultaneous imaging, near perpendicular) [3] and NEUTRA (off-line, parallel) [2], exemplified our pioneering approach. These developments have since inspired similar implementations at other beamlines, broadening the scope and impact of bimodal imaging [4, 5].

Early attempts to combine X-ray and neutron datasets relied on simple analytical tools such as bivariate histograms, which aimed to identify correlations between the datasets [6]. This method effectively highlights areas where the properties detected by X-ray and neutron imaging overlap or diverge, but it is limited in producing a nuanced, single representation of the material. Despite these limitations, bivariate histograms laid the groundwork for more sophisticated fusion techniques. Advanced fusion methods leverage more complex algorithms based on *pixel-level*, *feature-level*, and *decision-level* methods [7, 8]. These techniques, so far mainly exploited in 2D computer vision applications, allow for the integration of detailed spatial and spectral information from each modality, resulting in a comprehensive and coherent representation of the material's characteristics. They are tailored to capitalize on the strengths of each imaging modality, meticulously addressing challenges such as differences in penetration depth and sensitivity to different material constituents.

In this paper, we highlight three main procedures for image fusion or exploiting the complementary information from both modalities: (a) direct bivariate histogram-based fusion, (b) Laplacian Pyramid Fusion (LPF) [9], and (c) PCA-based fusion [10]. The first approach could be categorized as a *decision-level fusion*, as this method does not directly fuse the images into a single representation but rather uses statistical analysis to identify and highlight correlations between the datasets. This can then inform decisions about material properties and differentiation. While the second is an example of *pixel-level fusion*, as this technique integrates information from the neutron and X-ray images at the pixel level. It involves decomposing the images into multiple levels of resolution, fusing corresponding levels, and then reconstructing a single, high-quality image that combines features from both modalities. This approach enhances the visual and analytical quality of the resulting image by leveraging detailed spatial information from both sources. This technique excels at merging images with differences in focus or containing various details at different scales, which is a common scenario when fusing images captured with different detector systems and varying resolutions. Finally, the third method, represents an example of a *feature-level fusion*. It involves extracting relevant features (such as edges) from both neutron and X-ray images and then using Principal Component Analysis (PCA) to combine these features into a single fused image. PCA identifies the principal components that capture the most significant variance in the data, creating a fused image that maximizes the representation of this variance. This method enhances material differentiation by considering both intensity and structural features, providing a more comprehensive representation of the specimen.

These methods enhance material differentiation and facilitate subsequent segmentation and phase analysis, providing clearer and more detailed representations of complex specimens. By combining X-ray and neutron imaging, our research has significantly advanced materials characterization. Here, we show a few examples of such fusion, which can be implemented based on the requirements of each study, and we will report our recent developments in future publications.

2 Methods

Tomography and Registration: The neutron tomography was performed at the NEU-TRA beamline at the Paul Scherrer Institute (PSI). This dataset involved 625 projections over a 360° rotation, capturing a detailed neutron tomographic reconstruction of a heavily corroded archeological artifact. Neutron tomography provides high sensitivity to light elements such as hydrogen, making it a valuable tool for analyzing complex materials. The affine registration of the neutron and X-ray images was performed using the SimpleElastix library [11], based on mutual information (MI).

Image Fusion Methods: In this paper, we highlight three main procedures for image fusion or exploiting the complementary information from both modalities: direct bivariate histogram-based fusion, Laplacian Pyramid image fusion, and PCA-based feature-level fusion (Fig. 1). The first approach could be categorized as decision-level fusion. This method does not directly fuse the images into a single representation but uses statistical analysis to identify and highlight correlations between the datasets. To achieve this, we generate bivariate histograms by plotting the neutron tomography (NCT) values against the X-ray tomography (XCT) values for each voxel. The histogram is then analyzed to identify distinct clusters representing different material phases. This analysis highlights areas where the properties detected by X-ray and neutron imaging overlap or diverge, informing decisions about material properties and differentiation. The second method, Laplacian Pyramid image fusion, represents an example of pixel-level fusion. This technique integrates information from the neutron and X-ray images at the pixel level. It involves decomposing the images into multiple levels of resolution, fusing corresponding levels, and then reconstructing a single, high-quality image that combines features from both modalities. Specifically, the registered neutron and X-ray images are decomposed into multiple levels using the Laplacian Pyramid method. Corresponding levels from both modalities are then fused to integrate the information. Finally, the fused image is reconstructed from the pyramid, combining features from both neutron and X-ray images. This approach enhances the visual and analytical quality of the resulting image by leveraging detailed spatial information from both sources. The third method, PCA-based feature-level fusion, involves extracting relevant features, such as edges, from both neutron and X-ray images and then using Principal Component Analysis (PCA) to combine these features into a single fused image. PCA identifies the principal components that capture the most significant variance in the data, creating a fused image that maximizes the representation of this variance. The procedure begins with extracting edge features from the neutron and X-ray images using an edge detection algorithm, such as the Sobel filter. The images and their edge features are normalized to ensure they are within the same scale. These normalized intensity values and edge features are then stacked into a single feature matrix. PCA is applied to this combined feature matrix, and the first principal component is used to create the fused image. The fused image is reshaped to match the original image dimensions and normalized. This method enhances material differentiation by considering both intensity and structural features, providing a more comprehensive representation of the specimen.

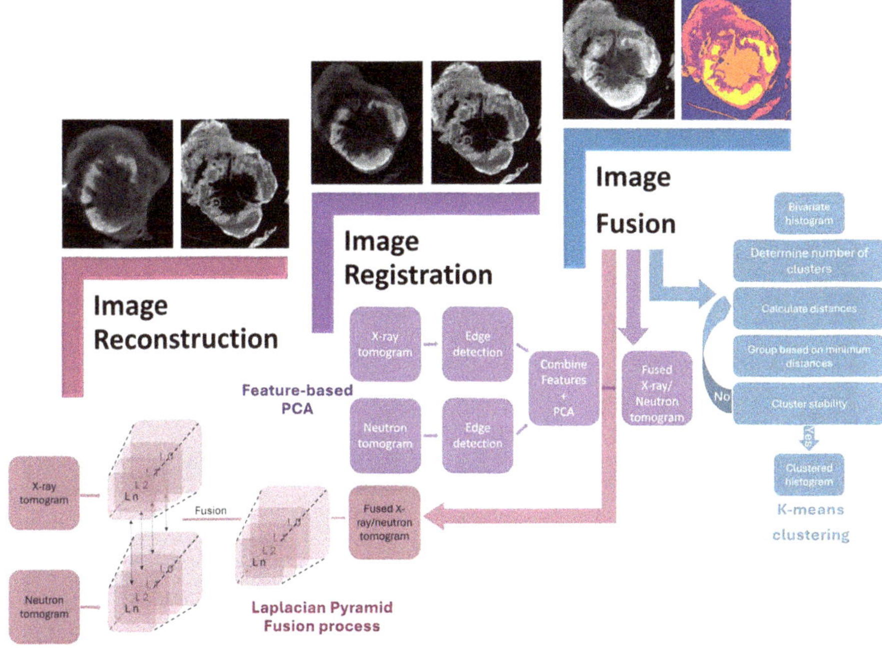

Fig. 1. Workflow for Bimodal Neutron and X-ray Imaging Fusion: This figure illustrates the workflow, consisting of three main stages: image reconstruction, image registration, and image/data fusion. The fusion stage highlights three methods: direct bivariate histogram-based fusion for material differentiation, Laplacian Pyramid image fusion for enhanced visual and analytical quality, and PCA-based feature-level fusion.

3 Results and Discussion

Figure 2 illustrates the results of using the bivariate histogram to analyze neutron and X-ray tomographic data. Panels (a) and (b) show longitudinal slices from the neutron (NCT) and X-ray (XCT) tomograms, respectively. Panel (c) shows a photograph of the original object. Panels (d) and (e) display X-Y cross-sections of the neutron and X-ray tomograms, respectively. The neutron tomogram is sensitive to light elements like hydrogen, useful for identifying hydrous phases, while the X-ray tomogram is sensitive to electron density variations and atomic number, highlighting metallic components and denser materials. Panel (f) presents a checkerboard super-positioning of these cross-sections, alternating small patches from both images. This visual representation provides a direct comparison of the information captured by each modality, helping to assess how well the datasets complement each other and identifying areas where they differ. Panel (g) shows the bivariate histogram of the neutron and X-ray tomographic data. Each point represents a voxel, with NCT intensity on one axis and XCT intensity on the other. Clusters in the histogram indicate regions where voxel intensities correlate between the modalities.

For example, a cluster might represent a specific material phase detected by both neutron and X-ray tomography. The color intensity in the histogram indicates the density

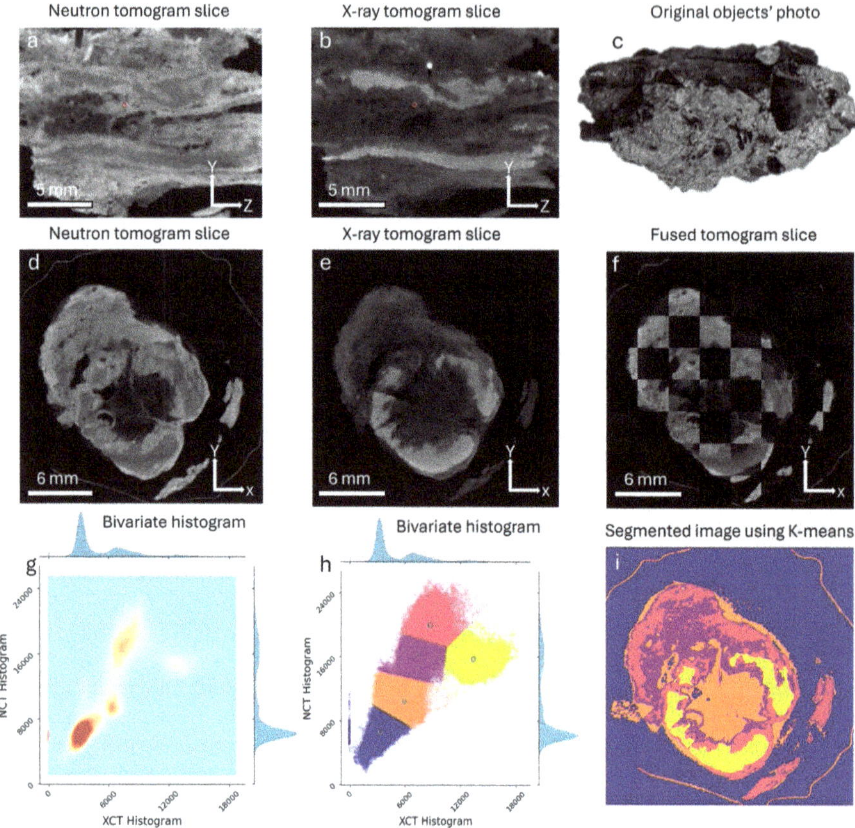

Fig. 2. Comparative analysis of neutron and X-ray tomography on an archaeological specimen. (a) and (b) display a longitudinal slice from neutron (NCT) and X-ray (XCT) tomograms, respectively, highlighting their differing material attenuations. (c) A photograph of the original object. (d) and (e) X-Y cross-sections. (f) Checkerboard representations of (d and e). (g) Bivariate histogram of the NCT and XCT. (h) highlighting the clusters and their centroids determined by K-means clustering algorithm. (i) Result of clustering mapped on the X-Y cross-section.

of data points, with brighter areas representing higher densities. Panel (h) illustrates the segmentation of the bivariate histogram using the K-means clustering algorithm. This clustering separates the histogram into distinct clusters, each representing a different material phase. Different colors in the histogram correspond to different clusters, which can be mapped back to the spatial domain to identify specific material components within the sample. Panel (i) maps the K-means clusters onto the X-Y cross-section, resulting in a segmented image that clearly delineates the different components of the specimen. Each color in the segmented image represents a different material phase, as identified by the clustering in the bivariate histogram. This segmentation provides clear visual differentiation of the material components, aiding in the analysis and interpretation of the specimen's internal structure and later statistical analysis. However, the direct bivariate histogram-based fusion method does not produce a single fused image that combines all the information from both modalities into a coherent representation. Instead, it relies on statistical analysis to highlight correlations and inform material differentiation.

Figure 3 presents the results of the Laplacian Pyramid image fusion method, which integrates neutron and X-ray tomographic data at the pixel level. This method enhances the overall image quality by combining detailed spatial information from both modalities. The fused image effectively highlights different material phases, providing a more comprehensive understanding of the sample's composition. Panels (a) and (b) display neutron and X-ray tomographic slices, respectively, while panel (c) shows the resulting fused image. The histograms in panels (d), (e), and (f) represent the frequency of pixel intensities for the neutron, X-ray, and fused images, respectively. The Laplacian Pyramid method decomposes each image into multiple levels of resolution, fuses corresponding levels, and then reconstructs a single, high-quality image.

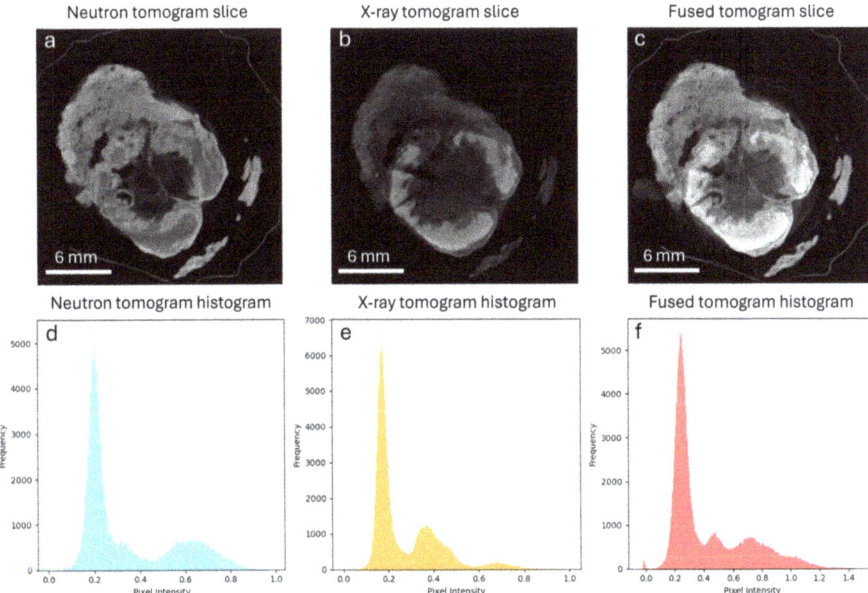

Fig. 3. Enhanced differentiation in a single representation showcased for an archaeological artifact, through image fusion. (a) and (b) NCT and XCT tomogram slices, respectively. (c) The resulting fused tomogram slice achieved through the Laplacian Pyramid image fusion method with 9 levels. The histograms (panels d, e, and f) display the frequency of pixel intensities, revealing the improved material contrast in the fused image. In panel (f), the histogram of the fused image highlights the enhanced ability to distinguish between metal components and other materials, demonstrating the effectiveness of the fusion technique in providing a clearer, more detailed representation of the artifact's composition.

This technique effectively combines the strengths of both neutron and X-ray imaging. In the fused image, the enhanced contrast is particularly notable for iron components, which are prominently highlighted due to the contributions from the X-ray tomogram.

This improved contrast facilitates the identification and differentiation of various material phases within the sample. The histogram of the fused image (panel f) shows distinct peaks corresponding to different materials, indicating successful integration of the complementary information from both modalities. The Laplacian Pyramid fusion method not only enhances visual clarity but also improves the analytical quality of the image. By leveraging detailed spatial information from both neutron and X-ray images, this approach provides a clearer and more detailed representation of the specimen. This is especially beneficial for advanced material characterization, where precise differentiation between material phases is crucial. Moreover, the method's ability to enhance material contrast and detail makes it a powerful tool for identifying specific components within complex structures. This is particularly useful in applications such as archaeology, where understanding the composition and structure of historical artifacts is essential.

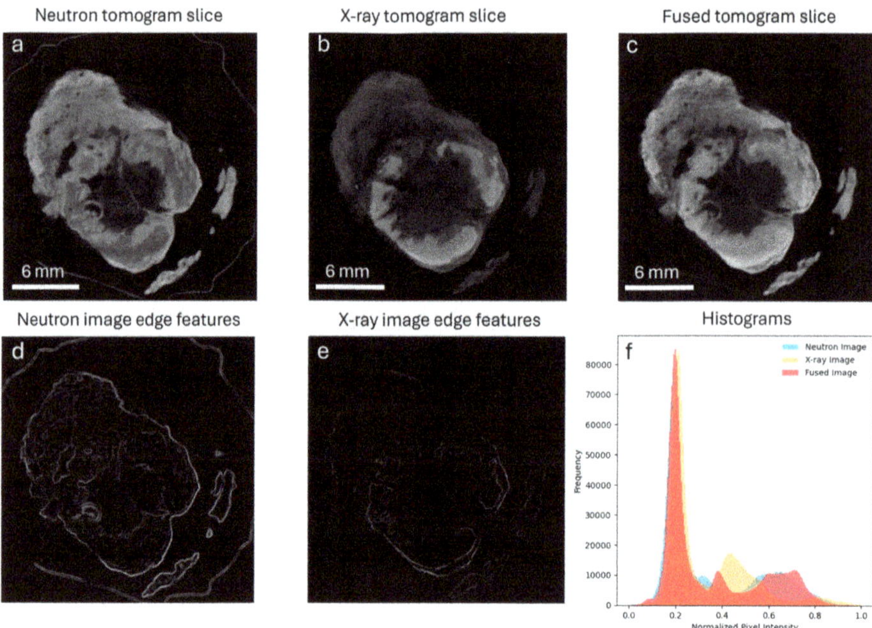

Fig. 4. PCA-Based Feature-Level Fusion: Panels (a) and (b) show neutron and X-ray tomographic slices. Panel (c) presents the fused tomogram using PCA. Panels (d) and (e) display edge features from neutron and X-ray images. Panel (f) compares histograms of neutron, X-ray, and fused images, showing enhanced material differentiation.

Figure 4 demonstrates the results of PCA-based feature-level fusion, which integrates intensity and structural features from neutron and X-ray tomographic data. This method effectively enhances material differentiation by capturing the most significant variance from both modalities, resulting in a comprehensive representation of the specimen. Panels (a) and (b) display the original neutron and X-ray images, respectively, while panel (c) shows the fused image obtained by combining these images using PCA. Panels (d) and (e) highlight the edge features extracted from the neutron and X-ray images using the Sobel filter, and panel (f) presents the histograms of the original images superimposed on the histogram of the fused image. The PCA-based fusion method begins by extracting edge features from both neutron and X-ray images. Edge detection is crucial for identifying structural boundaries within the images, providing additional information that enhances the fusion process. These edge features, along with the intensity values, are normalized and combined into a single feature matrix. PCA is then applied to this combined feature set to identify the principal components that capture the most significant variance in the data. This fused image maximizes the representation of significant variance, providing a detailed and nuanced understanding of the specimen's material composition.

4 Summary

In this study, we exemplified three distinct image fusion methods to integrate complementary information from neutron and X-ray tomography: direct bivariate histogram-based fusion, Laplacian Pyramid image fusion, and PCA-based feature-level fusion. Each method demonstrated unique advantages for material characterization. The bivariate histogram-based fusion highlighted correlations and divergences, identifying material phases via K-means clustering, though it relies heavily on statistical interpretation. Laplacian Pyramid fusion improved visual and analytical quality by enhancing material contrast, particularly for metallic components, making it invaluable for detailed spatial resolution needs. PCA-based fusion combined intensity and structural information, providing comprehensive representations and precise material differentiation. Future studies will focus on advanced methods like DeepFuse, a deep learning-based technique, to further enhance fusion quality and extend applications in engineering and archaeology. Enhanced imaging from these methods can lead to better material characterization, quality control, and artifact conservation, providing deeper insights and higher precision in various scientific and industrial fields.

References

1. Deschler-Erb, E., Lehmann, E.H., Pernet, L., Vontobel, P., Hartmann, S.: The complementary use of neutrons and X-rays for the non-destructive investigation of archaeological objects from Swiss collections. Archaeometry **46**(4), 647–661 (2004)
2. Lehmann, E., Thomsen, K., Strobl, M., Trtik, P., Bertsch, J., Dai, Y.: NEURAP—a dedicated neutron-imaging facility for highly radioactive samples. J. Imaging **7**(3), 57 (2021)
3. Kaestner, A.P., et al.: Bimodal imaging at ICON using neutrons and X-rays. Phys. Procedia **88**, 314–321 (2017)
4. Tengattini, A., et al.: NeXT-Grenoble, the Neutron and X-ray tomograph in Grenoble. Nucl. Instrum. Methods Phys. Res., Sect. A **968**, 163939 (2020)
5. LaManna, J.M., Hussey, D.S., DiStefano, V.H., Baltic, E., Jacobson, D.L.: NIST NeXT: a system for truly simultaneous neutron and x-ray tomography. In: Hard X-Ray, Gamma-Ray, and Neutron Detector Physics XXII, vol. 11494, pp. 61–71. SPIE (2020)
6. Angst, U.M., et al.: Chloride-induced corrosion of steel in concrete—insights from bimodal neutron and X-ray microtomography combined with ex-situ microscopy. Mater. Struct. **57**(4), 56 (2024)

7. Lip, C.C., Ramli, D.A.: Comparative study on feature, score, and decision level fusion schemes for robust multibiometric systems. Front. Comput. Educ., 941–948 (2012)
8. Lewis, J.J., O'Callaghan, R.J., Nikolov, S.G., Bull, D.R., Canagarajah, N.: Pixel-and region-based image fusion with complex wavelets. Inform. Fusion **8**(2), 119–130 (2007)
9. Nanavati, M.M., Shah, M.: Implementation and comparative study of pyramid-based image fusion techniques for lumbar spine images. Eng. Technol. Appli. Sci. Res. **13**(4), 11139–11145 (2023)
10. Kumar, S.S., Muttan, S.: PCA-based image fusion. In: Algorithms and Technologies for Multispectral, Hyperspectral, and Ultraspectral Imagery XII, vol. 6233, pp. 658–665. SPIE (2006)
11. Shakoorioskooie, M., Griffa, M., Leemann, A., Zboray, R., Lura, P.: Alkali-silica reaction products and cracks: X-ray micro-tomography-based analysis of their spatial-temporal evolution at a mesoscale. Cem. Concr. Res. **150**, 106593 (2021)

Limits in the Neutron Imaging Quantification of Hydrogenous Materials

Eberhard Lehmann[(⊠)] and Mahdieh Shakoorioskooie

Laboratory for Neutron Scattering and Imaging, Paul Scherrer Institute, Villigen, Switzerland
eberhard.lehmann@psi.ch

Abstract. The detection of hydrogen in neutron imaging is prevalent due to its high contrast relative to other structural materials present in or around samples. This presents a significant advantage over X-ray studies, where hydrogen contrast is almost negligible. In many applications, accurate quantification of hydrogen content is essential, warranting careful consideration of the approach's limitations. This paper, grounded in practical experiences, presents both successful and unsuccessful interpretations of neutron imaging data, highlighting the challenges and nuances of hydrogen detection in this context.

Keywords: neutron scattering · attenuation contrast · hydrogen content · calibration process · detection limits

1 Interaction of Slow Neutrons with Hydrogen

The interaction of hydrogen with (thermal, cold) neutrons is mainly via incoherent elastic scattering with a very high probability (cross-section). Therefore, hydrogen in a sample is well visible in neutron imaging, leading to many practical applications such as detecting explosives, tracking water migration in soil, studying fuel cells, other porous media and monitoring technical processes (fuel injection, gluing, lubrication, evaporation).

1.1 Cross-Section Data

Beer-Lambers's law is used for the quantification of the hydrogen content and its distribution, valid on first order. Using the attenuated neutron beam amount I in comparison to the undisturbed one I_0, the inverted law delivers the attenuation coefficient Σ, which corresponds to the microscopic cross-section σ via the nuclear density N of the material under investigation with a layer thickness d, e.g. hydrogen.

$$\Sigma = \sigma * N = \ln(I_0/I)/d \tag{1}$$

Because N is related to the material density ρ, the amount of the material can easily be determined (L = Avogadro's number; M = molecular weight):

$$N = \rho * L/M \tag{2}$$

A. E. Craft and H. Z. Bilheux (Eds.): WCNR 2024, SPPHY 348, pp. 19–24, 2026.
https://doi.org/10.1007/978-3-032-15003-5_3

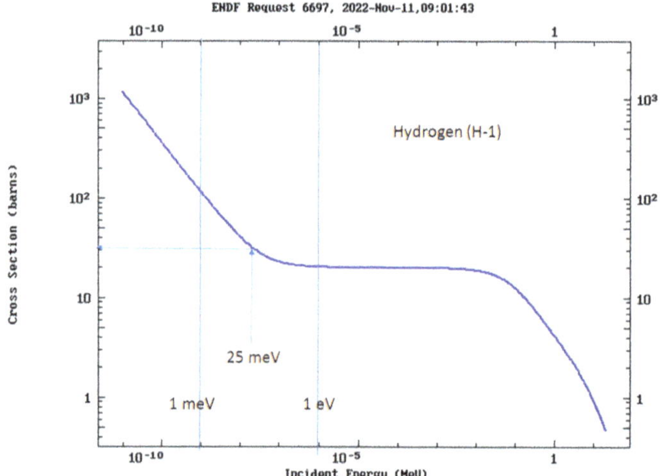

Fig. 1. The total cross-section of hydrogen (H-1) as derived from the ENDF library; relevant neutron energies are indicated separately. The logarithmic scales must be considered particularly in the low energy range!

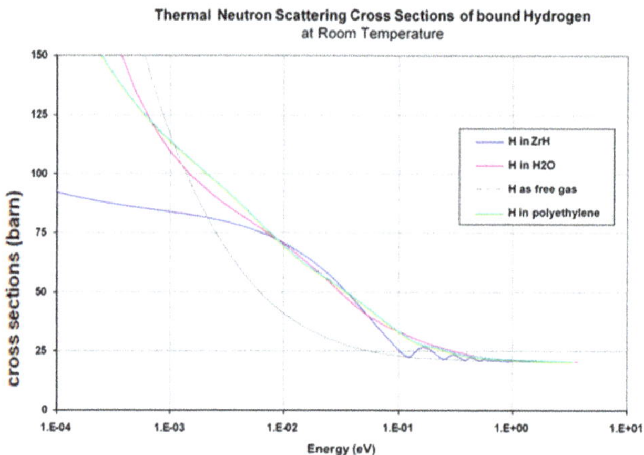

Fig. 2. Deviations from the "free gas" model, caused by the particular chemical bounds of the different hydrogenous materials.

The cross-section data for hydrogen are well-known and tabulated in cross-section libraries (e.g. ENDF/B) as shown in Fig. 1. This figure indicates a large slope towards lower neutron energies and describes the high sensitivity for spectral changes within the neutron beam.

On the other hand, these data are simplified according to the so-called "free gas model", in which the hydrogen molecules are handled as independent particles [6, 7].

Whereas this is applicable at higher neutron energies (above 1 eV), the particular chemical bound (e.g. with O in H_2O) has to be considered. Figure 2 shows examples of the deviating cross-sections, with high importance for the considered thermal and cold neutrons.

Taking both aspects into account: (1) the sensitivity for spectral changes by interaction within samples and (2) the uncertainty in the theoretical cross-section data by the chemical properties – a direct comparison to microscopic cross-section is hindered.

1.2 The Interaction Process

When the collimated beam of the neutron imaging facility hits a sample with a certain hydrogen content, the interaction is mainly by elastic neutron scattering. The total cross-section for thermal neutrons is given as about 30 barn, whereas absorption (0.3 barn) can be ignored on the first order. The scattered neutrons are distributed nearly isotopically around the sample. Because the attenuated signal is used in neutron imaging forward direction (see Eq. 1) for the quantification (an out-scattered neutron is considered as lost) and the total cross-section is taken for comparison.

However, a certain component of the scattered neutrons can contribute to the attenuated signal aside to the sample or even in the sample region. In this way, Beer-Lambert's law is not anymore valid, and the in-scattered neutrons must be considered separately.

This problem has higher importance for larger amounts (thicker layers) of hydrogen and for only small distances between sample and detector.

Although there have been several approaches to solve these problems by simulations of the scattering process [1, 2], we describe here an experimental way to limit the discrepancies.

2 Measurements with Hydrogenous Materials

The NEUTRA facility [3] at the spallation neutron source SINQ at PSI [4] was used for test measurements with hydrogenous materials of different thickness. For this purpose, thin layers of polyethylene were placed into the beam – either in direct contact with the detector (scintillator screen within a light-tight box, observed with a CCD camera detector) or in about 10 cm distance from it.

A typical projection (radiogram) of the sample is shown in Fig. 3 for the case "close contact". A sky shine becomes visible around the flat sample, caused by the amount of scattered neutrons from the sample's interaction. The profile in Fig. 3, right, describes the effect well, while the complementary case with the sample in 10 cm distance shows a flat profile and a lower amount of detected neutrons in the sample region.

It is quite difficult for the "close contact" case to derive the signal I and I_0, which are needed for the quantification according to Eq. 1, because of the clearly visible edge effects in Fig. 3.

However, the amount of hydrogen – the sample thickness of polymer foil in this case – also plays an important role in the quantification process. As shown in Fig. 4, thicker samples show smaller attenuation coefficients than thinner ones. This is the consequence of scattered neutrons, returning into the field-of-view. For thicker samples, the effect of

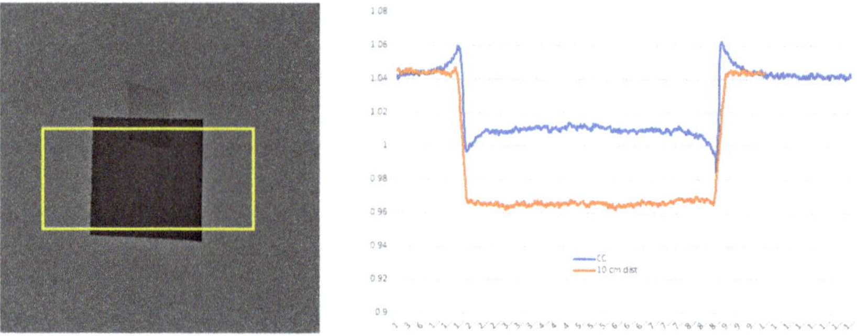

Fig. 3. Normalized transmission image of a 0.2 mm thick PE foils (obtained in close contact to the detector) and the profile as shown on the right side. The second profile was measured with the sample 10 cm away from the detector.

multiple scattering in the sample is increased, in addition to spectral changes. It is worth mentioning that the detector has a spectral response also due to its absorption probability in 1/v behavior (v = neutron velocity).

Another aspect in the quantification is the limits given by the detection system. Each detection system has its inherent noise level, given by different reasons, including the statistics of the neutron count reactions. Longer exposure in the data acquisition can help to reduce the noise level, even by stacking several exposures.

The detection limit regarding to the hydrogen content is reached if the disturbed and the undisturbed signal deviates just by two noise values. Taking into account also the electronic properties of the digital detection systems, the resulting noise level is on the order of a few percent [5].

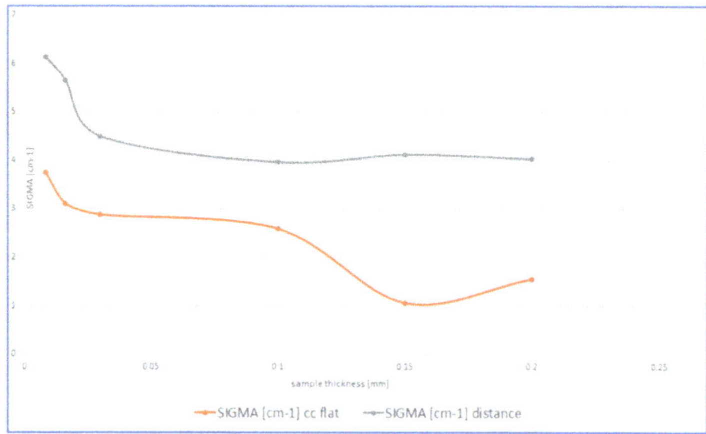

Fig. 4. Attenuation coefficients of thin PE foils with thickness between 0.01 and 0.2 mm obtained either for the "close contact" and the "10 cm distant" case, respectively.

3   Guidelines for Hydrogen Quantification

Because of the difficulties in the determination of the "disturbed" signal, which is needed for a quantification according to Eq. 1, we recommend a calibration process for the individual investigations. Samples with well-known hydrogen content and similar sample geometry should be measured under conditions, which will later be used in the practical approach.

The neutron spectra at different neutron imaging beam lines can differ enormously as shown for the three options at SINQ, PSI in Fig. 5. Therefore, it is important to perform the calibration process individually for each facility separately.

Because of the high amounts of scattered neutrons from the interaction with the sample material, which can disturb the measurement process in particular in low distances between sample and detector, we propose and advice a minimal distance of about 10 cm.

Even if the spatial resolution might be reduced in divergent beam a little, this should be tolerated in most cases. A higher spatial resolution can be obtained when no precise quantification is needed – as a second and separate study.

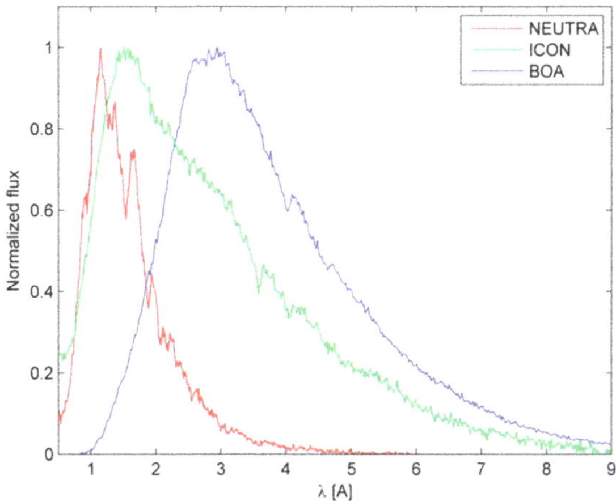

Fig. 5. Neutron spectra at the three beam lines at the spallation neutron source SINQ, used for neutron imaging. In transmission experiments, different effective attenuation coefficients are determined for each beam line.

4   Summary

Although determination of hydrogen amounts may seem easy at first glance, due to its high contrast and visibility, the process of neutron scattering in the sample and the particular detection process is much more complicated. It is necessary to understand all these aspects in detail before the best practices and evaluation strategy can be determined. As the scattering properties in hydrogen are much stronger than those of other materials, we must be especially careful here.

A direct comparison with data from cross-section libraries is very problematic because of the individual spectral conditions at the beam lines and within the samples, depending on their thicknesses or hydrogen content. Furthermore, the particular chemical bound and in some cases also the temperature has to be taken into account. Therefore, we propose performing a calibration procedure with samples of well-known content for better accuracy.

References

1. Hassanein, R., et al.: https://doi.org/10.1016/j.nima.2005.01.161
2. Raventos, M., et al.: https://doi.org/10.1016/j.phpro.2017.06.038
3. Lehmann, E., et al.: Nondestr. Test Eval. **16**, 191–202 (2001). https://doi.org/10.1080/105897 50108953075
4. Blau, B., et al.: https://doi.org/10.1080/10448630903120387
5. Grosse, M, et al.: https://doi.org/10.1016/j.phpro.2017.06.037
6. Perego, R.C., Blaauw, M.: Incoherent neutron-scattering determination of hydrogen content: Theory and modeling. J. Appli. Phys. **97**(12) (2005)
7. Barcellos, L.F.: Monte Carlo simulator project for neutron transport with continuous energy: shielding, criticality and spectral angular neutron flux analysis. PhD thesis UFRGS Brazil (2021) http://hdl.handle.net/10183/232701

How to Perform Cultural Heritage Research by Using Neutron Imaging? Considerations Based on PSI's Experiences

Eberhard Lehmann[(⊠)] and David Mannes

Laboratory for Neutron Scattering and Imaging, Paul Scherrer Institute, Villigen, Switzerland
eberhard.lehmann@psi.ch

Abstract. The observation of inner, hidden material distributions of objects with cultural heritage importance requires methods which enable a transmission of the used radiation. Because inspection with visible light is limited for non-opaque samples, only X-rays and neutrons can be used for such investigations. Alternative kinds of radiation like electrons, micro-waves are limited in the practical transmission and the required spatial resolution.

Besides a retrospective of successful project, we want to present ways how to collaborate and to emphasize the importance of personal commitment and contacts with relevant people in the museum's community.

To answer the question "how to present best our results", a library of image data is under preparation for ancient Tibetan bronze objects, studied with neutron imaging methods at the PSI facilities.

Keywords: image contrast · X-ray attenuation · neutron imaging · organic materials · visibility · detection limits · Buddhist bronze sculptures

1 Cultural Heritage Studies vs. the Art Market

We show in Fig. 1, there are two aspects for the study of samples of cultural heritage importance: the investigation for the general gain in knowledge about the object – and the art market. In this Figure, we point out that the interests are overlapping but also different in several aspects.

Whereas the society is interested in scientific facts about the samples of historical importance, the art marked is more privately driven and money oriented. Although most of our cultural relevant objects are stored and presented in museums, there are many private collections with a limited access by the public.

For the study of such samples in a non-invasive way, there are different methods available, as indicated in Fig. 2. Some of them can only be applied on the samples' surface, while X-ray and neutron imaging methods enable (under certain conditions) a full transmission and the study of the inner content in this way.

It must be stated, that some of the methods are already well established in larger museums and used for systematic investigations of objects situated there. This inhouse

© The Author(s) 2026

A. E. Craft and H. Z. Bilheux (Eds.): WCNR 2024, SPPHY 348, pp. 25–31, 2026.
https://doi.org/10.1007/978-3-032-15003-5_4

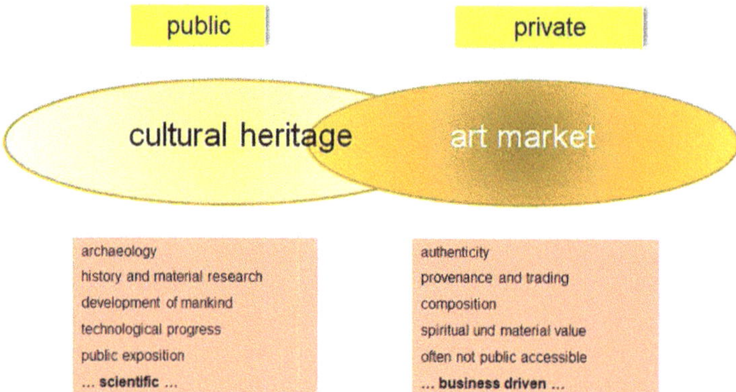

Fig. 1. The different aspects for studies of objects of cultural heritage importance

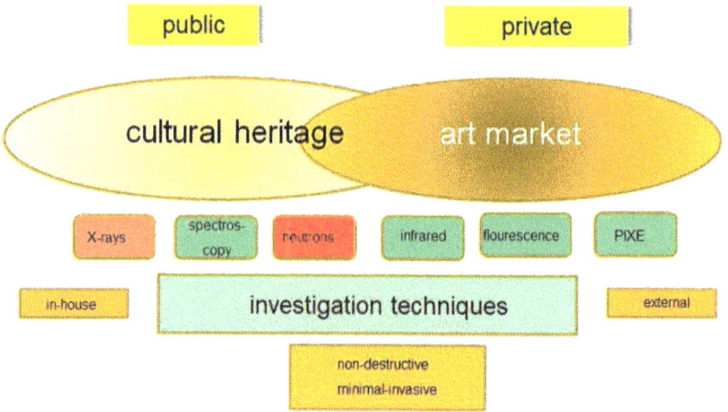

Fig. 2. Overview about methods for investigations of cultural heritage objects

research is performed by experts and driven by scientific requests. However, there are more sophisticated methods available, which can only be offered by so-called "large scale" facilities like accelerators, synchrotron light sources or nuclear reactors. In such cases, the objects must be transported to that facilities and the investigations are performed there.

This is the case particularly with neutrons because strong sources are immobile. This paper is dedicated to the use of neutron imaging methods for the study of cultural heritage objects with some link to X-ray methods – for comparison or/and combination. Because the effort for neutron imaging is quite higher than for X-ray studies, this method should be applied for cases when X-ray methods fail, or important additional information can be obtained by the neutrons. In chapter 3, we will present examples for such investigations that we performed successfully in the past.

2   What Kind of Materials/Objects Are Worth Studying with Neutrons?

Due to the high penetration power of slow neutrons for heavy elements, metals are preferred for inspection, in particular gold, lead, copper, … On the other hand, most light elements provide large contrast by their attenuation power – all hydrogenous materials can be detected in small amounts. Effects like corrosion, humidity distributions, conservation treatment and protection measures can be studied as processes on different time scales and contrast levels.

The combination of neutron and X-ray imaging provides a gain in information about the samples under investigation (see Fig. 3).

The physics behind this interaction behavior with matter of either neutrons or X-rays, respectively, is described in the literature [1–3] and needs not be repeated here.

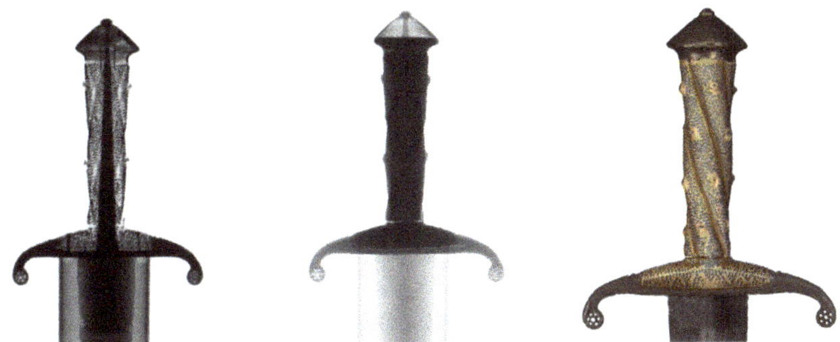

Fig. 3. Example of a combined study with X-rays (left), showing the metallic parts and neutrons (middle) indicating the remained wood parts: the sword from Lake Zug (Switzerland) [4]. The right image combines the information of both data sets into a complete picture of the sample.

3   Results of Previous Studies

With the basic knowledge about the higher transmission of metals by thermal neutrons, we have performed series of studies of relevant objects from different museums, Swiss based and also on international level at the neutron imaging facilities at PSI, Switzerland [5, 6].

The much higher contrast for organic materials has been demonstrated in a study, published in [7]. Here we show some results for typical sample materials of possible material fillings, see Fig. 4.

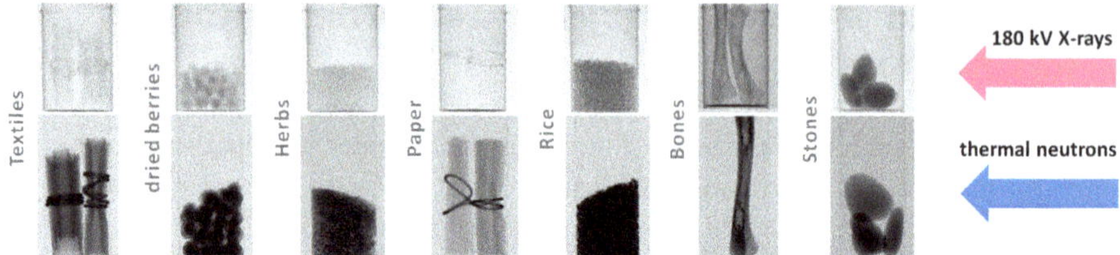

Fig. 4. Comparison of the attenuation contrast of different materials used for ritual fillings in Buddhist bronze statuettes, for 180 kV X-rays and thermal neutrons, respectively. The visibility for X-rays is often very limited, while neutrons provide a high contrast, already in thin material layers.

We summarize here some results of prominent investigations with their references:

- Renaissance bronzes from the Rijksmuseum Amsterdam, the Netherlands– investigation of the casting method and inner fillings [8]
- Roman bronzes from excavations within Switzerland – systematic tomography observation and determination of casting features [9]
- The bust of Marcus Aurelius, Avanches, Switzerland – determination of the wall thickness based on a full tomography run [10]
- The Roman gladius, Vindonissa museum, Brugg-Windisch, Switzerland, characterization of the object and the conservation status [11]
- Buddhist bronzes of the Rietberg Museum, Switzerland, determination of the holy fillings and their assembly [12]
- The violinist, sculpture from M. Gargallo, Barcelona, Spain, status of the inner lead corrosion, caused by interaction with the wooden core [13]
- Old brass instruments from 18th century, wall thickness determination and status of inner corrosion, playability [14]

Most of the studies were performed in order to describe the protection status of the samples and development of strategies for the further treatment and preservation. In some cases, also the wall thickness was derived non-invasively. Further, it was of important to know more details about the manufacturing process, e.g. casting and sealing.

4   A Fake ?

In selected cases, inspection results from neutron and X-ray studies have been used in auctions of renowned art sellers – with some success as additional information about the objects. The knowledge of inner features of art objects might be of essential interest to define their age, provenance and the cultural importance – and finally the price.

We pick out here one example, published by Sotheby's in 2016 [15] claiming an X-ray study and showing several features like in a tomography slice. It is shown in Fig. 4 together with the photo of the investigated object.

As stated above, we have experience and competence in the investigations of similar objects [16]. Therefore, we claim about the authenticity of these image data and believe to see a clever fake.

We have demonstrated in a dedicated other study [7], that copper layers of more than 2 mm will disable the visibility of any content due to the limited transmission of the X-rays. The 20 cm large sample under discussion has a much thicker bronze layer around the hidden filling. So, the claim to have an X-ray study is wrong in principle. Only neutrons are able to transmit such thick layers of the metal sealing.

However, the shown amount of different hydrogenous materials (paper, textile, wood, grains, …) would attenuate the neutron beam so much, that their visibility would be impossible. A possible gamma radiation scan cannot resolve organic materials in such clearance.

What is shown, can never be a view from the bottom because 20 cm of materials has to be transmitted. Moreover, the circular features at the picture edges indicates more a tomography slice near the bottom of the object with cropping artifacts by limited FOV. But there is so much of "sample material" to be analyzed that a separation is not possible in this quality.

Based on our practical knowledge and the arguments above, we tried to establish a dialogue with the art seller Sotheby's – but with no real dedicated feedback. This discussion ended in a newspaper article finally [17] (Fig. 5).

Fig. 5. Photo of a gilt-bronze figure depicting Ngawang Lobsang Gyatso, Dalai Lama 5th, Central Tibet, 17th century, sold by Sotheby's for 1.5 Mio US-$, based on the message: "Thanks to x-ray technology, here we can even take a glimpse of the contents of the interior of the bronze", shown on the right. Outer size is about 20 cm in length, height, depth, data from [15].

5   How to Present Our Results Best?

Results of the investigation of museums and art objects have been published either in journals of neutron and X-ray science, highlighting the power of both methods or in catalogues and presentations of the museums and collections. In all these cases, only a limited number of image data or other results can be presented and only a limited audience can be approached.

Therefore, we intend to use and apply much more modern methods and tools, which are organized on internet platforms.

This is in particular valid for a series of studies of Buddhist bronze statuettes from Asia and dated back centuries ago in their manufacturing. Until now, we have investigated by neutron imaging – the only valid method for this case – about 100 such samples successfully.

In the planned "Buddhist Data Base" (BDB) we will include next to photos and sample description the neutron radiography and neutron tomography data on first glance. Later, it might be possible to add also the projections from the tomography data acquisition, the volume data of the reconstructed tomography studies and selected slices through the objects. It might be, however, a handicap of these data to be on the order of several GBytes in size. The planned data base will be accessible by the public, however with restrictions for copy rights.

Maybe, our approach can become a template for other teams, museums and collections for the presentation of their objects and observation results.

6 Conclusions and Outlook

It has been shown that neutron imaging methods are very useful for the analyses of museums objects, in particular when thick metallic layers have to be transmitted and small amounts of hydrogenous materials have to be observed. Some other light elements as B, Li give also a high visibility by neutrons. In this way, corrosion products, moisture, conservation treatment and inner material distributions can be determined non-invasively. Therefore, neutron imaging complements and combines to X-ray studies, in the best way as a fused data set.

In order to present the valuable and often unique results of investigations to the public, an internet-based data base is in preparation, first of all for Buddhist bronze sculptures, which were studied with great success in the past years.

References

1. Kardjilov, N., Festa, G., (eds.): Neutron Methods for Archaeology and Cultural Heritage. Springer, (2017). https://doi.org/10.1007/978-3-319-33163-8
2. D'Amico, S., Venuti, V.: Handbook of Cultural Heritage Analysis. Springer Nature. https://doi.org/10.1007/978-3-030-60016-7
3. Strobl, M., Lehmann, E. (eds.), Neutron Imaging. IOP Publishing, Bristol. https://doi.org/10.1088/978-0-7503-3495-2
4. Mannes, D., et al.: Combined neutron and X-ray imaging for non-invasive investigations of cultural heritage objects. Phys. Proc. **69**, 653–660 (2015)
5. Lehmann, E., et al.: NEUTRA as European reference facility. Test Eval. **16**, 191–202 (2001). https://doi.org/10.1080/10589750108953075
6. Blau, B., et al.: Swiss Spallation neutron source SINQ. https://doi.org/10.1080/10448630903120387
7. Lehmann, E., Mannes, D., Trtik, P.: Study on the visibility of organic materials within metallic covers, the Pseudo-Buddha, these proceedings
8. van Langh, R., Lehmann, E., Hartmann, S., Kaestner, A., Scholten, F.: The study of bronze statuettes with the help of neutron imaging techniques. Anal. Bioanal. Chem. **395**, 1949–1959 (2009)

9. Deschler-Erb, E., Lehmann, E., Wöhrle, M.: Using neutron imaging methods for the non-destructive investigation of large ancient bronze artifacts. In: Deschler-Erb E, Ph. Della Casa (eds.) New research on ancient bronzes. Acta of the XVIIIth International Congress on ancient bronzes. Zurich studies in archaeology, vol 10, University Zürich, Zürich, pp 311–315 (2015)

10. De Pury-Gysel, A., Lehmann, E., Giumlia-Mair, A.: The manufacturing process of the gold bust of Marcus Aurelius: evidence from neutron imaging. J Roman Archaeol **29**, 477–493 (2016). https://doi.org/10.1017/S1047759400072275

11. Mannes, D.., et al.: The study of cultural heritage relevant objects by means of neutron imaging techniques. Insight **56**(3), 137 (2014)

12. Boehm, C.: Tibeto-Chinese gilt brass sculptures of the early Ming dynasty in the museum Rietberg. Art of Asia **49**(5), 84–97

13. Masalles, A., Lehmann, E., Mannes, D.: Non-destructive investigation of "The Violinist" a lead sculpture by Pablo Gargallo, using the neutron imaging facility NEUTRA in the Paul Scherrer Institut. Phys. Procedia **69**, 636–645 (2015)

14. Mannes, D., Lehmann, E.: Monitoring the condition of played historical brass wind- instruments by means of neutron imaging. In: Steiger, Av., Allenbach, D., Skamletz, M. (eds.) ROMANTIC BRASS – Präventive Konservierung, Material und Akustik. Symposien 4 und 5, Schliengen: Argus 2020 (Musikforschung der Hochschule der Künste Bern, Bd. 15)

15. https://www.sothebys.com/en/auctions/ecatalogue/2016/important-chinese-artn09541/lot.161.html

16. Lehmann, E., Hartmann, S., Speidel, M.: Investigation of the content of ancient tibetan metallic buddha statues by means of neutron imaging methods. Archaeometry **52**, 416–428 (2010)

17. Butin, H.: Des Dalai Lamas Kern,https://www.faz.net/aktuell/feuilleton/kunstmarkt/sotheby-s-das-roentgenbild-einer-bronzefigur-wirft-fragen-auf-17762033.html

Cr-coated Zircaloy-4 Performance During Ramping Conditions

Aaron W. Colldeweih[1]([⊠]), David W. Kamerman[1], Malachi Nelson[1], Aaron Craft[1], Nathan Capps[2], Caleb Massey[2], Mackenzie Ridley[2], Pavel Trtik[3], Michael Meyer[3], and Okan Yetik[3]

[1] Idaho National Laboratory, Idaho Falls, ID, USA
aaron.colldeweih@inl.gov
[2] Oak Ridge National Laboratory, Oak Ridge, TN, USA
[3] Paul Scherrer Institute, Villigen, AG, CH, Switzerland

Abstract. The development of accident tolerant fuel (ATF) concepts began as a result of the Fukushima nuclear accident in Japan. The primary objective of ATF cladding is to enhance the cladding corrosion resistance. The most prominent ATF cladding candidate is a chromium coated zirconium alloy. The material in this work involves a physical vapor deposition (PVD) Cr-coated Zircaloy-4 cladding. The test matrix included three sets of tubes that were all internally pressurized at elevated temperature (300 °C) to simulate in-reactor ramping conditions. This included pressurization induced biaxial loading conditions, namely high hoop strain in addition to axial strain. The second and third set of tubes were additionally subjected to high temperature, and high purity hydrogen environments to simulate an accelerated hydrogen pickup through potential flaws in the Cr-coating. The third set of tubes was then tested at the severe accident test station (SATS) at Oak Ridge National Laboratory (ORNL) where the tube was pressurized and heated at a rapid rate until bursting. Each set of tubes were then sectioned to 4.0 mm lengths and imaged along the axial direction of the tube using the Neutron Microscope detector (NM) at the ICON beamline at the Paul Scherrer Institute (PSI). Radiography results showed that even a severely strained Cr-coating was effective at preventing hydrogen uptake. Radiography also provided insight into the differences in deformation that took place in the pre-strained versus the as-received Cr-coating near the opening of a burst tube. These results highlight the capabilities and challenges of high-resolution neutron radiography for materials with thin coatings.

Keywords: Cr-coated cladding · zirconium alloy · hydrogen uptake

1 Introduction

Accident tolerant fuel (ATF) was developed over the last decade as a result of the Fukushima nuclear accident in Japan. The primary objectives of ATF cladding are to improve the oxidation resistance and increase the coping time during a loss-of-coolant accident (LOCA) [1, 2]. Cr-coated Zircaloy cladding is the leading ATF cladding concept

© The Author(s) 2026
A. E. Craft and H. Z. Bilheux (Eds.): WCNR 2024, SPPHY 348, pp. 32–39, 2026.
https://doi.org/10.1007/978-3-032-15003-5_5

benefitting from a relatively high level of technical readiness [3–6]. While the primary objective of ATF cladding is to improve performance during accident conditions, it is also important that ATF cladding maintains and improves the cladding performance during normal service conditions without severe neutronics and economical penalty.

Cr-coating could also reduce overall oxidation and hydrogen uptake of the cladding during normal operation. The reduction of oxidation and hydrogen uptake could extend the service lifetime of the cladding [7]. With an increased demand for electricity, extending the cladding service lifetime would support such needs. Other ways to support electricity demands include load following where the output power of a power plant is adjusted based on the needs of the electricity grid. However, nuclear power plants are traditionally designed and operated at constant power for baseload power generation. At constant power there are fewer and slower temperature transitions of the fuel occur compared to load following. During load following, also known as ramping, power must be increased or decreased in a shorter time span compared to traditional operation. The increased power cycling in a nuclear power plant results in increased pellet cladding interaction (PCI) as the fuel thermally expands faster than the cladding [8–13].

Load following and more frequent PCI inherently increases the hoop stress on the cladding, particularly the outer diameter. The increased hoop stresses could lead to an increased risk of cracking in the Cr-coating, given its brittle nature at operational temperatures [14], which may provide pathways for localized oxidation and hydrogen uptake. Further, it is important to understand the behavior of the cladding under accident conditions, namely LOCAs, after the potential coating fracture. Previous work demonstrated the effectiveness of neutron radiography at visualizing micro-cracks in zirconium alloys in addition to locally quantifying hydrogen around the tip of the crack [15].

In this work, coated Zircaloy-4 cladding tubes were internally pressurized to a controlled hoop strain at 300° C. A second set of tubes were subjected to hydrogen charging after the internal pressurization. Finally, a third set of tubes were burst at high temperatures. The objective is to evaluate the cracking characteristics of Cr-coated Zircaloy-4, hydrogen uptake protectiveness, and burst opening after straining.

2 Material and Methods

2.1 Materials

Samples consisted of cold-worked stress-relieved Zircaloy-4 (Zry-4) cladding conforming to ASTM B353 [16], with an outer diameter of 9.3 mm, a wall thickness of 0.57 mm, and cut to a length of 150 mm. The samples were coated with a dense chromium layer which was deposited using physical vapor deposition (PVD) to a 9 μm nominal thickness.

2.2 Mechanical Testing and Pre-characterization

To simulate ramping conditions, internal pressurization testing used the system to load samples with internal pressure [17, 18]. This testing system used feedback from a high temperature transverse (hoop) extensometer to control a venting pneumatic pressure regulator between a high-pressure reservoir and the test specimen. Samples were enclosed

34 A. W. Colldeweih et al.

in a furnace and heated to 300 °C in an argon atmosphere. The samples were loaded at 0.1%/min (16.7 x 10^{-6} /s) hoop strain rate to achieve the desired strain, held for a period of time, to measure the stress relaxation response. Following the strain events, the tubes were pre-characterized at radial cross sections of the rod adjacent to the ring section that would be imaged with radiography (see Fig. 1). The crack widths were measured respective of the hoop strain level and can be found in Table 1.

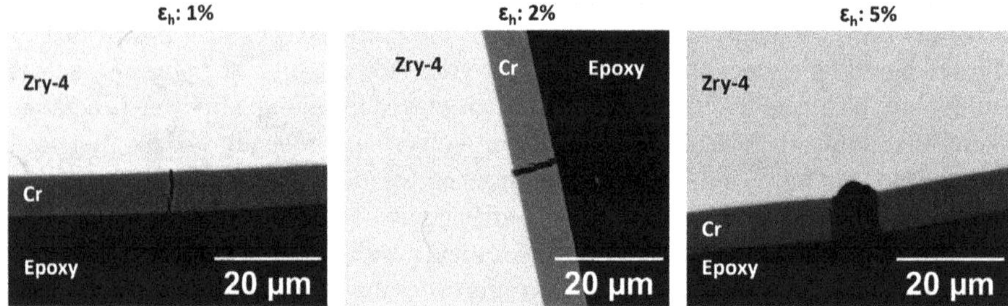

Fig. 1. Pre-characterization of the strained tubes. The backscatter electron (BSE) images are of representative cracks in the 1, 2, and 5% hoop strain samples. Micrographs provide insight into the features that are to be imaged with radiography.

Table 1. Crack width range for the respective hoop strain level.

Hoop Strain	Crack Width [µm]
1%	0.4–1.5
2%	0.5–1.7
5%	1.5–6.3

The second set of rods were internally strained as previously described and subjected to hydrogen charging. The hydrogen charging procedure was a gaseous type charging system. The system and procedure is well defined in literature [19] where high purity hydrogen fills inside a quartz tube inside a furnace. The zirconium alloy claddings were held within the charging chamber in a high temperature and high purity hydrogen environment, leading to the uptake of hydrogen in the cladding material. Kamerman reported for the specified procedure, several hundred wppm hydrogen would be absorbed either homogenously or in the form of a hydride rim, depending on the sample preparation [19]. Additional samples of the same base material were used to create standards for radiography measurements. The process of creating and imaging the samples for radiography, including hydrogen standards, are found in the following reference [15].

The third set of rods were strained through internal pressurization as described above and then subjected to LOCA conditions at the SATS facility. Pressure was applied at room temperature through the pressure line at the top of the sample. The sample was heated to 100 °C in static air. At 100 °C, steam flow was initiated and then heated to 300 °C and held for 5 min as a reference point for pseudo-normal operating temperatures. After the 5-min hold at 300 °C, the temperature was ramped by 5 °C/s until the cladding ruptured. The specimens were immediately air cooled after rupturing. The ruptures can be seen in the pre-characterization images in Fig. 2. Radial cross sections were then removed from the burst opening of each tube for radiography. The cut locations for the radial cross sections are designated by the red lines in Fig. 2.

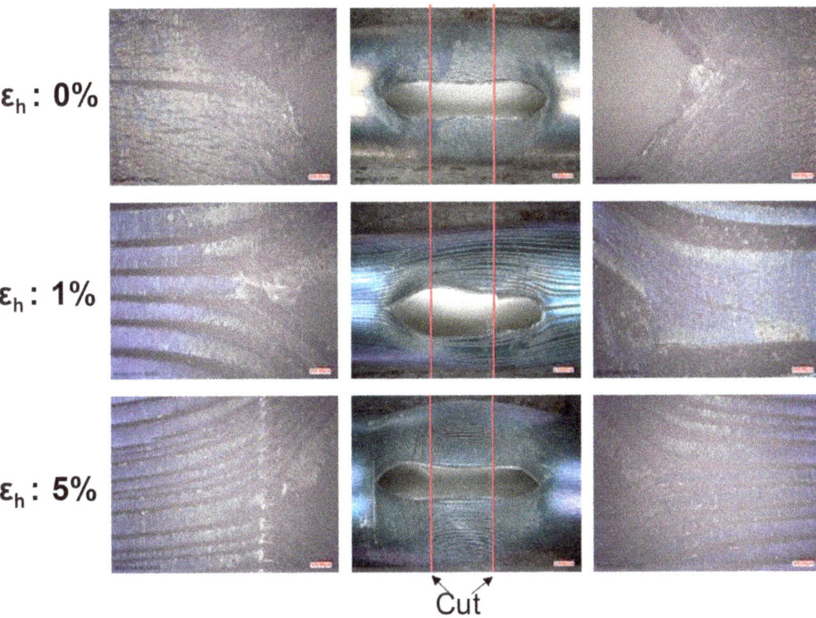

Fig. 2. Visual inspections of the burst openings from the SATS tested materials.

3 Radiography Results and Discussion

Neutron radiography was conducted at PSI's Neutron Microscope detector at ICON beamline, SINQ, PSI [20]. The microscope has been equipped with 16.9 MPixels sCMOS camera with 12 μm pixel size (Andor Balor). The 5:1 microscope magnification led to 2.4 μm isotropic image pixel size. The image acquisition was 60 s for each of the 40 sample projection and open beam images, as well as each of the 10 black body sample projections and black body open beam images. Further details of the image processing and calibration techniques can be found in the following reference [15]. Figure 3 presents the radiographs of strained samples and samples that were strained and subjected to the hydrogen charging. The strained samples did not present any details on the cracking in the coating. In addition to the strained samples, reference samples were imaged to create a calibration curve relating the image transmission intensity to hydrogen concentration.

A calibration curve was used to translate the average radial intensity of the cladding to hydrogen concentration as seen in Fig. 4. The radiographs and concentration curves showed that there was no measurable hydrogen pick up in the strained rods subjected to published hydrogen charging methods. It should be noted that the wall thickness in Fig. 4 may have been affected by the sample preparation, orientation of the sample in front of the scintillator detector, and slight dimensional variation in cladding lots. Observations of the cladding hydrogen uptake were additionally verified through hot gas vacuum extraction measurements to determine the hydrogen concentration. All cladding samples were measured between 10–20 wppm hydrogen as expected for as-fabricated Zircaloy cladding.

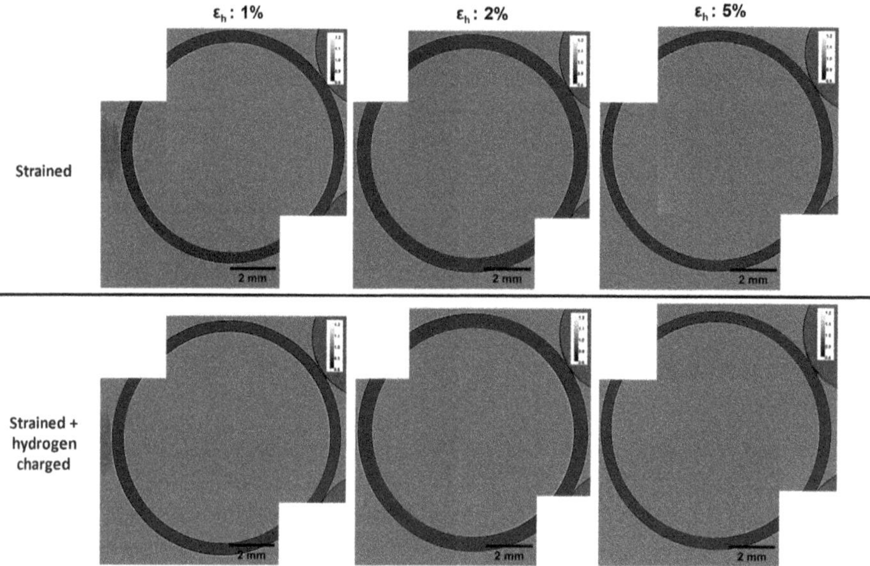

Fig. 3. Radiographs of the strained samples (top) and strained plus hydrogen charged (bottom).

Figure 5 shows the overall view of the burst opening of the sample burst at SATS. The figure also shows a higher magnification of an region of interest showing the extent of coating cracking. The backscatter electron (BSE) images below confirmed that the cracking did not extend into the zirconium matrix, but rather expanded in width into the Cr-coating. The BSE images highlighted that the radiographs were capable of visualizing small cracks that were roughly 9 μm deep and as little as 20 μm wide. The ability to visualize these cracks provides support for future development of this technique for the use of characterizing fractured ATF claddings that may be expanded to irradiated ATF claddings.

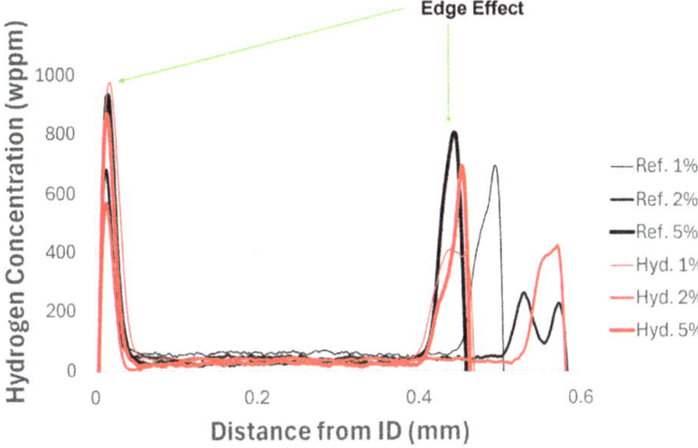

Fig. 4. Quantification results of the strained samples compared the strained and hydrogen charged samples.

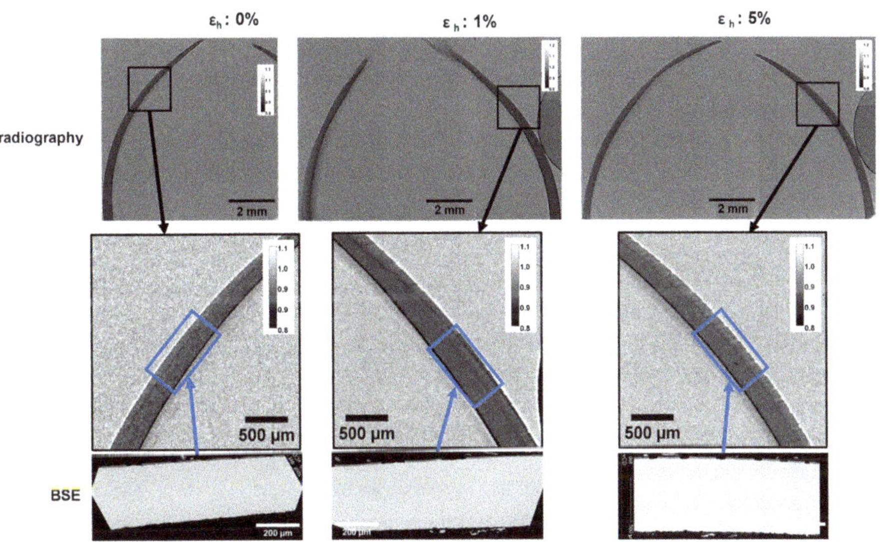

Fig. 5. Radiograph (top and middle) of the burst opening of the cladding. BSE images (bottom) show micrographs of representative cross sections of the magnified radiograph region.

4   Conclusion

Neutron radiography characterization was performed on internally pressurized Cr-coated Zry-4 tubes. Pre-characterization using light optical and scanning electron microxcopy showed coating crack widths ranging from ~ 0.4 – 6.3 µm. Such cracks were not detected through neutron radiography. However, through further characterization of hydrogen uptake, it was shown that protectiveness against hydrogen uptake of the cracked coatings remained after coating cracking at various levels. Hydrogen quantification provided quantitative information confirming that the samples subjected to hydrogen charging remained near as-received levels. Radiography of the samples burst at SATS showed that the coating deformation was distinguishable at crack depths and widths of about 9 μm and 20 μm, respectively. Additionally, the burst opening of the pre-strained cladding showed significantly different deformation characteristics than that of the reference material. The variation in deformation was qualitatively identified in radiography. The results demonstrate that neutron radiography is a viable method for analyzing flaws in thin coatings on cladding material and confirm protectiveness against hydrogen uptake of the strained Cr-coating.

Acknowledgments. Part of this work us based on experiments performed at the Swiss spallation neutron source (SINQ), Paul Scherrer Institute, Switzerland. The paper was submitted for the review in August 2024 and accepted for publication in January 2025.

References

1. Qiu, B., et al.: A review on thermohydraulic and mechanical-physical properties of SiC, FeCrAl and Ti3SiC2 for ATF cladding. Nuclear Eng. Technol., 1–13 (2020)
2. DOE, Development of Light Water Reactor Fuels with Enhanced Accident Tolerance. DOE, Washington D.C., USA (2015)
3. Lin, Y.P., et al.: Path Towards Industrialization of Enhanced Accident Tolerant Fuel
4. Bischoff, J., et al.: Development of Cr-coated Zirconium Alloy Cladding for Enhanced Accident Tolerance
5. Yeom, H., et al.: Improving deposition efficiency in cold spraying chromium coatings by powder annealing. Int. J. Adv. Manuf. Technol. **100**(5–8), 1373–1382 (2019)
6. Maier, B., et al.: Development of cold spray chromium coatings for improved accident tolerant zirconium-alloy cladding. J. Nucl. Mater. **519**, 247–254 (2019)
7. International Atomic Energy A, Waterside corrosion of zirconium alloys in nuclear power plants. IAEA-TECDOC-996 (1998)
8. Chan, K.S.: An assessment of delayed hydride cracking in zirconium alloy cladding tubes under stress transients. Int. Mater. Rev. **58**(6), 349–373 (2013)
9. Williams, C.D., et al.: Zircaloy-2 lined zirconium barrier fuel cladding. ASTM Special Technical Publication, vol. 1295 (1996)
10. Shimada, S., et al.: A metallographic and fractographic study of outside-in cracking caused by power ramp tests. J. Nucl. Mater. **327**(2–3), 97–113 (2004)
11. Sakamoto, K., Nakatsuka, M., Higuchi, T.: Simulation of outside-in cracking in boiling water reactor fuel cladding tubes under power ramp. J. ASTM Intl. (2011)
12. Arimescu, I., Karlsson, J.: Towards understanding beneficial effects of slow power ramps. J. Nucl. Sci. Technol. **52**(10), 1274–1280 (2015)
13. Karlsson, J.K.-H.: Ramp Testing at the Studsvik R2 Reactor 1969–2005. Top Fuel 2021 (2021)
14. Holzwarth, U, Stamm, H.: Mechanical and thermomechanical properties of commercially pure chromium and chromium alloys. J. Nuclear Mater. **300**, 161–177 (2002)
15. Colldeweih, A.W., et al.: Delayed hydride cracking in Zircaloy-2 with and without liner at various temperatures investigated by high-resolution neutron radiography. J. Nucl. Mater. **561**, 153549 (2022)
16. ASTM B353–12

17. Kamerman, D., Nelson, M.: Multiaxial plastic deformation of Zircaloy-4 nuclear fuel cladding tubes. Nucl. Technol. **209**(6), 872–886 (2023)
18. Kamerman, D., et al.: Development of axial and ring hoop tension testing methods for nuclear fuel cladding tubes, vol. 31, p. 2 (March 2022)
19. Kamerman, D., et al., Formation and characterization of hydride rim structures in Zircaloy-4 nuclear fuel cladding tubes. J. Nuclear Mater. **586** (2023)
20. Trtik, P. and E.H. Lehmann: Progress in high-resolution neutron imaging at the Paul Scherrer Institut-the neutron microscope project. J. Phys. Conf. Ser. **746**(1), 6 (2016)

Shapes of Mobile Interfaces in Tubes Revealed by Neutron Imaging: Sensitivity Analysis for Three-Phase High-Pressure Systems from Methane, *p*-xylene, and Water

Ondřej Vopička[1] , Tereza-Markéta Durďáková[1] , Petr Číhal[1] , Pierre Boillat[2,3] , Jongmin Lee[2] , Martin Melčák[1] , Jonatan Šercl[1] , Štěpán Tvrdý[1] , Jan Heyda[1] , and Pavel Trtik[2(✉)]

[1] Department of Physical Chemistry, University of Chemistry and Technology, Prague, Technická 5, Prague 6 166 28, Czech Republic
ondrej.vopicka@vscht.cz
[2] Laboratory for Neutron Scattering and Imaging, Paul Scherrer Institute, 5232 Villigen PSI, Switzerland
pavel.trtik@psi.ch
[3] Electrochemistry Laboratory, Paul Scherrer Institute, 5232 Villigen PSI, Switzerland

Abstract. The determination of interfacial energy (hence interfacial tension) from the shapes of mobile interfaces in cylindrical tubes is a rather viable though seldom utilized method despite the relative ease of its implementation. In this work, neutron imaging is used to reveal the shapes of interfaces in the system of perdeuterated *p*-xylene (p-C_8D_{10}) layered over water (10.88 mol.% of H_2O in D_2O) and exposed to pressurized methane (CH_4, from 1.0 to 101 bar) at 7.0 to 30.0 °C while enclosed in the titanium tubular cell. These non-tactile measurements are performed using relatively high spatial resolution (pixel size 20.3 μm). The sensitivity limits determining the uncertainty of thus derived interfacial tensions are analyzed. Interfacial tension for methane – *p*-xylene was determined at the experimental uncertainty of 2 mN·m^{-1}. Interfacial tension for *p*-xylene – water was estimated at the experimental uncertainty of 16 mN.m^{-1}. Water and *p*-xylene are severe formers of freeze-out from moist and/or hydrocarbons containing natural gas. We survey the conditions needed for the interfacial tension measurements of industrially very relevant systems with prospects of acceptable experimental uncertainties while utilizing the intrinsically straightforward and on the first principles based methodology.

Keywords: Interfacial tension · Uncertainty · Neutron Imaging

1 Introduction

1.1 Phase Interfaces

Interfacial tension determines the shapes of the mobile phase interfaces. The shape is described by the Young-Laplace equation [1–3], and is influenced by gravity. The general Young-Laplace equation simplifies greatly when the interfaces possess an axial

© The Author(s) 2026
A. E. Craft and H. Z. Bilheux (Eds.): WCNR 2024, SPPHY 348, pp. 40–46, 2026.
https://doi.org/10.1007/978-3-032-15003-5_6

symmetry, thus

$$z = \frac{\gamma}{\Delta\rho g}\left(\frac{z''}{\left(1+z'^2\right)^{3/2}} + \frac{z'}{x\left(1+z'^2\right)^{1/2}}\right) \tag{1}$$

The height coordinate of the mobile interface, $z(x)$, at the central plane can be calculated by numerically solving the boundary value problem. The principal radii of curvature can be expressed from Eq. (1), see the literature [1–3], $z' = dz(x)/dx$, $z'' = d^2z(x)/dx^2$. Variable x is the distance from the center, which ranges from zero to the tube inner radius, r, conditions are $z'(x = 0) = 0$ and $z'(x = r) = \cot(\theta)$, in which θ is the contact angle. Parameters include the density difference of the neighboring phases, $\Delta\rho$, the interfacial tension, γ, and the gravitational acceleration ($g = 9.8074$ m·s^{-2} in Villigen [4]). Although axially symmetric phase interfaces in wide-bore tubes oriented parallel to gravity can be easily prepared, their observations [5–7] appear much less common compared to those of (pendant) drops and bubbles. Besides the common pendant drop method [8, 9], interfacial tension can be measured by sensing capillarity [10–12], capillary waves [13, 14], *etc.* We elaborate here on the principles governing the sensitivity of the evaluation of the interfacial tensions from the shapes of mobile interfaces in tubes with focus on the interface shape sensitivity on the system parameters, and demonstrate these principles on the observation of the liquid-gas (methane as gas and supercritical fluid) and liquid-liquid interfaces using the state-of-the-art method of neutron imaging.

1.2 Investigated Systems

Water and BTEX (benzene, toluene, ethylbenzene, xylenes) compounds are impurities of natural gas that are relevant to the formation of freeze-out deposits during its production, transportation, and liquefaction [15–18]. For instance, natural gas hydrate forms at 172 bar and 276 K [18], solid p-xylene forms at 163 bar and 278 K [17]. Water and p-xylene, which has a much higher normal melting point (p-C$_8$H$_{10}$, 13.25 °C [19]) than the other BTEX compounds, are severe freeze-out formers. The use of isotopically labeled compounds enabled the adjustment of contrast among the phases in the studied system, the significant difference in neutron cross-section [20] between protium (H) and deuterium (D) was used. Namely, systems containing p-xylene (p-C$_8$D$_{10}$), water (10.88 mol.% of H$_2$O in D$_2$O) and methane (CH$_4$) were studied.

2 Results and Discussion

2.1 Phase Interfaces Revealed Using Neutron Imaging

The studied liquids were inserted into titanium vessels initially equilibrated at 1 bar of methane and then exposed to the methane pressure step (upper pressure ranged 45–101 bar). Spatially- and temporally-resolved neutron imaging was conducted at the NEUTRA beamline [21] at Paul Scherrer Institute at the measuring position No. 2 (L/D = 365) by MIDI-box detector system using a 30 μm-thick Gd$_2$O$_2$S:Tb scintillator screen and a sCMOS camera (Andor Neo) fitted with a 100-mm objective (Zeiss Makro-Planar). This experimental arrangement yielded the image of 2560 (W) × 2160 (H) pixels in size with an isotropic pixel size of 20.3 μm.

42   O. Vopička et al.

Mobile phase interfaces were detected in the central-plane tomographic reconstructions [22] derived from the radiographies collected using the previously reported methods and sample environment [5, 6, 23, 24].

Numerical solution of Eq. (1) was then fit to the observed interface using the Gauss-Newton method [25, 26], see Fig. 1, the optimized parameters were γ, θ, $z(0)$. Measures of the fit quality were Average Absolute Deviation (AAD) and the 95% confidence intervals of the parameters estimated using the Bonferroni method [25, 26], the latter covers the effects of fit deviations from measured data and sensitivity to the parameters. Higher AAD was observed for the liquid-liquid interfaces.

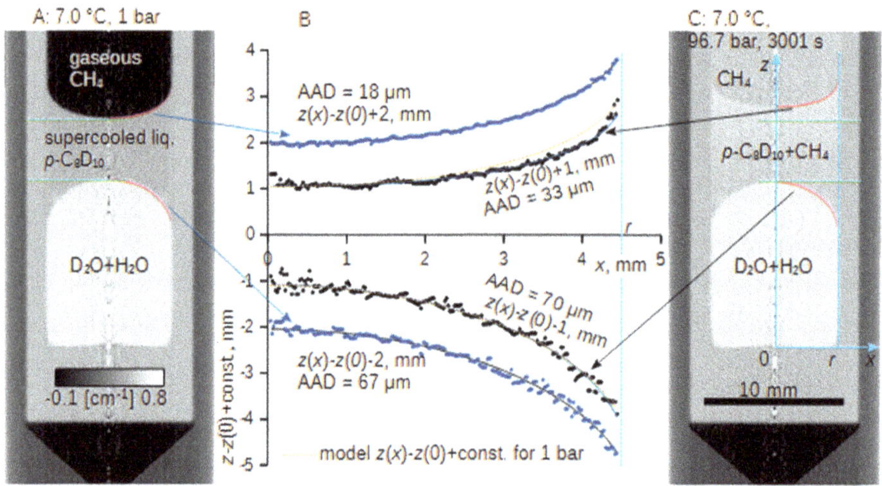

Fig. 1. Central-plane tomographic reconstructions with the solutions of Eq. 1 are shown in A and C (experimental conditions, such as temperature, pressure, tube radius, time after pressure step from 1 bar to 96.7 bar, scale of the linear attenuation coefficient are indicated). Cells are from titanium grade 5. Note that the perdeuterated p-xylene and its solution with methane are supercooled at the conditions in A and C, see the literature [5, 17] for details on the phase equilibrium conditions. Shapes of interfaces including their AADs and solutions of Eq. 1 are shown in B.

The model of the central-plane cut of the phase interface, *i.e.* solution of Eq. (1), is similar to a circle part at certain conditions, which is adverse for the interfacial tension measurement. Essential is the sensitivity of the interface shape to the change of $\gamma/\Delta\rho$, while this sensitivity decreases as $\gamma/\Delta\rho$ increases, and as θ increases (acute) or decreases (obtuse). The same is yielded for low gravitational acceleration, which is of low practical relevance here. The sensitivity increases with increasing r (Fig. 2). The $|\partial z(x)/\partial(\gamma/\Delta\rho)|$ at $z = 1/2 \cdot z(r)$ was chosen as a meaningful measure of sensitivity, which is expectedly meaningful also for observations of interfaces in very wide tubes, in which the interface becomes largely flat. The use of very wide tubes, however, seems unpractical for neutron imaging due to the attenuation.

To illustrate the above principles, the p-xylene-water interface had $\gamma/\Delta\rho \approx$ 2.10^{-4} m$^3\cdot$s^{-2}, while the methane-p-xylene interface had $\gamma/\Delta\rho \approx 2.5.10^{-5}$ m$^3\cdot$s^{-2}. Densities at the interfaces were calculated using equations of state for the pure components [27–29], and using the evaluation of concentration distribution, swelling, and interfacial tension reported previously [5, 6]. Therefore, the smaller sensitivity of the interface shape on $\gamma/\Delta\rho$ is yielded for the liquid-liquid interface, while AAD for Eq. (1) and experimental shape was higher. This renders higher uncertainty of the liquid-liquid interfacial tension. The uncertainty calculated based on the Bonferroni method yields from 13 to 16 mN$\cdot$m^{-1} for the p-xylene-water interface (Fig. 3), and 2 mN$\cdot$m^{-1} for the methane-p-xylene interface. Overall, high inner diameter of the sample tube, low $\gamma/\Delta\rho$, high spatial resolution [30] and contrast, and the materials choice leading best to complete wetting or complete non-wetting are the prerequisites for lowering the uncertainty of the interfacial tension measurement based on the mobile interface in tube oriented parallel to gravity.

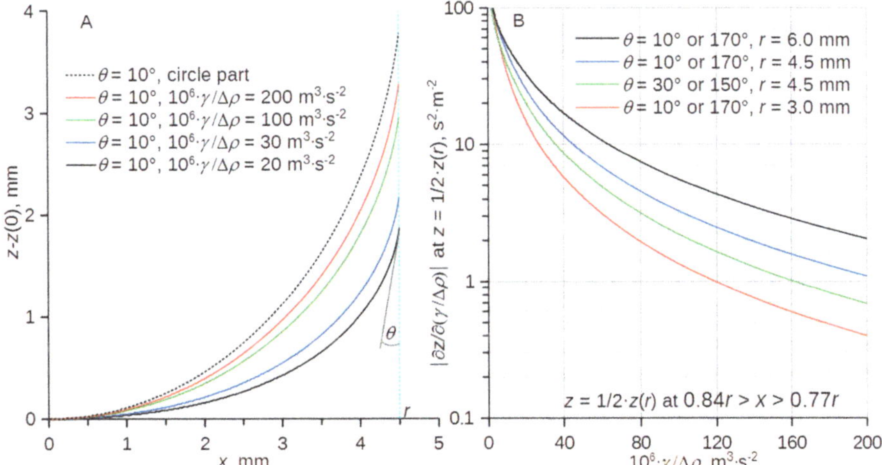

Fig. 2. Solutions of Eq. (1) for indicated model parameters (A). Sensitivity of the interface shape to interfacial tension at indicated conditions (B).

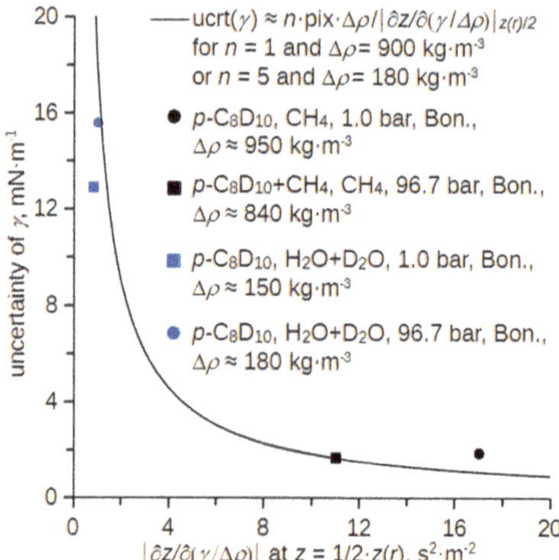

Fig. 3. The average uncertainty (95% confidence interval) of the interfacial tension estimated using the Bonferroni method (abbr. Bon.) for the confidence interval 95% plotted against the sensitivity of the interface shape to interfacial tension. Points are derived from experimental data, curve is the guide for the eye in which pix = 20.3 μm. Points were observed for the indicated systems and conditions for r = 4.5 mm.

3 Conclusions

The presented work provides overview of the first principles relevant to the measurement of interfacial tension based on menisci observed in cylindrical tubes, focus is on the experimental uncertainty. Despite the relative ease of the realization of such interfaces, this approach appears uncommon. While neutron imaging is a powerful tool enabling the observation of the interfaces even in pressurized systems in opaque (metal) tubes, the presented principles are general and foster the development of further methods of the interfacial tension measurement, which is demonstrated here for the industrially relevant system methane-p-xylene-water.

Acknowledgement. Authors acknowledge the financial support obtained from GACR and SNSF, project 23-04741K, and from the Ministry of Education, Youth and Sports within the specific university research grant A1_FCHI_2024_001.

This work is based on experiments performed at the Swiss spallation neutron source SINQ, Paul Scherrer Institute, Villigen, Switzerland.

The authors gratefully acknowledge the initial contributions of Dr. Vladimír Hynek and the late Assoc. Prof. Karel Trtík (both Universita třetího věku, CTU in Prague, Czech Republic).

This paper has been submitted for review on 30th September, 2024, and accepted for publication on 17th December, 2024.

References

1. Henriksson, U., Eriksson, J.C.: Thermodynamics of Capillary Rise: Why Is the Meniscus Curved? J. Chem. Educ. **81**(1), 150–154 (2004)

2. Adamson, A.W., Gast, A.P.: Physical chemistry of surfaces. Wiley, New York (1997)
3. Kralchevsky, P.A., Nagayama, K.: Particles at Fluid Interfaces and Membranes. Elsevier, Amsterdam (2001)
4. METAS Homepage. https://www.metas.ch/metas/en/home/dok/gravitationszonen.html, Accessed 30 Jan 2020
5. Vopička, O., Durďáková, T.-M., Číhal, P., Boillat, P., Trtik, P.: Absorption of pressurized methane in normal and supercooled p-xylene revealed via high-resolution neutron imaging. Sci. Rep. **13**, 136 (2023)
6. Vopička, O., et al.: One-pot neutron imaging of surface phenomena, swelling and diffusion during methane absorption in ethanol and n-decane under high pressure. PLoS ONE **15**(9), e0238470 (2020)
7. Bellur, K., et al.: Results from neutron imaging phase change experiments with LH2 and LCH4. Cryogenics **125**, 103517 (2022)
8. Sachs, W., Meyn, V.: Pressure and temperature dependence of the surface tension in the system natural gas/water principles of investigation and the first precise experimental data for pure methane/water at 25°C up to 46.8 MPa. Colloids Surfaces Physicochem. Eng. Aspects **94**(2–3), 291–301 (1995)
9. Satherley, J., Cooper, D.L., Schiffrin, D.J.: Surface tension, density and composition in the methane-pentane system at high pressure. Fluid Phase Equilib. **456**, 193–202 (2018)
10. Ramakrishnan, S., S. Hartland, S.: The determination of interfacial tension by differential capillary rise. In: Proceedings of The Indian Academy of Sciences (Chemical Sciences) **90**, 215–224 (1981)
11. Vinš, V., et al.: Possible anomaly in the surface tension of supercooled water: new experiments at extreme Supercooling down to −31.4 °C. J. Phys. Chem. Lett. **11**, 4443–4447 (2020)
12. Hadkar, U.B., Hadkar, A.S.: Interfacial tension between water and organic liquid heavier than water by capillary rise method. Asian J. Pharmacy Technol. **10**(1), 11–14 (2020)
13. Dechoz, J., Rozé, C.: Surface tension measurement of fuels and alkanes at high pressure under different atmospheres. Appl. Surf. Sci. **229**, 175–182 (2004)
14. Kerscher, M., et al.: Thermophysical properties of the energy carrier methanol under the influence of dissolved hydrogen. Int. J. Hydrogen Energy **48**(69), 26817–26839 (2023)
15. Netusil, M., Ditl, P.: Comparison of three methods for natural gas dehydration. J. Nat. Gas Chem. **20**(5), 471–476 (2011)
16. Stringari, P., et al.: Toward an optimized design of the LNG production process: Measurement and modeling of the solubility limits of p-xylene in methane and methane + ethane mixtures at low temperature. Fluid Phase Equilib. **556**, 113406 (2022)
17. Siahvashi, A., Al Ghafri, S.Z.S., Hughes, T.F., Graham, B.F., Huang, S.H., May, E.F.: Solubility of p-xylene in methane and ethane and implications for freeze-out at LNG conditions. Exper. Thermal Fluid Sci. **105**, 47–57 (2019)
18. Moudrakovski, I.L., McLaurin, G.E., Ratcliffe, C.I., Ripmeester, J.A.: J. Phys. Chem. **108**(45), 17591–17595 (2004)
19. DIPPR Project 801. https://app.knovel.com/hotlink/toc/id:kpDIPPRPF7/dippr-project-801-full/dippr-project-801-full, Accessed 30 Jan 2020
20. Dunning, J.R., Pegram, G.B., Fink, G.A., Mitchell, D.P.: Interaction of Neutrons with Matter. Phys. Rev. **48**, 265–280 (1935)
21. Lehmann, E.H., Kaestner, A., Grünzweig, C., Mannes, D., Vontobel, P., Peetermans, S.: Materials research and non-destructive testing using neutron tomography methods. Int. J. Mater. Res. **105**, 664–670 (2014)
22. Dasch, C.J.: One-dimensional tomography: a comparison of Abel, onion-peeling, and filtered backprojection methods. Appl. Opt. **31**(8), 1146–1152 (1992)
23. Boillat, P., et al.: Opt. Express **26**(12), 15769–15784 (2018)

24. Carminati, C., et al.: Implementation and assessment of the black body bias correction in quantitative neutron imaging. PLoS ONE **14**(1), e0210300 (2019)
25. Seber, G.A.F., Wild, C.J.: Nonlinear Regression. John Wiley & Sons Inc, New Jersey (2003)
26. Kubíček, M.: Numerické algoritmy řešení chemicko-inženýrských úloh. SNTL, Prague (1983)
27. Peng, D., Robinson, D.B.: A New Two-Constant Equation of State. Ind. Eng. Chem. Fundam. **15**(1), 59–64 (1976)
28. Wagner, W., Pruß, A.: The IAPWS Formulation 1995 for the Thermodynamic Properties of Ordinary Water Substance for General and Scientific Use. J. Phys. Chem. Ref. Data **31**, 387–535 (2002)
29. Cibulka, I., Takagi, T.: $P-\rho-T$ data of liquids: summarization and evaluation. 5. aromatic hydrocarbons. J. Chem. Eng. Data **44**, 411–429 (1999)
30. Trtik, P., Lehmann, E.H.: Progress in high-resolution neutron imaging at the paul Scherrer Institut - the neutron microscope project. J. Phys: Conf. Ser. **746**, 012004 (2016)

Measuring the Influence of Pulsed Injection on the Liquid Fraction Distribution in Foam Using Neutron Radiography

Friedrich Walzel[1], Muhammad Ziauddin[2], Leon Knuepfer[3], Tobias Lappan[3], Pavel Trtik[4], and Sascha Heitkam[2,3(✉)]

[1] Institute Charles Sadron, 67034 Strasbourg, France
[2] Technische Universität Dresden, 01069 Dresden, Germany
`s.heitkam@hzdr.de`
[3] Helmholtz-Zentrum Dresden-Rossendorf, 01328 Dresden, Germany
[4] Paul Scherrer Institute, 5232 Villigen, Switzerland

Abstract. Forced drainage configurations are often employed to control the liquid fraction in a foam column. In that, a constant flow of liquid is added at the top of the column, draining down through the foam, yielding a steady liquid fraction distribution. When the injected flow rate is subjected to pulsations, the liquid fraction pulsates as well. We performed time-resolved neutron radiography measurements to investigate the spatial and temporal distribution of these pulsations throughout the foam column. For high pulsation frequencies or low drainage flow rates, the pulsations are damped within a few millimeters from the injection point. Pulsation of lower frequencies or higher flow rates might spread several centimeters through the column. The results allow to estimate the influence of pulsating feed pumps on the liquid fraction distribution under forced drainage.

Keywords: foam drainage · pulsations · dampening · neutron radiography

1 Introduction

Foams play an important role in our daily life, e.g. in cushioning, insulation or food products. Also, industrial applications such as froth flotation [12] or foam fractionation [6] rely strongly on the existance of a well controlled froth zone.

An important property of a foam is the liquid fraction $\Phi = V_\mathrm{l}/V_\mathrm{f}$, which is the fraction of the volume of liquid V_l in a volume of foam V_f. It influences the stability, the permeability, the shear modulus and the topology of a foam [2]. Thus, the liquid fraction needs to be well controlled when performing foam experiments such as rheology measurements [10], foam stability measurements [5] or sensor trials [11]. To that end, typically forced-drainage setups are employed [5, 10, 11]. In these setups, a steady flow of liquid is added to the top of a foam column. The liquid drains down through the foam, maintaining a steady liquid fraction. If the liquid flow rate does not exceed a critical value [4], the foam remains static and a homogeneous liquid fraction can be expected.

© The Author(s) 2026

A. E. Craft and H. Z. Bilheux (Eds.): WCNR 2024, SPPHY 348, pp. 47–54, 2026.
https://doi.org/10.1007/978-3-032-15003-5_7

However, the liquid flow rate may be subjected to oscillations. They could result from the employed pump, in particular if a peristaltic, piston, or diaphragm pump is employed. Also, dripping from the liquid injection tube or accidentially trapped gas bubbles in the feed line may cause such oscillations. Verbist et al. [13] predict that in foam drainage no oscillations occur and that any oscillations in liquid fraction would be damped. However, the region of influence of such periodic flow rate oscillations is not yet clear.

We performed time-resolved neutron radiography experiments to measure the liquid fraction distribution in a draining foam subjected to pulsating liquid injection. In that way, we estimated the region of influence of such oscillations depending on liquid flow rate and pulsation frequency.

2   Material and Methods

Setup. Drainage experiments were performed in a closed flat chamber of $B = 100\,\text{mm}$ width, 200 mm height and $z_0 = 23.2\,\text{mm}$ thickness. Front and back wall consisted of neutron-transparent boron-free float glass. The upper half of the chamber was filled with foam. The lower half contained surfactant solution, i.e. distilled water with $6\,\text{gL}^{-1}$ sodium dodecyl sulfate (SDS). Liquid was extracted from the reservoir below the foam and injected into the foam by means of a thin needle. Figure 1 shows the setup of the experiment.

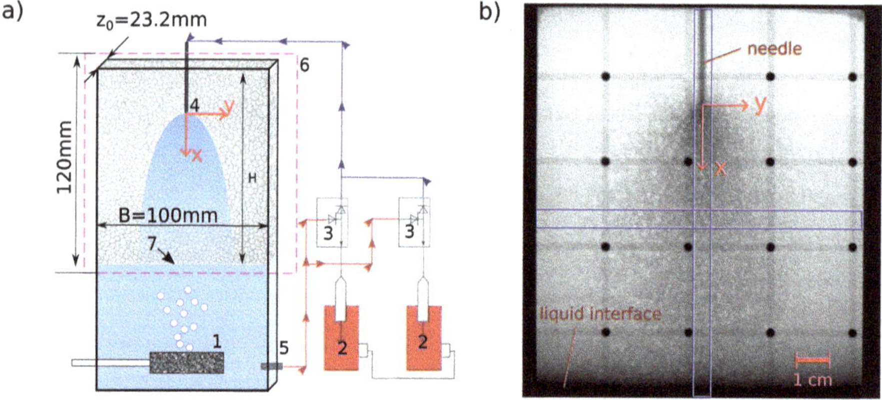

Fig. 1. Foam drainage setup a) with (1) porous media, (2) Aladin 1000 double syringe pumps, (3) valves, (4) needle for the liquid injection, (5) tube for extracting liquid, (6) neutron radiography field of view, (7) interface between foam and surfactant solution. Sample radiographic image b).

Air bubbles were produced by a submerged porous media charged with pressurized air of constant pressure from an air pump. Three different porous media were used, yielding bubbles with a Sauter mean radius of approximately $R_{3,2} = 1.0\,\text{mm}$, 1.5 mm and 1.8 mm, respectively. For all experiments except the study on bubble size influence in Fig. 5c, bubbles with a Sauter mean radius of $R_{3,2} = 1.5\,\text{mm}$ have been investigated.

During the experiment, two interconnected syringe pumps (Aladdin AL-1000) injected and removed liquid from the foam cell simultaneously, keeping the total amount of liquid in the cell constant. One pump injected and the other removed liquid at the same flow rate Q. The injection took place through a needle of 1.2 mm inner diameter. The needle was positioned in the center of the cell, 25 mm below the top wall.

Neutron Radiography. The experiment was performed in the neutron beam line NEUTRA [7]. The flux of thermal neutrons was about $13 \times 10^6 \mathrm{cm}^{-2}\mathrm{s}^{-1}$. The neutrons penetrated the measurement chamber and were partially attenuated. The transmitted neutrons hit a scintillator screen and generated visible light, which was recorded by a camera. The field of view of the beam line was 120 mm $\times$ 120 mm discretized by 1024 $\times$ 1024 pixel yielding a spatial resolution of approximately 0.12 mm per pixel. The temporal resolution was $10\,\mathrm{s}^{-1}$. Despite the high neutron flux in NEUTRA, such high frame rate resulted in a low signal-to-noise ratio. Thus, averaging over 50 $\times$ 50 pixels has been employed in the post-processing.

The Beer-Lambert law (1) gives a relation between incoming I_0 and transmitted neutron intensity I

$$I = I_0\, e^{-\mu\Phi z_0}. \tag{1}$$

The attenuation coefficient of water equals $\mu = 3.74\,\mathrm{cm}^{-1}$ and the thickness of the foam column is $z_0 = 23.2$ mm. The attenuation due to air in the foam can be neglected, because the attenuation coefficient of air is of magnitudes smaller than the one of water. A background image of the empty cell provides the incoming intensity I_0. This includes attenuation due to the cell and glass walls. From Eq. (1) one obtains the liquid fraction

$$\Phi = -\frac{ln(I/I_0)}{\mu z_0}. \tag{2}$$

Influencing factors that affect the accuracy of liquid fraction measurements are the dark current of the camera chip, scattered neutrons, and fluctuations in the neutron flux. The dark current was measured beforehand without any neutron flux and subtracted from all camera images. The intensity of scattered neutrons was measured for each image in the shadow of small neutron-opaque cylinders (see black dots in Fig. 4a) and interpolated over the field of view by a second order polynomials. The resulting scattered intensity was subtracted from the corresponding image. This has been done for the measurement of the background I_0 as well as for the actual measurement I [1,3]. To account for fluctuating beam intensity, each image was scaled with the instantaneous beam intensity, acquired from an undisturbed reference section in the top-left corner of the images [1,3].

Experimental Procedure. To investigate the pulsed liquid injection, two flow modes with different flow rates $Q(t)$ were used, sketched in Fig. 2. For the continuous injection flow mode, Q was constant in time t and thus, identical with the time-averaged flow rate $<Q>$. The measurement was started after the distribution of the liquid fraction reached a steady state. The stationary distribution of the liquid fraction was measured after 180 s for 10 s. Due to stationary liquid fraction distribution, the average over the 100 frames was considered. This reduces the statistical uncertainty significantly.

The pulsed injection mode had a steady periodic character. During a defined time T_{on} liquid is injected with flow rate Q_{on}. During the time T_{off} no liquid is injected. The total time of a period is T with $T = T_{on} + T_{off}$. In this work, the injection time and the pause time was equal $T_{on} = T_{off}$. This yields the time-averaged flow rate $< Q >= 0.5 Q_{on}$. The measurement was started after the liquid fraction reached a periodic distribution in time. This periodic distribution of the liquid fraction was measured after 90 s for $T_{measured} = 160$ s. The artificial fluctuation of the injection flow enables the investigation of the dynamic response of the foam drainage in respect to fluctuations of the injection flow. The damping character of a foam in relation to different flow rates $< Q >$ and pulse lengths T was investigated.

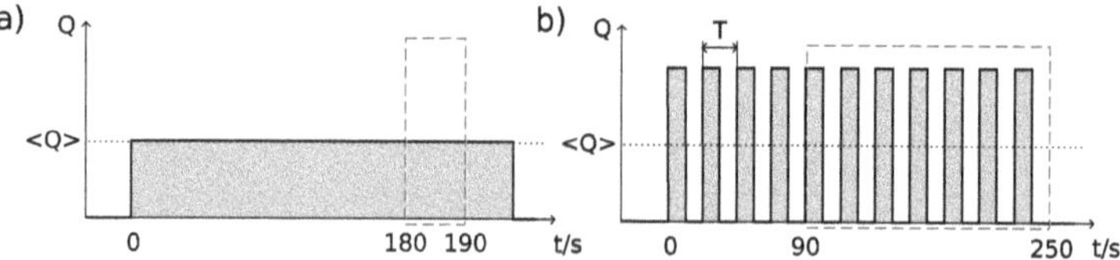

Fig. 2. Scheme for the continuous (a) and pulsed (b) injection mode. The instantaneous flow rate was Q and the averaged flow rate in time was $< Q >$. The red rectangle marks the measurement window.

Conditional Averaging. To tell noise and pulsation under pulsed injection apart, conditional averaging was employed. An averaged liquid fraction Φ_{pulse} was collected by averaging over a number P of consecutive pulses. All time instances t belonging to a pulse time instance $\tilde{t} \in (0, T)$ by $t = \tilde{t} + iT$ with $i \in \mathbb{N}$ were combined, yielding

$$\Phi_{pulse}(x,y,\tilde{t}) = \sum_{i=0}^{P} \frac{\Phi(x,y,t = \tilde{t} + iT)}{P}. \tag{3}$$

This conditional averaging works particularly well for small T, because a larger number of periods P could be recorded and evaluated. Figure 3 compares the instantaneous $\Phi(x,y,t)$ to the conditional average $\Phi_{pulse}(x,y,\tilde{t})$ for a region of 1 mm × 5 mm and for $P = 15$ and $T = 4$ s, at different distances to the needle.

For each position (x,y) the relative pulsation of the liquid fraction Φ was quantified by the relative standard deviation

$$s_P(x,y) = \frac{1}{< \Phi(x,y) >} \sqrt{\frac{1}{T} \int_0^T \left(\Phi_{pulse}(x,y,\tilde{t}) - < \Phi(x,y) > \right)^2 d\tilde{t}} \tag{4}$$

over one pulsation period $\tilde{t} \in (0, T)$.

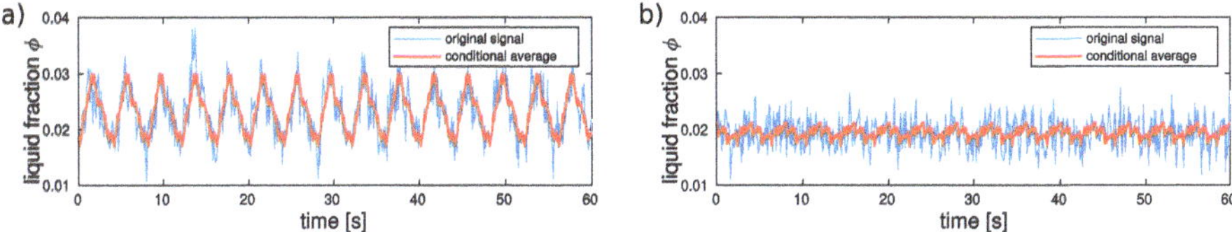

Fig. 3. Comparison of the original liquid fraction variation with the conditional-averaged variation for three different vertical positions, i.e. at a) $x = 10\,$mm and b) $30\,$mm below the needle. Liquid flow rate equals $< Q >= 3\,$mL/min and pulse width equals $T = 4\,$s.

3   Results

Stationary Liquid Injection. The liquid fraction distributions obtained from continuous injection experiments with different drainage flow rates are compared in Fig. 4. Underneath the needle position at $x = 5\,$mm, the liquid fraction is highest. In vertical direction the liquid spreads over an increasing cross section, yielding a decreasing liquid fraction below the needle. At the liquid interface at $x \approx 70\,$mm, a strong increase in liquid fraction is visible due to the capillary effect, sucking liquid from the reservoir into the foam. Horizontal and vertical profiles for different liquid flow rates are similar to each other, i.e. the horizontal penetration depth is not related strongly to the liquid flow rate.

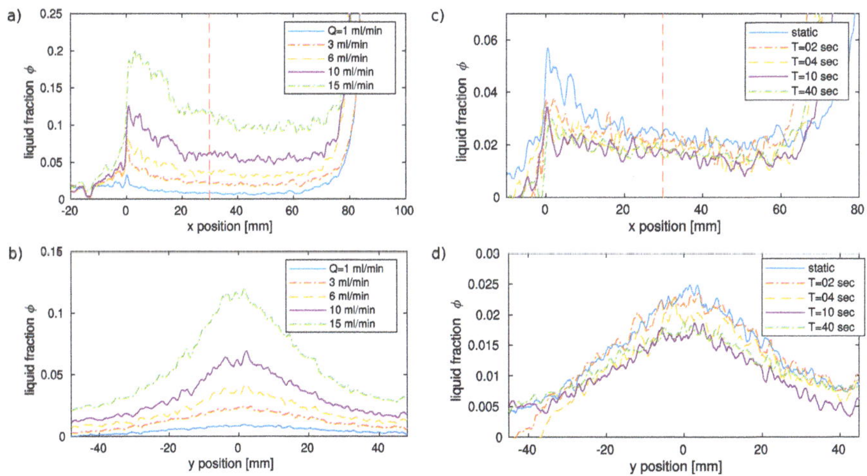

Fig. 4. Time-averaged liquid fraction profiles for static (a,b) and pulsating (c,d) injection at $< Q >$ of $3\,$mL/min. Horizontal profiles (b,d) intersect the vertical profiles (a,c) at the red broken line.

Pulsed Liquid Injection. In the first step, the time-averaged liquid fraction profiles are compared for continuous injection and pulsed injection with different injection pulse widths. Cases with identical time-averaged liquid flow rate $< Q >$ of $3\,$mL/min are shown in Fig. 4 c) and d). In the vicinity of the needle, the continuous injection leads to higher average liquid fractions. However, beyond 20 mm below the needle, the liquid fraction profiles for the static case and short pulse widths coincide well. Significantly lower liquid fractions were measured only for very long pulse widths. Additionally, near the liquid reservoir a strong deviation is visible. This is due to the different filling levels in different experimental runs and is not of relevance for this study.

In the second step, the amplitudes of the liquid fraction oscillation are investigated. Different pulse period times T, average injection flow rates $<Q>$ and bubble radii $R_{3,2}$ were considered. Figure 5 compares the relative standard deviations of the conditional averaged liquid fractions s_P according to Eq. (4), which are a measure for the pulsation amplitude of liquid fraction. The pulsation amplitude depends strongly on T, $<Q>$, and the distance to the needle. Below the needle, up to a distance of $x < 20$ mm, a significant pulsation is observed. With increasing distance to the needle the pulsation is damped. For higher pulsation frequencies, smaller drainage flow rates, and larger bubbles, the region of influence was reduced. Above the needle outlet, unphysically high oscillations were measured due to the filling and unfilling of the needle during pulsation.

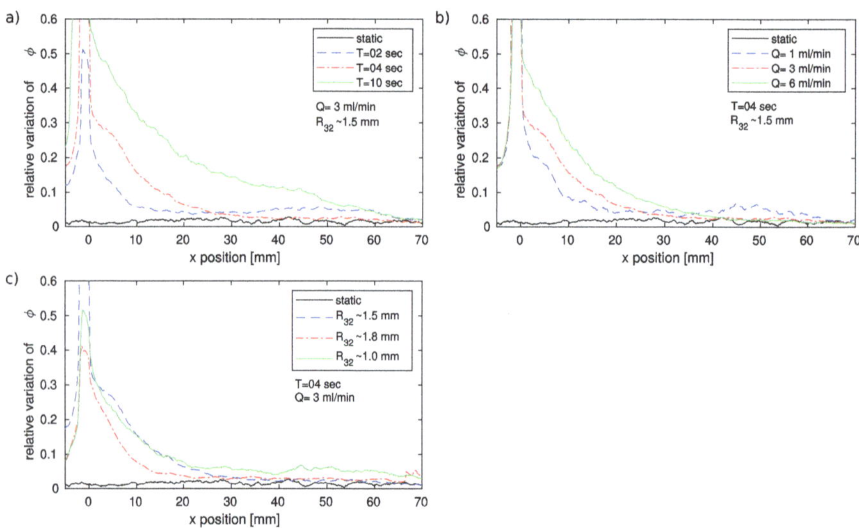

Fig. 5. Comparison of the relative standard variation of the liquid fraction (see Eq. (4)) under pulsed injection along the vertical position. Influence of a) pulse width, b) flow rate and c) bubble size is emphasized.

4   Discussion and Conclusion

Since neutron radiography is a new method for investigations of foams, especially foam drainage, the accuracy has to be considered critically. Our validation with the displacement of the foam-liquid interface yielded a systematic uncertainty of 15%. A possible explanation for this deviation could be a process similar to beam-hardening in X-ray radiography. Since the employed neutrons had a certain range in energy [9], low-energy neutrons are more likely to interact with the liquid than high energy neutrons. Consequently, with increasing penetration depth, the average energy of the transmitted neutrons increases and the effective attenuation coefficient decreases [9]. In [14], an adapted attenuation coefficient of $\mu = 3.64\,\text{cm}^{-1}$ has been used for water, yielding systematic deviations well below 5%. However, since this work is focussed on the magnitude of variations under pulsed drainage, a systematic relative uncertainty of 15 % in the liquid fractions does not falsify the conclusions from this work.

Figure 4 c) and d) show that $< \Phi(x,y,t) >$ is rather independent from the pulse durations T, provided an identical average injection flow rate $< Q >$ is imposed and T is not too large. For very large T, the non-linearity of the foam drainage equation becomes important. For SDS foam under forced drainage, the permeability α scales with [8] $Q \propto \alpha \propto \Phi^{3/2}$. During T_{on}, Q equals $2 < Q >$. Thus, Φ_{on} equals $2^{2/3} < \Phi >$. During T_{off}, Q and Φ are considered to be zero. Thus, the time-averaged liquid fraction equals $2^{2/3}/2 < \Phi > \approx 0.8 < \Phi >$. Accordingly, in Fig. 4, the continuous injection results in slightly higher liquid fractions than pulsed injection with long pulse widths. Verbist et al. [13] performed a linear stability analysis on foam drainage and concluded that no unstable modes exist. Thus, any pulsation of liquid fraction is damped. Accordingly, the pulses in liquid injection are damped and disappear after a certain vertical distance. When employing peristaltic or piston pumps, a certain pulsation in the injection is imposed. Figure 5 can help to decide to what extent pulsations have to be taken into account for given drainage flow rates and pulse durations. Typical pulse widths are below 4 s. Thus, at distances of more than 30 mm, the pulsation of the liquid fraction as well as the influence on the average liquid fraction can be disregarded for most application.

Acknowledgements. This work is based on experiments performed at the Swiss spallation neutron source SINQ, Paul Scherrer Institute, Villigen, Switzerland. The financial support by the Deutsche Forschungsgemeinschaft (HE 7529/3-1) is gratefully acknowledged.

This paper has been submitted for review in August 2024, and accepted for publication in December 2024.

References

1. Boillat, P., et al.: Chasing quantitative biases in neutron imaging with scintillator-camera detectors: a practical method with black body grids. Opt. Express **26**(12), 15769–15784 (2018)
2. Cantat, I., et al.: Foams: structure and dynamics. OUP Oxford (2013)
3. Carminati, C., et al.: Implementation and assessment of the black body bias correction in quantitative neutron imaging. PLoS One **14**(1), e0210,300 (2019)
4. Heitkam, S., Eckert, K.: Convective instability in sheared foam. J. Fluid Mechanics (2021)
5. Hutzler, S., Weaire, D.: Foam coarsening under forced drainage. Philos. Mag. Lett. **80**(6), 419–425 (2000)

6. Keshavarzi, B., et al.: Protein enrichment by foam fractionation: experiment and modeling. Chem. Eng. Sci. **256**(117), 715 (2022)
7. Lehmann, E.H., Vontobel, P., Wiezel, L.: Properties of the radiography facility neutra at sinq and its potential for use as european reference facility. Nondestructive Testing Evaluat. **16**(2–6), 191–202 (2001)
8. Lorenceau, E., Louvet, N., Rouyer, F., Pitois, O.: Permeability of aqueous foams. Euro. Phys. J. E **28**(3), 293–304 (2009)
9. Mannes, D., Josic, L., Lehmann, E., Niemz. P.: Neutron attenuation coefficients for non-invasive quantification of wood properties (2009)
10. Soller, R., Koehler, S.A.: Rheology of steady-state draining foams. Phys. Rev. Lett. **100**(20), 208301 (2008)
11. Staud, R., et al.: Minimal-invasive method for the evaluation of liquid fractions in foams with a point level sensor. Chem. Eng. Technol. **45**(8), 1397–1403 (2022)
12. Subrahmanyam, T., Forssberg, E.: Froth stability, particle entrainment and drainage in flotation–a review. Int. J. Miner. Process. **23**(1–2), 33–53 (1988)
13. Verbist, G., Weaire, D., Kraynik, A.: The foam drainage equation. J. Phys.: Condens. Matter **8**(21), 3715 (1996)
14. Ziauddin, M., et al.: Comparing wire-mesh sensor with neutron radiography for measurement of liquid fraction in foam. J. Phys. Condensed Matter (2022)

Application of Monte Carlo-Based Surrogate Model for Laser-Driven Neutron Radiography Source Optimization

D. P. Broughton[1(✉)], O. F. Erdem[1,2], C.-K. Huang[1], S. H. Batha[1], C.-S. Wong[1], R. E. Reinovsky[1], T. R. Schmidt[1], Z. Wang[1], B. T. Wolfe[1], M. Alvarado Alvarez[1], and A. Junghans[1]

[1] Los Alamos National Laboratory, Los Alamos, NM 87545, USA
david.brought@lanl.gov, oferdem@umich.edu
[2] University of Michigan, Ann Arbor, MI 48109, USA

Abstract. Short-pulse, high-intensity lasers offer a compact alternative to linear accelerators for driving high-flux neutron sources. One method to produce laser-driven neutrons is in a so-called "pitcher-catcher" setup, wherein deuterons and protons are accelerated into a converter material to produce neutrons. In this study, deuteron and proton spectra, along with angular ion divergence measured during experiments conducted at Omega EP (laser parameters: 500 J, 0.7 PS, 7 $\times\ 10^{19}$ W/cm^2), are used to optimize converter material and geometry for neutron radiography. The optimization process employs 3520 Monte Carlo N-Particle (MCNP®) simulation cases to build an Artificial Neural Network (ANN) surrogate model of neutron yield characteristics based on incident particle type (proton or deuteron), converter material (LiF or Be), and converter dimensions (thickness and radius of cylinders or cones). Using the surrogate model with the multi-objective genetic optimization algorithm we identify the Pareto front producing the optimal parameter combinations. The main parameters of interest are neutron yield and directionality, with an additional goal of minimizing converter volume. Relative errors between the Monte Carlo validation results and the surrogate model optimization results for each modeled objective range 0.1–2% in terms of mean absolute relative error. The resulting converter designs generate beams of up to 2.44×10^{10} neutrons per pulse, exhibiting the desired beam-like fast neutron angular emission profiles without requiring collimation. This successful demonstration of multi-objective optimization for source development can be extended from fast neutron beam development to thermal neutron source development.

Keywords: pulsed neutron source · surrogate model · multi-objective optimization

1 Introduction

Short-pulse laser-driven sources [1] offer a compact neutron pulse with potential for delivering higher instantaneous flux within a lower cost facility than traditional spallation-based neutron sources [2]. To date, experimental yields have exceeded 10^{10}

© The Author(s) 2026

A. E. Craft and H. Z. Bilheux (Eds.): WCNR 2024, SPPHY 348, pp. 55–63, 2026.
https://doi.org/10.1007/978-3-032-15003-5_8

neutrons per sr per pulse [1]. These neutron sources have numerous applications, the primary interest here is development of a neutron source for flash radiography of dynamic processes, including high-energy-density physics conditions.

Laser-driven neutron sources provide major technical benefits beyond cost and space savings. Lasers serve as a single driver for various radiation sources, enabling detailed test object characterization by pairing neutron radiography with laser-driven x-ray [3] or proton [4] radiography. The combination of the laser-pulse particle acceleration time in the tens of picoseconds and the sub-cubic centimeter neutron source size [1] improve time resolution compared to current spallation neutron sources [5]. While reproducible high-quality laser-driven flash radiography has not been realized (requiring $> 10^{12}$ n/sr); the upcoming multi-PW lasers are expected to increase peak neutron flux 10^3–10^6 × relative to today's facilities [6]. This work aims to demonstrate application of a generalized multi-objective optimization method on a neutron radiography source.

The highest yield short-pulse laser-driven neutron generation experiments have used a two-stage "pitcher-catcher" configuration (Fig. 1). In this setup, a high intensity laser, $>10^{18}$ W/cm^2, accelerates protons and deuterons within a thin deuterated-plastic film (100's of nm to several tens of um thick) referred to as the "pitcher". The ion beams interact via (p,n) and (d,n) reactions within a "catcher", typically Be or LiF, on the order of millimeters to centimeters thick [7].

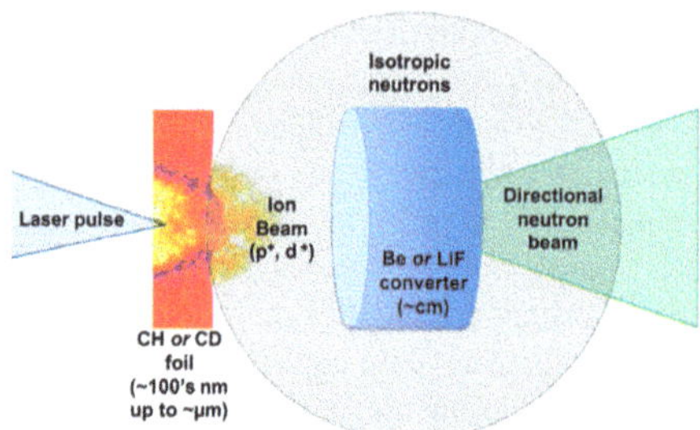

Fig. 1. Illustration of typical "pitcher-catcher" source configuration (not to scale).

This work builds on our previous simulation-based study focused on deuteron acceleration and the conversion to neutrons [8]. Here, experimentally measured deuteron and proton beams are used as the pitcher in simulations of the ion-to-neutron conversion. The absence of protons in our previous study is significant, as it omits an avenue of neutron production. Experimentally, we measured a significant proton source from what is believed to be surface contamination of the CD foil. Hydrogen may have attached to the CD foils surface during the process of floating the films onto washers while mounting them, or from ambient moisture and hydrocarbons present in even the near-vacuum experimental conditions. Laser-driven deuteron sources using foil targets may not exist without proton contamination. By using measured proton and deuteron sources, we increase the fidelity of neutron source simulations.

2 Methodology

The ion energy and angular distribution use experimental data collected at OMEGA EP at the Omega Laser Facility within the Laboratory for Laser Energetics. Figure 2(a) shows ion spectra measured from a 700 nm thick CD foil at 500 J in 0.7 PS using a Thompson Parabola spectrometer [9].[1] The relative ion yields correspond to the weighting factors used in modeling of 0.23 deuterons and 0.77 protons per ion.

Figure 2(b) shows the angular distribution of ions inferred from analysis of a radiochromatic film (RCF) separated 8 cm from the CD film. Since protons and deuterons are not distinguished on RCF, the averaged 2D Gaussian function fit from the experimental RCF is used for both simulated angular distributions. Laser-plasma acceleration produces a highly directional ion beam primarily focused within a narrow (~10–20°) cone, and an approximately Gaussian angular distribution within this range [10]. This resulting Gaussian fit, with $\sigma = 1.2642$ rad, was used as the ion source angular distribution in the MCNP simulations. Based on the laser's best focus spot size, ions were modeled as being emitted from a $\varnothing = 20$ µm disc, and CD film-to-converter separation was 1 mm.

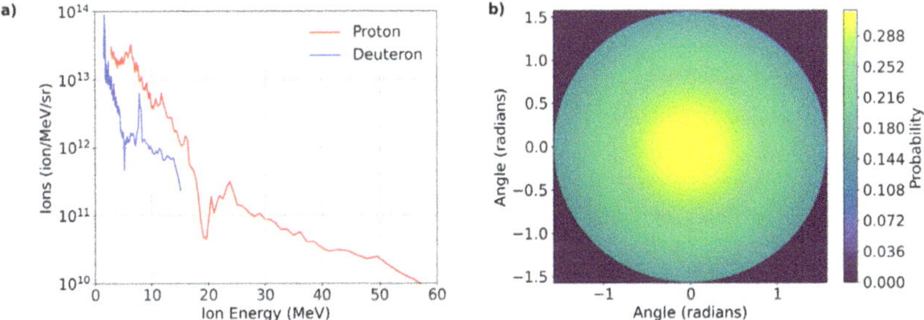

Fig. 2. (a) Experimentally measured proton and deuteron spectra (total of ~1.1 × 10¹³ p⁺ and ~3.3 × 10¹² d⁺), (b) angular ion distribution estimated from experimental data (contours indicate angular-probability bins used in MCNP simulations).

The relevant conversion reactions, Li(p,n) and Li(d,n), and Be(p,n) and Be(d,n), were modeled in MCNP6.3 [11] using validated cross sections for conversion of protons and deuterons to neutrons [12], and neutron scatter and absorption cross sections [13].

Converter geometry was either cylindrical or conical (the conical tip adjacent to ion source converges to a point), with the same heights and radii for both geometries during training data generation. A converter case matrix was generated depending on length, radius, material, and geometry. Based on the range of the highest energy protons (58.4 MeV) in Be, 20.34 ± 0.850 mm ($\pm$straggling) [14], converter length was sampled over 0–22 mm, and radius varied over 0–10 mm. Simulation input parameter sampling was evenly spaced.

---

[1] Based on scaling Thompson Parabola spectra using 0.078 sr (solid angle of cone with $h = 8$ cm, $\varnothing = 2.54$ cm) from an experiment at Omega EP under similar conditions.

Relevant neutron source objectives for radiography include, (1) directionality (i.e., ideally forward directed), (2) yield, (3) on-axis yield (within 5° cone angle), (4) beam-like emissions, and (5) converter volume. Corresponding numerical objectives were defined as, (1) the mean cosine of neutron emissions, (2) neutron yield per ion, (3) neutrons recorded within 5° width on-axis cone, and (4) beam-like nature (quantified as the forward flux, within the 5° on-axis cone, normalized by the full width half maximum (FWHM) of the neutron angular distribution), (5) volume.

The model training data used 3520 MCNP input files, each using 4×10^7 initial ions, to simulate 1760 converters, as each requires proton and deuteron cases. As this is small for an artificial intelligence dataset, the surrogate model applied a relatively simple Fully Connected Neutral Network (FCNN). There were four input and four output parameters. The hyperparameters chosen for this surrogate model are given in Table 1. Note, volume was calculated from the geometry and appended to the other model-predicted objectives prior to optimization.

Due to the limited dataset size, a model checkpoint was employed to prevent overfitting. The checkpoint method gathers the model with the lowest mean absolute error with respect to the validation data (224 simulated converters), even if the model overfits the training data (1272 simulated converters) after a given number of epochs. The metrics for each individual model output from the best model training checkpoint have been compared to the test data (264 simulated converters) and reported in Table 2 confirm its excellent predictive capability. This Artificial Neutral Network (ANN) [15] surrogate model is the basis for the fitness function in the optimization process.

Table 1. Selected ANN surrogate model hyperparameters.

Parameter	Value
Number of Hidden Layers / Neurons	4/150, 150, 150, 150
Initial Learning Rate	1e−3
Learning Rate Decay Rate/Patience	0.8/6 Iterations
Batch Size/Number of Epochs	1/200
Neuron Activation Function	Rectified Linear Unit
Input and Output Data Scaling Type	Linear Min. Max. (0–1)
Model Loss Function	Mean Squared Logarithmic Error
Model Checkpoint Monitor	Validation Data Mean Absolute Error
Early Stopping Patience	15 Iterations

The set of best converter designs was predicted by running the surrogate model through the multi-objective genetic optimization algorithm, Non-dominated Sorting Genetic Algorithm III (NSGA-III) [16], using the importance weights. The converter dimensions and materials were used to create a set of 100 MCNP simulations for validating performance of designs obtained using the surrogate model within the optimization algorithm.

The optimization was applied to maximize the first four objectives (yield and forward directionality) and minimize the last objective (volume) of the model. The geometric limits used in the training are also enforced on the optimization algorithm. The optimization was started with an initial population of 1000 converters. Each generation has 100 parent converters. 200 offspring converters are created with 75% cross-over probability and 25% mutation probability. Two-point cross-over was applied, and a 5% bit-flip mutation probability was used. The optimization ran for 100 generations.

Table 2. Test metrics corresponding to the best model checkpoint for each output parameter.

Metric	Output 1 (Mean Cosine)	Output 2 (Total Yield)	Output 3 (On-axis yield within 5°)	Output 4 (Peak angular flux-to-FWHM)
Mean Square Error	0.000295	0.000113	0.000795	0.000154
Mean Absolute Error	0.011474	0.007483	0.018413	0.008423
R^2	0.991221	0.999064	0.947715	0.991151

Selections were made to determine the best converter for various applications. The following describes the purpose, and, in parenthesis, the objectives being optimized for each case: (1) overall parameter optimization (all parameters), (2) fast neutron radiography beam (having maximum 5° on-axis yield and the maximum forward flux-to-FWHM), (3) input for thermal neutron beam development with moderation (maximum neutron yield), and (4) overall optimization for a LiF converter (all parameters). Importance weights applied in each case are given in Table 3.

Table 3. Importance weights applied to normalized objective values for each of the cases.

Case (optimizing)	Parameter Weights				
	Mean Cosine	Yield	5° Yield	Peak angular flux-to-FWHM	Volume
1 (all parameters)	1	1	1	1	−1
2 (neutron beam)	0	0	0.8	0.2	0
3 (neutron yield)	0	1	0	0	0
4 (all parameters, only LiF)	1	1	1	1	−1

3   Results and Discussion

The initial model training data consists of many simulations with large Monte Carlo statistical fluctuations impacting the ability to directly find the optimal converter. Applying a surrogate model reduces the impact of initial simulation statistics on the optimization. Unless overfitted, the ANN trained on the data represents the average objective performance among the statistical fluctuations. In Fig. 3(a) the surrogate model, trained on the lower particle MCNP runs, successfully eliminated these fluctuations. This is confirmed by the validation of dominating cases with higher number of particles, where the best performing results align with the best solutions suggested using the surrogate. Table 4 reports relative errors between MCNP validation results and surrogate model optimization results for each modeled objective, demonstrating strong overall performance, with predictions typically within 0.1–2%.

Table 4. Comparing surrogate model objective predictions against MCNP validation results.

Objective	Mean Cosine	Total Yield	5° Yield	Peak angular flux-to-FWHM
Mean Absolute Relative Error	0.0946%	1.9981%	1.6302%	1.2843%

The five-dimensional trade-off curves in this optimization problem are not easily visualized. The objectives of total yield, 5° on-axis yield, and peak angular flux-to-FWHM are used in Fig. 3(b) to show the spherical surface trade-off curve. As one objective increases, one or both others decrease on the Pareto front, creating a spherical surface.

The model optimization expectations are used to achieve the best results for the cases defined in Table 3. The best converter for each case is reported in Table 5, along with the corresponding performance metrics. The angular distributions shown in Fig. 3(c) for these converters highlights the beam-like nature of the optimized converters, where the peak on-axis flux (3×10^9 n/sr/pulse) is nearly double that greater than 45°. Figure 3(d) shows the similarity in the spectra between Be and LiF, where all the spectra peak just below 20 MeV with a low energy tail. It is noted that no cone converters were selected as optimal, likely due simply to fewer particles reach the converter compared to cylinder geometries reducing the efficiency.

These optimized converter results (Table 5) show the objective trade-offs for these best cases. To increase mean neutron cosine between case 1 and 2, total neutron yield and peak angular flux-to-FWHM values both decrease. To minimize converter volume by reducing converter radius between cases 2 and 3, the mean cosine, total yield, and 5° yield are all reduced, but the beam-like quality of neutrons increases, as indicated by the peak angular flux-to-FWHM value. If the converter must be LiF instead of Be, as is the change between cases 2 and 4, there is a decrease in total yield, 5° yield, and peak angular flux-to-FWHM, and an increase in the mean neutron emission cosine. Many application-specific optimal converter designs can be generated from the trade-off curve produced using the surrogate model within the genetic optimization algorithm.

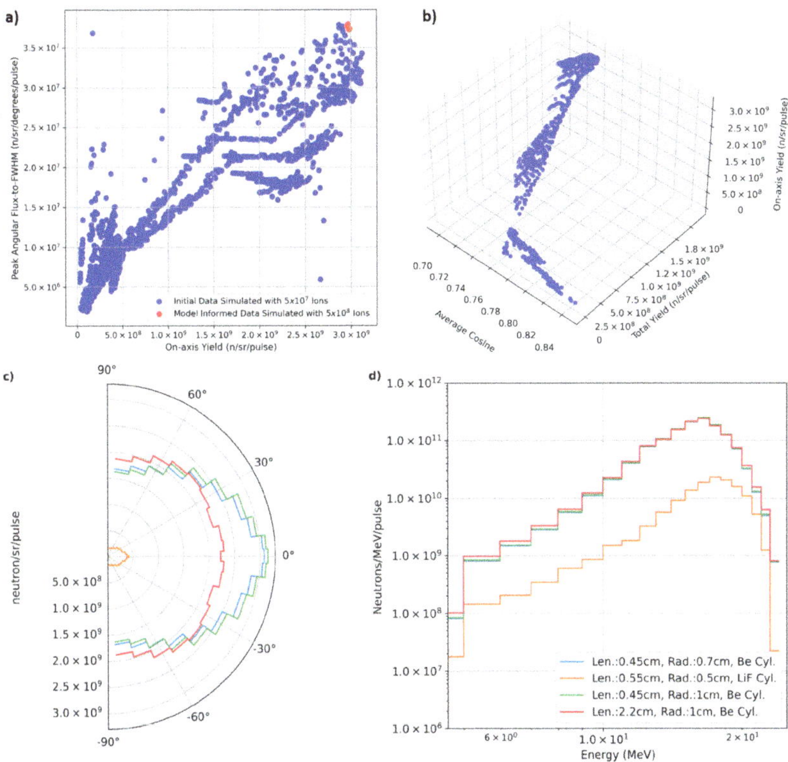

Fig. 3. (a) Comparison of initial MCNP data (run with 8×10^7 source ions) and model-informed best converter designs (run with 4×10^9 source ions) for the objectives peak angular flux-to-FWHM and on-axis 5° yield. (b) The optimal converters from the Pareto front balancing the objectives: total neutron yield, on-axis 5° neutron yield and peak angular flux-to-FWHM. For the converters detailed in Table 5 the outgoing neutron (c) spectra, and (d) angular distributions (same legend applies to (c) and (d)).

Table 5. Simulated performance metrics of the optimal converters chosen for the four distinct cases (note all optimal converters are cylinders).

Case (optimizing)	Mat.	Len. [cm]	Rad. [cm]	Vol. [cm^3]	Mean Cosine	Yield [$\times 10^9$ n/sr/pulse]	5° Yield [$\times 10^9$ n/sr/pulse]	Peak flux-to-FWHM [$\times 10^7$ n/sr/MeV/pulse]
1 (all)	Be	0.45	0.7	0.69	0.734	1.81	3.08	3.88
2 (beam)	Be	0.55	1	1.41	0.740	1.86	3.12	3.63
3 (yield)	Be	2.2	1	6.91	0.694	1.94	2.21	1.59
4 (all, LiF)	LiF	0.55	0.5	0.43	0.756	0.182	0.468	0.98

A clear extension of this work is simulating radiographic experiments using neutron sources developed here to evaluate image quality as a function of the number of laser pulses. To increase probability of single-shot radiography success, moderation of the maximum yield source, emitting 1.94×10^9 n/pulse, may be used as epi-thermal neutrons have much higher detection efficiency. The results, specifically from case 4, and the surrogate model developed in this work could be inputs for such a moderator design. The methods used here are suitable for optimizing moderator materials and thickness.

One factor limiting the precision of results presented here is accuracy of the ion spectra as a function of angle. The proven capability for identification of the objective trade-off curves here may justify implementing angular resolved measurements at Omega EP [10]. Also, as experimental chambers are typically constructed from Al, this could be included in the converter design to estimate the neutron production within the converter relative to the chamber. This would provide a signal-to-noise floor aiding experimental design. These improvements would notably enhance future studies.

Acknowledgements. This work was supported by the U.S. Department of Energy through the Los Alamos National Laboratory (LANL). LANL is operated by Triad National Security, LLC, for the National Nuclear Security Administration of the U.S. DOE (Contract No. 89233218CNA000001). This work was initiated under the LANL LDRD 20180732ER program and continued under OES ADP.

References

1. Yogo, A., et al.: Advances in laser-driven neutron sources and applications. Eur. Phys. J. A **59**, 191 (2023)
2. Vogel, S.C., et al.: Short-pulse laser-driven moderated neutron source. UCANS-8, Paris, France, vol. 231, p. 01008 (2020)
3. Broughton, D.P., et al.: Laser-driven flash x-ray radiography of a shocked metallic foil. AIP Adv. **15**, 055023 (2025)
4. Huang, C.-K., et al.: Characterization of laser-accelerated proton beams from a 0.5 kJ sub-picosecond laser for radiography applications. Phys. Plasmas **32**, 033107 (2025)
5. Zimmer, M., et al.: Non-destructive and isotope-sensitive material analysis using a short-pulsed laser-driven epi-thermal neutron source. Nat. Commun. **13**, 1173 (2022)
6. Chen, S.N., et al.: Extreme brightness laser-based neutron pulses for investigating nucleosynthesis in the laboratory. Matter Radiat. Extremes **4**, 054402 (2019)
7. Chichester, D.L.: Production and applications of neutrons using particle accelerators. In: Industrial Accelerators and Their Applications, pp. 243–305
8. Huang, C.-K., et al.: High-yield and high-angular-fluence neutron generation from deuterons accelerated by laser-driven collisionless shock. Appl. Phys. Lett. **120**, 024102 (2022)
9. Wong, C.-S., et al.: Multi-Probe 23A: post shot data and analysis. Los Alamos National Laboratory Report No. LA-UR-23-25028 (2023)
10. Jung, D., et al.: A novel high resolution ion wide angle spectrometer. Rev. Sci. Instrum. **82**, 043301 (2011)
11. Kulesza, J.A., et al.: MCNP® Code Version 6.3.0 Theory & User Manual. Los Alamos National Laboratory Report No. LA-UR-22-30006 (2022)
12. Nakayama, S., et al.: JENDL/DEU-2020: Deuteron nuclear data library for design studies of accelerator-based neutron sources. J. Nucl. Sci. Technol. **58**(7), 805–821 (2021)
13. Brown, D.A., et al.: ENDF/B-VIII.0: the 8th major release of the nuclear reaction data library with CIELO-project cross sections. Nucl. Data Sheets **148**, 1–142 (2018)
14. Ziegler, J.F., et al.: SRIM - The Stopping and Range of Ions in Matter. SRIM 2013. https://www.srim.org

15. Rumelhart, D.E., et al.: Learning representations by back-propagating errors. Nature **323**, 533–536 (1986)
16. Deb, K., Jain, H.: An evolutionary many-objective optimization algorithm using reference-point-based nondominated sorting approach. IEEE Trans. Evol. Comput. **18**(4), 577–601 (2013)

Lithium-Containing Semiconductors for Neutron Imaging

M. A. Benkechkache[1(✉)], E. Hoegberg[1], J. Gallagher[1], R. Golduber[1],
A. Kargar[2], H. Hong[2], J. Christian[2], M. Kanatzidis[3], E. Qian[3], K. Saheb[3],
S. Imam[3], and E. Lukosi[1]

[1] University of Tennessee, Knoxville, TN 37996, USA
`mbenkech@utk.edu`
[2] Radiation Monitoring Devices, Inc, Watertown, MA 02472, USA
[3] Northwestern University, Evanston, IL 60208, USA

Abstract. In the past decade, much research has explored the neutron-sensing properties and potential applications of lithium-containing semiconductors in neutron imaging. These semiconductors offer the possibility of achieving high spatial and temporal resolution, superior gamma discrimination, and excellent neutron detection efficiency. Such a combination of features is rare in neutron detection systems. In this paper, the promise of lithium-containing semiconductors to significantly advance neutron radiography, tomography, and energy-resolved imaging is demonstrated. The fundamental properties of lithium-containing semiconductors and their basic electronic pulse processing requirements are examined. Ongoing efforts to enhance the performance of ultrahigh-resolution neutron imaging systems, including improvements in imaging speed and the development of detectors with larger active areas for neutron diffraction facilities, are discussed.

Keywords: Lithium-containing semiconductors · LiInSe$_2$ ·
LiInP$_2$Se$_6$ · Neutron detector · Neutron imaging

1 Introduction

Neutron imaging has become a critical tool in applications ranging from scientific research to national security, offering unique insights that X-ray imaging cannot provide. By revealing otherwise invisible features and dynamic processes, it plays a vital role in advancing many fields. However, direct neutron detection remains challenging due to the weak interaction of neutrons with most matter [1]. Effective neutron detectors typically rely on target nuclides that decay into highly energetic charged particles after neutron capture, creating detectable signals through ionization [2].

In the realm of neutron imaging, the most widely used detectors are based on a combination of neutron-sensitive scintillator screens and imaging cameras (such as CMOS or CCD-based systems) [3,4]. These detectors are highly reliable for

A. E. Craft and H. Z. Bilheux (Eds.): WCNR 2024, SPPHY 348, pp. 64–75, 2026.
https://doi.org/10.1007/978-3-032-15003-5_9

applications where time-of-flight information is not required, offering advantages such as large-area coverage, flexibility, and low cost of implementation.

Certain isotopes such as ^{3}He, ^{6}Li, ^{10}B, and ^{157}Gd, with their large thermal neutron absorption cross sections, offer viable paths for the development of efficient neutron-sensitive scintillators and detectors [5–7]. Among the various detection technologies explored, lithium-containing semiconductors have emerged as promising candidates for compact and efficient neutron detection. Unlike scintillator-based detectors, semiconductor detectors are particularly attractive for applications requiring time-of-flight information or neutron counting, where each detected neutron's temporal and spatial information is critical. Materials such as $LiInSe_2$, $LiInP_2Se_6$, and $LiGaInSe_2$ are notable for their wide bandgap, enabling operation at room temperature [8–10].

In this paper, we explore the potential of lithium-containing semiconductors to revolutionize neutron imaging technologies. We examine their fundamental properties and basic electronic pulse processing requirements. Additionally, we highlight advances in crystal growth techniques aimed at enhancing charge-transport properties and optimizing crystal quality. Finally, we discuss ongoing efforts to develop novel detector architectures and geometries designed to achieve ultra-high-resolution neutron imaging down to the 5 μm range.

2 Lithium-Containing Semiconductors

Li-containing semiconductors directly undergo the ^{6}Li (n, α)^{3}H reaction when exposed to thermal neutrons, producing charged particles. This phenomenon is illustrated in Fig. 1 for ^{6}Li-containing semiconductor detectors. When a thermal neutron interacts with ^{6}Li, it yields tritium and alpha particles with a combined Q-value of 4.78 MeV. These charged particles have relatively short path lengths in the semiconductor material and quickly dissipate their kinetic energy, creating ionization and generating free electron-hole pairs. Under an applied electric field, these charge carriers flow and produce a measurable current, enabling a direct signal response to neutron absorption events in the semiconductor.

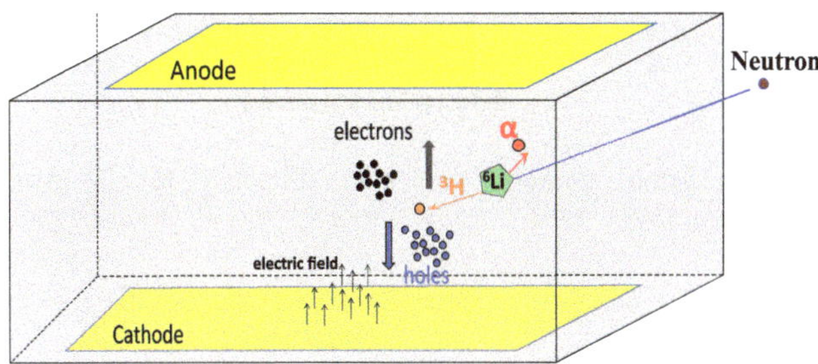

Fig. 1. Diagram of a Li-based neutron detector illustrating the ^{6}Li (n, α)^{3}H reaction, which generates charged particles, and the movement of electrons and holes toward their respective electrodes under an applied electric field.

Unlike conventional neutron detection systems, which often rely on a separate converter layer, such as ^{6}LiF, ^{10}B, or ^{157}Gd, lithium-based semiconductors integrate the neutron-sensitive isotope directly into the detection medium. In traditional systems, reaction products generated in the converter layer must escape and enter the adjacent semiconductor detector to be detected. However, the thickness of the converter layer poses a trade-off: while increasing the thickness enhances neutron absorption, it also leads to self-absorption, where reaction products are absorbed within the converter layer itself before reaching the detector, thereby reducing efficiency. By embedding ^{6}Li directly in the semiconductor, these limitations are eliminated, significantly improving neutron detection efficiency and simplifying manufacturing processes. Furthermore, direct signal generation in lithium-based semiconductors eliminates the need for photomultiplier tubes (PMTs) or silicon photomultipliers, enabling more compact designs with lower power consumption and enhanced reliability.

For a neutron detector based on lithium-containing compound semiconductors, the neutron absorption efficiency can reach very high values as detector thickness increases, but saturates below 100% due to competing reactions of thermal neutrons with isotopes other than ^{6}Li. These competing reactions contribute minimally to signal generation, resulting in negligible charge collection. The dependence of neutron detection efficiency on detector thickness is shown in Fig. 2, illustrating the relationship between the macroscopic thermal neutron absorption cross-section of the compound, its density, and the detector thickness, as previously outlined.

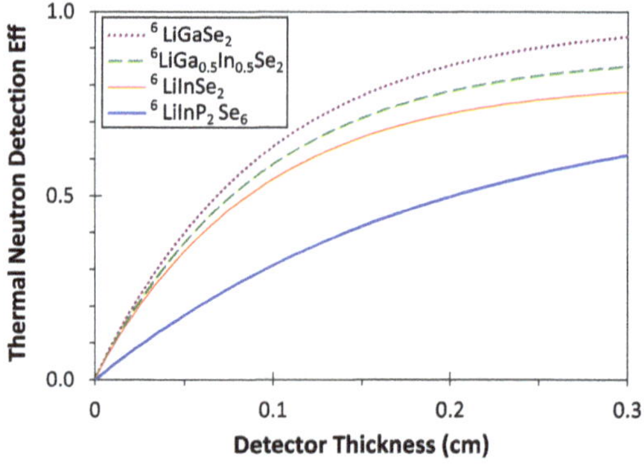

Fig. 2. Calculated thermal neutron detection efficiency for LiGaSe$_2$, LiInGaSe$_2$, LiInSe$_2$, and LiInP$_2$Se$_6$ [10].

2.1   LiInSe$_2$ Semiconductors

Lithium Indium Diselenide "LiInSe$_2$" is a promising material for neutron detection due to its unique properties and effective performance. LiInSe$_2$ wafers or crystals are grown using the Vertical Bridgman "VB" technique, ensuring the formation of high-quality crystals necessary for efficient neutron detection as a semiconductor with a band gap 'Eg' ranging from 2 to 3 eV. One of the distinguishing features of $LiInSe_2$ is its high neutron absorption efficiency, although it is limited to approximately 80% due to the presence of ^{115}In. $LiInSe_2$ also possesses excellent gamma-neutron discrimination capabilities. The large 4.78 MeV Q-value of the reaction ^{6}Li (n, α) ^{3}H contributes to this effective discrimination, making $LiInSe_2$ an exceptional candidate for neutron detection applications [8].

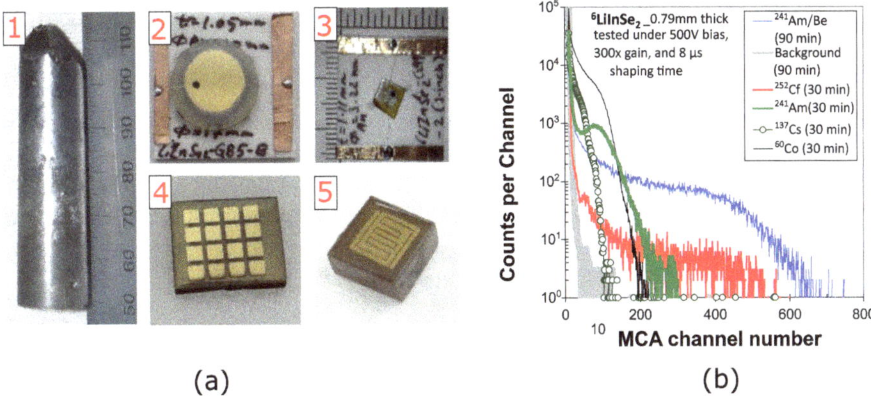

(a)   (b)

Fig. 3. (a) $LiInSe_2$ material with different electrode geometries with "1" a 17 mm VB $LiInSe_2$ crystal ingot "2" and "3" Planar $LiInSe_2$ detectors mounted on a 1-inch square ceramic substrate with Au contacts "4" a pixelated $LiInSe_2$ detector with Fe contacts "5" a 7.5×7.5 mm co-planar $LiInSe_2$ detector with Au contacts, and (b) ^{241}AmBe, background, ^{241}Am (alpha source), ^{60}Co, and ^{137}Cs pulse high spectra collected with 0.79 mm thick $LiInSe_2$ device with 3 mm diameter Au contacts. The channel number on the x-axis represents the output of the Multi-Channel Analyzer (MCA), which digitizes the voltage signal produced by the detector through an Analog-to-Digital Converter (ADC). This voltage is proportional to the energy deposited by the radiation.

The crystal boules of $LiInSe_2$ grown at Radiation Monitoring Devices, Inc. (RMD) are used to produce wafers for detector fabrication. RMD is consistently working on improving the crystal growth of Li-containing semiconductors. Purifying raw or synthesized materials to enhance charge transport properties and optimize crystal quality is crucial to achieving high-performance detectors. Detectors with various electrode geometries have been fabricated and tested to optimize detection capabilities and meet specific application requirements. Figure 3-a below shows an example of the $LiInSe_2$ growth boule and various wafers prepared for detector production with different electrode geometries.

The radiation performance of one of the $LiInSe_2$ detectors, with a thickness of 0.79 mm and 3 mm diameter Au contacts, is illustrated in Fig. 3-b. The plot shows the neutron response of the detector, biased at 500 V, evaluated with a moderated ^{241}Am/Be source. This measurement provided a characteristic thermal/cold neutron spectrum, producing a broad continuum response over the background, and demonstrated the ability of the LiInSe$_2$ material for gamma-neutron discrimination.

2.2 LiInP$_2$Se$_6$ Semiconductors

Another Li-containing material recently discovered by Northwestern University (NU) is Lithium Indium Phosphorus Selenide (LiInP$_2$Se$_6$), which has been grown using the Chemical Vapor Transport (CVT) technique. LiInP$_2$Se$_6$ is an effective material for neutron detection. As a semiconductor with a bandgap of approximately 2.06 eV, LiInP$_2$Se$_6$ has demonstrated both hole and electron transport properties, enabling high charge collection efficiency (CCE) in neutron detection applications. LiInP$_2$Se$_6$ has achieved the first-ever neutron-induced peak in a lithium-containing semiconductor, showcasing high neutron interaction efficiency and excellent gamma-neutron discrimination. Like the LiInSe$_2$ material, its large 4.78 MeV Q-value enhances its detection capabilities. Figure 4-a shows the LiInP$_2$Se$_6$ reaction vessel grown in NU and wafers prepared for detector production with different electrode geometries.

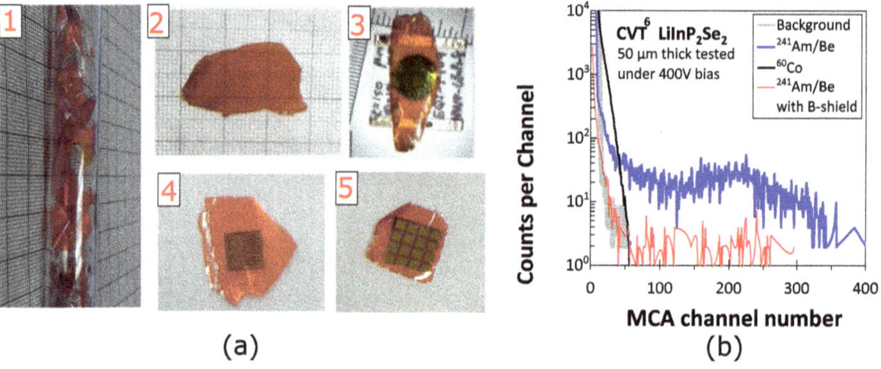

Fig. 4. (a) LiInP$_2$Se$_6$ material production with "1" the CVT reaction vessel "2" a single LiInP$_2$Se$_6$ crystal "3" Planar LiInP$_2$Se$_6$ detector mounted on a ceramic substrate with Au contacts "4" a LiInP$_2$Se$_6$ detector with a strip pattern "5" a pixelated LiInP$_2$Se$_6$ detector with Au contacts, and (b) The Energy spectra measured of LiInP$_2$Se$_6$ detector from ^{241}AmBe neutron source. The channel number on the x-axis represents the output of the Multi-Channel Analyzer (MCA), which digitizes the voltage signal produced by the detector through an Analog-to-Digital Converter (ADC). This voltage is proportional to the energy deposited by the radiation.

The radiation response of a LiInP$_2$Se$_6$ detector, made from a sample 50 µm thick and with 4 mm diameter Au contacts is illustrated in Fig. 4-b. The plot shows the neutron response of the detector, biased at 400 V, evaluated with a moderated ^{241}Am/Be source. This measurement provided a good neutron response over the background signal and demonstrated the capability of the $LiInP_2Se_6$ material. The ability of $LiInP_2Se_6$ to transport both holes and electrons contributes to a more uniform CCE distribution for a planar device, reducing the dependency on the depth of interaction. Consequently, the fabricated planar detector is not position-sensitive if both electrons and holes reach their corresponding electrodes, minimizing the effect of "tailing" in the pulse height spectrum. This allows threshold discrimination to remove gamma-ray contamination (<2 MeV) and form a peak on the neutron spectrum (Fig. 4-b), and enables the use of alternative readout techniques such as crossed strip designs.

The choice of a 50 μm thickness for the $LiInP_2Se_6$ detector is due to the current limitations in the crystal growth process using the 2D planar chemical vapor transport (CVT) technique, which is only suitable for growing relatively thin samples. While CVT can currently produce high-quality crystals up to a few hundred micrometers thick, further advancements are necessary to grow thicker materials and thus to achieve higher efficiencies. Efforts are underway to adopt the Bridgman growth technique, which offers the potential to grow much thicker crystals, up to a few millimeters.

These ongoing efforts at NU aim to enhance the quality of this promising semiconductor by producing larger and thicker crystals with improved neutron detection efficiency and charge transport properties, paving the way for broader applications in neutron detection technologies.

2.3   Radiation Tolerance of $LiInSe_2$ and $LiInP_2Se_6$

Li-containing semiconductors, particularly $LiInSe_2$ and $LiInP_2Se_6$, exhibit excellent radiation hardness in high-radiation environments. This is demonstrated through various material characterization techniques and direct testing in radiation sensing applications, which confirm their robustness under high radiation exposure conditions. Both materials were irradiated at Ohio State University (OSU) with neutron fluences up to 10^{15} n/cm^2 [11]. Despite additional wear from handling and treatment imposed by subjecting thin samples to several characterization tests, the results indicated remarkable resistance to radiation-induced damage, suggesting their potential use in high-radiation environments such as beamline instruments. Both materials continued to respond effectively to alpha radiation after neutron fluences up to 10^{15} n/cm^2. This is shown in Fig. 5, which illustrates the radiation response of both irradiated semiconductors, $LiInSe_2$ and $LiInP_2Se_6$, to neutron and alpha particles, respectively. This translates to a potential operational lifetime exceeding 1,000 d at neutron flux levels of 10^7 n/cm^2/s. Although systematic errors were present, they did not overshadow the primary finding that radiation exposure did not degrade the materials' performance, underscoring their reliability.

3 Lithium-Based Detection Systems

Li-containing semiconductors are promising materials for the development of neutron imaging detectors due to their excellent neutron detection capabilities, radiation hardness, and high-resolution imaging capabilities, as discussed in previous sections. Leveraging these unique properties, various detector designs and configurations have been built and tested using these direct neutron semiconductors, including a pixelated Timepix ASIC neutron detector and a cross-strip neutron imaging detector.

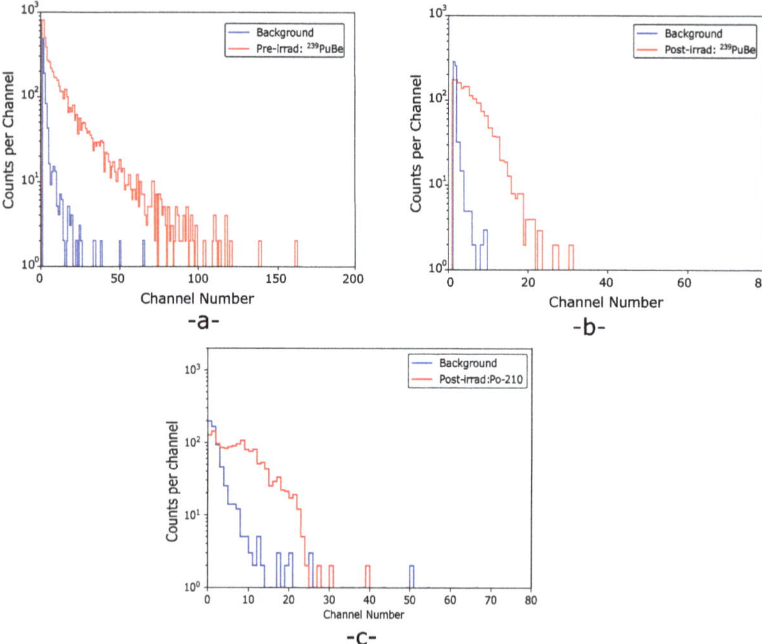

Fig. 5. (a) and (b) Pulse height spectra showing the response of LiInSe$_2$ to a moderated PuBe source and the corresponding background before and after irradiation, respectively, and (c) the response of LiInP$_2$Se$_6$ to an alpha source, ^{210}Po, and the corresponding background. The channel number on the x-axis represents the output of the Multi-Channel Analyzer (MCA), which digitizes the voltage signal produced by the detector through an Analog-to-Digital Converter (ADC). This voltage is proportional to the energy deposited by the radiation [11].

3.1 Li-Based Timepix Detector

Lithium Indium Diselenide has shown significant promise for high-resolution neutron imaging. Our group previously demonstrated this by developing a pixelated neutron imaging detector using a LiInSe$_2$ sensor with a 55 μm pixel pitch, coupled to the Timepix1 ASIC, as illustrated in Fig. 6a (adapted from [14]). The detector system was tested at the CG-1D Neutron Imaging Facility at HFIR, where it achieved a spatial resolution better than 200 μm using the PSI Siemens star mask (Fig. 6b). A feature size of 34 μm was resolved by imaging a 3D-printed University of Tennessee logo embedded in a 1cm^3 plastic block (Fig. 6c). This resolution was confirmed using a knife-edge test with a beam L/D ratio of 660:1 and analyzed through the Modulation Transfer Function (MTF) method, where the Edge Spread Function (ESF) was differentiated to obtain the Line Spread Function (LSF). The resolution was determined at the 10% MTF threshold, yielding 34 μm, approximately two-thirds of the pixel pitch. Further details on the detector design, experiment, and analysis can be found in the original publication.

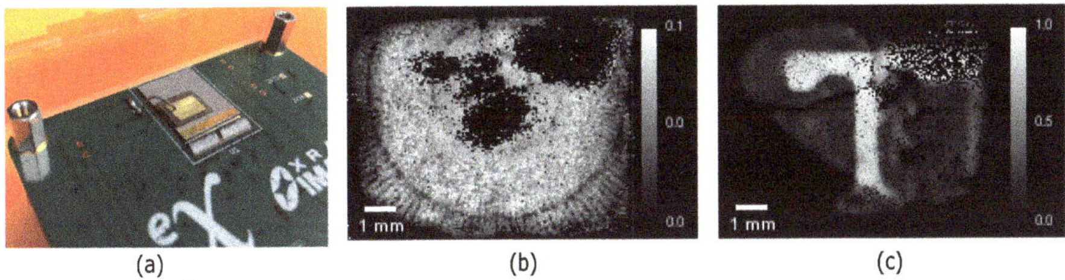

Fig. 6. (a) Bonded LiInSe$_2$-Timepix imager PCB, (b) neutron image of Seimens star on PSI resolution test mask, and (c) neutron image of a 3D printed "Power T" with a threshold of +24 mV (images adapted from [14], under CC BY 4.0.).

3.2 Double-Sided Strip Li-Based Detector

Another Li-based detector configuration successfully developed is the Double-Sided Strip Detector (DSSD) using Lithium Indium Phosphorus Selenide. The DSSD design leverages the excellent neutron detection and charge transport properties of LiInP$_2$Se$_6$, as discussed in previous sections, to enhance the neutron imaging capabilities. A key benefit of the DSSD design is the reduction of the readout channels from N^2 to 2N when directly coupled to an ASIC. This innovation reduces the complexity and power consumption of the readout process. Additionally, the DSSD design enables lower power processing electronics and a much-reduced data transfer rate, which is particularly beneficial for large-area neutron imaging applications.

The crossed-strip LiInP$_2$Se$_6$ detector we developed targets a spatial resolution of 5 μm, a temporal resolution of less than 1 μs, high neutron detection efficiency (>70%), and a strip pitch as low as 30 μm. To meet these objectives, a modified η-function is employed to analyze and optimize the charge sharing across the strips. This approach considers any interaction location and emission angle of the secondary charged particles (alpha and triton) resulting from neutron absorption by the Li-based semiconductor material. These specifications make it highly effective for enhanced neutron imaging.

The built detection system is based on a strip-patterned $LiInP_2Se_6$ sensor coupled to VATAGP7 Application-Specific Integrated Circuits (ASICs) manufactured by Integrated Detector Electronics AS (IDEAS), a Norwegian company specializing in ASIC solutions for advanced detection systems. These ASICs are hosted on a custom daughterboard, which connects to the main acquisition board. This main acquisition board incorporates a Field Programmable Gate Array (FPGA) and an Analog-Digital Converter (ADC), allowing for control of the readout process and communication with a PC [15,16]. Figure 7 shows the readout detection system developed for the DSSD $LiInP_2Se_6$ detector.

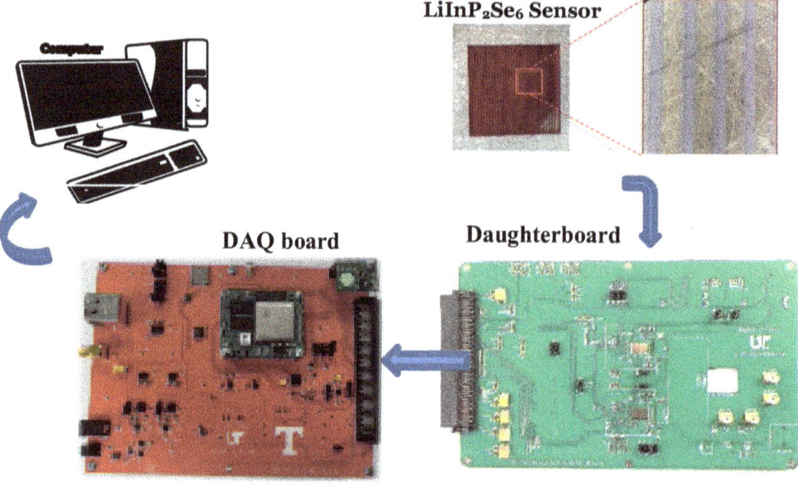

Fig. 7. A readout detection system developed for the DSSD $LiInP_2Se_6$ detector.

Develop Modified η-function. A study was conducted to develop a modified η-function coupled with machine learning techniques to enhance the spatial resolution of Li-based double-sided strip detectors (DSSD) for neutron imaging. By incorporating high-fidelity simulations and analyzing factors such as noise and charge collection efficiency, optimal strip width/pitch configurations of $10/15$ μm and $20/30$ μm were identified through simulations modeling charge sharing, electric fields, and charge collection dynamics. These configurations ensure sufficient charge sharing between adjacent strips, enabling machine learning models coupled with the η-function to reconstruct neutron interaction points with sub-5 μm spatial resolution. Smaller configurations like $10/15$ μm enhance charge sharing and spatial resolution, while the slightly larger $20/30$ μm configuration balances signal integrity with spatial precision. Validation using metrics such as RMSE and MTF demonstrated that these models maintained their accuracy under Gaussian noise and varying charge collection efficiency, ensuring their robustness for neutron imaging applications [17].

Figure 8-a presents data from a LiInSe$_2$ semiconductor detector, simulated with MCNP, used to evaluate the spatial-resolution-enhancing algorithm. Neutron interaction locations were determined using a thermal neutron beam on a gadolinium University of Tennessee-Knoxville Power T.

Figure 8-b shows a Coarse Tree Reconstruction of the University of Tennessee Power T, which refers to a machine learning-based regression model (Coarse Regression Tree) used to predict neutron interaction points. This model was trained on simulated data to map the interaction location based on the signals induced across multiple strip electrodes. Coarse trees are advantageous for initial reconstructions because of their simplicity and robustness in handling noise. The accurate simulation of the motion of the secondary charged particles in ^{6} Li neutron capture, including 4π emission and Bragg curves, was crucial. Neutron absorptions were binned into 4 μm pixels, representing a spatial resolution of 4 μm $\times$ 4 μm areas in the simulation grid. This fine binning ensures high granularity for analyzing neutron absorption patterns, which is critical for validating the sub-5 μm resolution goal. Signal formation was tracked using the Shockley-Ramo Theorem, which incorporates charge trapping across strip electrodes with a width/pitch of 20/30 μm.

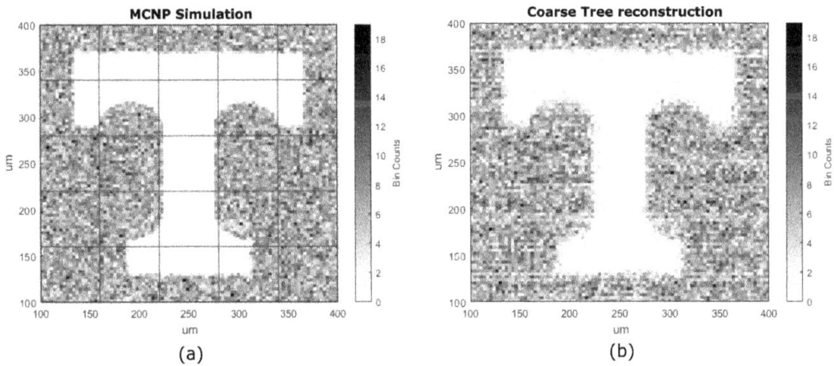

Fig. 8. (a) MCNP simulated University of Tennessee-Knoxville Power T used to evaluate the spatial-resolution-enhancing algorithm, red grid lines correspond to a sort of symmetry space required for the reconstruction process using pattern classification, and (b) A coarse tree reconstruction of the University of Tennessee Power T.

4   Conclusion

Li-containing semiconductors, such as LiInSe$_2$ and LiInP$_2$Se$_6$, offer significant potential for neutron detection due to their radiation hardness, high neutron absorption, and excellent gamma-neutron discrimination. In this paper, we presented new detector architectures to improve ultra-high-resolution neutron imaging. The LiInSe$_2$-Timepix system, previously developed by our group, achieved a spatial resolution of 34 μm, while the DSSD LiInP$_2$Se$_6$ detector, with advanced reconstruction algorithms, targets 5 μm resolution and less than 1 μs temporal resolution. These advancements further enhance neutron imaging for radiation-intensive applications, ensuring reliable performance and longevity.

Acknowledgements. This material is based on work supported by the U.S. Department of Energy, Office of Science, USA, Office of Basic Energy Sciences, USA, under Award Numbers DE-SC-0019446 and DE-SC0021554. Part of this work was conducted at the Micro-Processing Research Facility, a University of Tennessee Core Facility. A portion of this research used resources at the High Flux Isotope Reactor, a DOE Office of Science User Facility operated by the Oak Ridge National Laboratory. The beam time was allocated to CG-1D on proposal number IPTS-13833.1.

References

1. Dunning, J.R., et al.: Interaction of neutrons with matter. Phys. Rev. **48**(3), 265 (1935)
2. Knoll, Glenn F.: Radiation detection and measurement. John Wiley and Sons (2010)
3. Sakata, I., et al.: Three-dimensional observation of water distribution in PEFC by neutron CT. Nucl. Instrum. Methods Phys. Res. Sect. A **605**(1–2), 131–133 (2009)
4. Morgano, M., et al.: Neutron imaging options at the BOA beamline at Paul Scherrer Institut. Nucl. Instrum. Methods Phys. Res. Sect. A **754**, 46–56 (2014)
5. Trtik, P., Lehmann, E.H.: Isotopically-enriched gadolinium-157 oxysulfide scintillator screens for the high-resolution neutron imaging. Nucl. Instrum. Methods Phys. Res. Sect. A **788**, 67–70 (2015)
6. Kouzes, Richard T.: The ^{3}He supply problem. No. PNNL-18388. Pacific Northwest National Lab. (PNNL), Richland, WA (United States) (2009)
7. Tremsin, A.S., et al.: Optimization of high count rate event counting detector with microchannel plates and quad timepix readout. Nucl. Instrum. Methods Phys. Res. Sect. A **787**, 20–25 (2015)
8. Lukosi, E.: Lithium indium diselenide sensor for neutron transmission imaging. PSND Workshop, Forschungszentrum Julich (2014)
9. Chica, Daniel G., et al.: Direct thermal neutron detection by the 2D semiconductor ^{6}LiInP$_2$Se$_6$. Nature **577**(7790), 346–349 (2020)
10. Kargar, Alireza, et al.: Thermal neutron detection with lithium-based semiconductors. Hard X-Ray, Gamma-Ray, and Neutron Detector Physics XXIV, vol. 12241. SPIE (2022)
11. Golduber, Robert, et al.: Evaluation of neutron-radiation tolerance of lithium indium Diselenide semiconductors. Physica status solidi (a) **220**(9), 2300012 (2023)
12. McGregor, D.S., Bellinger, S.L., Shultis, J.K.: Present status of microstructured semiconductor neutron detectors. J. Crystal Growth **379**, 99–110 (2013)
13. Gueorguiev, A., et al.: Semiconductor neutron detectors. Hard X-Ray, Gamma-Ray, and Neutron Detector Physics XVIII, vol. 9968. SPIE (2016)

14. Herrera, Elan, et al.: Neutron imaging with Timepix coupled lithium indium Diselenide. J. Imaging **4**(1), 10 (2017)
15. Benkechkache, M. A., et al.: Front-end electronics development for a fine pitch AC Coupled double-sided $LiInSe_2$ strip detector. In: 2022 IEEE Nuclear Science Symposium and Medical Imaging Conference (NSS/MIC). IEEE (2022)
16. Benkechkache, M. A., et al.: Readout electronics and development of a double-sided strip design $LiInSe_2$ Neutron Imaging Detector. In: 2021 IEEE Nuclear Science Symposium and Medical Imaging Conference (NSS/MIC). IEEE (2021)
17. Gallagher, J., et al.: Neutron imaging with lithium indium diselenide semiconductors. In: 2020 IEEE Nuclear Science Symposium and Medical Imaging Conference (NSS/MIC). IEEE (2020)

Noise Reduction in Neutron Tomography

Qianru Zhan$^{(\boxtimes)}$, Mahdieh Shakoorioskooie, David Mannes,
and Anders Kaestner

Paul Scherrer Institut, 5232 Villigen - PSI, Switzerland
`qianru.zhan@psi.ch`

Abstract. This study focuses on noise reduction in neutron tomography, a non-destructive imaging technique with inherently low signal-to-noise ratio (SNR) due to the nature of neutron sources and detectors. This research explores the application of advanced denoising techniques, departing from iterative reconstruction methods, to improve image quality. Using a sample consisting of an aluminum cylinder filled with layers of copper balls and metal bars of various compositions, we reconstructed neutron tomography data at varying exposure times with the Filtered Backprojection (FBP) algorithm, followed by denoising using the Block Matching 4D (BM4D) algorithm. Results demonstrate that BM4D significantly enhances image quality by reducing normalized mean squared error (NMSE) and improving contrast-to-noise ratio (CNR). These findings suggest that BM4D can substantially reduce acquisition times in neutron tomography while maintaining detailed structural visibility and high image quality.

Keywords: Neutron tomography · noise reduction · BM4D · image quality

1 Introduction

Neutron imaging is a non-destructive technique used to unveil the internal structure of materials, complementing X-ray imaging. However, neutron images can suffer from a low signal-to-noise ratio (SNR) compared to X-ray images, primarily because of the compromises between spatial and temporal resolution. This low SNR arises because of the nature of neutron sources and the efficiency of neutron detectors, which often requires much longer exposure times to achieve acceptable image quality.

The challenge of low SNR in neutron tomography presents a significant obstacle, particularly when aiming to reduce acquisition times for practical and safety reasons. Reducing the scan time can reduce sample activation, which is particularly important for rare and sensitive objects e.g. archeological artefacts. The study of dynamic processes also aim at using short acquisition times to allow the detection of rapid changes like the interaction between the soil and roots.

This study aims to investigate the effectiveness of advanced denoising techniques on neutron tomography data acquired using different neutron doses. By

A. E. Craft and H. Z. Bilheux (Eds.): WCNR 2024, SPPHY 348, pp. 76–83, 2026.
https://doi.org/10.1007/978-3-032-15003-5_10

systematically applying denoising methods to datasets obtained with different exposure times, we seek to determine the limits of denoising and the potential for reducing acquisition times without compromising image quality. Various approaches have been explored to address low-dose tomography challenges in both X-ray and neutron imaging. These include sinogram denoising prior to reconstruction [1], model-based iterative reconstruction techniques [2], and volumetric denoising applied post-reconstruction [3,4,11]. Among these, volumetric denoising methods like BM4D have demonstrated significant potential in preserving structural details while effectively reducing noise, making them attractive for enhancing neutron tomography datasets obtained under reduced exposure conditions.

The focus is on using the Block Matching 4D (BM4D) algorithm [7], among others, to enhance the visibility of fine details and contrast in neutron tomographic images.

2   Materials and Methods

2.1   Sample Description

The sample consists of two parts (Fig. 1 (a)):

- **Upper Section:** An aluminum container filled with layers of copper balls of four various sizes (0.5 mm, 0.8 mm, 1 mm, and 2 mm) to study the resolution of the instrumentation setup.
- **Lower Section:** An aluminum cylinder inserted with six metal bars (Cu, Fe, Ti, Al, Pb, Ni) to study the contrast limits and detection capabilities.

2.2   Imaging Setup

The sample was imaged at seven different exposure times (0.5 s, 1 s, 2 s, 5 s, 10 s, 20 s, and 50 s), with a full scan comprising 1125 projections. An Andor iKON-L CCD camera with a Zeiss 100 mm f/2.0 macro lens and a 30 μm thick Gadox scintillator was used, leading to an effective pixel size is 37.4 μm.

2.3   Data Processing

The acquired dataset underwent pre-processing using the *spotsclean* and *ring removal* functionalities in Muhrec Software [6] to mitigate gamma hits and ring artifacts. Following pre-processing, the dataset was reconstructed using the Filtered Backprojection algorithm [5] implemented in Muhrec. The reconstructed volumes were subsequently denoised using the Block Matching 4D (BM4D) algorithm [4]. The denoised 50 s-exposure volume served as the reference for comparison.

The remaining volumes were evaluated relative to this reference. Both the ball packing in the upper part and the contrast sample in the lower part were assessed using the Normalized Mean Squared Error (NMSE). Additionally, the contrast-to-noise ratio (CNR) was computed for the contrast sample.

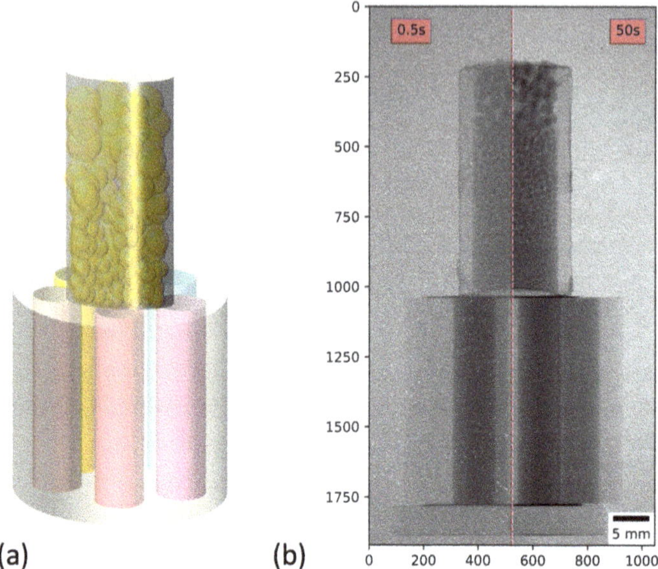

Fig. 1. (a) a 3D model of the sample; (b) Comparison of sample projection under 0.5 s and 50 s exposure times.

2.4   Volume of Interest (VOI) Selection

A Volume of Interest (VOI) was selected for each ball size present in the ball packing (0.5 mm, 0.8 mm, 1 mm, and 2 mm), corresponding to four distinct VOIs per volume. Due to the smaller dimensions of some balls, their layers exhibited more boundaries, increasing the likelihood of blurring introduced by the Filtered Backprojection (FBP) algorithm. To minimize the influence of boundary effects and ensure fair comparisons, VOIs were chosen to contain approximately similar amounts of balls across different sizes. This was achieved by first segmenting the copper balls from the aluminum container and selecting VOIs with varying heights: 50 slices for 0.5 mm balls, 80 slices for 0.8 mm balls, 100 slices for 1 mm balls, and 200 slices for 2 mm balls. Although this selection was based on practical observations rather than strict cubic scaling, the chosen VOIs were sufficient to avoid interference from neighboring layers and maintain consistency across evaluations.

2.5   Denoising with BM4D

The BM4D algorithm was utilized using a Python wrapper developed at the University of Tampere [7]. BM4D is a state-of-the-art denoising method that extends the principles of the Block Matching 3D (BM3D) algorithm [8] to four-dimensional data, where the fourth dimension represents the volumetric neighborhood. It operates by grouping similar blocks within the dataset, transforming them into a sparsified domain, applying shrinkage to reduce noise, and reconstructing the denoised data. The estimated noise sigma was calculated using a robust wavelet-based estimator of the noise standard deviation [9], and the Poisson noise was stabilized by the Anscombe transform [10]. Further details on the theoretical basis of BM4D can be found in the original publication by Mäkinen et al. [7], and the motivation behind using BM4D in our application was detailed in our previous work [11], where various filters were compared using synthesized tomographic data.

3   Results and Discussion

3.1   Reconstructed and Denoised Data

In the following figures (Figs. 2, 3) we provide a qualitative impression of the filtering effect.

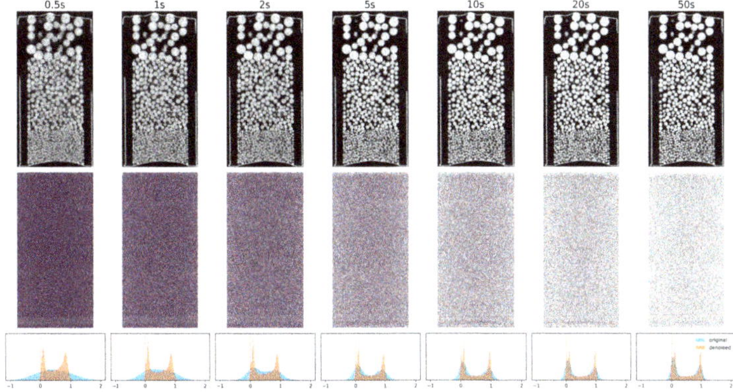

Fig. 2. Vertical slices through the ball packing sample after denoising with BM4D. The difference images below each slice show the differences between the originally reconstructed slices and the denoised ones. The histograms below illustrate the distribution of neutron attenuation coefficients before and after denoising.

3.2   Image Quality Assessment

To evaluate the effectiveness of the denoising process, we used two critical metrics: Normalized Mean Squared Error (NMSE) and Contrast-to-Noise Ratio (CNR). These metrics were calculated for the volumes before and after denoising across different exposure times, representing various SNR levels.

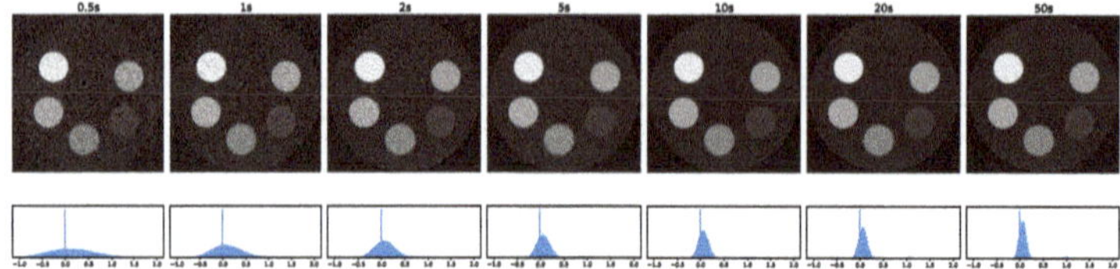

Fig. 3. Horizontal slices through the contrast sample reconstructed using various exposure times per projection. The histograms below show the distribution of grey levels in each volume.

NMSE. The NMSE measures the accuracy of the reconstructed volumes compared to a high-quality reference, defined as:

$$NMSE = \frac{\sum_{i=1}^{N}(I_i - R_i)^2}{\sum_{i=1}^{N} R_i^2}$$

where I_i represents the intensity values of the denoised or non-denoised image, R_i represents the intensity values of the reference image, and N is the number of pixels.

CNR. The CNR quantifies the visibility of different materials against the aluminum background and is calculated using the formula:

$$CNR = \frac{|\mu_1 - \mu_2|}{\sqrt{(\sigma_1^2 + \sigma_2^2)/2}}$$

where μ_1 and μ_2 correspond to the mean intensity of the signal and the background, σ_1 and σ_2 the standard deviation, respectively (Figs. 4 and 5).

From these figures, it is evident that denoising significantly enhances both CNR and NMSE across all materials, ball diameters, and exposure times. The CNR values of the denoised volumes are consistently higher, indicating improved visibility of the materials against the aluminum background. The NMSE values of the denoised volumes are consistently lower, indicating a closer match to the high-quality reference. The improvements are particularly noticeable at lower SNR levels (shorter exposure times), demonstrating the effectiveness of the BM4D denoising algorithm in enhancing image quality under challenging conditions.

Notably, the NMSE of the denoised volume at 5 s exposure is comparable to that of the non-denoised volume at 20 s exposure, demonstrating the potential to significantly reduce acquisition times without sacrificing image quality. Additionally, the CNR of the denoised volume at 0.5 s exposure is almost indistinguishable from that of the non-denoised volume at 50 s exposure, further highlighting the effectiveness of the BM4D denoising algorithm in maintaining high image quality under challenging conditions.

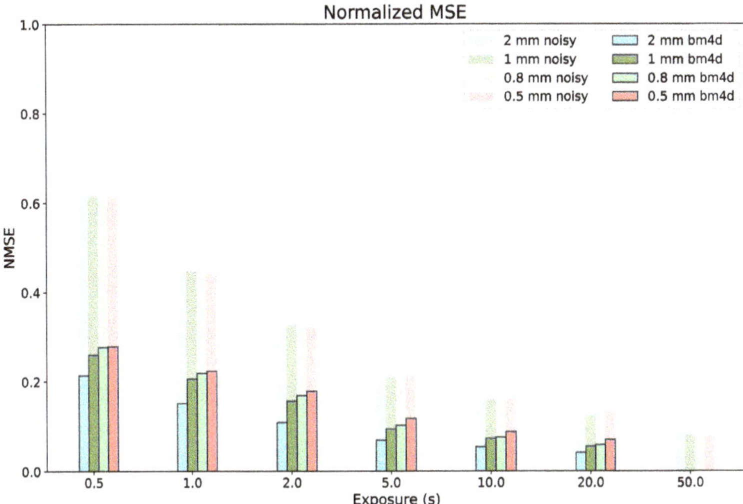

Fig. 4. Grouped bar plot of Normalized Mean Squared Error (NMSE) as a function of different exposure times for various ball diameters (2 mm, 1 mm, 0.8 mm, 0.5 mm). The NMSE values for non-denoised volumes are shown as semi-transparent bars, while the NMSE values for denoised volumes are shown as opaque bars of the same colors.

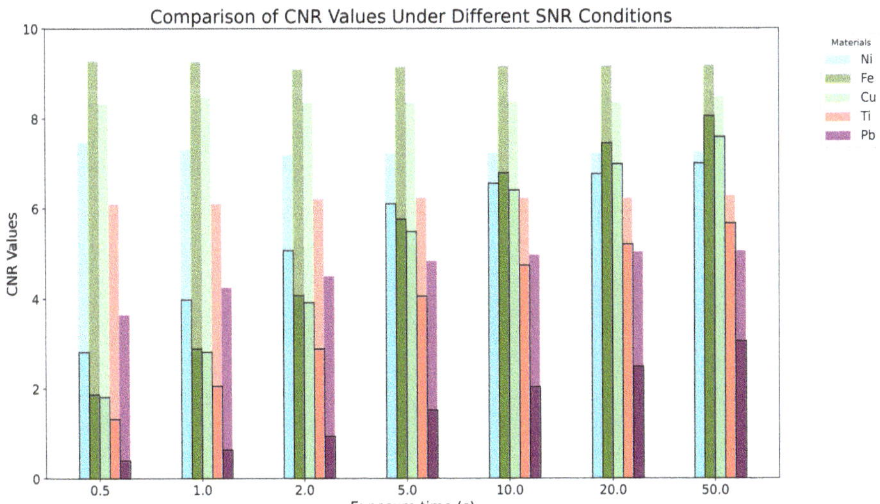

Fig. 5. Grouped bar plot of Contrast-to-Noise Ratio (CNR) as a function of different exposure times for various materials (Ni, Fe, Cu, Ti, Pb). The CNR values for non-denoised volumes are shown as opaque bars, while the CNR values for denoised volumes are shown as semi-transparent bars of the same colors.

4   Conclusion and Outlook

This study demonstrates the significant improvements in neutron tomography image quality achieved through the advanced denoising technique BM4D algorithm. By applying denoising to datasets acquired at various exposure times, we evaluated its effectiveness in enhancing contrast-to-noise ratio (CNR) and reducing normalized mean squared error (NMSE).

These results highlight BM4D's capability to improve image quality and reduce acquisition times in neutron tomography. Future work will involve leveraging this dataset for developing machine learning-based denoising techniques; and acquiring similar neutron imaging datasets to further optimize SNR studies.

References

1. Karimi, D., Deman, P., Ward, R., et al.: A sinogram denoising algorithm for low-dose computed tomography. BMC Med. Imaging **16**, 11 (2016). https://doi.org/10.1186/s12880-016-0112-5
2. Smith, E.A., et al.: Model-based iterative reconstruction: effect on patient radiation dose and image quality in pediatric body CT. Radiology **270**(2), 526–534 (2014). https://doi.org/10.1148/radiol.13130362.Epub 2013 Oct 29. PMID: 24091359; PMCID: PMC4228752
3. Diwakar, M., Kumar, P., Singh, A.K.: CT image denoising using NLM and its method noise thresholding. Multimedia Tools Appl. 14449–14464 (2018). https://doi.org/10.1007/s11042-018-6897-1
4. Maggioni, M., Katkovnik, V., Egiazarian, K., Foi, A.: Nonlocal transform-domain filter for volumetric data denoising and reconstruction. IEEE Trans. Image Process. **22**(1), 119–133 (2013). https://doi.org/10.1109/TIP.2012.2210725
5. Buzug, T.: Computed Tomography: from Photon Statistics to Modern Cone-Beam CT. Springer(2008). https://doi.org/10.1007/978-3-540-39408-2
6. Kaestner, A.P.: MuhRec - a new tomography reconstructor. Nucl. Instrum. Methods Phys. Res., Sect. A **651**(1), 156–160 (2011). https://doi.org/10.1016/j.nima.2011.01.129
7. Mäkinen, Y., Azzari, L., Foi, A.: Collaborative filtering of correlated noise: exact transform-domain variance for improved shrinkage and patch matching. IEEE Trans. Image Process. **29**, 8339–8354 (2020). https://doi.org/10.1109/TIP.2020.3014721
8. Dabov, K., Foi, A., Katkovnik, V., Egiazarian, K.: Image denoising with block-matching and 3D filtering. In: Proceedings of the SPIE 6064, Image Processing: Algorithms and Systems, Neural Networks, and Machine Learning, 606414 (17 February 2006). https://doi.org/10.1117/12.643267
9. Donoho, D.L., Johnstone, I.M.: Section 4.2 in Ideal spatial adaptation by wavelet shrinkage. Biometrika **81**(3), 425–455 (1994). https://doi.org/10.1093/biomet/81.3.425
10. Makitalo, M., Foi, A.: Optimal inversion of the generalized Anscombe transformation for Poisson-Gaussian noise. IEEE Trans. Image Process. https://doi.org/10.1109/TIP.2012.2202675

11. Zhan, Q., Shakoorioskooie, M., Mannes, D., Kaestner, A.: Image denoisers in tomography: enhancing quality of bimodal imaging for corrosion studies. In: Proceedings of the 13th Conference on Industrial Computed Tomography (iCT) 2023, 6 – 9 February 2024, School of Engineering, Wels Campus, Austria. e-Journal of Nondestructive Testing. https://doi.org/10.58286/29248

Spatially Resolved Evaluation of Texture Variations in Cold-Rolled Zircaloy-4 Cladding by Bragg Edge Imaging

Florencia Malamud[1]([✉]) [iD], Miguel A. Vicente Alvarez[2,3] [iD], Valentina Stella[2] [iD], Javier R. Santisteban[2,3] [iD], and Markus Strobl[1] [iD]

[1] Laboratory for Neutron Scattering and Imaging, PSI Center for Neutron and Muon Sciences, PSI, Forschungsstrasse 111, 5232 Villigen, Switzerland
`florencia.malamud@psi.ch`
[2] Laboratorio Argentino de Haces de Neutrones, CNEA, Buenos Aires, Argentina
[3] Consejo Nacional de Investigaciones Científicas y Técnicas, Buenos Aires, Argentina

Abstract. Bragg edge imaging experiments were carried out at the BOA beamline of the Paul Scherrer Institute on ring-shaped specimens cut at different axial positions of an unfinished, partially deformed, Zircaloy 4 tube. The height of the $(10\bar{1}0)$ and $(11\bar{2}0)$ Bragg edges was taken as indicative of the rotation of the grains around the c-axis during plastic deformation. Maps of these Bragg edge heights were constructed at each axial position to elucidate how this grain rotation changes along the hoop-radial plane. The results were consistent with neutron diffraction data, showing that the initial texture is dominated by the $(11\bar{2}0)//$ axial fiber, and during deformation the $(10\bar{1}0)//$ axial fiber became increasingly important at the expense of the first fiber.

Keywords: Bragg edge imaging · crystallographic texture · Zr alloy

1 Introduction

The production of Zirconium alloy cladding tubes (Zircaloy-2, Zircaloy-4, Zirlo, M5, E110) for various PWR (pressurized water reactors), BWR (boiling water reactors), and PHWR (pressurized heavy water reactors) nuclear power plants begins with a common intermediate product: an extruded, reduced, and recrystallized (TREX) tube. The process usually consists of 3–5 rolling passes in a cold-pilgering mill, with intermediate heat treatments to restore ductility by recrystallization, followed by a final stress relief heat treatment to achieve the desired mechanical properties and microstructural characteristics (grain size, texture) [1].

In particular, since the individual hexagonal close-packed (hcp) crystals of α-Zirconium are highly anisotropic, the relevant properties of polycrystalline reactor components (thermal expansion, irradiation creep, and growth, hydride orientation, corrosion rate, etc.) are influenced by the final texture of the specimen, significantly impacting their in-reactor performance. The basal poles (c-axes) in tubing are typically oriented in the radial-hoop plane, leading to irradiation growth in the axial direction of the

© The Author(s) 2026
A. E. Craft and H. Z. Bilheux (Eds.): WCNR 2024, SPPHY 348, pp. 84–92, 2026.
https://doi.org/10.1007/978-3-032-15003-5_11

tube. For cladding, a nearly radial basal pole texture is preferred, as it minimizes the risk of through-wall hydride cracking by encouraging the circumferential orientation of hydrides. Consequently, the texture of the cladding is meticulously controlled throughout the manufacturing process, and local variations in texture must be minimized.

The changes in texture through different stages of actual and potential manufacturing processes for Zircaloy tubes have been reported in several works [2–6]. Specifically, the changes in global texture during the manufacturing process of Indian Zircaloy-4 cladding tubes for PHWR reactors have been thoroughly studied. Overall changes in global texture at different stages of the full process are presented in [2], the texture evolution during annealing after the first cold pilgering in [7], and the local variations across the thickness in [8]. A more recent study characterized the through-thickness spatial variation of texture in detail for samples with different area reductions for the Argentinean process, ranging from the TREX tube to reductions up to 80%, corresponding to the tube geometry of the first rolling stage [8]. In samples extracted from different sections of a partially cold-rolled extreme, the texture was measured by neutron diffraction along the thickness for different axial positions, corresponding to different area reductions as well. This study revealed that the TREX tube exhibited a fiber texture with the $(11\bar{2}0)$ poles aligned along the axial direction and the c-poles distributed along the radial to hoop line. The texture gradients along the thickness of the TREX material are related to how the c-poles distribute along the fiber. Specifically, for the inner surface, these c-poles are found more frequently at an orientation $30°$ rotated from the radial direction, while for the outer surface, these poles primarily point along the hoop direction of the tube.

In recent years, Bragg edge imaging has gathered attention for its ability to provide valuable information on the spatial variations of texture with high spatial resolution [9]. This technique is based on analyzing the transmission attenuation of thermal neutrons as a function of their energy (or wavelength). In this experiment, the neutron beam passes through the sample, and the transmitted beam is collected by a position-sensitive detector. The attenuation of the beam intensity, determined as the ratio between the transmitted intensity with and without the sample, exhibits edges when a set of lattice planes is excluded from the diffraction process and the transmission increases. It was demonstrated that the height of the edge at a neutron wavelength $\lambda = 2d_{hkl}$, with d_{hkl} being the interplanar spacing of the (hkl) diffraction plane, is proportional to the number of grains with these (hkl) planes perpendicular to the beam direction. Moreover, it was also shown that for most structural materials, the shape of the attenuation as a function of energy in the 1 to 6 Å wavelength range is mainly affected by texture [9–12]. A complete description of the dependence of the attenuation coefficient with crystallographic texture can be found in [10, 13].

In this context, the transition from the $(11\bar{2}0)$ to the $(10\bar{1}0)//$ axial fiber texture as a function of deformation is an excellent case to study using Bragg edge imaging if the incident neutron beam is aligned along the axial direction of the tube. To achieve this, small ring samples of approximately 1 cm in length along the axial direction were sectioned from a partially deformed tube, and wavelength-resolved neutron images were obtained using the BOA instrument at the Paul Scherrer Institute in Switzerland. Bragg edge analysis would provide insights not only into the transition's dependence on the accumulated plastic deformation but also into how this transition varies along the radial-hoop plane.

2 Experimental

The last section of the cold rolled tube was cut after the first stage of pilgering rolling, including radial diameter changes from the original value (TREX, 0% deformation) to the final value with 81% area reduction (Fig. 1-a). Since each position of the tube along the axial direction represents a different value in area reduction, 10 small rings of approximately 1cm in length along the axial direction were cut from this piece, as schematically shown in red in Fig. 1-a, representing 0, 10, 29, 37, 47, 51, 58, 59, 80, and 81% of area reduction respectively.

The Bragg edge imaging measurements were performed at the cold neutron beamline BOA [14] at the Swiss spallation neutron source (SINQ) of the Paul Scherrer Institute (PSI) in Switzerland, employing the recently developed Frame Overlap Bragg edge Imaging (FOBI) technique [15] using a TPX3CAM detector system for event-based neutron detection [16]. We employed a chopper disk of 200 mm radius, with a pseudorandom angular distribution of 10 slits (1.8 mm width) repeated equally in 4 disk quadrants, using a rotation frequency 25 Hz, resulting in the 1.5 to 7 Å range wavelength range. The detector system was equipped with a 100 μm ^{6}LiF:ZnS scintillator positioned in the direct beam at 4.8 m distance to the chopper. Photons coming from the scintillator were reflected with a mirror positioned at 45° relative to the scintillator, such that the TPXCAM could be positioned outside the direct beam. To adjust the focus onto the scintillator surface and amplify the light coming off the scintillator, a 100 mm lens was coupled with a dual stage image intensifier. The optical setup provided a field-of-view of 10 cm × 10 cm with an effective pixel size of 90 μm after the event centroiding.

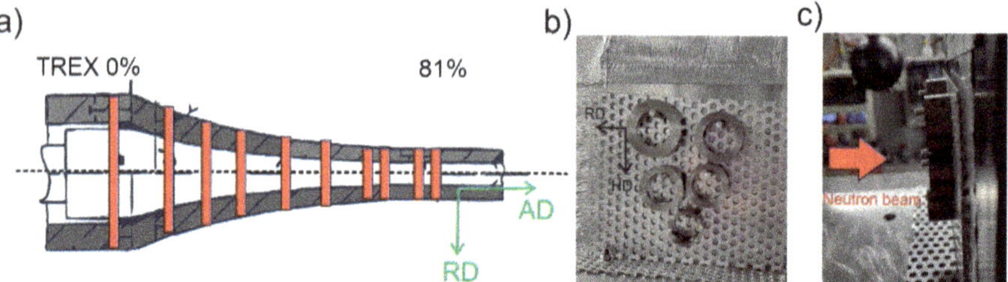

Fig. 1. a) Axial position of the samples extracted from the cold deformed tube, show-ing the radial direction (RD) and axial directions (AD), b) and c) sample arrangement in the sample holder at BOA, the red arrow indicates the direction of the incident beam, the detector is placed adjacent to the sample holder to its right.

To study the transition of texture from $(11\bar{2}0)$ // axis fiber texture of the TREX to $((10\bar{1}0))$ // fiber texture at 80%, the Bragg edge imaging measurements were collected by aligning the neutron beam along the axial direction of the tube, as shown in Fig. 1-b and

c. For each measurement, the samples were placed adjacent to the detector and exposed for 20 h in time-of-flight radiography mode, preceded by an open beam exposure of 10 hs and a pure iron powder measurement of 6 hs for calibration. The data were analyzed pixel by pixel employing the FOBI retrieval method and the signal processing algorithms presented in [15] with the corresponding chopper time delay scheme. After the FOBI retrieval, the time-of-flight to neutron wavelength calibration was performed by fitting the (110), (200) and (311) Bragg edges of a BCC iron powder standard with well-known lattice parameters, employing a non-linear function for the edge profile [17].

3 Bragg Edge Results

For each pixel and TOF value (or neutron wavelength λ), the transmission of the beam is evaluated by:

$$T(x, y, \lambda) = \frac{I(x, y, \lambda)}{I_0(x, y, \lambda)} \tag{1}$$

where I and I_0 are the intensity measured with and without the sample interposed into the beam, respectively. In the single scattering approximation, the transmission can be calculated as:

$$T(x, y, \lambda) = \exp\left(-\int_0^d \mu(\lambda, x, y, z)dz\right) \tag{2}$$

where d is the width of the sample along the direction of the beam and $\mu(\lambda, x, y, z)$ is the linear attenuation. In the case of a homogeneous material along the path of the beam, integration in Eq. (2) is straightforward and μ can be evaluated from the experimental transmission once the sample thickness along the beam direction d is known. The linear attenuation μ accounts for all mechanisms that take out neutrons from the beam, either by scattering or absorption. While texture effects manifest through the elastic coherent component, the other terms show a smooth dependence with neutron wavelength and can be calculated employing analytical functions as described in [18].

Figure 2-a shows the experimental linear attenuation as a function of wavelength for the different samples after averaging over all the pixels inside the rings to improve statistics. The blue curve (0%) corresponds to the TREX sample. In this case, the ($10\bar{1}0$) Bragg edge height is almost negligible, while the ($11\bar{2}0$) height is large. This is consistent with the fact that the texture of this sample is dominated by the ($11\bar{2}0$)// axial fiber with very low contribution from the ($10\bar{1}0$)// axial fiber. As deformation proceeds, the height of the ($10\bar{1}0$) Bragg edge increases while the height of the ($11\bar{2}0$) edge decreases. Figure 2-b shows the dependence of Bragg edge heights on area reduction, calculated as the difference in the attenuation coefficient at both sides of the edge. Error bars were estimated from the wavelength uncertainty due to the finite resolution of the instrument, $\Delta\lambda/\lambda \sim 10\%$. The largest $\left(10\bar{1}0\right)$ edge and the smallest ($11\bar{2}0$) edges are observed for the largest area reduction (81%). This is in agreement with neutron diffraction experiments [8] and reflect the grain rotation around the c-axis that occurs during the deformation process.

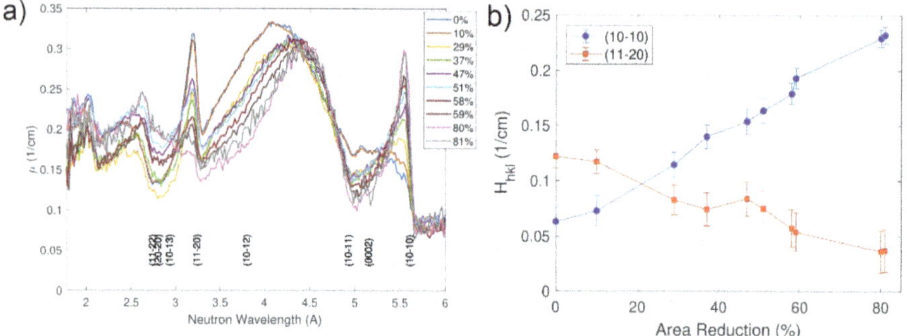

Fig. 2. a) Linear attenuation coefficient as a function of wavelength after averaging over the entire pixels corresponding to each of the rings. b) Evolution of the Bragg edge height as a function of area reduction. The error bar was calculated employing the resolution function of the instrument $\Delta\lambda/\lambda \sim 10\%$

A clear advantage of this imaging technique is the possibility to study the spatial variations of the transmission across the thickness and around the circumference, providing valuable information on the inhomogeneity of the deformation process. Figure 3 shows an image of the height of the $(10\bar{1}0)$ and $(11\bar{2}0)$ edges for five of the samples. The images have been produced using a macro-pixel of 270 µm, as a result of averaging over 3 adjacent pixels, to reduce the uncertainty in the edge height. For the samples corresponding to 29%, 47% and 59% area reduction, the region of the rings close to the outer and inner surfaces presents different values mid-thickness region. The samples for 29% and 47% area reduction also display variations around the circumference.

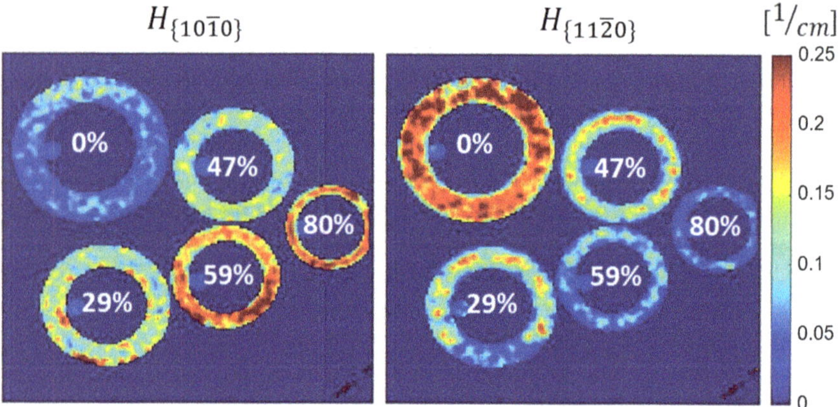

Fig. 3. Maps of Bragg edge height for selected samples. Tube thickness range from 14.6 mm to 4.6 mm. The spatial resolution is 270 µm.

To explore in detail the radial variation of the Bragg edge heights for the difference samples, Fig. 4-a represents the circumferential average of the $(10\bar{1}0)$ edge height measured at each radius, for the 10 samples studied. Figure 4-b presents the variation of such average height of both the $(10\bar{1}0)$ and $(11\bar{2}0)$ edges, versus a normalized thickness of the tube on the x-axis, going from the internal radius R_{INT} to the external radius R_{EXT} (Fig. 4-a) The data are presented from the internal surface to the external surface, as indicated by the red arrow in Fig. 4-a. From these figures, it is clear that initial TREX material does not present a clear trend of neither of the edge heights across the thickness. On the other end, both edges are clearly larger at mid- thickness than on the surfaces for area reductions larger than 80%. At lower degrees of deformation, between 10% and 50% area reductions, the $(11\bar{2}0)$ edge present a maximum at mid thickness, whilst the $(10\bar{1}0)$ edge displays a minimum. Such variations are likely due to the different shear forces experienced by the external material, in contact with mandrill and the die, and the internal region of the tube which is not. These shear forces are responsible for the rotation of grains around directions that are slightly deviated from the c-axis, producing a decrease in the number of grains with either $(10\bar{1}0)$ or $(11\bar{2}0)$ plane normal along the axial direction.

The circumferential variation observed in Fig. 3 was analyzed by dividing the ring into "pizza slices" or azimuthal sections and radially averaging over each section. Although variations does exist for areas reduction smaller than 60%, no clear trend around the circumference could be established. This can be explained by mentioning that during the pilgering process, the tube is deformed by two conical tools at the upper and lower position. After each deformation step, the tube moves forward 1mm and is rotated 51°. This process is repeated until the final shape of the tube is obtained. Each deformation step is highly anisotropic in terms of the azimuthal angle, but the effect of successive deformation steps each rotated 51° with respect to the previous one, finally blurs this anisotropy and corrects the tube ovality ending with an almost round shape tube. In the present experiment, each ring has a 10 mm length along the axial direction, meaning that at least 10 deformation + rotation steps were applied. In the present experiments, the circumferential variation seems to disappear after a 60% area reduction is achieved.

Both the through-thickness and circumferential inhomogeneities appearing across the section of the tube reported here could only be captured by the spatial resolution provided by Bragg edge imaging technique.

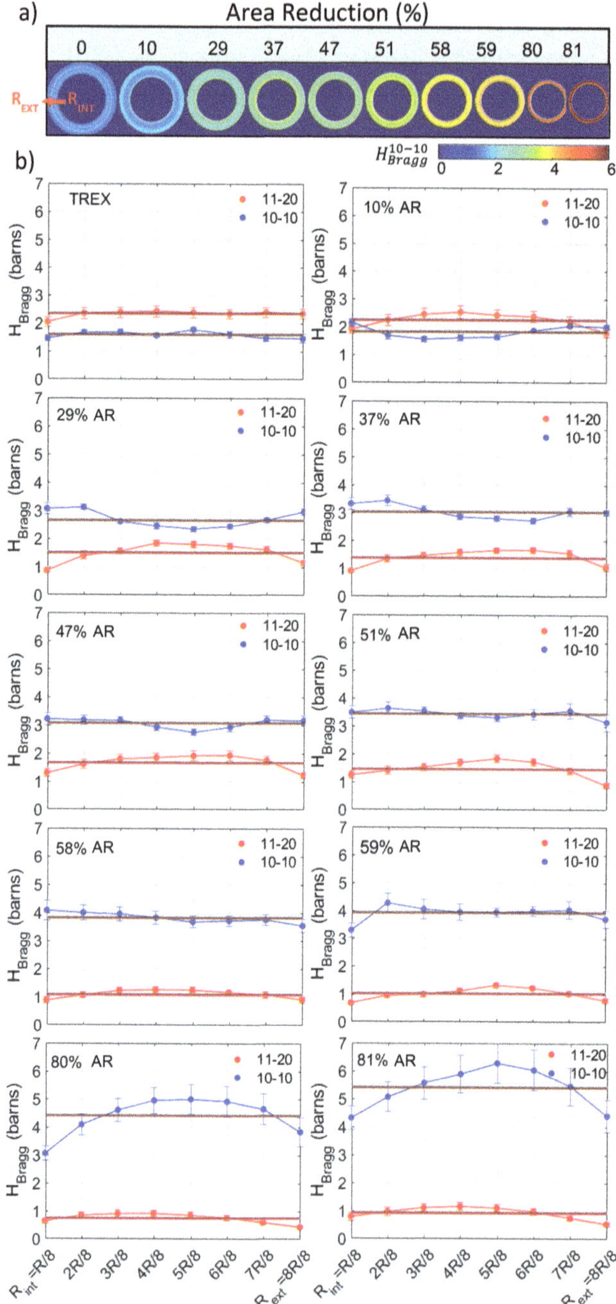

Fig. 4. Through-thickness variation of the height of the Bragg edges for all samples a) Graphical representation of the circumferential average of the $(10\bar{1}0)$ edge b) Quantitative variation for the $(10\bar{1}0)$ and $(11\bar{2}0)$ edges, versus normalized thickness for better comparison.

4 Conclusions

Bragg edge imaging revealed the spatial variation of the main changes in texture occurring on the first step of the cold rolling process used to produce nuclear fuel cladding. By analyzing the changes of specific Bragg edge heights, we were able to study the spatial variation of the number of grains of the corresponding lattice plane family oriented parallel to the incoming beam. In particular, the analysis of the $(11\bar{2}0)$ and $(10\bar{1}0)$ Bragg edges measured along the axial direction of 10 specimens representing different degrees of plastic deformation on the cold pilgering roll were interpreted in terms of two fibers textures that change in intensity as the tube cross section decreases. The results complement previous neutron diffraction studies, by revealing small through-thickness and circumferential variations in plastic deformation that result from the cold pilgering process. Circumferential variations are almost absent by the end of the first rolling step.

References

1. Lemaignan, C.: 2.07-Zirconium alloys: properties and characteristics. Comprehensive Nuclear Mater. **2**, 217–232 (2012)
2. Lebensohn, R.A., González, M.I., Tomé, C.N., Pochettino, A.A.: Measurement and prediction of texture development during a rolling sequence of Zircaloy-4 tubes. J. Nucl. Mater. **229**, 57–64 (1996). https://doi.org/10.1016/0022-3115(95)00210-3
3. Tomé, C.N., Lebensohn, R.A., Kocks, U.F.: A model for texture development dominated by deformation twinning: application to zirconium alloys. Acta Metall. Mater. **39**, 2667–2680 (1991). https://doi.org/10.1016/0956-7151(91)90083-D
4. Hiwarkar, V.D., et al.: Defining recrystallization in pilgered Zircaloy-4: from preferred nucleation to growth inhibition. J. Nucl. Mater. **412**, 287–293 (2011). https://doi.org/10.1016/j.jnucmat.2011.02.043
5. Linga Murty, K., Charit, I.: Texture development and anisotropic deformation of zircaloys. Progr. Nuclear Energy **48**, 325–359 (2006). https://doi.org/10.1016/j.pnucene.2005.09.011
6. Nagai, N., Kakuma, T., Fujita, K.: Texture control of Zircaloy tubing during tube reduction. In: Zirconium in the Nuclear Industry, ASTM International100 Barr Harbor Drive, PO Box C700, West Conshohocken, PA 19428–2959, pp. 26–38 (1982). https://doi.org/10.1520/STP 37044S
7. Singh, J., et al.: Texture development and plastic deformation in a pilgered Zircaloy-4 tube. Metall. Mater. Trans. A **46**, 1927–1947 (2015). https://doi.org/10.1007/s11661-015-2807-6
8. Juarez, G., Alvarez, M.A.V., Santisteban, J., Almer, J., Luzin, V., Vizcaino, P.: Global and local texture development during initial plastic deformation of cold-pilgered Zircaloy-4 tubing. J. Nucl. Mater. **558**, 153382 (2022). https://doi.org/10.1016/j.jnucmat.2021.153382
9. Woracek, R., Santisteban, J., Fedrigo, A., Strobl, M.: Diffraction in neutron imaging—a review. Nucl. Instrum. Methods Phys. Res., Sect. A **878**, 141–158 (2018). https://doi.org/10.1016/j.nima.2017.07.040
10. Malamud, F., Santisteban, J.R., Vicente Alvarez, M.A., Busi, M., Polatidis, E., Strobl, M.: An optimized single-crystal to polycrystal model of the neutron transmission of textured polycrystalline materials. J. Appl. Crystallogr. **56** (2023). https://doi.org/10.1107/S1600576722011323
11. Alvarez, M.A.V., Laliena, V., Malamud, F., Campo, J., Santisteban, J.: A novel method to obtain integral parameters of the orientation distribution function of textured polycrystals from wavelength-resolved neutron transmission spectra. J. Appl. Crystallogr. **54**, 903–913 (2021). https://doi.org/10.1107/s1600576721003861

12. Sato, H., Kamiyama, T., Iwase, K., Ishigaki, T., Kiyanagi, Y.: Pulsed neutron spectroscopic imaging for crystallographic texture and microstructure. Nucl. Instrum. Methods Phys. Res., Sect. A **651**, 216–220 (2011). https://doi.org/10.1016/j.nima.2011.01.063
13. Vicente Alvarez, M.A., Malamud, F., Santisteban, J.R.: Determination of crystallographic texture in polycrystalline materials from wavelength-resolved neutron transmission experiments: application to high-symmetry crystals. J. Appl. Cryst. **56** (2023). https://doi.org/10.1107/S1600576723008877
14. Morgano, M., Peetermans, S., Lehmann, E.H., Panzner, T., Filges, U.: Neutron imaging options at the BOA beamline at Paul Scherrer Institut. Nucl. Instrum. Methods Phys. Res., Sect. A **754**, 46–56 (2014). https://doi.org/10.1016/j.nima.2014.03.055
15. Busi, M., et al.: Frame overlap Bragg edge imaging. Sci. Rep. **10**, 14867 (2020). https://doi.org/10.1038/s41598-020-71705-4
16. Wolfertz, A., et al.: LumaCam: a novel class of position-sensitive event mode particle detectors using scintillator screens. Sci. Rep. **14**, 30495 (2024). https://doi.org/10.1038/s41598-024-82095-2
17. Santisteban, J.R., Edwards, L., Steuwer, A., Withers, P.J.: Time-of-flight neutron transmission diffraction. J. Appl. Crystallogr. **34**, 289–297 (2001). https://doi.org/10.1107/S0021889801003260
18. Malamud, F., Santisteban, J.R.: Full-pattern analysis of time-of-flight neutron transmission of mosaic crystals. J. Appl. Crystallogr. **49**, 348–365 (2016)

Monte Carlo Simulation of Prompt-Gamma Ghost Imaging: Proof of Concept

Cheng-i Chiang[1]([✉]), Alaleh Aminzadeh[1], Wilfred K. Fullagar[1], Ulf Garbe[2], Filomena Salvemini[2], Joseph J. Bevitt[2], Jeremy M. C. Brown[3], David M. Paganin[4], and Andrew M. Kingston[1]

[1] Department of Materials Physics, Research School of Physics, Australian National University, Canberra, ACT 2600, Australia
cheng-i.chiang@anu.edu.au
[2] Australian Centre for Neutron Scattering, Australian Nuclear Science and Technology Organisation, Lucas Heights, NSW 2234, Australia
[3] Optical Sciences Centre, Department of Physics and Astronomy, School of Science, Computing and Engineering Technologies, Swinburne University of Technology, Hawthorn, VIC 3122, Australia
[4] School of Physics and Astronomy, Monash University, Clayton, VIC 3800, Australia

Abstract. Neutron Activation Analysis (NAA) is a technique based on gamma rays emitted from neutron-nucleus interactions. Such interactions are unique for each element and isotope, and the measured gamma spectrum can identify the atomic composition of the target object. This proof-of-concept study explores the possibility of adding spatial information to NAA by employing ghost imaging (GI). In this article, the TOPAS Monte Carlo particle simulation tool was utilised to simulate the potential deployment of GI at ANSTO's DINGO neutron imaging beamline. In TOPAS the DINGO beamline was implemented to include the neutron source, collimator, and detector with an additional gamma-ray sensor. A GI experiment using a randomly structured Cd mask to pattern the neutron illumination was then simulated for 41 different mask positions. GI reconstruction enabled visualisation of the gold distribution in a test object. The simulation serves as a demonstration of the principle, and a means to better understand the appropriate experimental scenarios and acquisition times to successfully realise prompt-gamma GI in practice.

Keywords: neutron activation analysis · prompt-gamma emission · ghost imaging

1 Introduction

Neutron Activation Analysis (NAA) can determine the isotopic elemental constitution of a sample. It relies on measurement of gamma rays emitted from a sample after neutron irradiation. The emitted gamma spectrum is based on the neutron-nucleus interaction. Analysis of the peaks in the measured spectrum can identify the atomic composition of the target object. The gamma rays produced can be divided into two categories, namely "prompt gamma" rays emitted

© The Author(s) 2026
A. E. Craft and H. Z. Bilheux (Eds.): WCNR 2024, SPPHY 348, pp. 93–100, 2026.
https://doi.org/10.1007/978-3-032-15003-5_12

almost instantaneously after neutron interaction with a nucleus, and "delayed gamma" rays emitted from nuclei after a longer time scale. This proof-of-concept study explores the possibility of adding spatial information to prompt-gamma NAA through the use of emission ghost imaging. The x-ray analogue of this concept, namely x-ray fluorescence ghost imaging, has been demonstrated in the laboratory [1] and with synchrotron radiation [2].

Ghost imaging (GI) is a novel computational imaging method that does not require one to directly image the sample with a pixelated detector. Instead, the object is illuminated with a set of patterns and the total interaction (gamma emission in this case) is recorded with a single-sensor, or bucket detector. An image of the object is determined based on the correlation between the patterns and associated bucket measurements. The overall goal of this work is to explore the combination of these two methods, namely NAA and GI, to measure the spatial distribution of elements in a sample.

Such an investigation is motivated by potential applications that include (i) investigation of mining-related samples such as oil shale and gemstones, where one seeks to determine what minerals are present and where they are located or co-located, (ii) the study of elemental and isotope distributions in cultural artefacts, and (iii) fossil imaging. Regarding such possible future applications, there is the potential that the approach might be advantageous in comparison to x-ray fluorescence (XRF), since emitted rays have higher energy and are more penetrating, and can therefore scan larger and denser objects than XRF (which is limited to metals in biological samples). An additional practical motivation for this work is the possibility that it may, under certain circumstances associated with certain applications, give reduced dose in comparison to existing conventional imaging approaches.

A first step to realising prompt-gamma ghost imaging (PG-GI) is its simulation with a particle-physics simulation tool. This will both enable the demonstration of the principle and provide valuable information towards the development of appropriate experimental scenarios and protocols to successfully realise PG-GI experimentally. The TOPAS Monte Carlo particle simulation tool (topas-mc.org [3]) was utilised to simulate the physical interaction process; this software is based on the GEANT4 Monte Carlo radiation transport modelling toolkit [4–6]. A model of ANSTO's DINGO neutron imaging beamline was implemented including the neutron source, collimator, and transmission detector. A GI experiment was then simulated using a randomly structured Cd mask to pattern the neutron illumination, with a gamma-ray sensor also being incorporated into the model. Ghost-image reconstruction tools were then used to visualise the distribution of gold in a test object.

The principles of neutron activation analysis and the ghost-imaging technique are described in Sect. 2. Section 3 outlines the Monte Carlo model of ANSTO's DINGO neutron imaging beamline for prompt-gamma ghost imaging. The experimental setup and measurement protocol are detailed in Sect. 4 along with the recovered prompt-gamma images and a brief analysis and discussion of the results. Finally, Sect. 5 offers a number of concluding remarks and potential future research directions.

2   Background

2.1   Neutron Activation Analysis

When a neutron collides with an atomic nucleus, the nucleus has the potential to undergo a transformation that may result in the creation of new isotopes, i.e., neutron activation. These newly formed isotopes are often unstable, so distinctive gamma rays are one mechanism by which excess energy is released. Different spectral features represent different elements and isotopes in the target material. NAA involves analysing the spectrum of gamma rays emitted to discover the constituent elements of the target material and their relative proportions. This compositional information alone can be very useful. However, the spatial distribution of the elements within the object could provide valuable additional information that is currently inaccessible. This work explores the possibility to add spatial information to NAA by employing emission ghost imaging techniques.

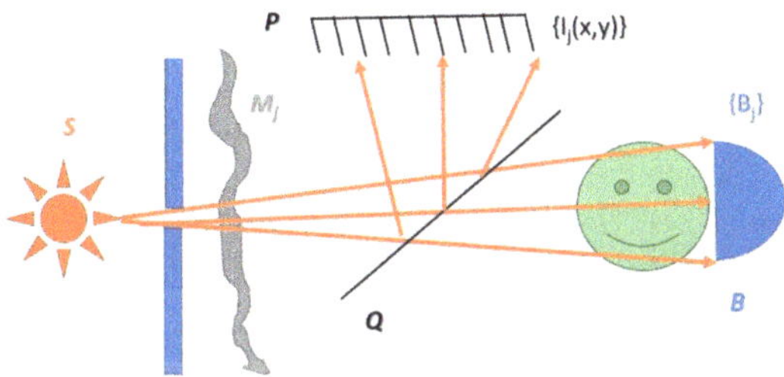

Fig. 1. Classical GI experimental setup; S: light source; M_j: illumination pattern number j; P: pixelated detector; Q: beam splitter; B: bucket detector.

2.2   Ghost Imaging

Transmission Ghost Imaging: Classical ghost imaging is a relatively recent and unconventional imaging technique where the photons or particles that interact with the object do not need to be recorded with position information [7]; a single sensor is sufficient (noting that the in/out aperture of a spectrometer can serve this role). Image formation relies on the correlation between two patterned light beams or particle streams, one of which interacts with the object, with the other acting as a reference beam (see Fig. 1). Different illumination patterns can be generated by translating a structured mask, M. A set of illumination patterns $I_j(x, y)$ is measured for each mask position M_j, along with the "bucket values," B_j, representing object interaction (e.g., transmission in Fig. 1). An image of the object can be recovered from these data based on the correlation between the pairs of measurements, $\{I_j(x, y), B_j\}$, as demonstrated in Ref. [8].

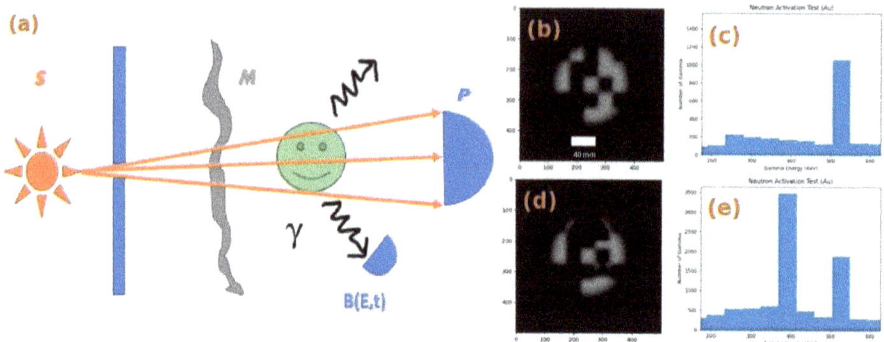

Fig. 2. (a) GI setup to add spatial information to NAA. (b) Apertured mask pattern measured at pixelated detector with no object. (c) Associated gamma spectrum at bucket detector. (d) Image at pixelated detector including Au object. (e) Gamma spectrum now with Au peak at 411 keV.

Emission Ghost Imaging: Figure 2 depicts how GI can be employed to add spatial information to NAA. For J measurements: (i) translate mask to a known position; (ii) collect neutron images of the aperture, mask, and sample (Fig. 2(d)); (iii) collect the prompt-gamma spectrum (Fig. 2(e)). The gamma spectrum's peak height for each element relative to the background (e.g., Fig. 2(c)) serves as a bucket value that is correlated with the set of mask images to recover its spatial distribution. Note that Ref. [2] describes how images of the mask only (such as in Fig. 2(b)) can be determined from the set of mask and sample images.

3   Monte Carlo Modelling of Prompt-Gamma Ghost Imaging

The ANSTO DINGO beamline at OPAL is simulated using the TOPAS [3] Monte Carlo tool for particle simulation built on GEANT4. This section outlines the details of each individual component that makes up the overall model. First, the neutron source modelled as a two-stage process will be described and then the different detectors used in the measurement introduced. Next, the mask information is elaborated and, finally, the test object described. All particle transport was simulated with the following physics list of GEANT4 constructors: G4HadronElasticPhysicsHP, G4DecayPhysics, G4EMStandardPhysics_option3, HadronPhysicsQGSP_BIC_HP, G4RadioactiveDecayPhysics. Note that this is the default TOPAS physics list devised in Ref. [9], with the exception that option 3 was employed for EMStandardPhysics rather than option 4. This does not have atomic de-excitation enabled but is a faster implementation and is expected to model sufficient physics for this proof-of-concept demonstration.

Neutron Source: The DINGO neutron imaging beamline at ANSTO uses a thermal neutron source with a mean energy of 36 meV (or 1.5 Å), energy spread (standard deviation) of 30% and has a flux of 2.15×10^7 neutrons $\mathrm{cm}^{-2}\mathrm{s}^{-1}$. The simulations were done in two stages, to reduce computation time and the geometrical configuration mimicked the methodology outlined in Jakubowski *et al.* [10]. Stage 1 is completed at the surface coloured red in Fig. 3 where the particle type, position, momentum, and energy of the particles at the surface were registered in a phase-space file. In stage 2, a virtual neutron source is simulated at the output of the helium flight tube with the same properties as those measured in stage 1.

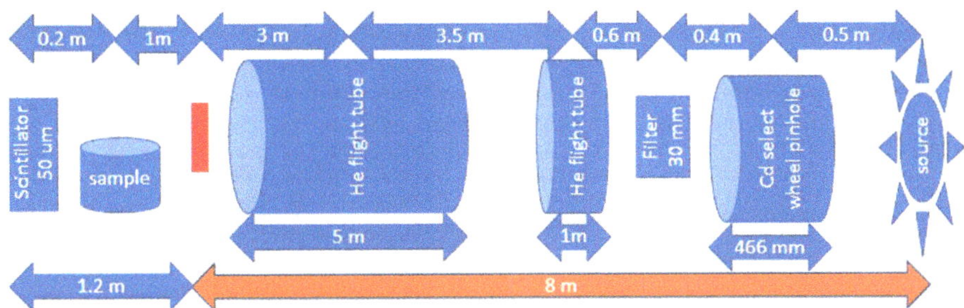

Fig. 3. DINGO beamline simulation details. The red feature is the phase-space box which records the properties of the output downstream of the He flight tube.

Neutron Transmission Detector: The detector at DINGO is a 550 mm × 550 mm ^{6}LiF/ZnS:Cu scintillator detector 1150 mm downstream of the virtual neutron source (blue square in Fig. 4). For simplicity the detector was modelled as a 500 × 500 pixel detector with ideal spatial, energy, and timing resolution. The properties of the particles interacting with the scintillator detector were registered in a CSV file, which includes the position and fluence information.

Gamma Emission Detector: To minimise simulation times, a large gamma spectrometer (200 mm × 200 mm sensor) was modelled and placed 200 mm above the test object (pink square in Fig. 4). Again the sensor was modelled with ideal spatial, energy, and timing resolution. The properties of the particles escaping the surface of the gamma spectrometer were registered in a phase-space file. The scored properties included the particle type, position, momentum, and energy.

Aperture and Mask for Patterned Illumination: A 2 mm-thick, 25 × 25 pixel Cd mask with a pixel size of 20 mm was modelled and placed 400 mm downstream from the virtual neutron source. The mask has a random binary pattern for patterned illumination. A 20 mm-thick Cd aperture was also inserted (red disc in Fig. 4) with diameter 100 mm, placed 200 mm downstream from the virtual neutron source.

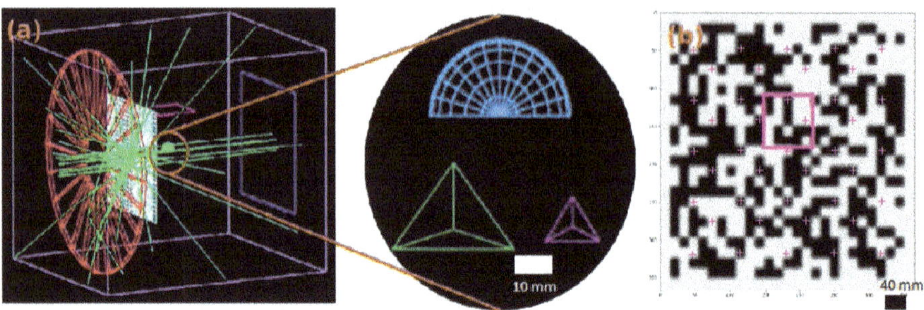

Fig. 4. (a) 3D TOPAS simulation setup showing the Cd aperture (red), Cd mask (white), neutron scintillator detector (blue), gamma sensor (pink) and zoom in to the Au object. (b) An image of the 25×25 pixel random Cd mask with 5×5 field-of-view (pink) scanned over the mask for each measurement.

Test Object: A simple Au test object composed of a 60 mm hemisphere and two tetrahedra with side lengths of 25 mm and 15 mm, was placed 1 m downstream from the virtual neutron source.

4   Prompt-Gamma Ghost Imaging Monte Carlo Experiment

Experimental Method: The field-of-view for each measurement is 5×5 Cd mask pixels (pink square in Fig. 4). One million neutron histories were simulated for each measurement with 41 different mask positions. The mask is translated in a diamond pattern with the centre points of each position marked with pink crosses in Fig. 4c. For each position, neutron images were recorded of the combined aperture, mask, and test object along with each corresponding gamma spectrum. The 411 keV Au emission peak height above background was extracted to give GI bucket values (blue plot in Fig. 5c). The mean peak height was around 3000 photons. The estimated noise determined from several runs with the same mask position is around 2%, i.e., Poisson as expected. The equivalent time on DINGO for each measurement in our simulation is 0.12 ms. This must be scaled to account for realistic neutron/gamma-ray detection efficiencies and a smaller gamma sensor.

Experiment Validation: Based on the neutron radiograph of the target object (Fig. 5a), the PG-GI data were digitally simulated from a binary 2D object (Fig. 5b) with a mean of 3000 photons assuming a Poisson distribution. This process generated "expected" bucket values for each mask position (orange plot in Fig. 5c). Compared with the Monte Carlo measurements, in blue, a similar trend can be seen with a different amplitude. This implies that (i) the prompt-gamma detection rate is proportional to the object projection area illuminated by the random pattern mask, and more importantly, (ii) the signal-to-noise ratio for

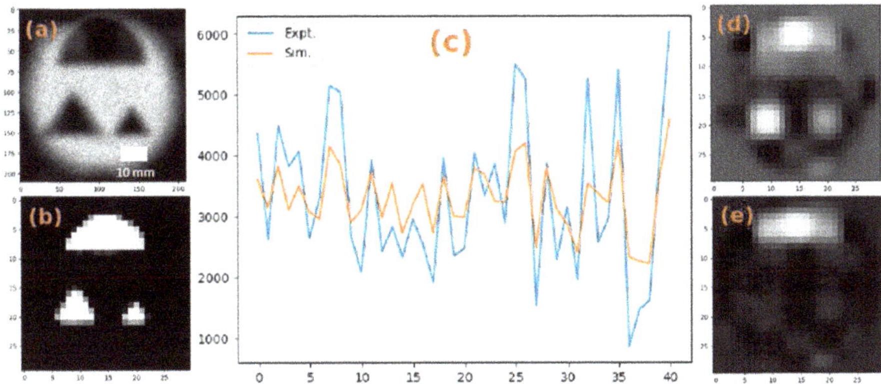

Fig. 5. (a) Monte Carlo neutron radiograph of Au target object with Cd aperture; the same scale bar is used for a, b, d and e; (b) binarisation of object in (a) with reduced resolution; (c) comparison of Monte Carlo 400 keV Au PG emission photon count for each mask position (blue) with simulated values from (b) with realistic noise (orange); (d) ghost image of target object reconstructed from simulated data in (c); (e) ghost image recovered from the Monte Carlo experiment data in (c). Note: panels (d-e) were reconstructed using 10,000 Landweber iterations.

the Monte Carlo data should be sufficient. The discrepancy in amplitude between the two plots indicates that the digital validation data is possibly oversimplified.

Image Reconstruction Results: The GI reconstructed from the oversimplified validation data (orange plot in Fig. 5c) is shown in Fig. 5d. This resembles the target object up to the resolution of the mask pixels, as expected. The PG-GI recovered from the Monte Carlo data (blue plot in Fig. 5c) is shown in Fig. 5e. It has recovered the Au hemisphere at the top of the image; however, the two Au tetrahedra are not visible at the bottom.

Since the PG spectrometer was located above the object, the discrepancy between the simplified validation result (Fig. 5d) and the full PG-GI Monte Carlo result (Fig. 5e) is understood to be from self-absorption by the test object. Specifically, the large Au hemisphere blocks the PG emitted from the Au tetrahedra. The gamma extinction length of Au at 411 keV is 2.5 mm, so transmission through 60 mm is 10^{-11}. Therefore it can be concluded that the Monte Carlo model was indeed able to capture the physics correctly and that the distribution of Au as visible by the gamma spectrometer was successfully reconstructed. This numerical simulation thereby establishes proof-of-concept.

5   Conclusion and Future Work

This simulation-based work outlined a proof-of-concept for prompt-gamma ghost imaging at ANSTO's DINGO neutron imaging beamline using Monte Carlo radiation transport simulation. It was shown that the distribution of Au (as seen by the gamma spectrometer) in a test object was able to be reconstructed with existing GI reconstruction techniques. These results are encouraging as they show that the developed simulation platform is able to mimic a simple realisation of the physics of radiation transport of this application reasonably well. However, these results also show that sample choice is important, and that further work is required to refine the experiment design. To further refine the model and improve on these initial expectations, more physically realistic features will need to be added to the simulation platform. These features would include an upgraded PG detector model, inclusion of time stamping to distinguish between PG and delayed-gammas, and adding beam-on/off modulation to determine background signal and clean up the PG measurement. Once this is all in place, it will be possible to optimise the experiment configuration and protocols for future physical experiments.

References

1. Klein, Y., Sefi, O., Schwartz, H., Shwartz, S.: Optica **9**(1), 63 (2022)
2. Manni, M., et al.: Opt. Lett. **48**(23), 6271 (2023)
3. Perl, J., Shin, J., Schümann, J., Faddegon, B., Paganetti, H.: Med. Phys. **39**(11), 6818 (2012)
4. GEANT4-Collaboration. Nucl. Instrum. Methods Phys. Res. Sect. A **506**(3), 250 (2003)
5. GEANT4-Collaboration. IEEE Trans. Nucl. Sci. **53**(1), 270 (2006)
6. GEANT4-Collaboration. Nucl. Instrum. Methods Phys. Res. Sect. A **835**, 186 (2016)
7. Pelliccia, D., et al.: IUCrJ **5**(4), 428 (2018)
8. Kingston, A.M., et al.: Phys. Rev. A **101**(5), 053844 (2020)
9. Arce, P., et al.: Med. Phys. **48**(1), 19 (2021)
10. Jakubowski, K., et al.: Sci. Rep. **13**(1), 17415 (2023)

TRIXIE – Development and Construction of a New Neutron Imaging Instrument in the Czech Republic

Jana Matouskova[1]($\boxtimes$), Nikolay Kardjilov[2], Burkhard Schillinger[3], and Lubomir Sklenka[1]

[1] Czech Technical University in Prague, Prague, Czech Republic
`jana.matouskova@fjfi.cvut.cz`
[2] Helmholtz-Zentrum Berlin (HZB), Berlin, Germany
[3] Heinz Maier-Leibnitz Zentrum (FRM II), Technische Universität München, Munich, Germany

Abstract. The Czech Technical University in Prague is currently building a new neutron imaging instrument named TRIXIE at the research reactor LVR-15. The LVR-15 is a 10 MW reactor located in Rez, near Prague, in the Czech Republic. The TRIXIE project started in the spring of 2022, establishing a conceptual design of the major components. This is the result of a collaboration between Czech Technical University in Prague, Helmholtz-Zentrum Berlin and Heinz Maier-Leibnitz Zentrum (FRM II). The TRIXIE instrument is intended for thermal neutron radiography and tomography techniques. This instrument comprises components such as fast neutron and gamma filters, a beam collimator with an aperture for thermal neutrons and primary and secondary shutters. The experimental area is a large shielding structure with a detection system, sample positioning tables and other equipment placed approximately 5 m away from the source. The TRIXIE instrument is expected to be completed by the beginning of 2025, and it will provide access to national and international user communities in various research disciplines. This should help to fill the gap between high users' requests and insufficient capabilities, which existing neutron imaging instruments provide nowadays in Europe.

Keywords: Neutron Imaging Instrumentation · Instrument Development · Neutron Radiography · Neutron Tomography · Thermal Neutron Imaging

1 Introduction

In 2022, the Czech Technical University in Prague successfully built and commissioned a facility for neutron imaging at its very low-power research reactor VR-1 [1]. Based on the experience gained from the development and testing of the neutron imaging facility at the VR-1, CTU decided to use an opportunity of an unused beamline at the research reactor LVR-15 near Prague, in the Czech Republic and design and build a new state-of-the-art neutron imaging instrument. Several factors motivated the construction of the new neutron imaging instrument, namely, the decreasing number of research reactors in

A. E. Craft and H. Z. Bilheux (Eds.): WCNR 2024, SPPHY 348, pp. 101–108, 2026.
https://doi.org/10.1007/978-3-032-15003-5_13

Europe, e.g., the BER II reactor at HZB [2], which was permanently shut down in 2019, problems with unexpected extended shutdowns, e.g. the FRM II reactor at MLZ [3] and high user demand on a very limited number of neutron imaging instruments - most of the instruments are overbooked by a factor ranging from two to seven. Moreover, even though there were several attempts to build instruments for neutron imaging in the past [4], the LVR-15 reactor does not have one, which makes it one of the last, if not the last, high-power reactors in Europe to be equipped with one.

The project of development of a new imaging instrument at the LVR-15 reactor is based on a collaboration between Czech Technical University in Prague, Heinz Maier-Leibniz Zentrum, Technische Universität München, and Helmholtz-Zentrum Berlin für Materialien und Energie GmbH, Berlin. Based on the collaboration, the instrument has been named TRIXIE, aka TRICERATOPS, which is an acronym for TRI-CEnter RAdiography and TOmograPhy inStrument.

2 TRIXIE Neutron Imaging Instrument

2.1 Research Reactor LVR-15

The research reactor LVR-15 is operated by Research Centre Rez (CVR) near Prague in the Czech Republic. The reactor was first commissioned in 1957 as a VVR-S reactor, with a nominal power of 2 MW [5]. Between 1988 and 1989, the reactor VVR-S underwent a major reconstruction and became known as the LVR-15 reactor, with an increased nominal power of 10 MW [5]. The reactor was designed as a state-of-the-art instrumentation for research in reactor physics and material irradiation. At present, the LVR-15 reactor is used for a wide range of applications, including industrial and medical radioisotope production [6], material research [7], neutron transmutation doping of silicon, both prompt and standard neutron activation analysis [7] and also beam experiments, such as neutron diffraction, small angle neutron scattering and neutron depth profiling [8].

The LVR-15 is a tank-type reactor cooled and moderated by light water. It uses an ITR-4 type fuel containing low-enriched uranium [8]. The nominal power of the reactor is 10 MW [8]. The reactor is equipped with several experimental facilities, such as nine horizontal beamlines, several irradiation facilities, in-core irradiation positions, experimental loops, and a thermal column [8]. All the operational beamlines are used by external users, while beamlines 3, 4, 6, 8 and 9 are used by the Nuclear Physics Institute of the Czech Academy of Sciences [8] for various activities like neutron diffraction, SANS and neutron depth profiling. Beamline 2 is used by the Czech Technical University in Prague for neutron diffraction [9]. The rest of the beamlines, namely 1, 5 and 7, are currently unused, in the case of beamlines 5 and 7, due to lack of space. The new neutron imaging instrument TRIXIE is currently under construction at beamline 1. The layout of the nine horizontal beamlines is given in Fig. 1 [adapted from 8].

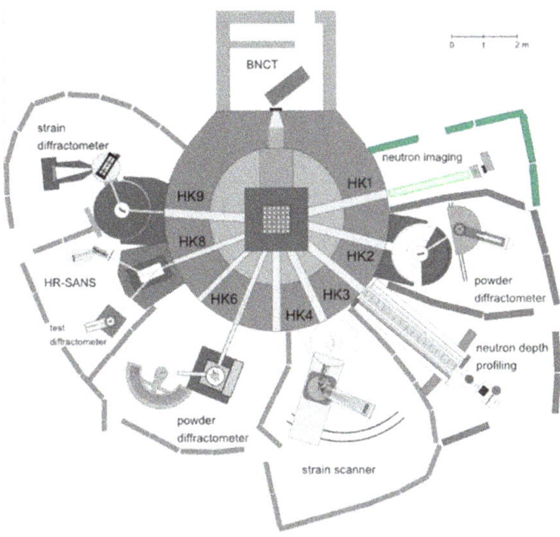

Fig. 1. The layout of horizontal beamlines at the LVR-15 reactor [adapted from 8].

2.2 Instrument Layout

The TRIXIE neutron imaging instrument will be a state-of-the-art facility for research in various fields. The instrument will provide a medium-intensity thermal neutron beam both for in-house research and also for users as a part of open access. The individual parts of the instrument are then described in detail in Sects. 2.3–2.5. The instrument layout is shown in Fig. 2.

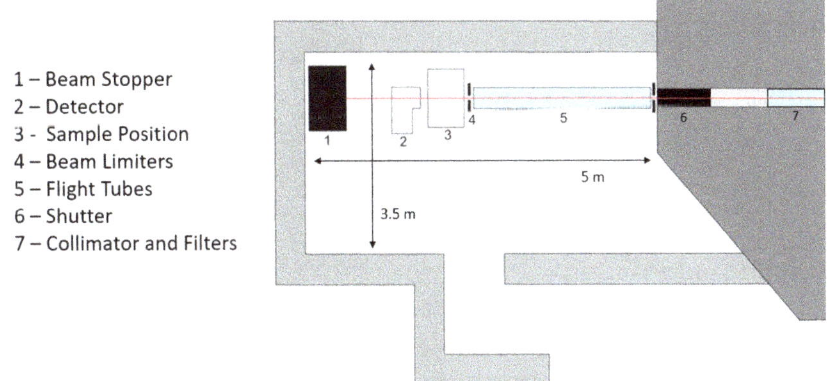

Fig. 2. The overview layout of the TRIXIE instrument

The basic parameters of the TRIXIE instrument are given in Table 1.

Table 1. The basic parameters of the TRIXIE instrument

Parameter	Value
Neutron spectrum	Thermal
Collimation ratio L/D	150–250
Max. flux at the sample	$1 \cdot 10^7$ n/cm^{2}s
Max. field of view	15×15 cm^2

2.3 Beamline

The instrument is located at beamline 1 of the LVR-15 reactor. To modify the collimation of the neutron beam, a conical collimator consisting of a combination of lead and borated aluminium was designed. The smallest diameter D of the beam is 20 mm, and the length of the beamline L is variable from 3 to 5 m, resulting in a collimation ratio L/D between 150 and 250. Since it is a radial beamline, it was necessary to not only modify the collimation of the neutron beam but also the neutron and gamma spectrum. The spectrum of an empty beamline roughly copies the reactor spectrum, with a high ratio of fast neutrons and gamma rays. Filters were designed to adjust the spectrum. To find suitable filter material and size, calculations using Monte Carlo Calculation code MCNP in version MCNP 6.1 with ENDF/B-VII.1 library were performed [10]. The final configuration of the beam includes three sapphire monocrystals with a total length of 30 cm for filtering fast neutrons and one bismuth monocrystal with a length of 10 cm for filtering gamma rays, all placed in an aluminium casing. A model of the beamline with the collimator and filters is shown in Fig. 3.

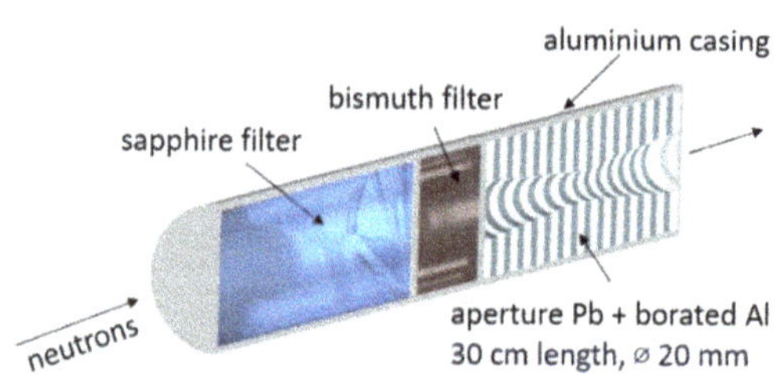

Fig. 3. A model of the beamline with the collimator and filters

The calculated flux at the detector position (4 m), with the collimation ratio L/D = 200, after filters and collimator, was calculated using the MCNP code [10] with a value of $1 \cdot 10^7$ n/cm^{2}s. The neutron spectrum was also calculated using the MCNP code. The calculated spectrum peaks around 32 meV, corresponding to a wavelength of 1.6 Å, as shown in Fig. 4. The ratio of thermal-to-epithermal/fast in the beam is more than 98%.

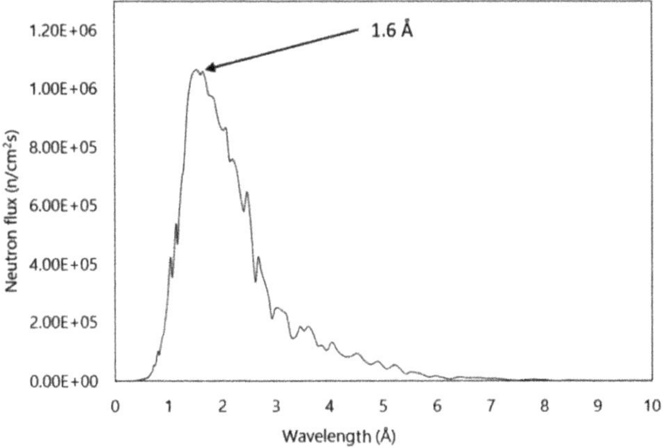

Fig. 4. The calculated neutron spectrum at the TRIXIE instrument

2.4 Detection System

Currently, there are two detectors available at the TRIXIE instrument with variable field of view. These detectors are a standard neutron imaging detection system consisting of a CMOS camera in combination with a scintillator screen, a mirror and optics. The first detector uses a ZWO ASI294MM Pro camera with 14-bit digitisation, an 8288×5644 pixel array and a $4.64\ \mu m$ pixel size [11]. The second detector uses a ZWO ASI2600MM camera with 16-bit digitisation, a 6248×4176 pixel array and a $3.76\ \mu m$ pixel size [12]. Both cameras can mount different lenses, including 25 mm, 35 mm, 50 mm and 75 mm C-mount lenses. Both cameras are also placed in light-tight boxes with interchangeable parts for the scintillator screen, which allows the change of the field of view from $15 \times 15\ cm^2$ to $5 \times 5\ cm^2$ [13]. The used scintillator screen is $^6LiF/ZnS:Cu$ type of different thicknesses, 100 μm and 200 μm [14]. In the future, new types of scintillators will be added to the set of available scintillators, for example, Gadolinium Oxysulfide (Gadox) scintillators, etc. The detector box is equipped only with small shielding in front of the camera to cut direct gammas from the scintillation screen and sample. An external shielding must be placed around the detector box for protection.

A special high-resolution detector based on a Heliflex lens, currently under development, will also be added to the set of available detectors [15].

2.5 Shielding

The instrument's shielding is based on heavy concrete blocks. Individual walls are 40 cm thick and made of heavy concrete with a density of 4 t/m^3. Heavy concrete shielding blocks were part of the central neutron guide shielding of the BER II reactor at HZB and after its permanent shutdown, they were donated to CTU. The internal side of the walls will be covered with plates of borated polyethene with a thickness of 5 mm. The shielding model is shown in Fig. 5.

Fig. 5. Model of the shielding of the instrument

2.6 Other Components

The instrument will also be equipped with components including a positioning table for adjusting the position of the sample against the detector and for performing computed neutron tomography measurements, beam limiters for reducing the size of the beam to avoid unnecessary activation and irradiation of components or flight tubes with highly flexible concept, modular and movable (will be added to the instrument in the later stages of operation). The primary/safety shutter is already installed at the beamline of the reactor, however, it will be supplemented with a secondary shutter, which will be used to close the beam between individual images during tomography and to reduce the activation of the sample. A beam stopper, designed mainly to stop thermal neutrons and remaining fast neutrons and gammas, consisting of heavy concrete and borated polyethene, will be placed at the end of the entire beamline.

The instrumentation control system of the TRIXIE instrument is based on the Networked Instrument COntrol System (NICOS) [16] developed at MLZ and used to control the full instrument suite at FRM II as well as at other large-scale facilities [17]. Communication to the hardware was implemented using Entangle [18], an implementation of TANGO [19] developed at MLZ. This enables the state-of-the-art, user-friendly control of the instrument, including scripting of measurement procedures and a graphical user interface.

3 Conclusions

A new neutron imaging instrument, TRIXIE, is currently under construction at the 10 MW research reactor LVR-15 near Prague, Czech Republic. The project of development of this instrument is based on collaboration between CTU, MLZ and HZB. This instrument will be a state-of-the-art facility for thermal neutron radiography and tomography techniques. The major components, such as filters, collimators, detectors, etc., are designed and manufactured and will be installed at the beamline after the installation of the shielding. The filters are designed to eliminate high ratios of fast neutrons and gamma in the neutron spectrum, resulting in a thermal neutron beam. The collimator is a fixed pinhole made of a combination of lead and borated polyethylene and will provide a collimation ratio of $L/D = 150\text{--}250$. The detection systems will comprise a standard neutron imaging detection system and also a high-resolution detection system. The shielding is based on heavy concrete shielding blocks in combination with borated

PE desks. Shielding is expected to be installed in the summer of 2024, later followed up by the installation of other components. The first test measurements should be performed at the beginning of 2025.

Starting in mid-2025, the instrument will be used for in-house research in various research fields and also for national and international user communities as a general-purpose instrument. This should help to fill the gap between high user requests and insufficient capabilities that existing neutron imaging instruments provide nowadays in Europe.

References

1. Matouskova, J., Schillinger, B., Sklenka, L.: New neutron imaging facility NIFFLER at very low power reactor VR-1. J. Imaging **9**(15) (2023)
2. HZB, Decommissioning Research Reactor BER II. https://www.helmholtz-berlin.de/projects/rueckbau/index_en.html. Accessed 24 July 2024
3. TUM, Research Neutron Source Heinz Maier-Leibnitz (FRM II). https://www.frm2.tum.de/en/frm2/about-us/news-media/press/press-releases/article/neutrons-from-garching-again-as-of-summer-2024/. Accessed 24 July 2024
4. Soltes, J., et al.: The new facilities for neutron radiography at the LVR-15 reactor. J. Phys. Conf. Ser. **746** (2016)
5. CVR, Research and Services. https://www.cvrez.cz/cs/vyzkum-a-sluzby/provoz-reaktoru/vyzkumny-reaktor-lvr-15-12065. Accessed 24 July 2024
6. Koleška, M., et al.: Capabilities of the LVR-15 research reactor for production of medical and industrial radioisotopes. J. Radioanal. Nucl. Chem. **305**(1), 51–59 (2015). https://doi.org/10.1007/s10967-015-4025-5
7. Koleška, M., Kysela, J., Marek, M., Vsolak, R., Zlamal, O.: LVR-15 reactor and fusion related activities in material research and technology at CV Rez. IAEA TECDOC No. 1724, pp. 41–51, ISSN 1011-4289 (2013)
8. NPL, NPL instruments at the research reactor LVR-15. https://www.ujf.cas.cz/en/departments/department-of-neutron-and-ion-methods/instruments/lvr15/. Accessed 24 July 2024
9. Kucerakova, M., Vratislav, S., Kalvoda, L., Trojanova, Z.: Investigation of textures of α-zirconium and zirconium alloy by neutron and x-ray diffraction. J. Surface Investig. X-ray, Synchrotron Neutron Tech. **14** (2020)
10. Pelowitz, D.B.: MCNPX User's Manual – LA-CP-11-00438. Los Alamos Nat. Lab., Los Alamos, NM, USA (2011)
11. ASI, ASI294 Pro Series. https://www.zwoastro.com/product/asi294/. Accessed 24 July 2024
12. ASI, ASI2600 Pro Series. https://www.zwoastro.com/product/asi2600/. Accessed 24 July 2024
13. Schillinger, B.: An affordable image detector and a low-cost evaluation system for computed tomography using neutrons, x-rays, or visible light. Quant. Beam Sci. **21**(3) (2019)
14. RC TRITEC, 6Li-based scintillators. https://www.rctritec.com/en/isotopes/li-f-based-scintillators-2. Accessed 24 July 2024
15. Schillinger, B., et al.: Flexible camera detector box design using 3D printers. J. Phys. Conf. Ser. **2605** (2023)
16. NICOS, NICOS open source scientific instrument control software. http://www.nicos-controls.org. Accessed 24 July 2024
17. Schillinger, B., Craft, A., Kruger, J.: The ANTARES instrument control system for neutron imaging with NICOS/TANGO/LiMA converted to a mobile system used at Idaho National Laboratory. In: Neutron Radiography, 11th World Conference on Neutron Radiography (WCNR-11)

18. TANGO, TANGO base class specification. https://forge.frm2.tum.de/entangle/defs/entangle-master/. Accessed 24 July 2024
19. TANGO, TANGO controls. http://www.tango-controls.org. Accessed 24 July 2024

Neutron Generators: From Logical Assumptions to totally different behavior

Burkhard Schillinger[1(✉)], Tomas Bily[2], and Jana Matoušková[2]

[1] Heinz Maier-Leibnitz Zentrum (FRM II), Technical University of Munich, Munich, Germany
Burkhard.Schillinger@frm2.tum.de
[2] Czech Technical University, Prague, Czech Republic

Abstract. Two kinds of tabletop neutron generators, D-D and D-T, were used for thermal and for fast neutron imaging, both at a confidential facility that did not want to be named, and at the Czech Technical University in Prague. The first setup with water and polyethylene shielding close to the neutron beam produced unexpected effects by fast neutron scattering and gamma generation in the channel walls that made many images unusable, while a tabletop setup at CTU without any shielding close to the generators produced better, but still insufficient results for nondestructive testing.

Keywords: Neutron Imaging · Neutron Radiography · Fast Neutrons · Neutron Generators · Facility Design

1 Introduction

1.1 Preface

This article is explicitly NOT meant to discuss the usefulness of larger accelerator-driven neutron sources, of which several successful installations exist, but it is limited to so-called tabletop neutron tubes. Quantitative data was not available for these provisionary setups within two days of measurement.

1.2 Motivation

Tabletop neutron generators are often advertised as simple and affordable neutron sources for testing and training, even nondestructive testing of materials. A D-D fusion generator produces fast neutrons based on the $D(d,n)^3He$ reaction at 2.6 MeV neutron energy, while a D-T generator produces fast neutrons based on the $T(d, n)\alpha$ reaction at neutron energy 14 MeV, with accompanying gamma radiation around 70 kV. The generators have an emission cone of about 40°, slightly tilted to the perpendicular to the axis.

Such generators are mostly used for prompt gamma activation analysis called 'active neutron interrogation' [1]. Direct imaging with fast neutrons is possible using special fast neutron scintillation screens, but cross sections for all materials are very low at these

© The Author(s) 2026
A. E. Craft and H. Z. Bilheux (Eds.): WCNR 2024, SPPHY 348, pp. 109–118, 2026.
https://doi.org/10.1007/978-3-032-15003-5_14

high energies, and also the detection efficiency is low, so useful technical images are hard to achieve.

Imaging with thermal neutrons is only possible using a moderator on the output (typically a 2 cm thick piece of PE), where fast neutrons are downscattered to thermal energies, but in first approximation, at least in the center of gravity system, moderated neutrons are scattered into 4π solid angle, so the flux directed at the detector is orders of magnitude lower than the fast flux.

Two table-top neutron generator facilities were used to test the possibility of neutron imaging. The generator brands of the first facility were not given, but it was supposedly the ThermoFisher P 385 [2] as at the second facility, which is available both for D-D and D-T reactions. This generator produces in the order of 3 x 10^8 n/sec in total, which are emitted in a cone angled about $70°$ to the generator axis. The official manual [3] is not available to non-customers, so no information about the cone size is given, and no flux measurements were available, so according to the intensities on the detector, we can only speculate about 10^3–10^4 n/cm^{2}s fast neutrons and 10^2–10^3 n/cm^{2}s thermal neutrons (after moderation) in half a meter distance for the D-T generator, and 100 times less for the D-D generator. This estimate is derived from typical intensities measured with this kind of detector at low-power reactor sources, and the contrast seen in the images.

At a confidential facility – they did not want to be named, but gave permission to publish the results – both types of generators were set up surrounded by shielding of polyethylene (for D-D) or water canisters (for D-T) that provided beam channels towards the detector. An optional cadmium liner can be inserted into the channels to block thermal neutrons from the channel walls. The CCD camera detector has an L-shape that can be inserted into the channel so that the neutron screen is deep within the channel.

At CTU, both generators were simply set up on a table in a well-shielded basement room, which was evacuated (by people) for the measurements. There was no shielding directly surrounding the generators or the detector.

1.3 Measurements at the First Facility with the D-D Generator

The D-D generator was set up vertically in a polyethylene shielding, with the output directed into a horizontal channel. The channel had an optional Cd liner. There was no moderator placed directly on the generator output, but fast neutrons were moderated in the channel walls and in the wall behind the generator. The camera detector has an L-shape with the scintillation screen extended about 20 cm into the channel. Images were generated with a thermal neutron screen (400 μm plastic emulsion LiF + ZnS) and a 3 mm plastic screen (ZnS) by Scintacor [4], and a 200 μm LiF + ZnS powder screen and a 2 mm plastic screen by RC Tritec [5] (Fig. 1).

The first measurements were performed with a 400 μm thermal plastic emulsion screen at 300 s exposure time. Figure 2 left shows the very common test for thermal neutrons – a Cd sheet placed on the detector, which gives good contrast and seems to prove the presence of thermal neutrons. The next image in Fig. 2 shows an image of a 3.5″ hard drive (center top) with very little contrast. For the next radiography, the Cd liner (center bottom) is inserted into the channel, which gives even less contrast, but still some contours of the spin motor and the reading arm drive mechanism (upper right).

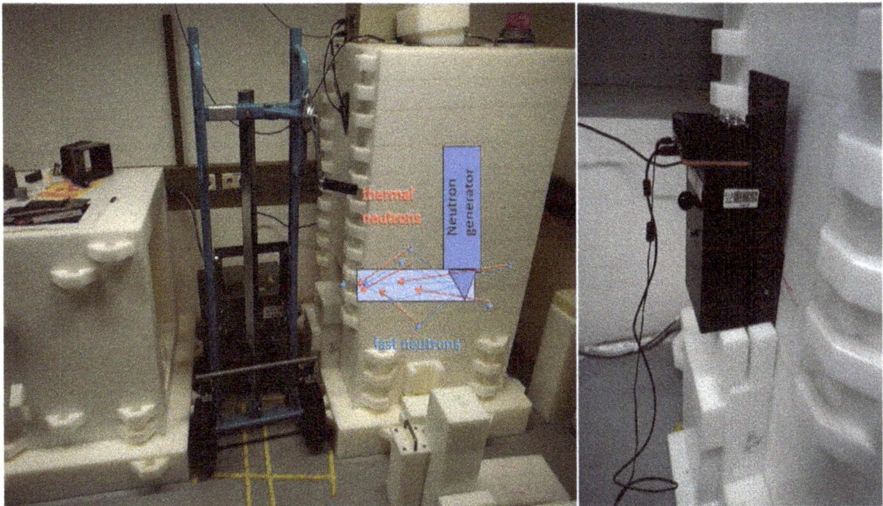

Fig. 1. Setup of the D-D generator with a horizontal channel (left). The detector is inserted into that channel (right).

Fig. 2. Measurements with thermal Scintacor screen: A Cd sheet, a hard drive (thermal), photo of the hard drive and Cd liner, and an image of the hard drive with the Cd liner inserted.

In Fig. 3, a 5 cm thick PE Siemens star is measured. No contours are visible, which proves that there are no neutrons (thermal or fast) along the channel axis, all contrast is caused by radiation scattered in the channel walls. The left image is recorded without the Cd liner, and seems to be a thermal image, the next image is recorded with the Cd liner inserted – still showing significant contrast, but it cannot be thermal. The image on the right is recorded with a 200 μm RC Tritec thermal powder screen with the Cd liner inserted, and shows no image at all. This leads to the conclusion that the Scintacor plastic emulsion screen is also sensitive to fast neutrons (via recoil protons), and that the powder screen is insensitive to fast neutrons, but is illuminated by gamma radiation only. No quantitative data could be obtained from the setups of both facilities, so quantification of the signals is not possible here – where measurement of high-energy gammas and high-energy neutrons would pose a problem all of its own.

Then, the Cd sheet was mounted in 4 cm distance to the screen, and its image disappeared completely, which proves that the image was ONLY generated by gamma radiation (in part from the generator, but mostly from the walls), to which both screens are also sensitive to some extent.

Fig. 3. Measurement with the thermal Scintacor screen of a 5 cm thick PE Siemens star without and with the Cd liner inserted. On the very right is a measurement with the RC Tritec powder screen with the Cd liner inserted.

For the next measurements, the RC Tritec 2 mm fast neutron plastic screen was used, and the detector was mounted OUTSIDE the channel so the screen was at the channel exit, not inside. In this position, the screen was not exposed to radiation coming from the channel walls from behind and from the sides, only to what comes out of the channel in a more or less straight path. The plastic screen is slightly transparent for optical light and gave an image even with the neutron generator turned off, until we covered it with aluminium foil. Figure 4 shows measurements outside the channel with the Siemens star directly on the screen (left) and in 5 cm distance (next), which delivers a considerably sharper image in spite of the increased distance, but with much less blurring from scattering. The next image in the row shows a 5 cm PE brick (top) and a 5 cm lead brick with similar contrast, which hints at a mixed fast neutron – gamma-image. The last image in Fig. 4 shows the Siemens star behind a 5 cm lead filter – this is mostly a fast neutron image.

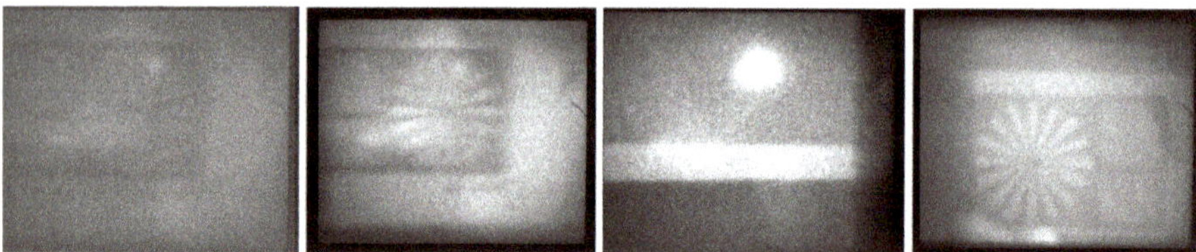

Fig. 4. Fast neutron images with RC Tritec screen for fast neutrons. Siemens Star on the screen and in 5 cm distance, then 5 cm PE and lead brick (mixed neutron-gamma image), and the Siemens Star behind a 5 cm lead filter (mostly fast neutrons).

1.4   Measurements at the First Facility with the D-T Generator

Figure 5 shows the setup of the D-T-generator with a vertical beam channel made of PE, surrounded by water canisters. In the basic configuration, there is no moderator on the output of the generator, moderation happens only in the channel walls. The moderated thermal neutrons can be blocked by a Cd insert. In the new configuration, a 2.5 cm PE brick is placed on the generator output as moderator.

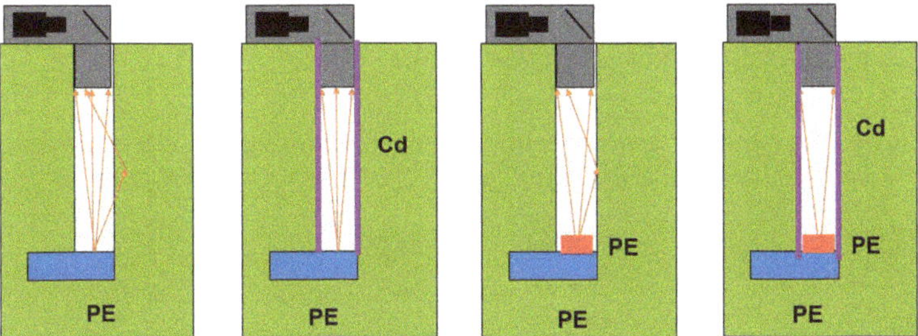

Fig. 5. Vertical beam channel made of PE for the D-T generator without and with moderator on the generator output, with optional Cd insert to block thermal neutrons from the channel walls.

Figure 6 (left) shows the Siemens star block mounted at a 5 cm distance from the 400 μm Scintacor thermal screen, with the Cd liner inserted, and a 5 cm lead filter. No image is visible, the screens seems to be illuminated by gammas only. Figure 6 (right) shows a boron nitride mask on the 400 μm screen plus the 200 μm Tritec screen in the center, with the Cd liner removed, and a 5 cm lead filter. There are thermal neutrons present which are moderated in the channel walls. Calculating spatial resolution from this mask does not make much sense, since the incoming radiation, much of it from the walls, is highly divergent, and mask channels are only visible if the mask is directly put on the screen, for a relatively thin mask. Any value derived from such a calculation would have no meaning for larger objects. The same applies for the plastic Siemens star shown below – the visibility depends on the thickness of the star, and decreases for a thicker object, when no neutrons on an inclined path can penetrate the trenches anymore, and for even thicker objects, the image is wiped out by scattering in the plastic star.

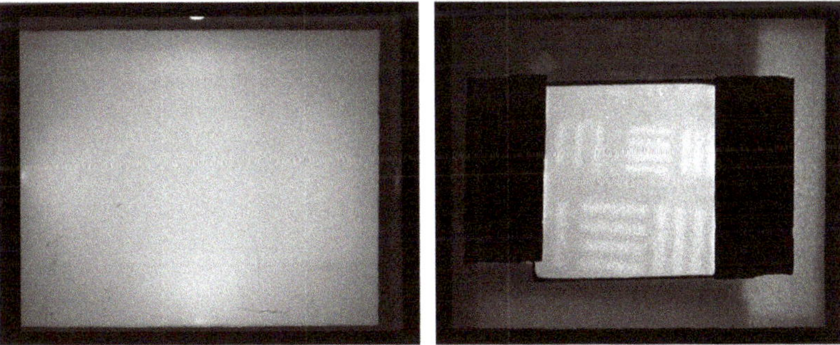

Fig. 6. Siemens star with 5 cm distance from detector with 5 cm lead filter. Thermal 400μm screen with the Cd liner inserted (left), Boron nitride mask on screen with 5 cm lead filter on 400 μm Scintacor screen plus 200 μm Tritec screen with the Cd liner removed (right).

Figure 7 (left) has the Cd liner removed and is slightly brighter, which proves that there are still fast neutrons penetrating into the channel walls and being moderated before they contribute to the image. Figure 7 (right) shows the image with the Cd liner inserted, so thermal neutrons can only come from the moderator on the generator exit, not from the channel walls. The images are scaled for display, the grey scales not absolute. Using 5 cm PE as a moderator produced similar results.

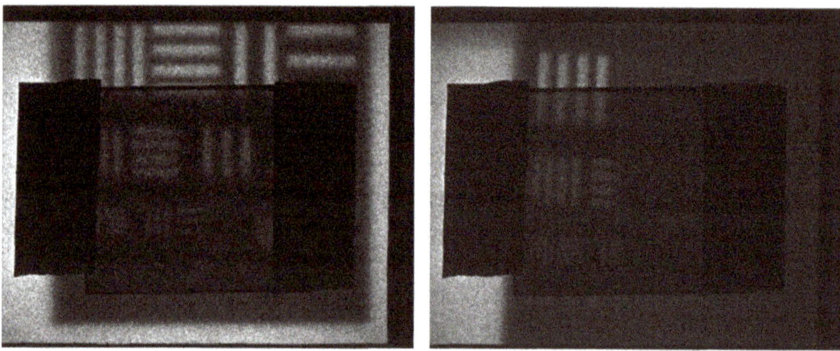

Fig. 7. Boron nitride mask with 2.5 cm PE moderator on the generator exit with the Cd liner removed (left) and inserted (right).

Without additional moderator, both generators produce thermal neutrons only by moderation in the channel walls. Screens are also sensitive to gammas, which are mostly produced in the shielding. A Cd liner lets the thermal image disappear. On the D-T-generator, adding 2.5 – 5 cm moderator directly on the generator output produces a directed beam of thermal neutrons.

1.5 Measurements at CTU

At CTU in Prague, the D-T neutron generator was simply placed on a table in a basement room with thick walls, without shielding directly surrounding the setup. For measurements, all persons left the basement room. Figure 8 shows the setup at CTU with 2.5 cm PE as moderator on the generator output, the detector with borated PE shielding next to the camera, optional lead foil shielding (too thin to stop high energy gammas from the target) and the sample directly on the screen.

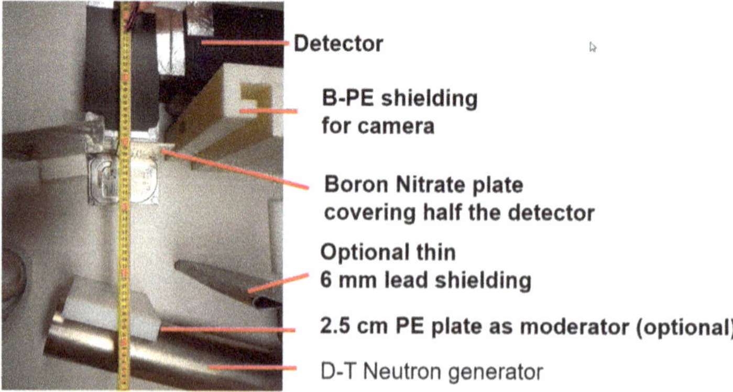

Fig. 8. D-T generator setup un table top at CTU.

The emission characteristics of the generator is slightly inclined to its axis, so the generator was tilted.

Figure 9 shows measurements of a boron nitride plate with the RC Tritec fast neutron screen without and with 6 mm lead inserted.

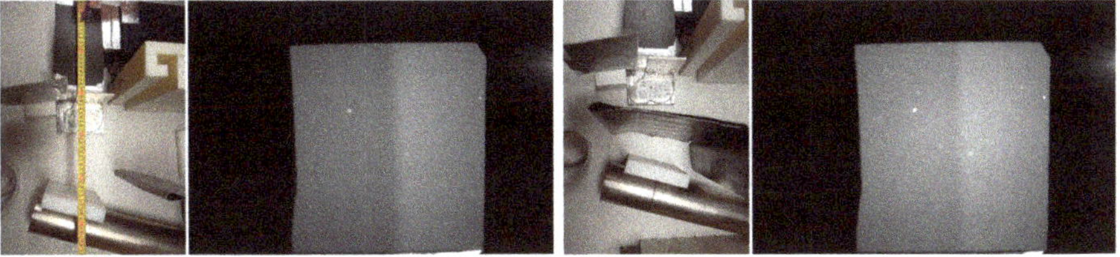

Fig. 9. Fast neutron screen measurement with BN mask without and with 6 mm lead inserted. The PE moderator was removed for these images.

With the lead, although too thin to stop ALL gammas, the contrast decreased, which proved that the first image was partially a gamma image. In Fig. 10, a hard drive was placed directly on the fast neutron screen (center) and in 10 cm distance (right). The very faint contrast shows that technical applications of fast neutrons are difficult. The image on the right is magnified by the cone beam geometry of the neutron generator.

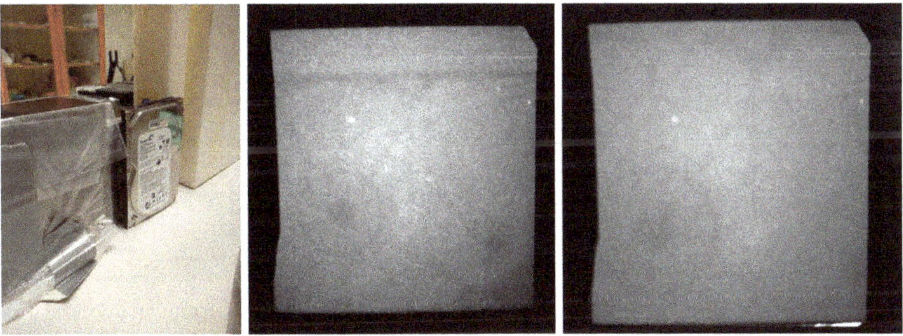

Fig. 10. Fast neutron screen measurement of a hard drive in direct contact (center) and in 10 cm distance from the screen (right). The contrast is very faint. The image on the right is magnified due to cone beam geometry.

Figure 11 shows a fast neutron measurement of a 5 mm and 50 mm thick PE Siemens star with the 400μm thermal neutron plastic emulsion screen with NO moderator inserted. The screen is clearly sensitive to fast neutrons. Surprisingly, the visible star on the left, looking from detector to generator, is the 5 mm thick star, which was verified several times to be certain. For the gaps in the 50 mm star, the beam collimation is insufficient to have neutrons fly straight through, and the PE scatters the neutron beam.

Fig. 11. 400μm thermal neutron screen measurement of a 5 mm and 50 mm PE Siemens WITH-OUT moderator. The plastic emulsion thermal screen is clearly sensitive to fast neutrons. The visible star on the left is the 5 mm star.

Figure 12 shows the effect of the lateral size of the PE neutron moderator on the thermal neutron intensity, measured with the 400 μm thermal screen. On the left, there is no moderator inserted, then PE plates of 2.5 cm thickness, and 3 cm, 6 cm and 9 cm width. The intensity increases with the moderator size, proving that the whole surface emits thermal neutrons after moderation by multiple scattering. But of course the beam collimation decreases with the size of the moderator as effective source, and images of the BN mask remain sharp only because it is placed directly on the screen. Further experiments should show the effect of smaller moderators on intensity and beam collimation, using samples placed in some distance of the detector.

Fig. 12. 400μm thermal neutron screen measurement with no moderator, and 3cm, 6 cm and 9 cm wide PE plates as moderator.

1.6 Conclusions

Again, this article is explicitly *not* meant to discuss the general possibilities of using accelerator-driven neutron sources, of which some very successful installations exist, it is limited to so-called table-top neutron tubes.

If table-top neutron generators are employed for neutron imaging, they should be used without shielding or channels directly next to their output to avoid moderation and gamma production in channel walls. Instead, they should be surrounded by shielding at some distance.

Thermal neutrons can be produced with a PE plate as a moderator, but the collimation in 30 cm distance is very low.

Due to the low overall intensity, it is difficult to use neutron generators for technical applications. Any thermal or fast neutron screens are also gamma sensitive to some extent, and one must be careful to distinguish the gamma content. Some thermal plastic screens are also sensitive to fast neutrons.

Such small table-top neutron generators have very limited use for technical applications besides detector tests (as opposed to larger accelerator facilities). Several organisations have made efforts to use this kind of generators for field-portable fast neutron imaging in nuclear security applications (non-imaging) and for portable thermal neutron imaging systems. Seeing our own limited results, further reference goes beyond the scope of this paper. The measurements shown above were performed with test masks or thin samples placed directly on the detector. A moderator for the generation of thermal neutrons should, on one side, be rather large to create a large emitting surface, but should, on the other hand, be small to improve collimation. Small pieces of moderator produced even less intensity, further limiting the usability of the device for technical examinations. Strongly absorbing test masks, as shown above, allow for a simple yes/no contrast, but as the images of a hard drive have shown, a distinction of different attenuations and gray scales is hard to achieve with the available flux.

A setup for neutron generators should have the generator free-standing in the center of shielding walls, to avoid moderation, scattering and gamma production in shielding too close to the generator. In the case of CTU Prague, a basement room with thick walls was used. A suggested setup for a neutron generator is shown in Fig. 13.

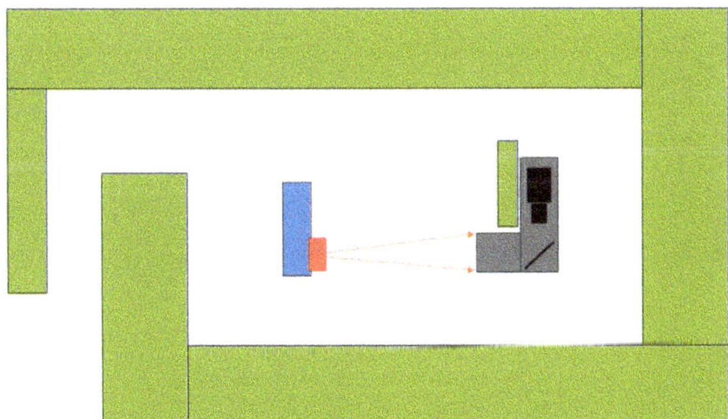

Fig. 13. Suggested setup for a neutron generator, free-standing in the center of shielding walls, to avoid moderation, scattering and gamma production in shielding too close to the generator.

References

1. Chichester, D.L., Simpson, J.D., Lemchak, M.: Advanced compact accelerator neutron generator technology for active neutron interrogation field work. J. Radioanal. Nucl. Chem. **271**, 629–637 (2007)
2. Bily, T., Huml, O.: Characterization of neutron emission during pulse mode of low output electronic neutron generator. EPJ Web Conf. **253**, 04003 (2021). https://doi.org/10.1051/epjconf/202125304003
3. Thermo Fisher Scientific, 2010. P 385 Neutron Generator Operation Manual, 2010. Technical report, Manual P/N 120006-A 062907

4. https://scintacor.com/products/neutron-detection-screens
5. https://www.rctritec.com/en/isotopes/li-f-based-scintillators-2

How to Avoid Going Wrong in Neutron Imaging Facility Design

Burkhard Schillinger[1(✉)], Aaron Craft[2], and Nikolay Kardjilov[3]

[1] Heinz Maier-Leibnitz Zentrum (FRM II), Technical University of Munich, Munich, Germany
`Burkhard.Schillinger@frm2.tum.de`
[2] Idaho National Lab (INL), Idaho Falls, USA
[3] Helmholtz Zentrum Berlin (HZB), Berlin, Germany

Abstract. Neutron imaging had a breakthrough when electronic camera detectors became available 30 years ago, which allowed for digital image processing, but also for sacrificing intensity for higher resolution and sharper images. Detectors have become so sensitive that neutron computed tomography can be performed at 500 W (!) reactor power, and installations are now possible at low-power research reactors that were previously regarded as providing insufficient neutron flux, but such requires an optimal design of the neutron beam for imaging. This article provides an overview about lessons learned from experiences designing many neutron beam lines and facilities for imaging. It will not save a new user from looking up many references, but it highlights hidden but crucial details that are essential for a well-performing neutron imaging facility.

Keywords: Neutron Imaging · Neutron Radiography · Neutron Beam · Neutron Imaging Facility · Facility Design

1 Introduction

1.1 Motivation and Beam Collimation

Early installations with film-based detectors simply opened a big channel for neutrons from some source (e.g. nuclear reactor, accelerator) to reach a sample adjacent to a converter and film. Images took long exposure times and were unsharp, so they were considered not very useful for technical and scientific applications. Digital camera detectors enable digital image processing, and offer higher sensitivity with much shorter exposure times, which also simplified tests with modifications of beam lines [1]. Such results from different neutron beam lines soon made clear that beam collimation, or local parallelism is the most important parameter for high-quality images. As a property of the facility, this parameter is characterized by the ratio of distance L between source and detector, and effective diameter D of the source, which is typically a pinhole in the beam (sometimes referred to as an aperture or diaphragm). For characterizing the actual projection sharpness, the source-to-object distance is used, which is nearly the same for a flight path of several meters and a sample-to-detector distance (l) of a few cm. Looking from a

© The Author(s) 2026
A. E. Craft and H. Z. Bilheux (Eds.): WCNR 2024, SPPHY 348, pp. 119–128, 2026.
https://doi.org/10.1007/978-3-032-15003-5_15

point on the sample towards the source, the extended source is seen under a certain angle from edge to edge, emitting neutrons that potentially hit the point on the sample under that maximum angle. After that imaginary point on the sample, individual neutron rays from the source area spread out again over the distance l to project that point as a circle or disk of diameter d on the detector, leading to unsharpness (d). The equation is given as L/D = l/d, so one can calculate the estimated geometric unsharpness (d) by dividing the sample-to-detector distance (l) by the L/D ratio. Even in a beam spreading to a large area, the angle under which the source is seen is mostly constant, defining *local* beam parallelism.

2 Collimators

2.1 Pinhole Projection of the Real Source, Collimator Shaping

The effective source is usually a pinhole that forms the smallest diameter of an incoming, converging collimator. The *real* source is usually the end or the nozzle of the beam tube. The converging collimator should be shaped in such a way that the whole source diameter is visible from all points on the pinhole, i.e. not cut off the sight to the outer parts (Fig. 1). This determines the main properties of the beam like divergence and intensity distribution. A simple collimator does *not* influence the spectrum. A diverging collimator after the pinhole only limits the size of the projected beam and does *not* influence the beam properties in the center. The diverging collimator should be designed to at least attenuate the penumbra region of the beam, or even further limit its size if the whole possible diameter is not needed. Any unused beam produces unnecessary neutron activation and gamma-rays in detector materials and the beam stop.

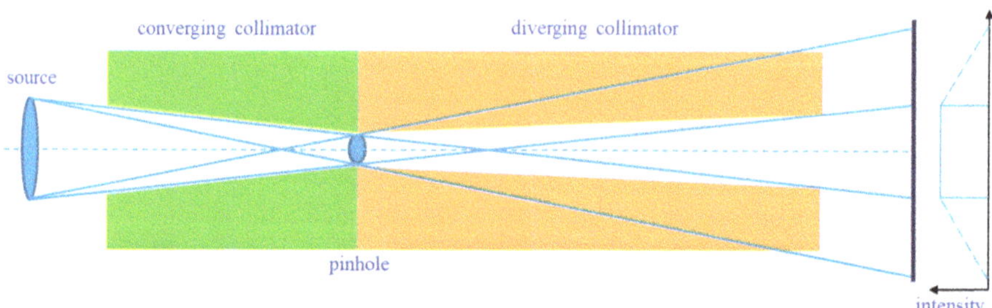

Fig. 1. Theoretically optimal collimator design that acts like a pinhole camera projecting the source onto the detector. The uniform fully illuminated area in the center is the area that can see the whole source. The penumbra region that sees less of the source may be cut off by the shape of the diverging collimator.

The aperture acts as a pinhole camera that projects the source onto the sample and detector. Any existing intensity distribution of the source will be projected as is, and cannot be influenced by the collimator. However, a poor design of the collimator can prevent the direct projection of the source distribution. Some facilities inserted a straight tube section between the converging and diverging collimators. In such a configuration, a point on the beam axis sees a larger area of the source than a point off the axis (Fig. 2). Overall, this leads to a beam profile with lower peak intensity and a smaller area with maximum uniform intensity at the center, quickly dropping towards the edges of the beam profile.

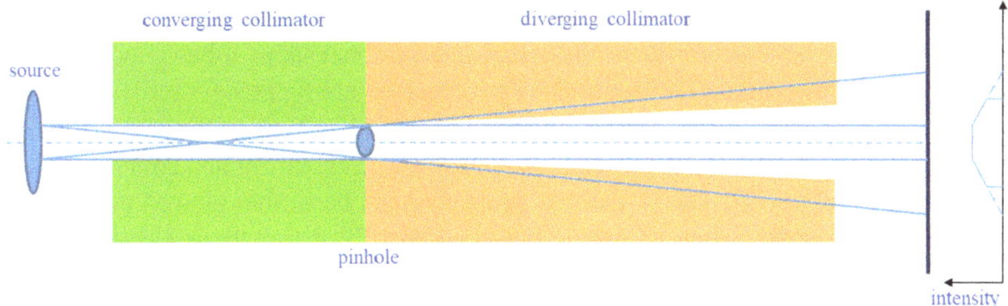

Fig. 2. A straight tube limits the visibility of the source and leads to a conical beam profile, wasting most of the beam.

2.2 Filters

Any filters (bismuth or sapphire) should be placed into the incoming collimator *before* the narrowest section (Fig. 3) so that the smallest diameter of the beam path is still physically limited by the collimator shape. Typical filter lengths are 10 cm of bismuth for gammas, which attenuates about 90% of the thermal neutron intensity, and 10 cm of sapphire for fast neutrons, which attenuates about 30% of the thermal neutron intensity. For very low power reactors, the installation of such filters is not recommended. The detectors (e.g. scintillator screens) are, in first approximation, insensitive to gammas and fast neutrons; it is more of a shielding issue.

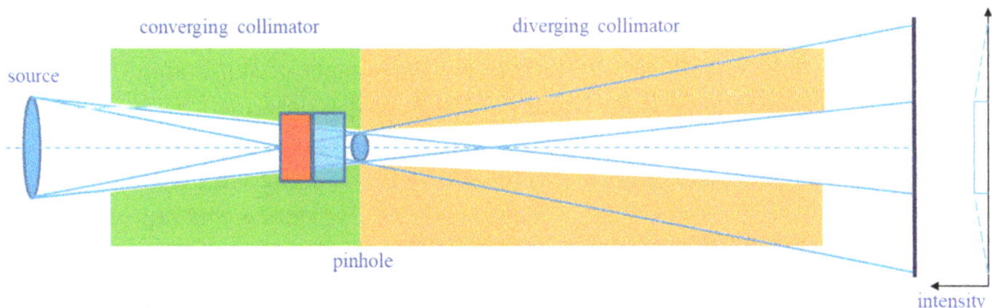

Fig. 3. Any filters should be placed in the converging collimator BEFORE the smallest diameter.

2.3 Collimator Material

A great material for collimators is borated steel, which hardly activates compared to steel without any boron content. Steel always contains traces of cobalt and nickel, which builds up long-term activation products such as ^{60}Co, and the boron in borated steel reduces the production of ^{60}Co. As a practical solution, many alternate layers of polyethylene and lead are often used.

3 Shielding, Beam Tubes and Beam Catcher

3.1 Shielding Material for Walls

Reactor shielding must shield both gammas and fast neutrons. For large assemblies like measurement chambers, lead and polyethylene layers would often be too massive and too expensive, so they are replaced by other shielding materials, mostly heavy concrete. Since lead is mostly transparent for neutrons, iron is chosen instead, as it has good shielding properties for both gammas and neutrons. The typical shielding process for fast neutrons needs hydrogen to slow down fast fission neutrons from 1–2 MeV to the 100 keV range, where iron shows resonance absorption, i.e. it absorbs fast neutrons and emits them with much lower energies [2]. The re-emitted neutrons must be further slowed down to the thermal range again by hydrogen, where boron will finally absorb them. Heavy concrete usually contains colemanite or magnetite, minerals that contain iron, but also a lot of crystal water as moderator. Magnetite $Fe^{2+}Fe^{3+}_2O_4$ must get additions of ferroboron, FeB, while colemanite $Ca[B_3O_4(OH)_3]\cdot H_2O$ already contains boron. (Please note that Barite, often used for X-ray shielding, does not contain boron, and no extra crystal water.) Technical University of Munich has developed a more efficient shielding material that contains no cement or other non-shielding materials. The material is a mixture of steel shots, ferroboron and paraffin oil. The material maintains the consistency of wet sand that must be filled into a steel container. The mixture hardly activates and can be re-used by filling it into new containers. A former patent has expired; the material composition is free now [2].

3.2 Beam Tubes

After the collimator, a beam tube made of aluminium should be placed that is either evacuated or filled with helium. Steel tubes must be avoided because of strong gamma creation and activation, but also the aluminium tube diameter should be sufficiently large that the beam never hits the tube walls. Although the capture cross section of aluminium is rather small, a tangential beam will effectively see a meter of aluminium (Fig. 4) and create 8 MeV prompt gammas on capture. This is easily prevented by stripper disks made of B_4C or BN inside the beam tube that absorb the penumbra beam while emitting only 478 keV gammas from capture in ^{10}B. Avoid any coating with cadmium, which would be a very strong source of high-energy (1.2 MeV) gammas that are very hard to shield from the detector and the whole measurement chamber.

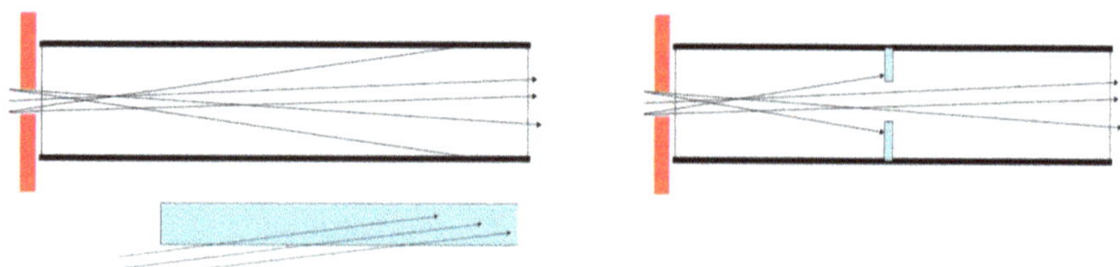

Fig. 4. Any neutrons reaching the beam tube tangentially will create a huge amount of capture gammas. This is easily remedied with B_4C stripper disks inside the beam tube.

3.3 Beam Catcher

A beam catcher should be placed not too close to the detector because of capture gamma creation that influences the detector by creating white spots from gammas that directly hit the camera chip, best more than 1.5 m away. A flat wall would backscatter neutrons into the measurement chamber and eventually let them reach the detector, and create more capture gammas. Any beam catcher should therefore be shaped like a cup to reduce backscattered neutrons, wide enough to catch the full beam including its penumbra region (Fig. 5). The cup should be about as deep as it is wide, so multiply scattered neutrons can mostly run dead inside. The cup should therefore be coated with borated PE on the inside, with lead or concrete on the outside, seen from the beam. Calculations have been done for beam catchers that would be split into many smaller individual cups [2], but the improvement showed to be less than 10%, and is not worth the effort.

Coating the inner walls of the measurement chamber with cadmium would be a mistake, as it was previously done for indirect film radiography, which is, unlike electronic detectors, insensitive to gammas. A cadmium coated wall would make the use of electronic detectors nearly impossible.

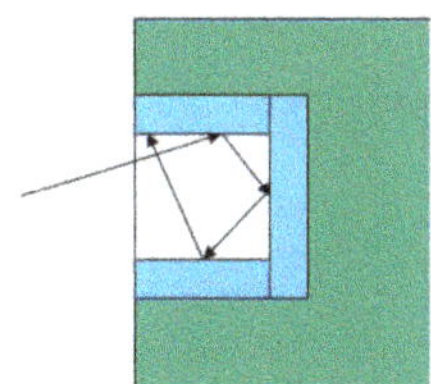

Fig. 5. Beam catcher, coated with B_4C on the inside (blue) surrounded by lead or concrete (green). Backscattered neutrons are more likely to be attenuated with a cup approach compared to a flat back wall.

4 Neutron Guides

4.1 Imaging at a Neutron Guide

When imaging is done at a neutron guide end position, a dedicated converging and/or diverging collimator is not required, as bent neutron guides do not carry a background of fast neutrons and gammas. For thermal and cold neutrons only, a selector wheel or plate with pinholes in absorber material like B_4C can be installed. The selected pinhole must be followed by a flight tube of several meters length to provide sufficient collimation just like a standard collimator above. The divergence directly at a neutron guide exit is too large to create sharp images of larger objects without the pinhole and flight tube, which again define divergence in the classical way.

Again, the pinhole projects the 'source' onto the detector. The use of a neutron guide brings a new problem, as the 'source' is the flux distribution inside the neutron guide, which is constructed of short segments adhered together. The joints of short sections do not reflect and appear as dark stripes in the flux distribution. The smaller the pinhole, The smaller the pinhole, the sharper the projection of these black stripes will be, where the intensity can drop nearly to zero (Fig. 6).

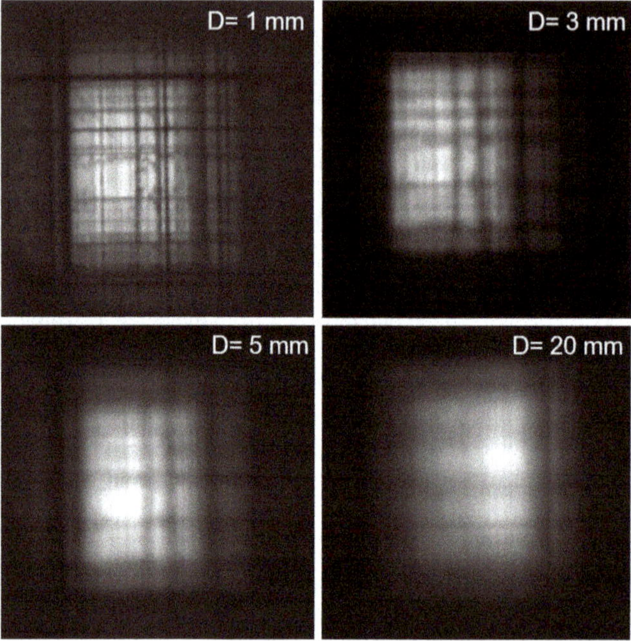

Fig. 6. Beam profile after a 5 m flight tube after different pinholes on a neutron guide for 1 mm, 3 mm, 5 mm and 20 mm pinholes. The smaller the pinholes, the sharper the projection of the stripes and the rectangular shape of the guide.

This can be mitigated by installing a 5–20 mm thick aluminium container with graphite powder just at the guide exit before the pinhole. The graphite powder acts as a diffuse scatterer that homogenizes the beam by 'filling' up or smearing out the black stripes, with some losses from smearing out. Figure 7 shows the intensity dips in a beam from a neutron guide with 3 cm and 1 cm pinhole, without and with the graphite filter. The homogenized intensity here is a factor of two less than the peaks of the unfiltered beam.

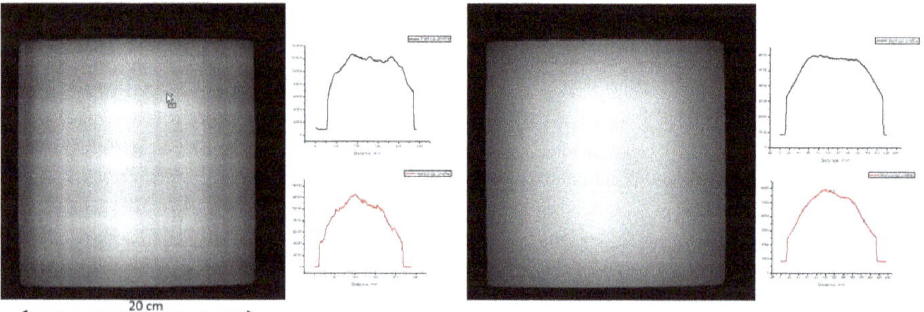

Fig. 7. Beam profile after a 9 m flight tube after a neutron guide without (left) and with (right) a 2 cm graphite diffusor and 3 cm pinhole. Beam profiles are shown in red (horizontal) and black (vertical).

Neutron guides are usually built with a curvature that eliminates direct sight to the reactor source, so fast neutrons and gammas cannot reach the instrument. The curved section is usually followed by a shorter straight section of guide that helps to homogenize the beam profile – most neutrons are reflected at the outside of the curvature, which leads to an inhomogenous distribution at the end of the curved section.

The curvature of the guide defines a cutoff wavelength for the beam, where shorter wavelengths cannot be reflected through the curved guide, usually in the range of 2 Å. This wavelength defines the maximum spread of the beam with a homogenous spectrum, which is *not* equal to the beam collimation. A typical ^{58}Ni guide reflects thermal/cold neutrons under a maximum angle of 0.1° per Å of the wavelength; for supermirror guides, this is multiplied by the factor m given by the specific supermirror coating. This means that from the pinhole of the guide end, neutrons of different wavelengths spread in cones of different widths, depending on the wavelength, with larger cones for larger wavelengths. Outside the cone of the shortest (i.e. cutoff) wavelength, the beam does not contain the full initial spectrum anymore, so the attenuation coefficient for sample materials increases with the lower neutron energy spectrum, and a homogenous tomography is no longer possible. It is therefore recommended to limit the beam at the sample position to the area containing the full spectrum.

Neutron guides are employed at large scale facilities with neutron guide halls, with typical lengths of 20–30 m. For small and low-power reactors, it is not worth building a neutron guide for neutron imaging.

5 Special Beam Tubes

5.1 Imaging at a Through-Going Tangential Beam or at a Radial Beam

With the availability of low-cost electronic detectors, neutron imaging is being installed at many older reactors at beam ports that had not originally been designed for it. A radial beam, with direct line of sight to the reactor core, transports a considerable amount of fast neutrons and gammas [7] that create unwanted background at the detector position, which can be mitigated by installing crystal filters as described previously. Sometimes, radial beam tubes are installed very close to the core to optimize for epithermal neutrons [7]. In such cases, the water gap between the core and the beam tube nozzle is not optimal to create the maximum possible thermal flux. The thermal flux can be increased by inserting a plug with additional moderator (wax or PE) into the beam tube; in one case, a factor of 1.4 was achieved with a 4 cm plug [8].

For higher reactor power, heating and embrittlement of such a plug becomes an issue, and it should be encapsulated and sealed in aluminium (Fig. 8).

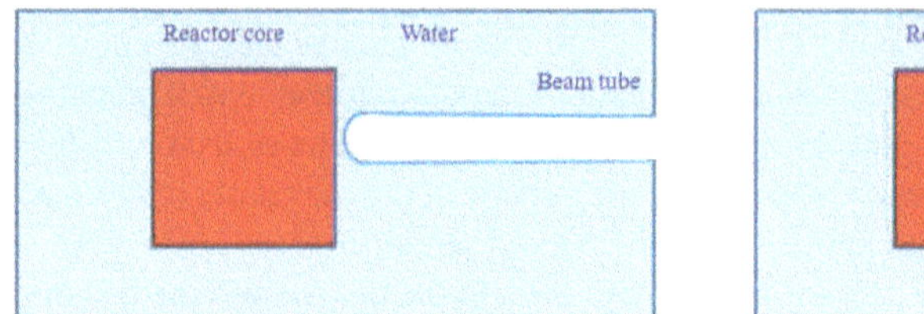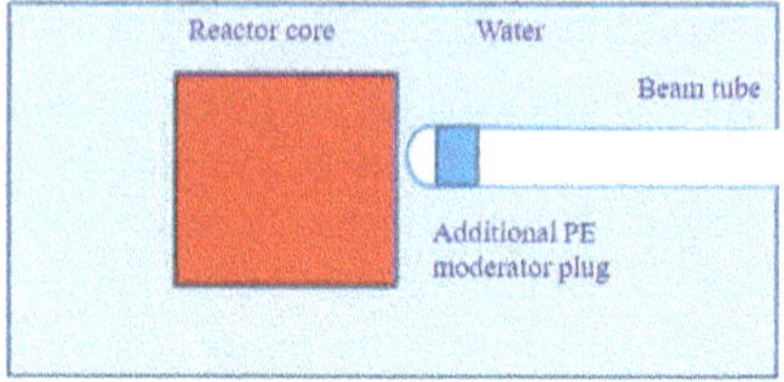

Fig. 8. Radial beam tube nozzle very close to the core, without and with extra moderator plug.

A tangential beam ending in the reactor pool is to be preferred because of drastically reduced fast neutron and gamma flux. Some reactors have a through-going tangential beam tube that comes out of the moderator tank both sides. If the tube is empty, neutrons can only enter the tube tangentially under small angles, and there will be no neutrons that fly parallel to the beam axis. Such a beam has much reduced intensity and a large average divergence, without any neutrons flying straight. To remedy the situation, a moderator plug (e.g. graphite, wax or polyethylene) must be inserted into the beam tube next to the core (Fig. 9).

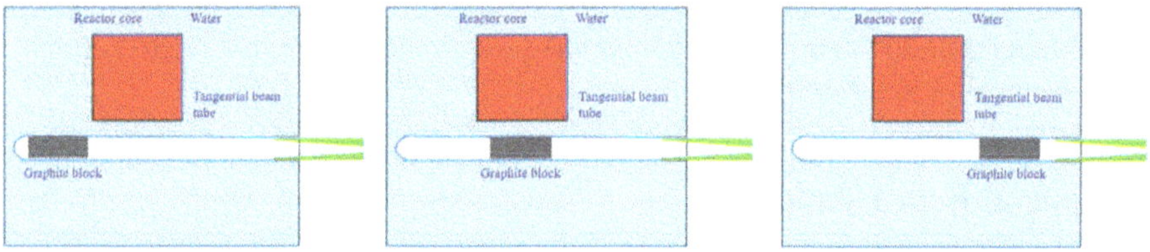

Fig. 9. Moderator plug in a tangential beam tube – behind the core, next to it and between core and beam tube exit.

Future work will consider multiple positions for the moderator. The position next to the core is expected to be best as it is surrounded by the highest flux. The position behind the core is expected to produce the next highest flux, as it is still in the thermal flux and can emit unhindered into the tube. The position in front of the core is expected to be the worst solution, as the higher flux closer to the core will be attenuated in the plug again. Detailed simulations must be run for each individual situation.

6 Detectors

6.1 Camera Detectors

Camera detectors for neutrons typically use a neutron-sensitive screen, a mirror and a cooled CCD or CMOS camera, e.g. from [9, 10]. The camera views the screen via a 45° mirror to get out of the direct beam. Multiple publications exist about such detectors, so only a few vital conditions should be mentioned: The camera is sensitive to gamma rays from the sample or structural materials that directly hit the camera chip itself. Local charge clouds are created in the chip by absorption of gamma photons, which appear as white spots or streaks in the image. Sometimes, permanent damage is caused in the crystal lattice, likely from higher-energy neutron interactions [11]. Therefore, the camera must be shielded against gamma rays and scattered neutrons, at the very least by a lead brick blocking the direct line of sight from the sample to the camera chip, through all housings and detector walls. A layer of borated rubber may be added to stop scattered neutrons. In [12, 13], a low-cost system using cooled astronomy cameras is described. The system has the camera box separated from the box containing the screen and mirror, so the latter can be swapped out for different sizes and fields of view. The 3D-printed camera box itself contains a split lead brick in front of the camera, but is so small that it relies on externally stacked shielding, which makes it much more flexible. A complete tomography system with controller and software is also described, with the plans available for free [14].

7 Conclusion

With new low-cost astronomy CCD and CMOS cameras, neutron radiography and computed tomography have become easy to do on nearly any beam line. However, high-quality imaging requires a dedicated beam line and collimators set up for neutron imaging. Detectors and sample manipulators can be easily replaced, but the neutron imaging beam line including collimators, beam tubes and shielding is the core of an imaging facility, and best efforts should be made to create them best as possible according to the rules shown above. There are many more considerations, and a lot of literature research will be required to build a new instrument. This paper covered some of the most important aspects in the basic design of an imaging facility.

Finally, the golden rule in creating an imaging facility is: "*First, you grab all available space. If somebody else wants some of it, he will complain. If you don't do this, you will regret it later.*" [Elbio Calzada, private communication]

References

1. Schillinger, B: Improved radiography and 3D tomography due to better beam geometry. In: Palmer, S.B. (ed.) Nondestructive Testing and Evaluation vol. 16, no. 2–6, pp. 277–286. Gordon & Breach Publishers (2001)
2. Grünauer, F, Schillinger, B: Optimization of the beam geometry and radiation shieldings for the neutron tomography facility at the new neutron source in Munich. In: 4th International Topical Meeting on Neutron Radiography, PennState University (2001). J. Appl. Rad. Isotopes **61**(4 Oct. 2004), 479–485
3. Calzada, E., Grünauer, F., Schillinger, B., Türck, H.: Reusable shielding material for neutron- and gamma-radiation. Nuclear Instrum. Methods Phys. Res. Sect. A: Accelerat. Spectrom. Detect. Assoc. Equip. **651**(1), 77–80 (2011). ISSN 0168-9002. https://doi.org/10.1016/j.nima.2010.12.239
4. Schillinger, B.: Estimation and measurement of L/D on a cold and thermal neutron guide. In: Palmer, S.B. (ed.) Nondestructive Testing and Evaluation, vol. 16, no. 2–6, pp. 141–150. Gordon & Breach Publishers (2001). https://doi.org/10.1080/10589750108953071
5. Schillinger, B., Bleuel, M., Böni, P., Steichele, E.: Design and simulation of neutron optical devices plus flight tube for neutron radiography. IEEE Trans. Nuclear Sci. **52**(1) (2005)

6. Kardjilov, N., et al.: Neutron tomography instrument CONRAD at HZB. Nucl. Instrum. Methods Phys. Res., Sect. A **651**(1), 47–52 (2011)
7. Giegel, S.H., Pope, C.L., Craft, A.E.: Determination of the neutron energy spectrum of a radial neutron beam at a TRIGA reactor. Nucl. Instrum. Methods Phys. Res., Sect. B **454**, 28–39 (2019)
8. Matouskova, J., Schillinger, B., Sklenka, L.: New neutron imaging facility NIFFLER at very low power reactor VR-1. J. Imaging **9**(1), 15– (2023). https://doi.org/10.3390/jimaging9010015
9. www.zwoastro.com
10. www.qhyccd.com
11. Pugliesi, R., Andrada, M.L.G., Dias, M.S., Siqueira, P.T.D., Pereira, M.A.S.: Study of pixel damages in CCD cameras irradiated at the neutron tomography facility of IPEN-CNEN/SP. Nuclear Instrum. Methods Phys. Res. Sect. A **804**(59–63) (2015)
12. Schillinger, B.: An affordable image detector and a low-cost evaluation system for computed tomography using neutrons, X-rays or visible light. https://www.mdpi.com/2412-382X/3/4/21
13. Schillinger, B., Geerits, N., Jünger, T., et al.: Flexible camera detector box design using 3D printers. J. Phys. Conf. Ser. **2605**(1), 012008 (2023). https://doi.org/10.1088/1742-6596/2605/1/012008
14. https://forge.frm2.tum.de/wiki/projects:mobile_tomography:index

A Vertical Neutron Beam Device Used to Examine Liquids

Burkhard Schillinger[1(✉)], Stephan Sponar[2], Alessandro Tengattini[3],
and Clemens Trunner[2]

[1] Heinz Maier-Leibnitz Zentrum (FRM II), Technical University of Munich, Munich, Germany
Burkhard.Schillinger@frm2.tum.de
[2] Atominstitut der TU Wien, Vienna, Austria
[3] Institut Laue-Langevin, Grenoble, France

Abstract. Neutron beam lines are mostly horizonal, with some exceptions at specialized reactors as the McClellan reactor (Sacramento, USA) that is built underground and has inclined beams, or the small reactor at Kingston college, Canada. These inclined beam lines have never been used to examine thin liquids with at least a vertical beam component. We have built and tested a new device with one or two crystal reflections on pyrolytic graphite to deflect a horizontal neutron beam to the vertical, and examined thin liquids. Initial tests were performed at the TRIGA reactor of Atominstitut Vienna (ATI). Further measurements were conducted at ILL Grenoble, France. Air bubbles in a horizontal CPU water cooler were examined in real time.

Keywords: Neutron Imaging · Neutron Radiography · Bragg Crystal · Pyrolytic Graphite · Vertical Neutron Beam · Examination of Liquids

1 Introduction

1.1 Motivation for a Vertical Beam

The use of pyrolytic graphite (and other) crystals for monochromatization or simple beam deflection is nothing new – many imaging installations use a double crystal monochromator to tune the wavelength of their horizontal beams. But to our knowledge, nobody has used crystals to create a vertical beam for imaging before. The sole motivation for a vertical beam is to examine horizontally spread thin liquids that do not mix easily, or melts in direction parallel to gravity. A killer application has not been found yet.

The purpose of the initial experiments was to determine whether such a beam generated on a weak thermal spectrum would deliver sufficient intensity for imaging, and sufficient contrast on thin liquids. Simulations have not yet been performed.

1.2 Practical Setup for a Vertical Beam

For a horizontal beam, double crystal monochromators are often used, with the sample in rather large distance, so the incoming beam is locally rather parallel, mostly given

© The Author(s) 2026

A. E. Craft and H. Z. Bilheux (Eds.): WCNR 2024, SPPHY 348, pp. 129–139, 2026.
https://doi.org/10.1007/978-3-032-15003-5_16

130   B. Schillinger et al.

by the Length L between sample and crystal, and the diameter D of the crystal. For a vertical beam, distances of 10–20 m are hardly possible, a first setup must work on less than a meter distance, with equipment that can be transported in a car or in a suitcase. On top of this, the first experiments had to be done on a thermal beam at Atominstitut (see below).

For the first installation, large pyrolytic graphite crystals were available with 50×50 mm^2 size and 5 mm thickness. Their mosaicity was rather large with 3°. The idea was to get a sufficiently large reflected beam on a very short distance. Using Bragg's law $n\lambda = 2d\sin\theta$, with λ the incident wavelength, d the lattice constant and θ the incident angle, and with d = 3.48 Å for the 001 reflection we find λ = 4.71 Å for 45° incident angle and thus 90° total deflection angle. The wavelength 4.71 Å is well in the cold range of a neutron spectrum, and is nearly non-existent in a thermal beam. However, the 90° deflection also works for half the wavelength, or the 002 reflection at 2.36 Å, which is well within the thermal range. Typically, the second order (n = 2), has about 40% of the relative intensity of the base or first order (n = 1) wavelength, but in a typical cold spectrum, the intensity at 2.36 Å is two to three times as high as at 4.71 Å. The next higher order (n = 3) at 1.57 Å is in the hot/thermal range of a direct beam, but will be cut off in a neutron guide.

The first test setup consisted of a framework carrying the imaging detector looking down on a crystal holder (Fig. 1).

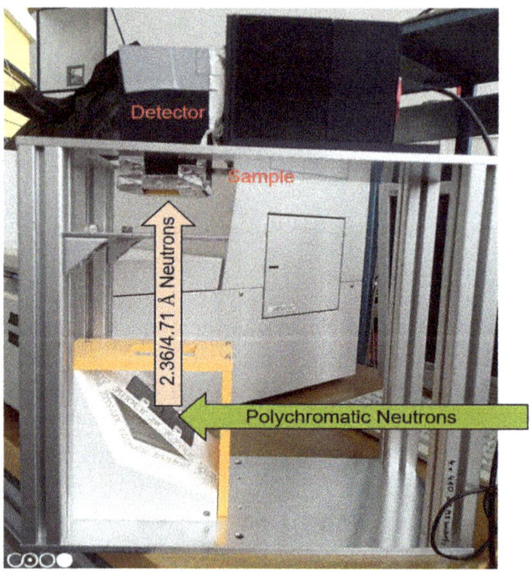

Fig. 1. First vertical beam test setup. The framework holds the camera detector looking down on a crystal holder. The sample (a test pattern) is taped close to the neutron screen.

The crystal holder has adapters for two crystals set up at 45° for a single reflection of 4.71 Å and second-order (n = 2) 2.36 Å neutrons (Fig. 2a) or for a double reflection of 2.36 Å neutrons on a total of four crystals at 22.5° and 67.5° (Fig. 2b). The double reflection setup was explicitly tested for a thermal spectrum that does not contain cold neutrons. As expected, due to the large mosaicity of the crystals, the losses on the double first order ($n = 1$) reflection for the thermal wavelength were greater than for the second-order ($n = 2$) single reflection, and the 45° setup delivered more intensity. The grey and orange windows consist of borated plastic to limit the incoming and outgoing beams. Because of the tilt of the crystals, the effective cross section in the beam is multiplied by the sine of the tilt angle. To increase the effective cross section and the resulting beam size, two crystals are used in this first approach to create a beam as large as possible, although this creates more beam unsharpness.

In a typical radiography setup, the sample is mounted in distance L from a pinhole with diameter D (Fig. 3). Beam collimation (and sharpness) is defined as the ratio L/D. With a crystal, the beam is not just mirrored, but due to its mosaicity, the crystal acts as a secondary source, with its projected size the new diameter D (Fig. 4). In our test setup, L is only about 40 cm, with $D = 10$ cm $\times \cos 45° = 7.1$ cm in beam direction, 5 cm across, rendering $L/D = 5.6$ resp. 8 (to be exact, L varies along the tilt of the crystal), corresponding to an angle $\theta = 5.05°$, giving very bad collimation, and thus unsharp images. Samples had to be installed very close to the detector. However, at this short distance, this estimate is not correct – the 3° mosaicity of the crystals does not allow such a large geometrical spread for a parallel incoming beam; the collimation at Atominstitut is $L/D = 125$, corresponding to 0.46°, and at ILL, it is $L/D = 315$, corresponding to 0.18°. Combining these with the mosaicity would roughly give 3.46° and 3.18° spread, corresponding to roughly $L/D = 16.5$ and 18. For much larger vertical distances, L/D is solely given by the crystal size. The mosaic spread of the crystal also causes a wavelength spread as described in (1), but this is irrelevant for these crude experiments.

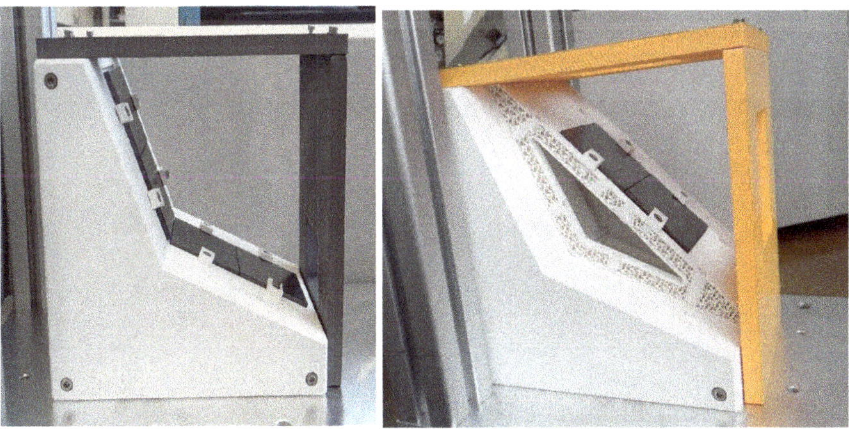

Fig. 2a (left) and 2b (right): Crystal setup for double reflection at 22.5° (left) and for single reflection at 45°. Due to mosaicity losses, the single reflection setup delivered more intensity. The grey and orange windows consist of borated plastic.

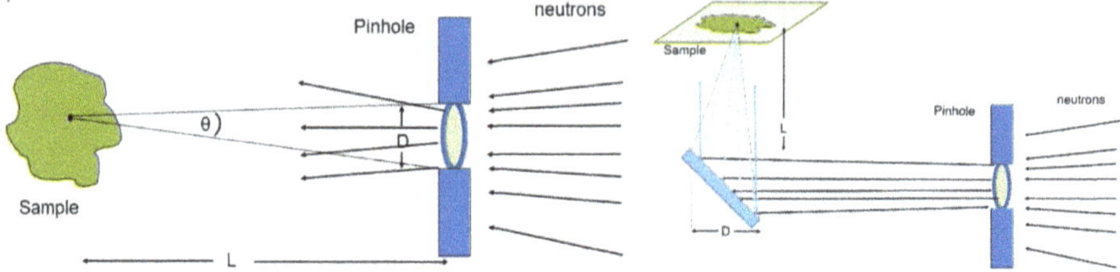

Fig. 3. (left) and 4 (right): Classic image geometry and crystal as secondary source: A point on the sample sees the source area under an angle θ that determines collimation. Due to the crystal mosaicity, a crystal acts as a secondary source, with its projected size the new diameter D. This is, however, not exact on short distances, as the collimation of the incoming beam must be taken into account, and the purely geometrical L/D may be smaller than the one given by the crystal mosaicity. Note that Fig. 4 is only a schematic drawing – the thin blue lines are assumed to be reflections from different angles of the mosaicity, and different wavelengths, and L should be drawn much longer.

2   Measurements at Atominstitut

2.1   Installation

The vertical beam device was installed at the radiography beam port of the 250 kW TRIGA reactor of the Atominstitut of Technical University Vienna, which looks at the thermal column and delivers a perfect thermal spectrum. Figure 5 shows the scaffold installed on the left, with the beam shutter and beam port visible to the right. Figure 5 on the right shows the detector installed, with a lead brick underneath to protect the camera against gamma radiation. The sample holder can be moved vertically, and close to the detector. The incoming intensity was in the order of 10^5 n/cm2s, with the reflected intensity estimated a hundred times less.

Fig. 4. First vertical beam test setup. The framework holds the camera detector looking down on a crystal holder. The sample (a test pattern) is taped close to the neutron screen.

2.2   Measurements at ATI

First measurements were performed at the imaging installation on the thermal column of the 250 kW TRIGA reactor of Atominstitut Vienna (2).

Figure 6 shows a Gadolinium line pair gauge at 5 min and 30 min measurement time, in near-direct contact to the detector. The width of the pattern is 30 mm, a line pair is 10 mm long. Intensity was very low.

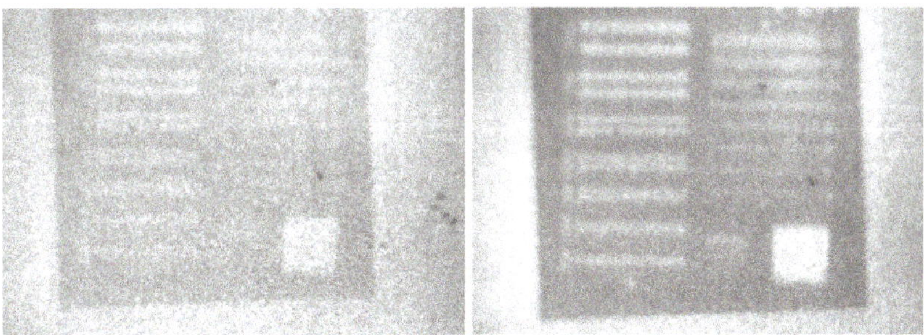

Fig. 5. Gadolinium line pair gauge at 5 min and 30 min measurement time.

Figure 7 shows two oil drops floating on ~ 1.5 mm of water in a container of aluminium foil with ~ 4 cm diameter. Due to the container size, the water was about 2.5 cm away from the detector, giving an unsharp image.

Figure 8 shows two drops of wax floating on water in a 4 cm diameter container.

Figure 9 shows an air level with air bubble, ca. 2.5 cm size, with the original image in the center, and the image on the right normalized to an open beam image.

To simulate voids or bubbles, Fig. 10 shows glass beads in about 5 mm of water, again in a 4 cm wide container.

Exposure times for these measurements were very long, in the order of 20–30 min per image, but these preliminary results helped us to get beam time at the NeXt beam line at the ILL reactor in Grenoble, with much higher intensity.

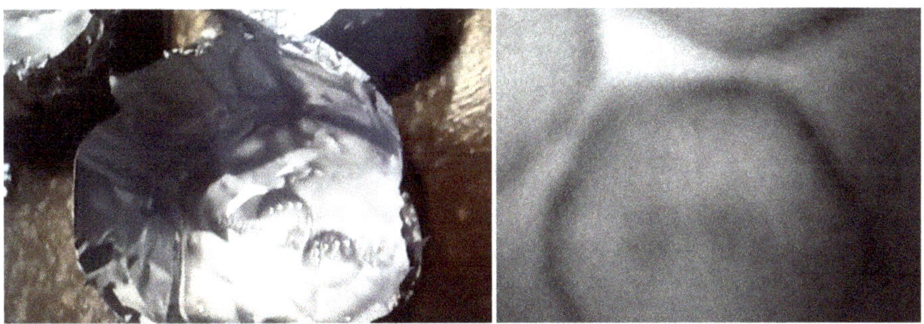

Fig. 6. Two oil drops floating on ~ 1.5 mm of water in a 4 cm wide container.

Fig. 7. Two drops of wax floating on water in a 4 cm container.

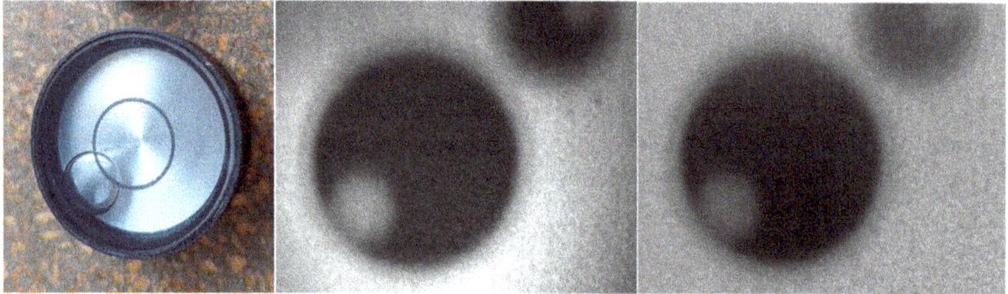

Fig. 8. Air level with air bubble, ca. 2.5 cm size, normalized on the right

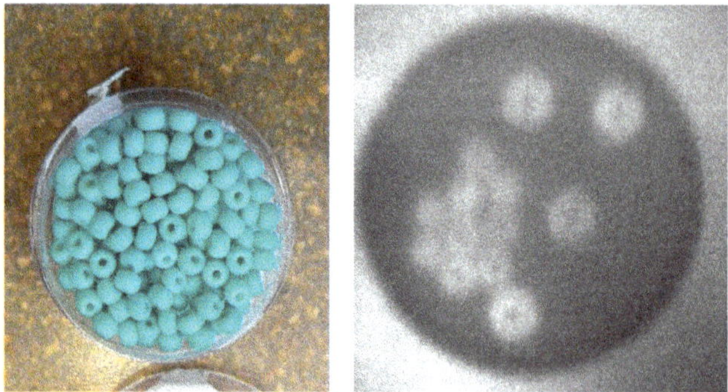

Fig. 9. Glass beads in water. The left image is the storage container to show the beads, the neutron radiography on the right is a again a 4 cm diameter container with water.

3   Measurements at NeXT at ILL

Measurements with 10^3 times higher intensity were performed at the imaging facility NeXT at ILL Grenoble [3].

3.1 Static Measurements at ILL

Figure 11 shows the repetition measurement of the air level, with higher initial intensity and collimation, but no better quality than the measurement at ATI – which proves that due to the mosaicity of the crystal, the beam collimation is determined by the crystal size and distance to the sample.

Fig. 10. Repetition of air level at higher flux and initial collimation.

Figure 12 shows water drops first floating on the little hairs of a crude cardboard, then they touched the surface and spread out.

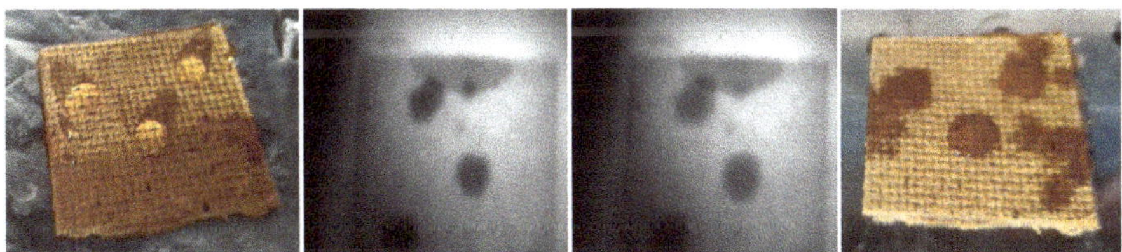

Fig. 11. Water drops first floating on the little hairs of a crude cardboard, then spreading out.

Figure 13 shows the Gadolinium line pair gauge in 3 cm distance, in direct contact with the detector, and in 1 cm distance from the detector when an oval 10 mm pinhole in borated rubber was placed over the crystal to reduce the virtual source size – which significantly increased the collimation and image sharpness.

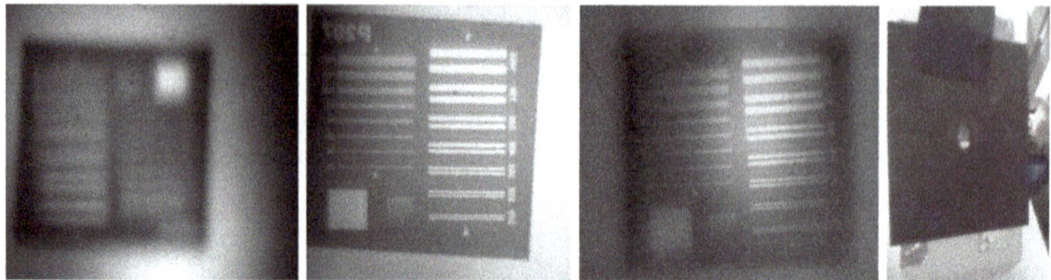

Fig. 12. Gadolinium line pair gauge (total size 30 mm, one line pair is 10 mm long) in 3 cm distance, in direct contact with the detector, and in 1 cm distance from the detector with an oval 10 mm pinhole in borated rubber (right image) placed over the crystal.

3.2 Dynamic Measurements at ILL

For dynamic measurements, a bubble machine was created using an aquarium air pump that blew air bubbles into a funnel and tube inside a closed water container that was connected by a tube to another, open container placed at a higher level, via a horizontal CPU water cooler made from aluminium. The collected fine bubbles in the funnel united as larger bubbles of random size and moved through the tube and the CPU cooler as a two-phase flow. Another tube connected both containers for pressure equalization (Fig. 14).

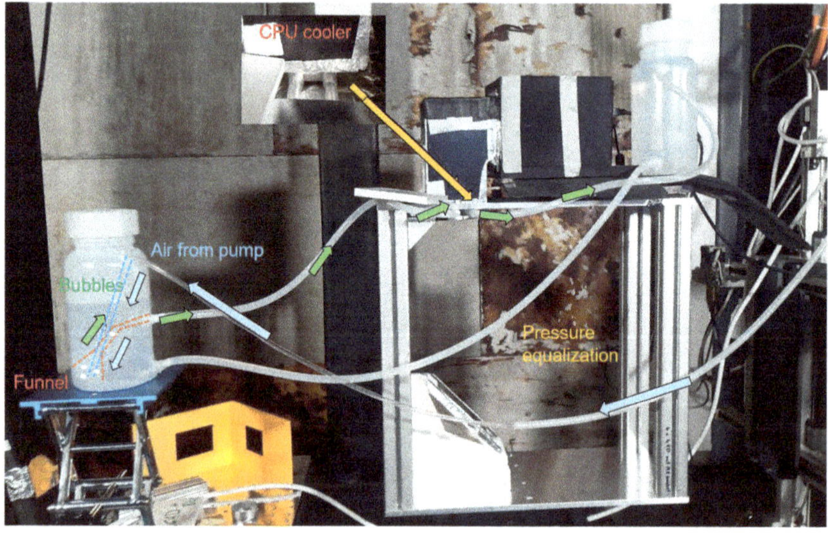

Fig. 13. The bubble machine installed with the vertical beam device and detector. A horizontal CPU water cooler is mounted in contact with the detector.

This setup was solely built to test time resolution and contrast, the bubbles in this cooler could also have been examined with a horizontal beam, and the cooler standing upright. The intensity even of the reflected beam was sufficient to record images with 300 ms, 200 ms, 100 ms and even 50 ms time resolution. Although the images are noisy due to insufficient neutron statistics, the flow of bubbles was clearly visible – which can only be depicted in static images here (Figs. 15).

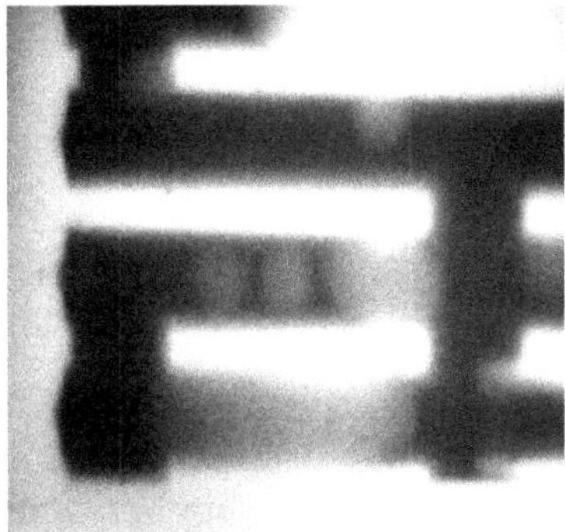

Fig. 14. Static image of water and bubbles in the CPU cooler.

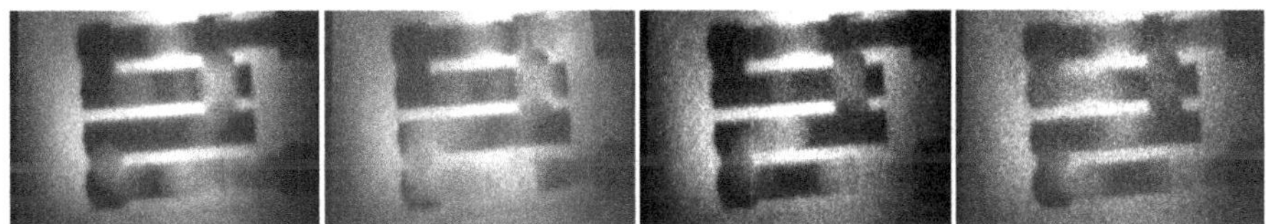

Fig. 15. Dynamic images of water and bubbles at 300 ms, 200 ms, 100 ms and 50 ms single exposure time.

3.3 Improvements of the Resolution

Since the size of the crystal acts as a secondary source with diameter 'D', the resolution can be improved by either reducing D or extending L. Further tests were done with a 15 x 15 mm^2 sized crystal with 0.7° mosaicity, which improved the resolution, at the cost of a smaller beam. Interesting enough, the intensity of the beam in the center was more than with using the 50 x 50 mm^2 3° mosaicity crystals – because the initial beam collimation was L/D = 315 (L = 35m, D = 35 mm), which corresponds to an incoming source angle of 0.18°, which is smaller than the 3° mosaicity – like the beam at Atominstitut with L/D = 125, corresponding to 0.46°, the ILL beam could not even illuminate the full mosaicity of the crystals.

The initial measurements were done with equipment that had to fit into a suitcase, hence the small scaffolding that resulted in such a small collimation. Further measurements were conducted with the detector on telescopic bars, increasing the distance L which also increased the collimation and resolution. For future experiments, it is easy to extend the detector distance L to two meters or more to get better collimation and resolution.

3.4 'Real' Use of the Vertical Beam

The experiments shown here were only for testing the method, and demonstration. Real scientific applications may be in the examination of melts, and two-phase flow. For thermal neutrons, copper crystals may be employed. ILL is currently planning to integrate a vertical beam option into the future thermal imaging facility at ILL.

Another possible application is water management in tilted fuel cells, at variable angles not necessarily 90 degrees – an electric car must be able to go uphill.

But a real 'killer application' has not been found yet.

Acknowledgements. Special thanks goes to Daniel Clemens of Helmholtz Zentrum Berlin for the crystals.

References

1. Al-Falahat, A.M., et al.: Energy-selective neutron imaging by exploiting wavelength gradients of double crystal mon-ochromators—Simulations and experiments. Nucl. Inst. Meth. A, vol. 943. https://doi.org/10.1016/j.nima.2019.162477
2. Hameed, F., Zawisky, M., Rohatsch, A.: Neutron tomography imaging of restored original samples from Austrian Cultural Heritage. In: Conference: 9th International Symposium on Image and Signal Processing and Analysis (ISPA 2015). https://doi.org/10.1109/ISPA.2015.7306042
3. Tengattini, A., Lenoir, N., Andò, E.G., B, Atkins, D., Beaucour, J., Viggiani, G.: NeXT-Grenoble, the Neutron and X-ray tomograph in Grenoble. Nucl. Inst. Meth. A, 968 09/2020, https://doi.org/10.1016/j.nima.2020.163939

Future from the Past: Long-Term Corrosion of Iron Analyzed Using Bimodal Neutron and X-ray CT

Ocson Cocen[1,3]($\boxtimes$) (iD), Mahdieh Shakoorioskooie[2] (iD), Elodie Granget[1] (iD),
David Mannes[2], Anders Kaestner[2] (iD), Jean-Marie Drezet[3], and Laura Brambilla[1] (iD)

[1] Haute Ecole Arc Conservation-restauration, HES-SO University of Applied Sciences and Arts Western Switzerland, Espace de l'Europe 11, 2000 Neuchâtel, Switzerland
ocson.cocen@he-arc.ch
[2] Laboratory for Neutron Scattering and Imaging, PSI Center for Neutron and Muon Sciences, Forschungsstrasse 111, 5232 Villigen, Switzerland
[3] Tribology and Interfacial Chemistry Group, Institute of Materials, SCI-STI-SM, École Polytechnique Fédérale de Lausanne, Station 12, 1015 Lausanne, Switzerland

Abstract. Advancements in analyzing iron archaeological artifacts are helpful in many applications, i.e., elucidating corrosion phenomena experienced by nuclear waste repositories and lifetime prediction of still-buried objects, such as pipelines. This paper investigates the corrosion state of iron-based archaeological objects through bimodal neutron and X-ray computed tomography, SEM-EDX analysis, and Raman spectroscopy. The nail BdC2, excavated from the archaeological site Bois de Châtel in Switzerland, is used as a representative case study. The workflow, starting from securing the sample from the archaeological site, beamtime experiment parameters, and data processing steps until segmentation results validation, is described in this paper. For the segmentation validation, SEM-EDX and Raman spectroscopy analyses were performed. The K-means segmentation algorithm produced satisfying results despite a mis-segmentation of one gap space between two parts of the nails. The work described in this paper is part of a larger analytical system development that aims to enable 3D physical/geometrical reconstruction and chemical composition identification and distribution within ferrous objects via bimodal neutron and X-ray computed tomography experiments, eliminating the need for invasive characterization methods.

Keywords: neutron computed tomography · X-ray computed tomography · multimodal imaging · iron corrosion · heritage conservation

1 Introduction

Environmental changes affect the corrosion of ferrous objects buried in the ground over centuries or millennia [1–3]. The resulting complex corrosion products are irreproducible through relatively short, simplified in-laboratory experiments [4, 5]. An alternative is to study corrosion phenomena experienced by iron archaeological artifacts (IAA) that

A. E. Craft and H. Z. Bilheux (Eds.): WCNR 2024, SPPHY 348, pp. 140–148, 2026.
https://doi.org/10.1007/978-3-032-15003-5_17

are analogs to long-term corrosion of other types of buried ferrous objects. However, even though many extensive (post-excavation) studies have been performed [6, 7], the long-term corrosion phenomena of IAA have not been comprehensively elucidated.

Post-excavation and invasive characterizations of IAA might capture non-pertinent information, as the corrosion layers might be reactive when exposed to a new environment, e.g., open-air post-excavation, and therefore, their physicochemical characteristics might change. Thus, there is a need for method development to capture objects' corrosion state in their buried condition, and the long-term corrosion phenomena should be elucidated based on these conditions.

Mannes et al. demonstrated capabilities of bimodal neutron and X-ray computed tomography (CT) analysis on a sword that allow discernment of organic parts, i.e., wooden, from metal parts, detailed 3D visualization of each of sword components, and distinguishment of corroded areas in the sword [8]. Further capabilities of bimodal neutron and X-ray CT were highlighted by Jacot-Guillarmod et al. They correlated and addressed possible corrosion products present in an archaeological iron nail by utilizing several oxides also oxyhydroxides as reference [9]. By combining complementary neutron-X-ray detection characteristics, bimodal neutron [10] and X-ray [11] computed tomography (CT) is an appropriate non-invasive technique that provides and displays physical and (a priori) chemical information of the corroded objects.

2 Material and Methods

2.1 Material

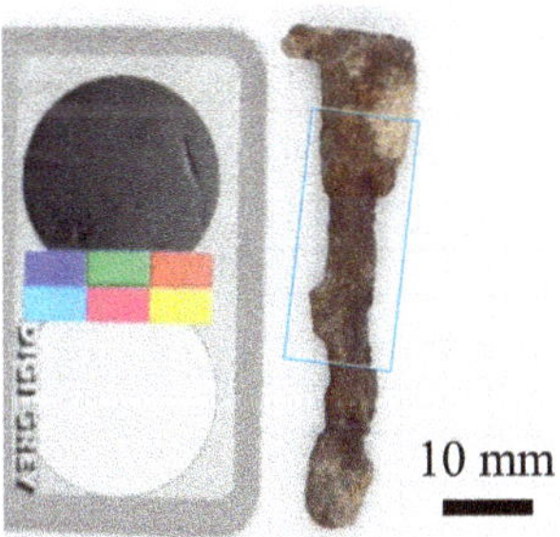

Fig. 1. Appearance of BdC2 sample. The blue rectangle indicates the sample part imaged by neutrons and X-rays.

The sample named BdC2, which is shown in Fig. 1, is one among the excavated IAAs from Bois de Châtel, one of the Site et Musée romains d'Avenches (SMRA) archaeological sites in Switzerland. All excavated IAAs were stored and transported anoxically, as described in Fig. 2: the IAA, found with the help of a metal detector, is collected using an aluminum tube to secure it together with original surrounding soil; both tube ends are then wrapped with aluminum foil before being inserted into a transparent, nitrogen-purged, sealed ESCAL bag; finally, inside a glovebox under nitrogen atmosphere, the IAA is transferred into a sealed, nitrogen-purged storage device.

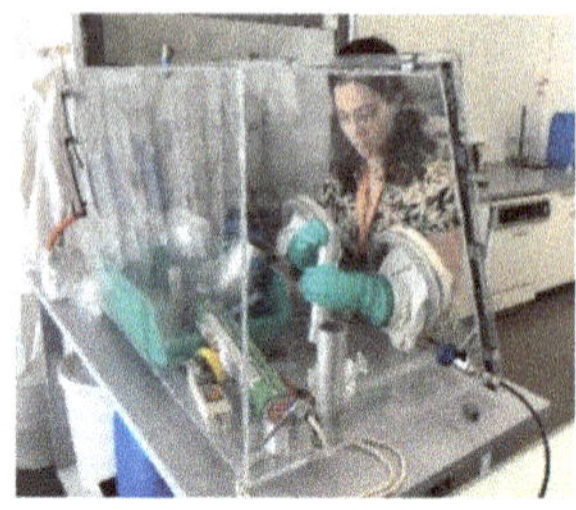

Fig. 2. Sample preparation workflow: (left) detectorist sought for possible IAAs, (middle) insertion of an aluminum tube into the ground to collect IAA and surrounding soil, (right) IAA was transferred into a bespoke storage device in a glove box under nitrogen atmosphere

Just before the CT experiments, the BdC2 sample was taken from the storage device, wrapped in aluminum foil, and put in a sample holder, as shown in Fig. 3.

2.2   Neutron and X-ray Computed Tomography Experiments

Fig. 3. Sample positioning on the ICON beamline

All CT experiments were performed at the Imaging with COld Neutron (ICON) beamline at Swiss Spallation Neutron Source (SINQ) at Paul Scherrer Institute in Switzerland [12]. Two tomography experiments, i.e., neutron and X-ray CTs, on BdC2, were taken in series, with some critical parameters listed in Table 1. The field of view was 27.6 mm x 27.6 mm.

Table 1. Neutron and X-ray CT experiment parameters

	Neutron CT	X-ray CT
Number of projections	1125/360°	1125/360°

(continued)

Table 1. (*continued*)

	Neutron CT	X-ray CT
Exposure time	50 s	15 s
Resulting pixel size	$(13.5)^2\ \mu m^2$	$(12.55)^2\ \mu m^2$

2.3 CT Data Processing and Validation

The neutron and X-ray tomography images were reconstructed by means of the MuhRec software [13] using the filtered back-projection algorithm for the parallel beam and the Feldkamp-Davis-Kress (FDK) algorithm, respectively. The FDK algorithm accounts for the cone-beam geometry and corrects the associated artifacts. The reconstructed neutron tomograms were then registered to X-ray tomograms using the SimpleElastix image registration library, applying an affine transform based on the mutual information similarity metric [14]. The resulting voxel size of the registered tomograms is $(12.55)^3\ \mu m^3$.

The segmentation is based on the K-means algorithm [15]. Various numbers of clusters were first used in the segmentation, i.e., four to eight clusters. We deemed that the five-cluster segmentation is a good example for BdC2 sample, thus, only the five-cluster segmentation results are presented in this paper. The 3D visualization was generated using Dragonfly software [16]. After CT experiments, BdC2 was cut for destructive analysis, and its cross-sections were examined via scanning electron microscopy (SEM), energy dispersive X-ray (EDX), and Raman spectroscopy.

3 Results and Discussion

Registration of reconstructed images of one modality to the other (neutron–X-ray) is one of the most pivotal milestones in our image processing workflow. Proper registration is an excellent basis for generating high-precision segmentation. As a qualitative registration checkpoint, a checkered image consisting of interlacing images from the registered neutron and X-ray tomograms was created, as shown in Fig. 4. The visible features from each modality are seamlessly interconnected, indicating that the registration was successful. These registered images were then segmented.

The segmentation produced five clusters: one dedicated to the background and four to the BdC2 bulk. The four clusters are named after the terminology described in the literature [17]: *M* for the remaining metallic substrate of BdC2; *DPL 1* and *DPL 2* for the dense product layers, the corrosion products contained in BdC2; *TM* for the transformed medium, a mixture of corrosion products and soil elements.

Figure 5 shows a slice of the BdC2 sample seen across various modalities. From the X-ray image, we could distinguish the metallic part much better (X-rays $\rightarrow$ *M*), while from the neutron image, the corrosion products and soil organic materials contrast more (neutrons $\rightarrow$ *DPL* and *TM*). After the beamtime experiment, a small chunk of the sample fell off; thus, only one big part of BdC2 was embedded in the resin, as shown in Fig. 5, instead of two parts as imaged in neutron and X-ray images.

TM is the outermost layer of the BdC2, as it is the interface between the environment (the soil) and the object itself. A mis-segmentation is observed (refer to red arrows in Fig. 5): A void between the two BdC2 parts is segmented into the *TM* cluster.

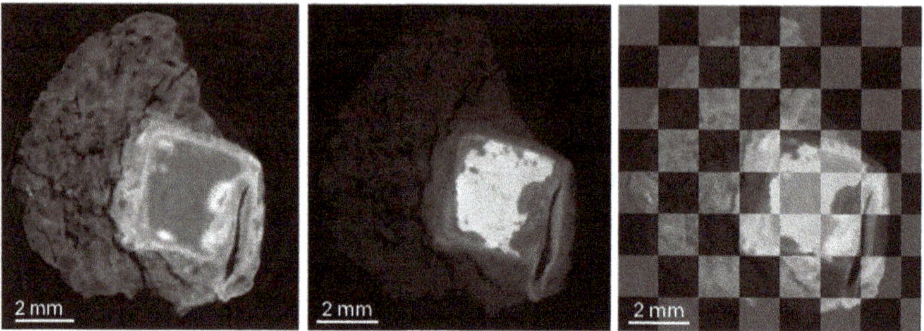

Fig. 4. Checkered image (right) consists of neutron image (left) and X-ray image (middle).

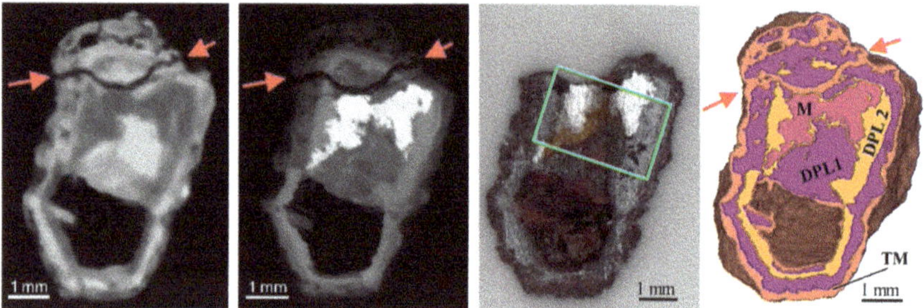

Fig. 5. Views of the same BdC2 cross-section in neutron image (left), X-ray image (middle left), camera photo (middle right, the green rectangle indicates the area analyzed by SEM-EDX), and segmented image (right). Red arrows point to a mis-segmentation; void was segmented as TM.

Figure 6 shows the SEM image and EDX maps with the distribution of three elements: iron, oxygen, and chloride. The top left corner in Fig. 6 is the embedding resin. Iron corrosion products often appear as compounds containing oxygen in the form of oxides, oxyhydroxides, or carbonates, and some may contain chloride—hydrogen and chloride attenuate neutrons more than X-rays. Therefore, the chloride distribution map explains why a bright area is observed in the BdC2's center area (*DPL1*) in the neutron image in Fig. 5.

Based on the oxygen distribution map in Fig. 6, we can categorize three different regions, namely: (1) oxygen-rich – intense blue, (2) low oxygen – mild blue, and (3) oxygen-depleted – black-colored. The oxygen-depleted region corresponds geometrically with the intense red area in the iron (Fe) distribution map, confirming that it is metallic iron (Fe^0), indicated as *M* in the segmentation. One of the two oxygen-containing regions exhibits more chloride, suggesting at least two different corrosion products are present: compounds with and without (or low) chloride, thus two *DPL*s in the segmentation.

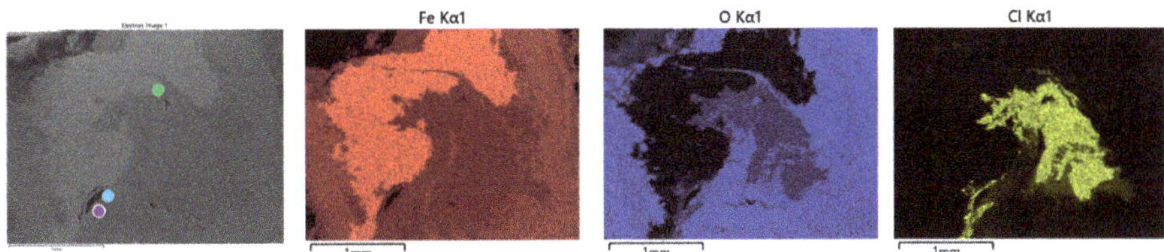

Fig. 6. SEM image (left, colored dots indicate spots analyzed using Raman spectroscopy) and EDX map on BdC2 cross-section: iron (middle left), oxygen (middle right), and chloride (right)

After SEM-EDX analysis, Raman spectroscopy was performed on the same cross-section. Three spot analyses are indicated by the three colored dots in Fig. 6 in the leftmost image. The same color legend is used for the curves in Fig. 7, which displays the Raman spectra and each matching compound from the Raman spectral database. All raw spectra were normalized for comparison displays, but no further adjustments were performed. Raman spectroscopy results indicate that the corrosion product in the chloride-rich region is akaganeite (β-FeOOH), and in the low-chloride region is magnetite (Fe_3O_4). The presence of hydrogen in akaganeite further explains the brighter area in the chloride-rich region in the neutron image.

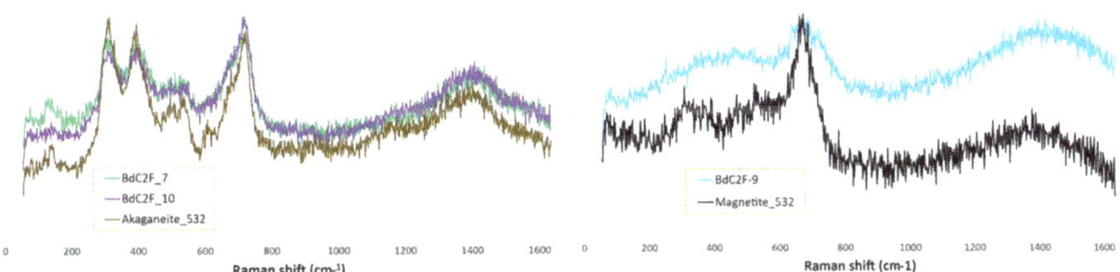

Fig. 7. Raman spectra on BdC2 cross-section

By calculating the mass percentage of Fe in akaganeite and magnetite, then looking at the distribution of iron in the EDX map, the segmentation of *M*, *DPL 1*, and *DPL 2* produced consistent results (refer to Fig. 5, Fig. 6, Fig. 7, and Table 2). The most intense, red-colored area in the EDX map is *M*. By omitting the embedding resin and cracks, the least intense Fe region belongs to akaganeite (*DPL 1*), and the second-least intense region belongs to magnetite (*DPL 2*).

Table 2. Mass percentage of iron in akaganeite and magnetite

Iron corrosion compound	wt.% of Fe
FeOOH (akaganeite, β-FeOOH)	62.9
Fe_3O_4 (magnetite)	72.4

4 Summary and Outlook

The archaeological nail, BdC2, was analyzed non-invasively utilizing bimodal neutron and X-ray tomography and invasively via SEM microscopy, EDX, and Raman spectroscopy. A complete insight into the object is obtained by exploiting the complementary nature of neutrons and X-rays. X-ray data emphasizes the remaining metallic iron in the object's center, while neutron data discerns more of the soil elements and iron corrosion products. Thus, every part of the object can be visualized and virtually accessed afterward. The invasive analyses supported and validated the observations derived from the neutron and X-ray tomography images.

The segmentation process needs further optimization, as there was a mis-segmentation. Such activity is envisaged for the future by trying other k numbers of clusters in the K-means segmentation algorithm and attempting segmentation with different methods/algorithms, e.g., the Gaussian Mixture Model.

This paper presents some parts of a larger process workflow shown in Fig. 8 with the dashed rectangle. The present data will feed a machine-learning model. Other works are ongoing in parallel, such as workflow automation for better reproducibility, gathering ground truth, and recording iron corrosion compounds' bimodal material signatures for calibration purposes. The overarching goal is to acquire 3D physical and chemical information without invasive analyses, only with bimodal neutron and X-ray computed tomography analysis.

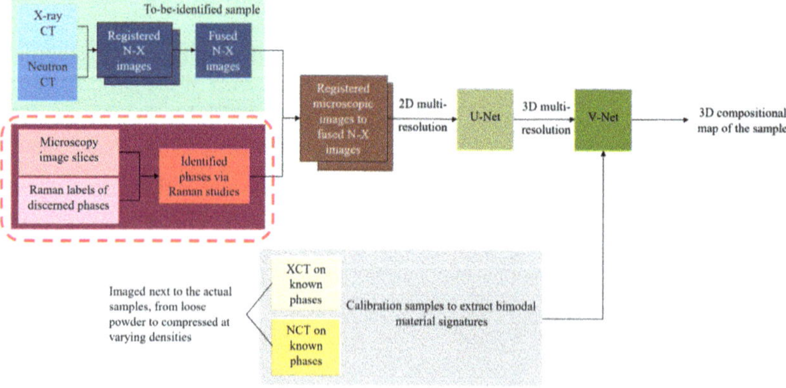

Fig. 8. Simplified version of the complete process workflow based on [18]

Acknowledgment. The authors thank Pierre Blanc and Myriam Krieg from the Site et Musée romains d'Avenches (SMRA), who helped provide all archaeological samples. We acknowledge the support of the Swiss National Science Foundation (SNSF) through the Sinergia Project CORINT (grant number 205883) in this work. This work is based on experiments performed at the Swiss Spallation Neutron Source (SINQ), Paul Scherrer Institute (PSI), Villigen, Switzerland.

Competing Interest. The authors declare that they do not have any competing interests.

References

1. Neff, D., Tidblad, J., Watkinson, D., Grassini, S.: Corrosion challenges towards a sustainable society: Protection of cultural heritage. Mater. Corros. **73**(11), 1745–1746 (2022). https://doi.org/10.1002/maco.202213140
2. Gerwin, W., Baumhauer, R.: Effect of soil parameter on the corrosion of archaeological metal find. Geoderma **96**, 63–80 (2000). https://doi.org/10.1016/S0016-7061(00)00004-5
3. Salem, Y., Oudbashi, O., & Eid, D. Characterization of the microstructural features and the rust layers of an archaeological iron sword in the Egyptian Museum in Cairo (380–500 A.D.). Herit. Sci., 7(1), 1–12 (2019). https://doi.org/10.1186/s40494-019-0261-2
4. Saheb, M., Neff, D., Foy, E., Gallien, J.-P., Dillmann, P.: An analytical methodology for the study of the corrosion of ferrous archaeological remains in soils. Conservation Manage. Archaeological Sites **14**(1–4), 16–27 (2012). https://doi.org/10.1179/1350503312z.0000000002
5. Terryn, H.: Corrosion challenges towards a sustainable society: Finite element modelling and artificial intelligence for corrosion prediction the way to go **73**(11), 1746–1747 (2022). https://doi.org/10.1002/maco.202213140
6. Neff, D., Saheb, M., Monnier, J., Perrin, S., Descostes, M., L'Hostis, V., et al.: A review of the archaeological analogue approaches to predict the long-term corrosion behaviour of carbon steel overpack and reinforced concrete structures in the French disposal systems. J. Nucl. Mater. **402**(2), 196–205 (2010). https://doi.org/10.1016/j.jnucmat.2010.05.003
7. Kibblewhite, M., Tóth, G., Hermann, T.: Predicting the preservation of cultural artefacts and buried materials in soil. Sci. Total. Environ. **529**, 249–263 (2015). https://doi.org/10.1016/j.scitotenv.2015.04.036
8. Mannes, D., Schmid, F., Frey, J., Schmidt-Ott, K., Lehmann, E.: Combined neutron and X-ray imaging for non-invasive investigations of cultural heritage objects. Phys. Procedia **69**, 653–660 (2015). https://doi.org/10.1016/j.phpro.2015.07.092
9. Jacot-Guillarmod, M., Schmidt-Ott, K., Mannes, D., Kaestner, A., Lehmann, E., Gervais, C.: Multi-modal tomography to assess dechlorination treatments of iron-based archaeological artifacts. Heritage Sci. **7**, 29 (2019). https://doi.org/10.1186/s40494-019-0266-x
10. Neutron imaging at the spallation source (SINQ) information for potential users and customers. Paul Scherrer Institute (2016). https://www.psi.ch/sites/default/files/import/niag/ImagingBrochureEN/Neutron_Imaging_User_2016.pdf
11. Bernabale, M., Cognigni, F., Nigro, L., Rossi, M., de Caro, T., De Vito, C.: A comprehensive strategy for exploring corrosion in iron-based artefacts through advanced Multiscale X-ray Microscopy. Sci. Rep. **12**(1), 6125 (2022). https://doi.org/10.1038/s41598-022-10151-w
12. Kaestner, A.P., Hartmann, S., Kühne, G., Frei, G., Grünzweig, C., Josic, L., et al.: The ICON beamline – a facility for cold neutron imaging at SINQ. Nucl. Instrum. Methods Phys. Res., Sect. A **659**(1), 387–393 (2011). https://doi.org/10.1016/j.nima.2011.08.022
13. Kaestner, A.P.: MuhRec—a new tomography reconstructor. Nucl. Instrum. Methods Phys. Res., Sect. A **651**(1), 156–160 (2011). https://doi.org/10.1016/j.nima.2011.01.129

14. Shakoorioskooie, M., Griffa, M., Leemann, A., Zboray, R., Lura, P.: Alkali-silica reaction products and cracks: X-ray micro-tomography-based analysis of their spatial-temporal evolution at a mesoscale. Cem. Concr. Res. **150**, 106593 (2021). https://doi.org/10.1016/j.cemconres.2021.106593
15. Ikotun, A.M., Ezugwu, A.E., Abualigah, L., Abuhaija, B., Heming, J.: K-means clustering algorithms: A comprehensive review, variants analysis, and advances in the era of big data. Information Sciences, 622, 178–210 (2023). https://doi.org/10.1016/j.ins.2022.11.139
16. Dragonfly [computer software], version 2022.2 for Windows. Comet Technologies Canada Inc., Montreal, Canada; software available at https://www.theobjects.com/dragonfly
17. Neff, D., Vega, E., Dillmann, P., Descostes, M., Bellot-gurlet, L., Béranger, G.: Contribution of iron archaeological artefacts to the estimation of average corrosion rates and the long-term corrosion mechanisms of low-carbon steel buried in soil. In: Dillmann, P., Béranger, G., Piccardo, P., Matthiesen, H. (eds.) Corrosion of Metallic Heritage Artefacts, pp. 41–76 (2007). Woodhead Publishing. https://doi.org/10.1533/9781845693015.41
18. Granget, E., Cocen, O., Shakooriokooie, M., Zhan, Q., Lumongsod-Thompson, M. N., Kaestner, A., et al.: Developing a quantitative multimodal and multi-scale, fully non-destruc-tive technique for the study of iron archaeological artefacts. Acta IMEKO, 13(2), Article 2 (2024). https://doi.org/10.21014/actaimeko.v13i2.1799

2D and 3D Visualization of Localized Plastic Deformation

Khanh Van Tran[1,2,3], Thawatchart Chulapakorn[4,5(✉)], Nikolay Kardjilov[2], Henning Markötter[2,6], Stephen A. Hall[4], Ingo Manke[1,2], and Robin Woracek[4(✉)]

[1] Technische Universität Berlin, Straße des 17. Juni 135, 10623 Berlin, Germany
[2] Helmholtz-Zentrum Berlin, Hahn-Meitner-Platz 1, 14109 Berlin, Germany
[3] Thuyloi University, 175 Tay Son, Dong Da, Hanoi, Vietnam
[4] Division of Solid Mechanics, Lund University, 221 00 Lund, Sweden
{thawatchart.chulapakorn,robin.woracek}@ess.eu
[5] European Spallation Source ERIC, 221 00 Lund, Sweden
[6] Bundesanstalt für Materialforschung und -Prüfung (BAM), Unter Den Eichen 87, 12205 Berlin, Germany

Abstract. This study introduces a novel contrast modality in neutron imaging, enabling the visualization of plastically deformed zones within ARMCO® iron. Utilizing differences in transmitted neutron intensity, we exploit variations in Bragg diffraction strength between deformed and non-deformed areas. Our findings reveal that deformed regions exhibit stronger diffraction due to extensive sub-grain divisions and angular spreading caused by plastic deformation. This contrast effect, attributed to extinction phenomena, is particularly evident in ARMCO® iron. We demonstrate the applicability of this technique through detailed 2D and 3D imaging of tensile and deep drawing samples. In tensile samples, the contrast effectively highlights regions of localized plastic deformation and Lüders band formation. For deep-drawing samples, the technique captures significant deformation patterns around the punch area and differentiates between the inner and outer surfaces of the sheet metal. The contrast modality is another step forward in fully exploiting and unraveling the rich information in transmission based neutron imaging and enables full-field characterization for samples up to tens of centimeters in size.

Keywords: Plastic deformation · Plastic localization · ARMCO · Neutron tomography · Extinction

1 Introduction

The deformation of metallic materials and the study of deformation microstructures have a long history in metallography and remain critical for improving structures and developing new materials. During plastic deformation, line defects (dislocations) are introduced into the lattice, organizing themselves into ordered patterns that govern most mechanical, functional, and kinetic properties of metallic alloys. Line profile analysis, based on X-ray or neutron diffraction [1] and synchrotron based 3D techniques [2–7] have provided capabilities for *in-situ* analysis of deformation structures.

© The Author(s) 2026
A. E. Craft and H. Z. Bilheux (Eds.): WCNR 2024, SPPHY 348, pp. 149–157, 2026.
https://doi.org/10.1007/978-3-032-15003-5_18

In this paper, we report experimental observations that reveal plastically deformed areas in centimeter-sized samples made from ARMCO® technically pure iron. The underlying contrast mechanism, that we attribute to extinction (and multiple scattering) [8], will be discussed elsewhere. Here, we focus on exploiting this contrast feature to expand the characterization of deformation microstructures to macroscopic and representative length scales of 'real world' components. The latter refers, for example, to sample sizes that are typical for standardized mechanical tests, such as tensile, compression, and bending tests, commonly performed in industrial and research laboratories. These tests often involve specimens with dimensions on the order of centimeters, conforming to international standards like ASTM or ISO, which ensure consistent and comparable mechanical properties across different materials and applications. Neutron imaging can capture key features of deformation, including strain localization (via Bragg edge analysis) and microstructural changes/dislocation patterns (introduced in this work), within these larger test specimens. This capability bridges the gap between microscale investigations, using electron or synchrotron X-ray methods, and the bulk behavior of materials as observed in engineering-scale components. We propose the term 'NIODS' (Neutron Imaging of Deformation Structures) for this application. This technique is straightforward to implement at most neutron imaging facilities. It leverages intensity difference in transmitted neutrons due to diffraction intensity variations, which depend on the level of sub-grain divisions due to plastic deformation.

We show how the NIODS technique can be used to follow the appearance and development of deformation structures, specifically the dense network of dislocations during tensile deformation. In this context, the onset of plastic deformation and the formation of Lüders bands can be followed *in-situ*. Also, we demonstrate the application of NIODS to sheet metal deformed by the Erichsen deep-drawing (Erichsen cupping) test, which is used to determine the stretch-forming capacity of sheet metals in accordance with DIN EN 10139 and DIN EN 10130.

Following the work of H. Sato *et al.* [9], which addresses extinction as well as forward-scattering artifacts arising from the sample-to-detector distance, the present work raises further awareness of this effect and paves the way for its application in spectral neutron imaging [10]. This includes potential combinations with diffraction contrast [11] or inelastic scattering contrast, which can be utilized for temperature mapping [12–14].

2 Observations in Plastically Deformed Tensile Samples Made from ARMCO

Tensile samples with a notched geometry were prepared from commercial ARMCO® iron (composition in wt %: 0.01% C, 0.01% Si, 0.059% Mn, <0.01% P, <0.010% S, 0.02% Cr, <0.005% Mo, 0.038% Ni, 0.013% Al, balance Fe) using water jet cutting. The three samples presented in this section were normalized at 950 °C for 30 min under nitrogen atmosphere and then furnace-cooled at a rate of 5 °C/min. The macroscopically observable upper yield point was 165 MPa, while the lower yield point was 145 MPa. These three samples were deformed under tensile loading in a custom-built tensile machine [15] under the following conditions:

- Sample (i): Maximum macroscopic stress of 160 MPa, before reaching the upper yield point, with a deformation rate of 1.2 µm/min.
- Sample (ii): Maximum macroscopic stress of 160 MPa, before reaching the upper yield point, with a deformation rate of 200 µm/min.
- Sample (iii): Maximum deformation of 5500 µm (stress of 280 MPa) with a deformation rate of 200 µm/min.

Neutron radiography was performed at the CONRAD-2 beamline [16, 17], formerly located at Helmholtz Zentrum Berlin (HZB), using a 1 cm diameter pinhole, corresponding to a L/D value of 1200. The radiography was conducted at two sample-to-detector distances (0 cm and 33 cm) with an exposure time of 60 s. The shorter distance was selected to achieve optimal spatial resolution, while the larger distance was chosen to enhance polychromatic in-line phase contrast. The latter has been demonstrated in the past to be effective in revealing localized plastic deformation resulting from armourers' marks in metallic cultural heritage objects [18].

Neutron transmission images of all three samples, taken at both sample-to-detector positions, are presented in Fig. 1. The figure also includes the von Mises stress distribution predicted by a Finite Element Model (FEM) for the two deformation states. The experimental data shows a clear and distinct contrast variation across all samples and both distances. Neutron transmission is reduced in areas subjected to higher mechanical stress, such as the notched gauge section. It should be noted that sample (iii) exhibits higher transmission in the notched gauge section that can be attributed to visible necking (thinning) of the sample.

In addition to the notched gauge section, samples (i) and (ii) also display darker contrast around the holes where the samples were secured with hardened pins. Stress concentration is expected around the pins, as indicated by the FEM, with exact localization dependent on the contact interface between the pin and hole. The darker contrast in the notched region for samples (i) and (ii) extends beyond the notch, corresponding to the FEM-predicted stress distribution. The experimental data also reveals an inclined angle, which can be attributed to the formation of Lüders bands, a typical feature that can be expected for ARMCO® when transitioning from elastic to plastic deformation.

The darker region observed in the upper grip area of sample (ii) is due to the presence of aluminum tape used to secure the sample during measurement. Samples (i) and (ii) were loaded to the same target stress at different deformation rates and display similar contrast, with only slight differences in the exact shape of the Lüders bands and grip areas. Sample (iii), which was subjected to more severe deformation, shows darker contrast throughout its entire length. Aside from the thinned center section, only certain regions near the grips remain in brighter contrast. These presumably non-plastically deformed areas align with the regions of lowest stress, as predicted by FEM.

The images of samples measured at the two sample-to-detector distances appear marginally different at first, with bright-to-dark contrast dominating in both cases. As expected, the sample edges appear darker due to refraction (commonly referred to as 'edge enhancement' [19]) at the larger distance. The interface regions between dark and bright areas become more distinct at the larger distance, showing a line of higher transmission at the interface. This effect is even more pronounced in the sample discussed in the next section, where bright lines at the interfaces between dark and bright areas are distinctly visible, indicating the boundary between plastically deformed and non-plastically deformed zones. This contrast has been used to study cultural heritage objects, often made from iron or low-carbon steel [18]. The results presented here aim to further support the interpretation of similar investigations in the future.

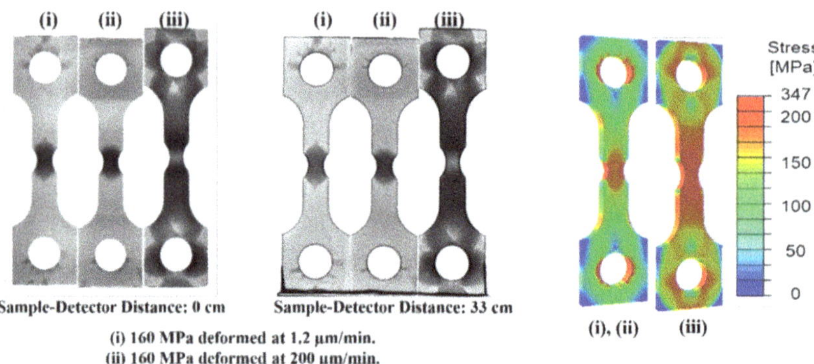

Fig. 1. Neutron radiographs of samples loaded to: (i) 160 MPa with a deformation rate of 1.2 μm/min, (ii) 160 MPa with a deformation rate of 200 μm/min, and (iii) a displacement of 5500 μm with a deformation rate of 200 μm/min. The left and middle panels correspond to sample-to-detector distances of 0 cm and 33 cm, respectively. The right panel shows von Mises stress distributions obtained from Finite Element modeling, corresponding to a macroscopic stress of 160 MPa and more severe plastic deformation.

3   In-Situ Tensile Deformation of Samples Made from ARMCO

To investigate the previously observed contrast and its development in greater detail, an in-situ tensile deformation experiment was conducted. The tensile sample was prepared from ARMCO® iron in the same manner as described earlier, though this sample was not annealed beforehand. The same custom-built tensile machine was used with a deformation rate of 200 μm/min. At selected intervals, the sample was unloaded to the nominal seating load of 20 MPa, as depicted in the stress-deformation curve in Fig. 2. Neutron transmission images were captured at eleven specific deformation states, with exposure times of 60 s, at a sample-to-detector distance of 33 cm on the CONRAD-2 beamline, and an L/D ratio of 1200 as in the previous section. The 11 neutron images are shown in Fig. 3, where the top row presents images taken after the applied load was released following each deformation step.

At 120 MPa, which is still well below the macroscopic elastic limit, a region in the center gauge section becomes less transparent (Step 2). Since this contrast persists when unloading to 20 MPa (Step 3), it can be inferred that the contrast is due to plastic deformation. As the sample is further loaded, the contrast becomes more pronounced, eventually filling the narrow gauge section (Steps 6 and 7). This observation aligns with the expected stress concentration predicted by the Finite Element Model (FEM), where the sides of the notched gauge section initially experience the highest stress, serving as starting points for plastic flow. The dark contrast then expands outward from the gauge area, forming a distinctly inclined band that we interpret as a Lüders band, as discussed in the previous section. On the contrary to the two inclined angles observed in the previous samples, here a single band is seen. It should also be noted that this band remains in the same position between deformation Steps 8 and 10, after which the sample begins to neck.

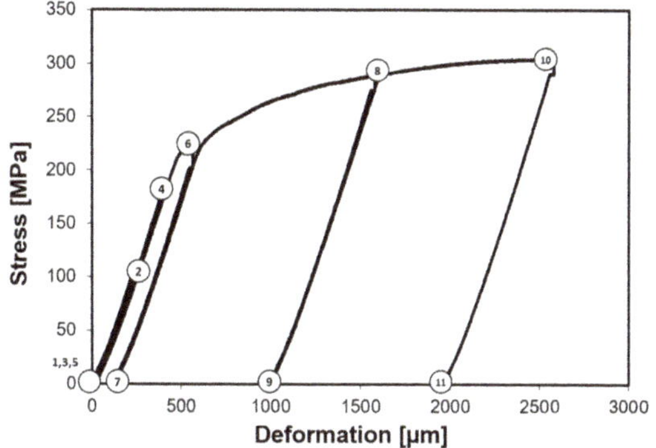

Fig. 2. Stress-deformation curve for the non-annealed ARMCO® sample presented in Sect. 3. Neutron radiographs were captured at the eleven indicated points along the curve.

A striking observation is the difference between the measurements taken with and without an applied elastic load on the sample. The unloaded results reveal the previously described bright line between the interfaces of plastically deformed and non-plastically deformed zones. However, this line is not visible when the elastic load is applied (e.g., comparing Step 6 vs. 7, Step 8 vs. 9, and Step 10 vs. 11). This finding warrants a detailed follow-up study.

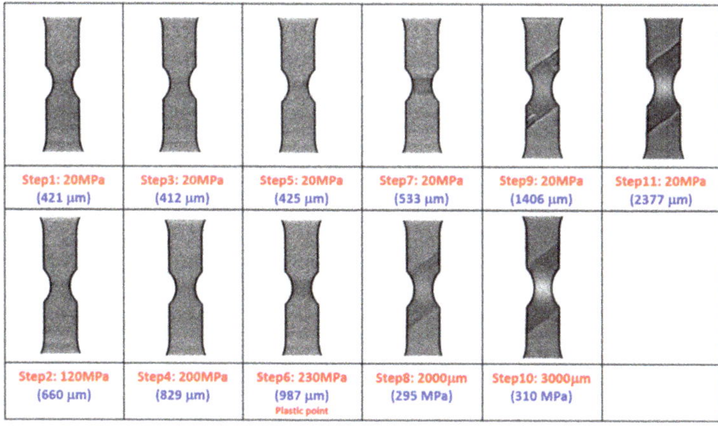

Fig. 3. Neutron radiographic images during interrupted *in-situ* tensile-loading.

4   3D Visualization of Plastically Deformed Deep–Drawing Samples

To study the applicability of the NIODS contrast, three sheet samples of ARMCO® iron with a thickness of 4 mm were deformed using the Erichsen cupping test. The samples, referred to as T04, T02, and T03, were deformed under the following conditions:

- Sample T04: Normalized at 950 °C for 30 min and then furnace-cooled; subjected to the Erichsen cupping test with a maximum load of 32 kN, resulting in a displacement of 18.6 mm.
- Sample T02: Normalized at 950 °C for 30 min and then furnace-cooled; subjected to the Erichsen cupping test with a maximum load of 35 kN, resulting in a displacement of 15.0 mm.
- Sample T03: Not annealed prior to deformation; subjected to the Erichsen cupping test with a maximum load of 31 kN, resulting in a displacement of 18.3 mm.

The deformed samples are shown in Fig. 4. Neutron tomography was performed at the CONRAD-2 beamline using a 3-cm pinhole (L/D = 400). A total of 400 projections were recorded, with exposure times of 10 s for each projection. The samples were positioned 6 cm from the detector, and all samples were stacked on top of one another and rotated in the horizontal plane. The effective pixel size of the reconstructed images is 135 μm/pixel. The reconstructed tomography data is presented in Figs. 5–7.

Figure 5 displays the surface of the samples along with their cross-sectional views. The circumferential cracks near the die in samples T04 and T03 are clearly visible. It is observed that the region around the die exhibits the highest neutron attenuation (indicated by a red tone). The strength of the attenuation varies with the actual punching loads.

The largest deformation typically occurs in the region directly beneath the punch, particularly around the area of the sheet metal that is stretched into the die. This region, known as the punch nose radius and the bottom of the cup, experiences the most significant biaxial tensile stress, resulting in the highest degree of plastic deformation. The area under the punch exhibits higher neutron attenuation, indicating larger plastic deformation.

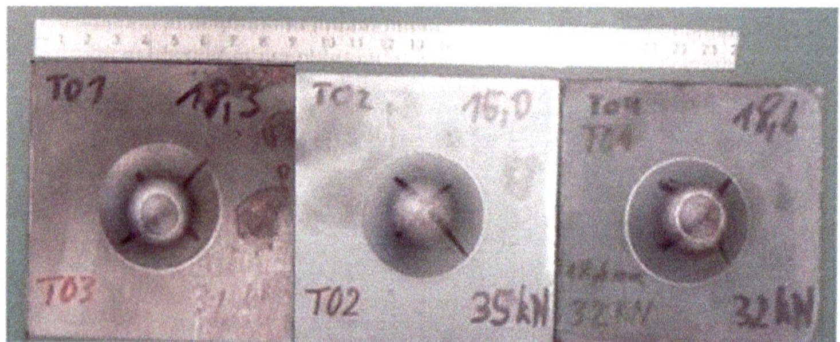

Fig. 4. Deep drawing samples: T03, T02, and T04 after loaded at 32, 35, and 31 kN, respectively.

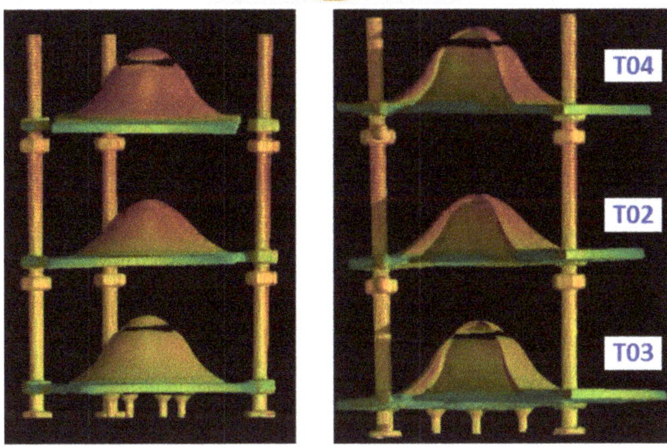

Fig. 5. 3D reconstructed images for the T04, T02, and T03 samples, displayed in both surface view and cross-sectional view.

Figure 6 presents top-view slices from various heights of the deep–drawing samples. Just outside the immediate contact area, within the transition zone, the metal is subjected to significant tensile forces, contributing to the overall deformation. However, the degree of deformation decreases toward the flange, which remains relatively flat and clamped. This general trend is also reflected in the neutron data, indicating reduced plastic deformation in these regions.

Notably, there is a distinct difference between the inner surface (in contact with the punch) and the outer surface (in contact with the die) of the sheet metal. During the Erichsen cupping test, the outer surface experienced more tensile stress and stretching, while the inner surface faced compressive stresses and frictional forces. Figure 7 provides a close-up view of the center cross-section of the three samples, facilitating detailed observation of the described zones.

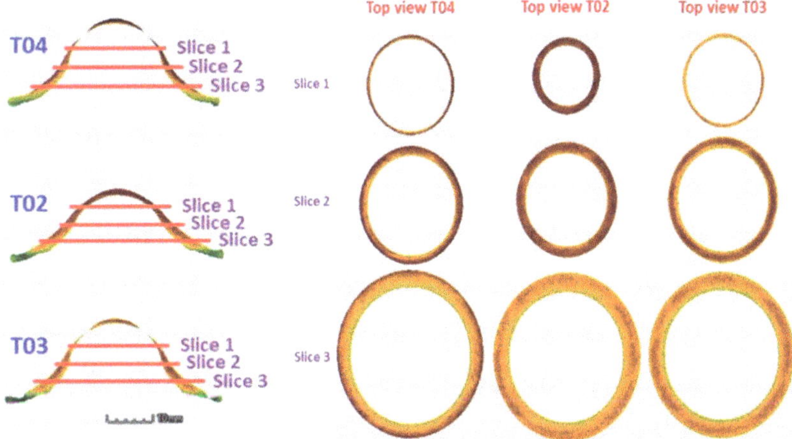

Fig. 6. Cross-section view from top at different heights between die and flanges.

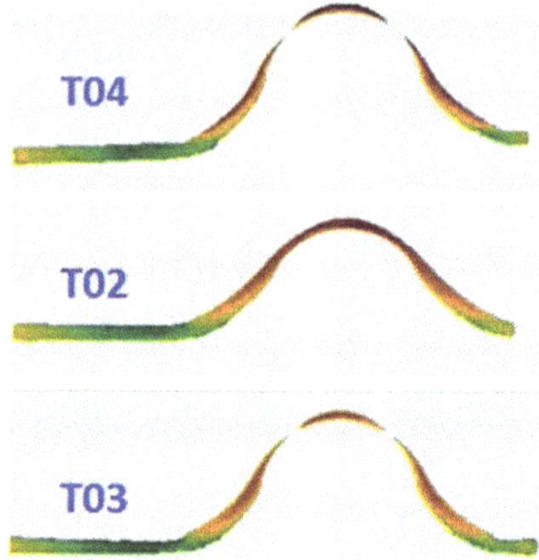

Fig. 7. The three samples virtually cut in the center.

5 Conclusion and Outlook

We have demonstrated a largely overlooked contrast modality in neutron imaging that enables the visualization of plastically deformed zones within ARMCO® iron. ARMCO® iron, characterized by its relatively large grains (average grain size of 70 μm), undergoes significant sub-grain divisions with angular spreading during plastic deformation. We attribute the observed differences between non-deformed and deformed areas to variations in the strength of Bragg diffraction, influenced by the extent of sub-grain divisions and angular misorientations induced by plastic deformation. Specifically, deformed areas exhibit stronger diffraction compared to their non-deformed counterparts, a phenomenon already confirmed by accompanying measurements and additional experiments conducted by the authors. We hence attribute the contrast variation primarily to extinction (note that extinction refers to the strength of the diffraction strength, not the transmitted intensity), a behavior previously reported for ARMCO® iron in earlier studies [8].

It is important to note that forward scattering effects, sample-to-detector distance, and related quantification methods in neutron imaging are well-documented in the literature [9, 20, 21]. While these factors do not explain the specific differences observed in transmitted neutron intensities, they will be essential for developing a quantitative description of the observed contrast. Qualitatively, however, this contrast modality enables effective 2D and 3D visualization of areas undergoing plastic deformation, demonstrating its applicability to both tensile and deep-drawing samples made from ARMCO® iron.

Potential applications related to ARMCO® iron include studies of its magnetic properties, which are directly influenced by its microstructure and residual stress [22]. Further investigations will also explore the applicability of the NIODS technique to other materials and microstructures.

Acknowledgements. We would like to thank B. Pfretzschner and the Bundesanstalt für Materialforschung und -Prüfung (BAM) for their assistance in preparing samples using the Erichsen cupping test. K.V. Tran expresses gratitude to Thuyloi University for their support throughout the project. The authors also wish to extend their appreciation to J. Santisteban, S. Vogel, D. Di Julio, and I.C. Noyan for their fruitful discussions and valuable suggestions regarding the topic of extinction in neutron imaging experiments.

References

1. Ungár, T., Ott, S., Sanders, P., Borbély, A., Weertman, J.: Acta Mater. **46**, 3693–3699 (1998)
2. Larson, B., Yang, W., Ice, G., Budai, J., Tischler, J.: Nature **415**, 887–890 (2002)
3. Jakobsen, B., et al.: Science **312**, 889–892 (2006)
4. Simons, H., et al.: Nat. Commun. **6**, 1–6 (2015)
5. Jakobsen, A., et al.: J. Appl. Crystallogr. **52**, 122–132 (2019)
6. Simons, H., et al.: Nat. Mater. **17**, 814–819 (2018)
7. Wang, L., Li, M., Almer, J., Bieler, T., Barabash, R.: Front. Mater. Sci. **7**, 156–169 (2013)
8. Weiss, R.: Phys. Rev. **86**, 271 (1952)
9. Sato, H., et al.: Phys. Procedia **88**, 322–330 (2017)
10. Tran, K.V., et al.: Materials Today Advances **9**, 100132 (2021)
11. Woracek, R., Santisteban, J., Fedrigo, A., Strobl, M.: Nuclear Instruments and Methods in Physics Research Section A: Accelerators. Spectrometers, Detectors and Associated Equipment **878**, 141–158 (2018)
12. Sato, H., Miyoshi, M., Ramadhan, R.S., Kockelmann, W., Kamiyama, T.: Sci. Rep. **13**, 688 (2023)
13. Jamro, R., et al.: J. Phys: Conf. Ser. **2605**, 012026 (2023)
14. Al-Falahat, A.M., et al.: J. Appl. Crystallogr. **55**, 919–928 (2022)
15. Woracek, R., et al.: J. Appl. Phys. **109** (2011)
16. Kardjilov, N., Hilger, A., Manke, I., Woracek, R., Banhart, J.: J. Appl. Crystallogr. **49**, 195–202 (2016)
17. Kardjilov, N., et al.: J. Imaging **7**, 11 (2021)
18. Williams, A., Edge, D., Grazzi, F., Kardjilov, N.: Stud. Conserv. **64**, 10–15 (2019)
19. Lehmann, E., Schulz, M., Wang, Y., Tartaglione, A.: Phys. Procedia **88**, 282–289 (2017)
20. Kang, M., et al.: Nuclear instruments and methods in physics research section a: accelerators. Spectrometers, Detectors and Associated Equipment **708**, 24–31 (2013)
21. Carminati, C., et al.: PLoS ONE **14**, e0210300 (2019)
22. AKSteel, ARMCO® PURE IRON - Product Data Bulletin (2024)

In-Situ Neutron Radiography with Hydrogenated Tensile Samples in the INCHAMEL Facility

Sarah Weick[1]([✉]), Mirco Grosse[1], Conrado Roessger[1], Martin Steinbrueck[1], and Anders Kaestner[2]

[1] Institute for Applied Materials - Applied Materials Physics, Karlsruhe Institute of Technology, 76344 Eggenstein-Leopoldshafen, Germany
sarah.weick@kit.edu
[2] Laboratory for Neutron Scattering and Imaging, Paul Scherrer Institute, 5232 Villigen, Switzerland

Abstract. In Light Water Reactors, fuel rod cladding tubes made of zirconium alloys undergo an oxidation process where a certain fraction of hydrogen may penetrate into the cladding. Hydrogen precipitates as hydride in the metallic matrix during cooling in the post-operation phase. An applied tensile force in the elastic range increases the hydrogen solubility and diffusivity and thus may initialise a reorientation process of hydrides that could be destructive for the metal. During long-term dry storage conditions, the stress influence in the elastic range on the zirconium claddings seems to be relevant, but not quantified yet.

This paper presents in-situ neutron radiography experiments of hydrogenated zirconium tensile samples under the influence of different elastic tensile stresses. The experiments were conducted during the commissioning of the In-situ Neutron radiography CHAmber for tests under MEchanical Load facility, a tensile testing machine with an inductive heating system. The stress influence on hydrogen movements in the elastic range for short time frames (hours) was shown to be negligible. Only under the influence of plastic tensile stresses, hydrogen movements from regions with lower to higher stress were observed. Nevertheless, the first results show that the new facility can be successfully used for investigations of the influence of elastic stresses on the hydrogen diffusion and solubility in zirconium. However, the duration of the experiments and/or the resolution during the neutron imaging have to be optimised.

Keywords: Neutrons · Imaging · Hydrogen · In-situ · Radiography · Diffusion · Stress · Zirconium

1 Introduction

Zirconium alloys that contain more than 98% zirconium are used as cladding tube material for the nuclear industry because of their low neutron absorption, a good corrosion resistance and strength during and after nuclear operation, which ensures a secure covering of the pellets. In Light Water Reactors (LWR), cladding tubes are in direct contact

© The Author(s) 2026
A. E. Craft and H. Z. Bilheux (Eds.): WCNR 2024, SPPHY 348, pp. 158–167, 2026.
https://doi.org/10.1007/978-3-032-15003-5_19

to the surrounding cooling water, which leads to a formation of oxides due to the oxidation of zirconium by water. During the oxidation, a certain fraction of the produced free hydrogen may penetrate into the cladding material where it remains in solution as long as the terminal solid solubility limit for precipitation (TSSp) of hydrogen is not exceeded [1]. When the TSSp is reached, hydrogen precipitates in form of zirconium hydrides (Fig. 1), which have a detrimental effect on the mechanical properties of the metal. The actual impact depends on several factors like the hydride orientation, length, connectivity, etc. Further, cladding tubes are exposed to strain from within, because of the pressurization of fuel rods by helium and fission gases that increase the inner pressure.

One of the most promising analytical methods to study the hydrogen diffusion in zirconium and zirconium containing cladding tubes is neutron imaging, as intensively used in many works, e.g. from Gong et al. [2] and Grosse et al. [3, 4]. Multiple advantages come along with this analytical method. It is a non-destructive method, even small amounts of hydrogen are detectable, it is spatially resolvable, the sample's hydrogen content is three-dimensionally resolvable with neutron tomography, and in case of metals with low neutron cross sections it can visualize the diffusion dynamics in-situ. Because of the very low neutron cross section of zirconium, the metal is nearly invisible for neutrons in contrast to hydrogen that scatters neutrons strongly and thus appears darker in neutron images.

The experimental aim of this study is to investigate the hydrogen diffusion into zirconium in dependence of temperature and time under the influence of an elastic strain. The experiments help to gather information about the long-term behaviour of spent nuclear fuel during interim dry storage. During this phase, the cladding tubes are exposed to several stresses like elastic stresses, thermal stresses, and creep. Internal stresses within cladding tubes result from inner pressures of 70 to 150 bar that are caused mainly by the formation of gaseous fission products.

Fig. 1. Optical micrograph showing predominantly circumferential oriented zirconium hydrides in a Zircaloy-4 cladding tube [5].

2 Materials and Measurements

2.1 Sample Preparation

Flat zirconium tensile samples (Fig. 2) with a varying rectangular cross section (from 2 x 3.5 mm^2 to 6 x 3.5 mm^2) in axial direction were eroded from a 3.5 mm thick Zr-plate (99.2%). Their geometry provides an internal stress gradient if a tensile force is applied. Two holes at the sample's heads are used to mount the sample to the sample holder extension. Only the sample's bar remains visible for neutrons and thus can be monitored in-situ during the experiment.

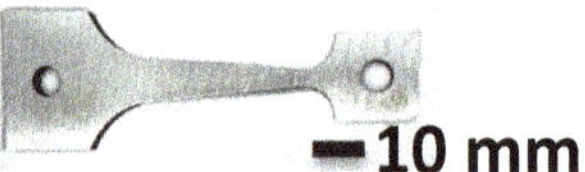

Fig. 2. Zirconium flat tensile sample with its smallest cross section of 2 x 3.5 mm^2 and its largest cross section of 6 x 3.5 mm^2.

2.2 Hydrogenation

The samples are ground with SiC300 for 1 min and then homogeneously hydrogenated in the SICHA (Sieverts chamber for hydrogen absorption) [6] at 450 °C and pressures below 1 bar from the gas phase until the sample contains either 80 or 160 wt.ppm hydrogen. At the beginning, the inset gas warms up from room temperature, which leads to a short pressure increase. Thereafter, the pressure drops non-linearly with a slowing rate as visualized in Fig. 3. After 100 and 550 h the samples were hydrogenated with 80 and 160 wt.ppm hydrogen.

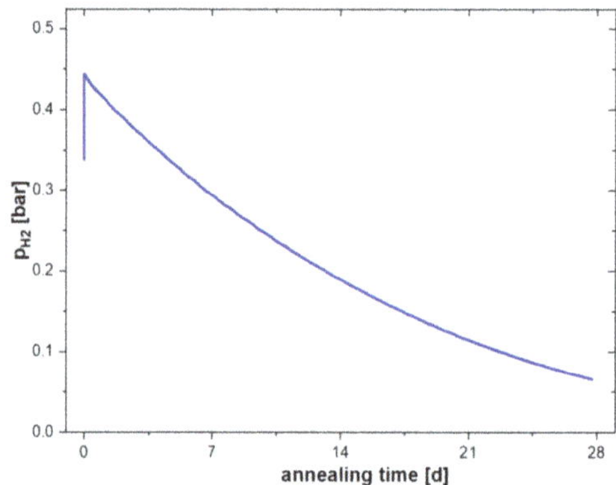

Fig. 3. Time-dependent pressure drop of a zirconium sample at 450 °C with hydrogen gas in the SICHA during 550 h.

2.3 Tensile Testing

For in-situ tensile tests with these pre-hydrogenated zirconium samples at defined temperatures, the KIT's portable INCHAMEL (in-situ neutron radiography chamber for tests under mechanical load) facility [5] is mounted at the ICON (imaging with cold neutrons) beamline [7] at the Paul Scherrer Institut, Switzerland. The samples are always placed in the exact same position and with the same distance to the scintillator screen with the sample holder extension shown in Fig. 4.

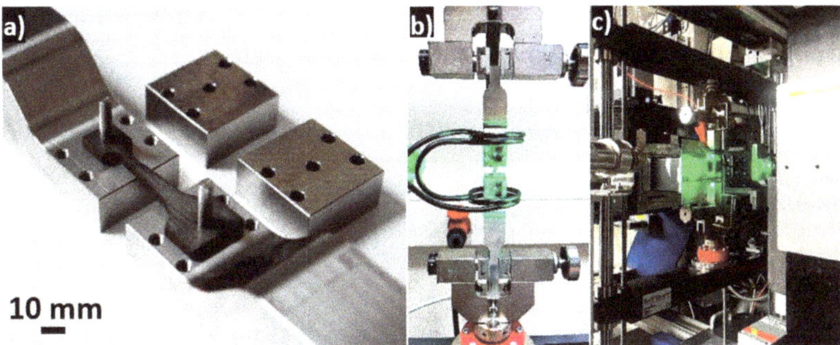

Fig. 4. a, b) INCHAMEL facility with a flat sample and a sample holder extension, c) placed in front of the scintillator screen of the ICON beamline [7] of the Paul Scherrer Institute in Villigen, Switzerland [5].

The samples are heated up homogeneously with a constant heating rate of 10 K/min. After reaching target temperatures of 320, 375 or 450 °C, the samples are elastically strained for 43, 43 or 24h, respectively, while applying a constant external force, F, of 725, 660 or 530 N. Thereby, the applied tensile stress is kept below the yield strength. While applying these forces, the samples' stress distribution can be calculated with $\sigma = F/A$ [N/mm^2] and differs between the largest and the smallest cross section by 93, 85, or 69 MPa (Fig. 5).

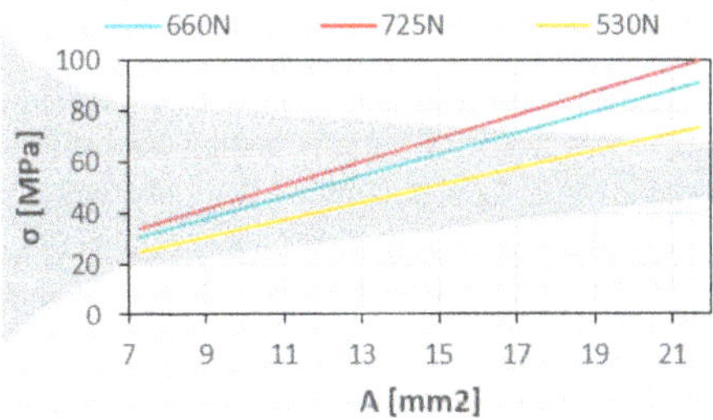

Fig. 5. Smallest and largest sample cross section of the flat tensile samples with the non-niform stress distribution within the samples at a nominal force of 530, 660 or 725 N.

2.4 Neutron Radiography Analysis

During the in-situ experiments, neutron radiographs are taken continuously with a CCD camera (Andor iKon L) by using illumination times of 55s. The collimation was 342 in terms of L/D (L: distance between pinhole and samples; D: pinhole diameter), and the neutron flux 1.3×10^7 cm^{-2} s^{-1}. A sample-detector distance of about 100 mm resulted in a lateral resolution of about 300 µm. A 30 µm Gadox scintillator screen with a pixel size of 0.0133 mm was applied.

The dark current intensities (I_{DC}) and the intensities with (I) and without samples (I_0) are measured during the experiments. With the measured intensities, the local neutron transmissions T, the total macroscopic cross section Σ_{total} distributions and the hydrogen concentrations in the sample were calculated. The factor D_0/D adjusts the attenuation term to normalize for differences in neutron flux or exposure.

$$T = \frac{I - I_{DC}}{I_0 - I_{DC}} = \exp(-\Sigma_{total} \bullet s \bullet \frac{D_0}{D}) \tag{1}$$

3 Results

The first commissioning neutron radiography measurements with flat zirconium samples in the INCHAMEL facility at the ICON beamline [7] were made under the influence of a constant temperature (320 - 500 °C) and an applied tensile stress – in the elastic or the plastic range. During these experiments, neutron radiographs that show the hydrogen distributions within the samples during several experimental stages were taken and Carrier Gas Hot Extraction (CGHE) measurements and metallographic analysis were made for each sample after the experiment.

A constant and homogeneous temperature of the sample was measured by two independent pyrometers of the INCHAMEL facility and a thermal imaging camera. The emissivity was set to 0.31 for zirconium.

By raising the temperature to 500 °C and applying an external force of 530 N, the sample creeped and failed after about 100 min at the area of the smallest cross section.

The hydrogen concentration within the samples changed during the experiments with the INCHAMEL facility. Before the experiments, the hydrogen concentration was homogeneous within the samples, as found in Table 1.

Table 1. Zr tensile samples used for the experiments. Sample 1 was used for various testing conditions and observed in-situ, whereas sample 2 was observed ex-situ.

	c_{H0} [ppm]	F [N]	T [°C]	t [h]
Sample 2	160	725	320	43
		530	450	24
		530	500	1,7
Sample 1	80	660	375	43

The results from Neutron Radiography (NR) and CGHE show that the hydrogen concentration changed significantly during the application of a plastic tensile stress. Thus, plastic stresses may induce hydrogen movements in zirconium. In general, hydrogen moves from regions with lower to higher stress regions and thus, the hydrogen in solid solution migrated towards the area with the lowest cross section and thus the highest strain. The highest hydrogen concentration is measured in the area of the smallest sample cross section. Nevertheless, sample 1 shows a cotrary behaviour, as visualized as results of the CGHE measurements in Fig. 6. The yellow marked parts of the samples were additionally analysed by NR and are of interest, whereas the rest was additionally measured, but refers to the sample heads. The application of a plastic tensile stress

(530 N, 500 °C) lead to creep and after 100 min to necking and breaking of sample 2. Its cross section in the necked zone was minimized and thus shows the effect of a stress influenced hydrogen redistribution. The other sample's (1) condition after the experiment differs strongly from sample 2. The area with the smallest cross section (length position 39 mm) shows the lowest hydrogen concentration, but the highest at length 30. The optical micrograph of sample 1 with a hydrogen concentration of 160 wt.ppm shows randomly distributed small zirconium hydrides. They further show no preferred hydride orientation. This optical micrograph is representative for the whole sample, because it shows an optically homogeneous hydride distribution.

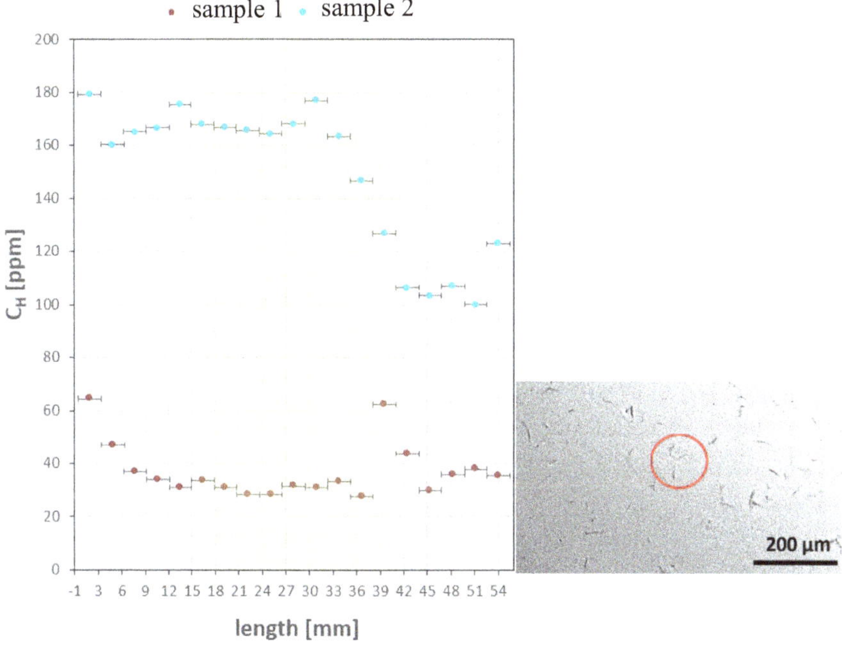

Fig. 6. *Left*: post-experimental measured (CGHE) hydrogen concentration of the flat tensile samples in dependence of their sample length. *Right*: post-experimental optical micrograph with zirconium hydrides (e.g. in the red circle) in sample 1.

After the neutron radiography for sample 2 was completed, its hydrogen concentration was calculated by referencing the neutron transmission to the pre-test neutron transmission of the sample. Figure 7 shows the hydrogen concentration of sample 2 during the tensile test with three different experimental settings. Setting 1 refers to a two-day tensile test in the elastic range with 320 °C and 725 N. During these two days, no observable hydrogen movements occurred, because the hydrogen concentration at the beginning (1a), after 16 h (1b), 32 h (1c) and 48 h (1d) are principally similar. Setting 2 refers to a one-day tensile test in the elastic range with 450 °C and 530 N. The hydrogen concentrations are shown at the beginning (2a), after 8 h (2b), 16 h (2c) and 24 h (2d). After 24 h a slight change in the hydrogen concentration is visible in the form of less fluctuations in the concentration. This trend stops with setting three that refers to a 100 min tensile test in the plastic range with 500 °C and 530 N. The hydrogen concentrations are shown at the beginning (3a), after 30 min (3b), after 1 h (3c) and after 100 min (3d) – right before the sample's break. Further, the relative hydrogen concentration during this setting remains.

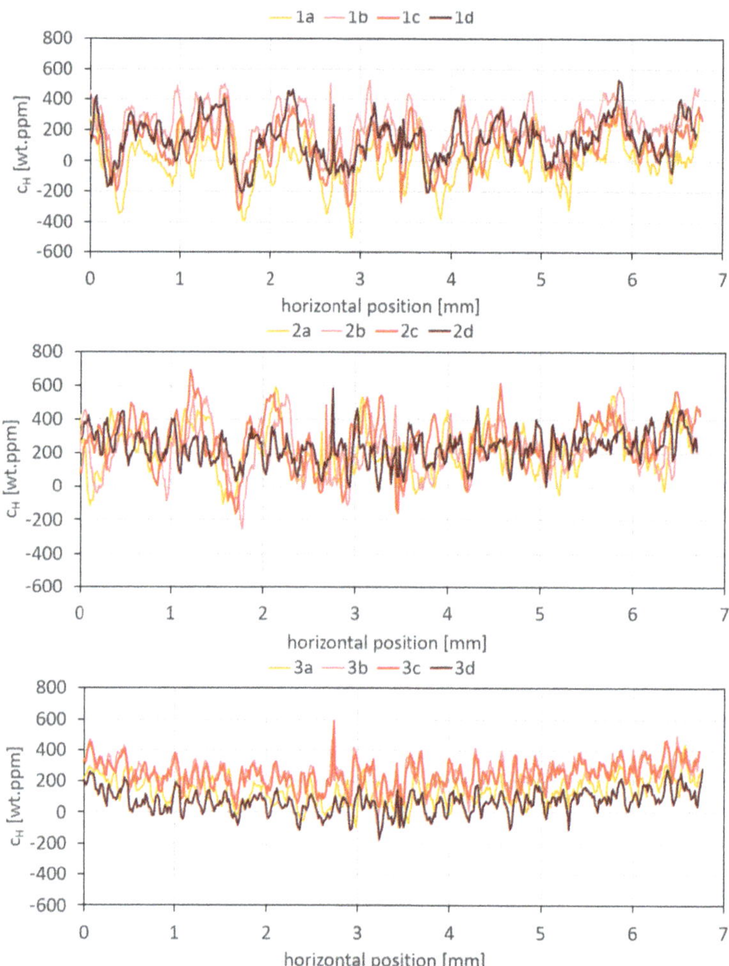

Fig. 7. With NR measured hydrogen concentrations of sample 2 during four stages (t_0, 1/3 t, 2/3 t, t) of three different experimental settings (setting 1: 320 °C, 725 N; setting 2: 450 °C, 530 N; setting 3: 500 °C, 530 N). Position 0 equals the largest cross section of the sample. The last setting includes a plastic tensile strain, whereas the other two settings comprise an elastic tensile strain..

Because the optical micrographs show very small hydrides after the experiment, the authors state that locally shorter hydrides had formed during cooling after the test. This can be underlined with Fig. 8, where the evaluated hydrogen concentrations are shown for sample 2 at 450 °C and 530 N with different counting statistics. Because no matter if the pixel intensities were averaged over 20 or 1300 images, the relative fluctuations in the hydrogen concentration stay similar. Thus, the consequence of inaccuracy caused by the counting statistics is negligible. Instead, locally great fluctuations in concentration are caused by the precipitation of large hydrides and the hydrogen depletion between them. Referring to Fig. 8, it shows that the fluctuation is less caused by the counting statistics. The smaller fluctuation of the hydrogen concentration at the end of the test compared to that before applying the mechanical load gives hints that larger hydrides resulting in a locally high hydrogen concentration dissolute with an applied elastic tensile stress and reprecipitate in form of local equal distributed smaller hydrides. Further, the average hydrogen concentration through the sample remains unchanged. Negative values arise due to the large fluctuations and the referencing to the initial sample's state.

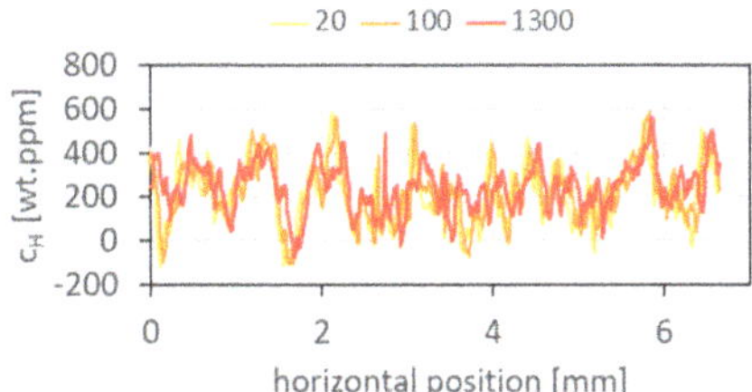

Fig. 8. With NR measured hydrogen concentrations of sample 2 showing that counting statistics (with frames 20, 100, 1300) have less influence on the measurements. It refers to Fig. 7, setting 2 with 450 °C and 530 N.

4 Discussion

Because the introduced neutron radiography beamtime was used mutually as a first commissioning tests of the INCHAMEL facility, the timeframe for multiple experiments was limited. Nevertheless, the first results are presented and discussed in this paper. Generally, the neutron radiography and the results from CGHE reveal that the stress influence on hydrogen movements in the elastic range for short time frames of hours is negligible, but plastic stresses do indeed influence the hydrogen movements within zirconium.

The experiments at the ICON beamline were made with homogeneously hydrogenated zirconium samples under constant temperatures (320 - 500 °C) and an applied tensile stress in the elastic or the lower plastic range. The hydrogen level was chosen to stay below and slightly above its solubility limit. The constant and homogeneous conditions exclude the possibility of other driving forces of the hydrogen movement besides the stress gradient. A concentration and temperature gradient influences the hydrogen movement stronger than a stress gradient and thus was tried to be excluded before the experiment. The flat tensile zirconium samples' shape with varying gauge was selected in order to obtain a stress gradient within the sample. Thus, the hydrides within the samples were expected to dissolve and then to move towards the area of greatest stress. Nevertheless, in the elastic range, these expectations were not confirmed. Instead, neutron radiographies revealed in-situ that the large fluctuations in hydrogen concentration through the samples diminished. For sample 1, we even measured less hydrogen at the region of the highest stress than in the rest of the sample after the experiment. In this context, it can not be excluded that a slight temperature concentration through the sample during the experiment was causing this behaviour, since the other sample shows in its hydrogen distribution a direct response to the plastic stress and thus might have had a slight temperature profile throught the sample, too. The thermal heating camera showed a homogeneous temperature profile, but a possible temperature gradient through a sample might not be visible for very small temperature differences, once the sample temperature is measured with reference to the room temperature background. The optical micrographs show very small hydrides after the experiment. It is a hint that locally shorter hydrides had formed during cooling after the test.

The lowest experimental temperature was 320 °C, even if no significant hydrogen uptake at and below 375 °C is expected. Nevertheless, it was checked if there is a visible hydrogen redistribution at this temperature due to the stress influence. The results show that at least for the small experimental period, no hydrogen redistribution is recognisable. Thus, these temperature and strain conditions shall be applied for future samples for longer time scales (weeks or months). Even though Kammenzind [8] could not observe significant hydrogen movements within longer time scales (from 6 to 60 days), he only made ex-situ experiments. Thus, minor hydrogen movements could possibly be detected in-situ, independendly from the final experimental stage. Further, if these future results exclude a hydrogen redistribution under these conditions, it is planed to continue with experiments only at temperatures above 375 °C. From these first results it can be derived that during the SNF phase of interim dry storage, a relevant elastic stress/strain driven hydrogen redistribution within zirconium containing cladding tubes might not take place when a temperature of 375 °C is reached during the cooling process. Nevertheless, future experiments with cladding tubes made of zirconium alloys at the same conditions should be taken into account, since they contain supplementary elements and a strong texture due to manufacturing, which might either facilitate or impede the hydrogen movement within the material, depending on the strain direction. In this case, only tensile stresses were applied on the samples. In reality, tensile hoop stresses occur.

5 Conclusion

In-situ and ex-situ NR experiments in the manner of the commissioning of the tensile testing machine INCHAMEL were made for Zr flat tensile samples. The samples were hydrogenated with concentrations relevant to dry storage conditions of SNF and then elastically strained at temperatures relevant to dry storage. The results of the two samples tested at different conditions lead to the following conclusions:

- The stress influence on hydrogen movements in the elastic range for short time frames (hours) is negligible.
- Plastic stresses induce hydrogen movements within zirconium.
- Hydrogen moves from regions with lower to higher stress regions.
- The authors assume that the stress gradient is of minor significance compared to temperature and concentration gradients.

Acknowledgements. This work is part of the SPIZWURZ project and funded by the Federal Ministry for the Environment, Nature Conservation, Nuclear Safety and Consumer Protection (FKZ 1501609B). The authors thank the PSI for the beamtime at the ICON beamline.

References

1. Northwood, D.O., Kosasih, U.: Hydrides and delayed hydrogen cracking in zirconium and its alloys. Int. Metals Rev. **28**, 92–121 (1983)
2. Gong, W., Trtik, P., Valance, S., Bertsch, J.: Hydrogen diffusion under stress in Zircaloy: high-resolution neutron radiography and finite element modeling. J. Nucl. Mater. **508**, 459–464 (2018)
3. Grosse, M., van den Berg, M., Goulet, C., Kaestner, A.: In-situ investigation of hydrogen diffusion in Zircaloy-4 by means of neutron radiography. J. Phys.: Conf. Ser. **340**, 12106 (2012)
4. Grosse, M., Santisteban, J.R., Bertsch, J., Schillinger, B., Kaestner, A., Daymond, M.R., Kardjilov, N.: Investigations of the hydrogen diffusion and distribution in zirconium by means of Neutron Imaging, KT **83**, 495–501 (2018)
5. Weick, S., Grosse, M., Steinbrueck, M., Seifert, H.J.: The INCHAMEL facility – a new device for in-situ neutron investigations under defined temperatures with applicable mechanical load, ITMNR-9 Proceedings (2022)
6. Weick, S., Grosse, M., Steinbrück, M., Seifert, H.J.: The SICHA apparatus – a cheap & advantageous device for hydrogen loading of zirconium alloys. In: 5th Symposium on Extended Dry Storage of Spent Nuclear Fuel (SEDS) (2021)
7. Kaestner, A.P., et al.: The ICON beamline – A facility for cold neutron imaging at SINQ. Nucl. Instrum. Methods Phys. Res. Sect. A **659**, 387–393 (2011)
8. Kammenzind, B.F., Franklin, D.G., Duffin, W.J., Peters, H.R.: Hydrogen pickup and redistribution in alpha-annealed Zircaloy-4 (1993)

Boron-Based Neutron Scintillator Screen Characterization with X-rays and Neutrons

William Chuirazzi[1]([✉]) [ID], Burkhard Schillinger[2] [ID], Steven Cool[3] [ID], Aaron Craft[1] [ID], Zoltan Kis[4] [ID], and Laszlo Szentmiklosi[4] [ID]

[1] Idaho National Laboratory, Idaho Falls, USA
william.chuirazzi@inl.gov
[2] Heinz Maier-Leibnitz Zentrum (FRM II), Technische Universität München, Munich, Germany
[3] DMI/Reading Imaging, Reading, MA, USA
[4] Budapest Neutron Centre, HUN-REN Centre for Energy Research, Budapest, Hungary

Abstract. Recent work on boron-based neutron scintillator screens suggests these screens can offer superior performance when compared to commonly used screens. Borated neutron scintillator screens perform well in terms of light output (5–6 times greater than a standard Gadox screen) and detection efficiency (greater than standard LiF + ZnS screens). However, previously manufactured boron-based screens have exhibited non-uniform surface coating and poor mixture uniformity between phosphor and converter particles. The objective of this work was to evaluate newly fabricated scintillator screens to determine if enhanced fabrication methods produced a more homogeneous distribution between neutron converter and scintillation phosphor particles. Uniformity of scintillator material deposition was also inspected. This new iteration of screens appeared more uniform than previous generations with the new coating method improving surface chemistry and scintillator material homogeneity. Additionally, a new methodology for screen characterization, involving the correlation of a neutron image taken with a borated scintillator screen to X-ray computed tomography of that same screen, was demonstrated to elucidate a relationship between scintillator screen thickness and relative light output of the screen under neutron exposure. This method suggested that the ideal thickness of combined scintillator and boron-based converter material for this specific screen chemistry and fabrication method was ~ 150 µm to maximize light output of the screen.

Keywords: Neutron scintillators · neutron computed tomography · X-ray computed tomography · neutron imaging · scintillator evaluation

1 Introduction

Boron-based neutron scintillator screens have recently been studied as a suitable neutron detector in digital neutron imaging systems [1–6]. In digital imaging systems, improving the scintillator screen can directly increase the spatial resolution and dynamic range of the imaging system while reducing the imaging time, in turn increasing instrument throughput. Additionally, when compared to ^{6}LiF, a commonly used neutron converter

© The Author(s) 2026
A. E. Craft and H. Z. Bilheux (Eds.): WCNR 2024, SPPHY 348, pp. 168–179, 2026.
https://doi.org/10.1007/978-3-032-15003-5_20

in commercial neutron imaging screens, ^{10}B has a neutron absorption cross section over four times greater, half the reaction energy, and shorter daughter product ranges [4, 5], all of which lead borated neutron scintillator screens to offer the potential for improved light output and neutron detection.

In this work, the scintillator screens consist of a mixture of phosphor and converter particles with small median particle sizes, as opposed to separate layers of converter and phosphor. The development of boron-based neutron scintillators has undergone several iterations to optimize the neutron converter and phosphor scintillator stoichiometry, particle sizes, and scintillator surface chemistry [4–6]. Fine milling of ^{10}B-containing converter material to improve the escape probability of the conversion reaction's daughter products [7] is essential to producing sufficient light output to produce a useable scintillator screen. The previous generation of borated screens employed this finely milled material, however testing of these screens revealed a lack of consistency and homogeneity in the distribution between boron-based neutron coFnverter particles and phosphor scintillators [6]. This can be observed in Fig. 1 where a neutron image taken with a borated scintillator screen displays dark circles (converter particles) surrounded by white rings (phosphor) instead of a uniform intensity of light output.

Fig. 1. Neutron image from a previous version of borated neutron scintillator screen. The range of sizes of particles and particle agglomerations within the scintillator can be seen by the dark clumps (converter particles/agglomerations) surrounded by bright rings (phosphor particles) [6].

Before these boron-based scintillators can be deployed for regular use in digital neutron imaging systems, the homogeneity of the scintillator mix as well as the uniformity of the surface chemistry must be improved to create a consistent scintillator. A new mixing and deposition method for scintillator fabrication was piloted in an effort to correct these issues. The goal of this work was to characterize a new generation of borated scintillator screens to determine if the new fabrication technique improved screen quality. This manuscript details the testing and characterization results of the latest effort to produce boron-based neutron scintillators.

2 Methods

2.1 Scintillator Screens

Based on the results from previous testing campaigns [4–6] ten new scintillator screens were fabricated with the goal of improving screen homogeneity, minimizing large grains in the scintillator coatings, creating a smoother surface, and reducing the uneven distribution between the converter and phosphor materials. Scintillator material was deposited on the substrate using two different methods. Four screens had material deposited using the traditional method used to produce screens in the previous studies while a new deposition method, which sought to more evenly distribute the converter materials around the phosphor powders, was utilized in the fabrication of six screens.

Once the scintillator material was deposited to a uniform thickness of ~ 200 μm, the scintillator was scraped to remove material and approximate a "wedge" of different thicknesses to enable characterization of performance property as a function of thickness[8]. Note that this screen characterization method neither assumes nor requires a perfect wedge, but rather determines the thickness for each pixel from the neutron imaging measurements. As long as a range of thicknesses are available with a given screen, then the neutron capture efficiency and relative light output can be measured as a function of thickness for each pixel.

In addition to the different scintillator material deposition techniques, the ten new scintillator screens represent different phosphor materials and molecular ratios between converter atoms and phosphor molecules. Screen parameters are displayed in Table 1 while optical photographs of each screen are shown in Fig. 2 (figures throughout this manuscript use red text to denote screens created using the previously employed fabrication methodology). Note that the screens were initially uniform after scintillator deposition but the scraping method for the purpose of this test resulted in "damaged" screens as shown in the optical images.

Table 1. Parameters of tested scintillator screens

Screen ID	Converter/Phosphor	Converter Atoms: Phosphor Molecules	Fabrication Method
D2-01a	^{10}B-B_2O_3/ZnS:Cu	1:1	New
D2-01b	^{10}B-B_2O_3/ZnS:Cu	1:1	New
D2-02a	^{10}B-B_2O_3/ZnS:Cu	2:1	New
D2-02b	^{10}B-B_2O_3/ZnS:Cu	2:1	New
D2–03	^{10}B-B_2O_3/ZnS:Cu	1:1	*Standard*
D2–04	^{10}B-B_2O_3/ZnS:Cu	2:1	*Standard*

(continued)

Table 1. (*continued*)

Screen ID	Converter/Phosphor	Converter Atoms: Phosphor Molecules	Fabrication Method
D2–05	^{10}B-B$_2$O$_3$/ZnS:Ag	1:1	New
D2–06	^{10}B-B$_2$O$_3$/ZnS:Ag	2:1	New
D2–07	^{10}B-B$_2$O$_3$/ZnS:Ag	1:1	*Standard*
D2–08	^{10}B-B$_2$O$_3$/ZnS:Ag	2:1	*Standard*

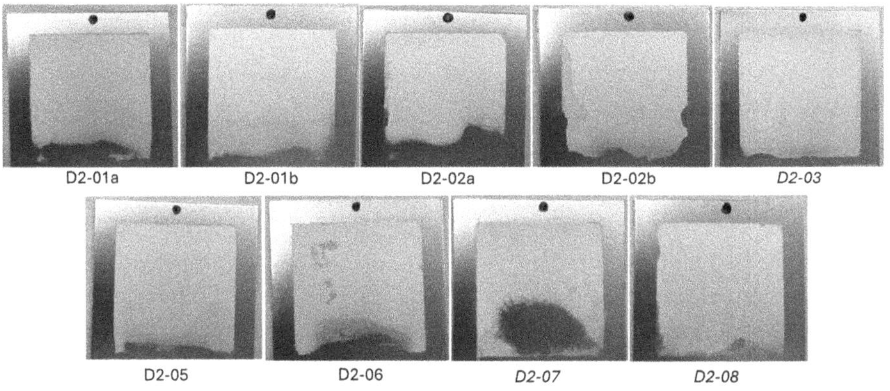

Fig. 2. Optical images of the scraped scintillator screens.

2.2 Budapest Neutron Centre Beamline Experiments

Neutron measurements in this work were conducted at the NORMA beamline at the Budapest Neutron Centre (BNC) [9]. This facility is an integral part of the combined NIPS-NORMA measurement station situated at the end of a cold neutron guide, where NIPS is for elemental analysis and NORMA is for imaging [10]. The largest cross-sectional area of the neutron beam in the sample chamber is 43×43 mm^2, and the maximum thermal equivalent flux at the center of the beam is 2.7×10^7 cm^{-2}s^{-1}. The measured collimation ratio L/D is 233.

Two neutron imaging setups were deployed for these experiments. The first setup consisted of a larger field-of-view (FOV) detector box with a cooled scientific CMOS camera (ASI 178MM cool, now out of production) with a standard 50 mm lens and mirror to capture the entirety of the screens, producing a spatial resolution of ~ 9.2 µm/pixel. The acquisition time for this setup was 60 s per radiograph and the images were open-beam corrected using a 60 s image of a 10 µm thick gadolinium oxysulfide (Gadox) screen. Slight movement occurred between subsequent images of screens caused by a misfit between the screens and sample holder.

A Heliflex 90-degree lens was used in the second configuration to provide higher spatial resolution (~2.53 µm/pixel) at the expense of FOV. The heliflex lens of 55 mm focal length and 90-degree deflection captured a FOV of 15 mm diameter and projected this to infinity, which was captured by the CMOS camera with a 50 mm lens in a 1.1:1

projection. A Siemens star was projected onto the image plane and this projection was used to focus the imaging system. The smaller FOV prevented imaging of the entire screen so the top, middle, and bottom sections were imaged separately with acquisition times of 60 s per image. No open beam images were obtained as the tested screens were both reference and sample, and any other screen like Gadox would have its own inhomogeneities. However, these images were still useful in determining screen homogeneity even if quantitative analysis was precluded.

2.3 X-ray Microscope

X-ray computed tomography data for screen D2–03 was acquired using a ZEISS Xradia 620 Versa X-ray microscope (Carl Zeiss X-ray Microscopy Inc., Dublin, CA). This instrument is located at the Irradiated Materials Characterization Laboratory (IMCL) at Idaho National Laboratory and is used to nondestructively examine a variety of irradiated and unirradiated energy materials [11]. In this work, an X-ray energy spectrum with a maximum energy of 40 kVp was utilized in conjunction with a low-energy beam filter supplied by the manufacturer. The lower X-ray energy spectrum generated better contrast through the relatively low-Z scintillator material. The flat panel detector was employed with no detector binning to produce a spatial resolution of 18.47 μm/voxel. A total of three separate scans, each consisting of 4801 projections which were exposed for 24 s each, were collected across 360 degrees of rotation. Each of the three datasets was then reconstructed and the reconstructed data was stitched together to capture the entirety of the scintillator screen.

2.4 Data Processing and Analysis

Data Processing.
Full FOV neutron imaging data was open beam corrected by dividing the open beam Gadox image into the neutron image of the scintillator screen. X-ray images were dark and open beam corrected using the commercial ZEISS Scout and Scan Control System Reconstructor software. Corrected images were used for subsequent analysis.

Once the XCT data was obtained, a previously developed analysis technique [12] was applied to calculate the scintillator material thickness. First, the three-dimensional data (3D) volume was cropped to reduce edge effects while maintaining the entirety of the scintillator volume. A standard deviation filter was then applied to the data to emphasize the scintillator volume in contrast to the aluminum substrate. Manual grayscale thresholding was employed to binarize the image before performing connected component analysis [13] to select the region of scintillator material.

Morphological closing was then implemented to generate a complete binary mask of this region. A Euclidean distance transform [14] was then applied to the inverse of the masked grayscale data to generate the thickness of each scintillator layer in pixels, converting the 3D dataset to a two-dimensional (2D) map of scintillator thickness. This result was multiplied by the spatial resolution to generate screen thicknesses in metric units.

Data Analysis.

First, the neutron image of screen D2–03 and the 2D thickness map generated from the X-ray data were spatially aligned using the landmark registration method in the Thermo Scientific™ Avizo software (Thermo Fisher Scientific, Waltham, MA). The neutron image was then resampled using Lanczos interpolation [15] to match the resolution of the X-ray thickness data. The aligned neutron and X-ray thickness map allowed for light output (from the neutron image) and screen thickness (from the X-ray thickness map) to be correlated with one another. The average value of light output for each discrete thickness value was then plotted as a function of scintillator screen thickness.

3 Results

Neutron Imaging Results.

The neutron images taken with the screens in the full FOV arrangement are shown in Fig. 3. The purpose of these measurements is to observe the relative light output across the entirety of the wedged screen as well as to generate data to correlate with the XCT screen thickness measurements. These images show that while light output is more uniform in some screens than others, screens still contain defects and other inhomogeneities that produce varied light output, even when accounting for the wedged shape of the screens. Despite the varied light output between screens, screens such as D2-01b and D2-02b exhibit regions of strong uniformity. Qualitative assessment of these results suggest the new fabrication method may offer improved, if inconsistent, scintillator deposition. Note that the slight movement between images of the screens compared with the original position of the screen used for the open beam measurements, caused by a misfit between the screens and sample holder, is most likely the cause of the cross-hatched pattern seen in the images in Fig. 3. The screen used for the open beam measurement was created using a screen-printing method and the misalignment of images highlights the 90° printing of the screen layers. Additional method development may be necessary to improve the repeatability of this new deposition technique.

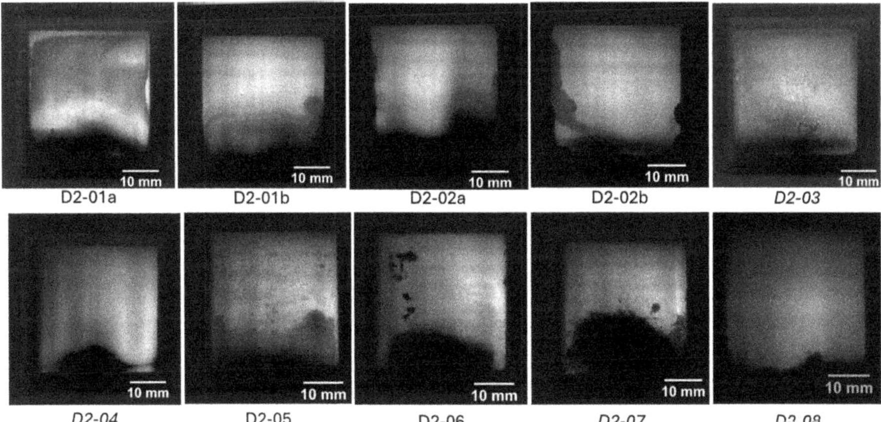

Fig. 3. Open-beam corrected neutron images taken with scraped borated scintillator screens in the full FOV configuration. These images show the light output of the respective screens.

174   W. Chuirazzi et al.

Heliflex lens images, displayed in Fig. 4, were used to examine the uniformity of the screens. A Siemens star was imaged with this set-up and the images shown are all taken at the middle regions of the screens. At this resolution, some screens are seen to be more homogenous than others. The spatial resolution achievable varies from screen to screen, most likely due to a combination of the different screen coating as well as potential differences in the mixing between converter and phosphor particles. However, the estimated spatial resolution is just less than 100 μm/lp, based on the innermost ring representing 50 μm/lp and the next ring outside that representing 100 μm/lp. Generally, the homogeneity of the screens is improved over those from previous fabrication efforts, an example of which is shown in Fig. 1. Additionally, these results suggested the new scintillator deposition method may result in a more even mixture of converter and phosphor particles as well as a more uniform deposition of the scintillator layer.

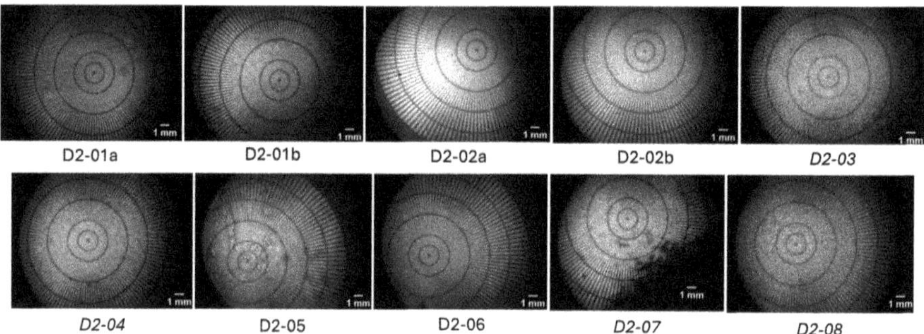

Fig. 4. Neutron images taken with borated scintillator screens with the Heliflex lens. These images are of the center image of each screen, with the goal being to examine the homogeneity of the scraped screens.

X-ray Computed Tomography Results.

Based on the optical and neutron image results, screen D2–03 was selected to undergo X-ray computed tomography for method demonstration on correlating XCT measurements of scintillator screen thickness to relative light output under neutron exposure. Screen D2–03 was chosen for X-ray examination due to its uniform scintillator deposition, lack of obvious defects, and its relatively consistent light output under neutron exposure. Although D2–03 was selected to demonstrate this method, this technique could be applied to any of the screens. Time limitations restricted XCT to a single screen instead of the entire set.

Results of XCT for D2–03 are presented in Fig. 5. While the optical image (Fig. 5a) shows a relatively uniform deposition of scintillator material, segmentation of the XCT data (Fig. 5b) reveals some thinner layers (shown as holes) where the scintillator thickness was less than the spatial resolution of the scan. Once the Euclidean distance transform is applied to this data, the 3D dataset is reduced to a 2D map of the thickness of each region of the scintillator (Fig. 5c). This data can then be viewed as a histogram (Fig. 5d) to determine the distribution of thicknesses in the screen.

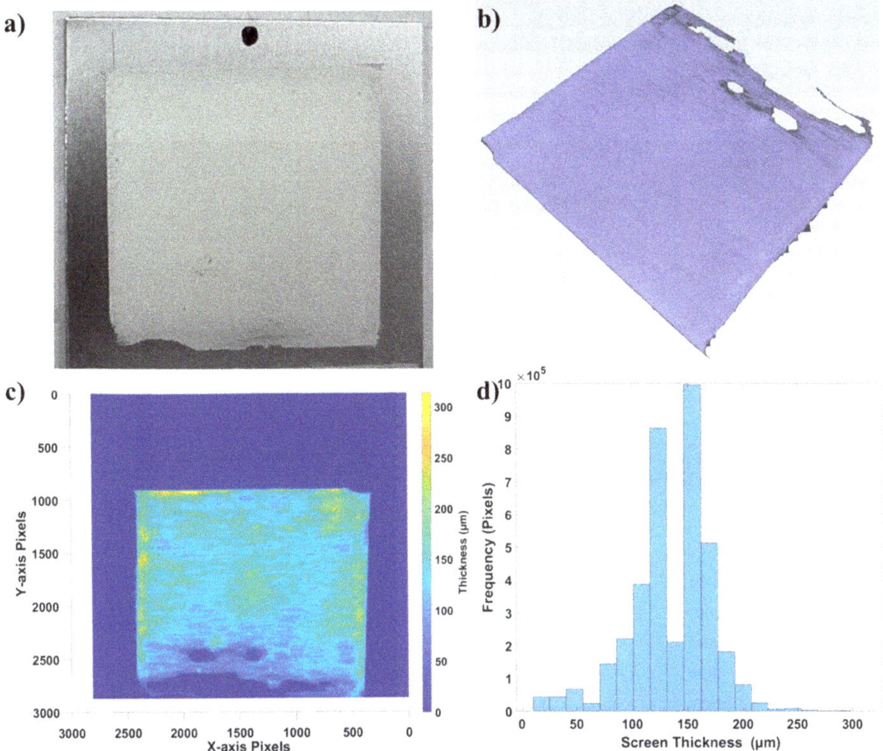

Fig. 5. XCT results for screen D2–03. a) An optical image of the screen can be compared to the b) segmentation of the scintillator material from the XCT measurements. c) Screen thickness as a function of position can be visualized as d) a histogram to show the screen thickness distribution.

While the screens had a nominal maximum thickness of ~ 200 μm, a few areas near the edges were slightly thicker than the target maximum, since screens were coated more thickly in anticipation of scraping. Additionally, the most common thickness appeared to be around 150 μm. An ideal wedged screen would have had the same number of thickness occurrences from the maximum value down to zero. However, these results reveal that there are a variety of thicknesses present, approximating a wedged shape, which is all that is necessary to correlate screen performance to scintillator thickness. One advantage to the pixel-by-pixel measurement is that even if the screen is nonuniform or has irregular thickness variations, these datapoints are still useful in generating light output as a function of screen thickness.

Data Correlation.

Once the neutron image and X-ray measurement of screen thicknesses were completed, the two images (displayed in Fig. 6) were correlated to spatially relate the data from one image to the other. This enabled a pixel from the neutron image, which contains light output information, to be spatially correlated to a corresponding pixel from the X-ray image, which contains information on the screen thickness at that given location. The correlated results, exhibited in Fig. 7, suggest that the maximum light output occurs around a scintillator thickness of 150 μm for this particular coating composition and converter atom to phosphor molecule ratio for screen D2–03, before self-shielding effects reduce the light output.

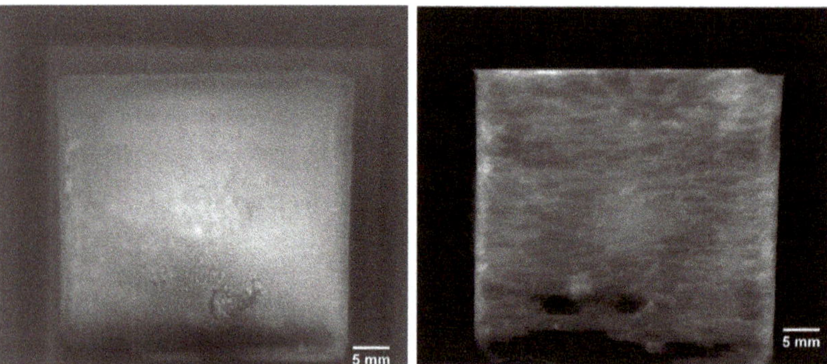

Fig. 6. (Left) Neutron image and (Right) X-ray thickness measurement map of screen D2–03.

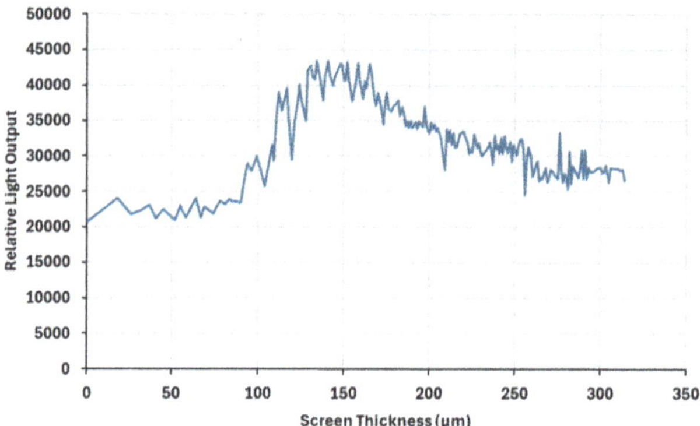

Fig. 7. Results of relative light output as a function of scintillator screen thickness. Maximum light output is achieved at an approximate thickness of 150 μm.

4   Discussion

The focus of this study was to observe the homogeneity of the newest generation of boron-based neutron scintillator screens compared to previous measurements of earlier generations of screens. The purpose of producing these new screens was to generate a more homogeneous mix between neutron converter (boron-based) and phosphor (ZnS-based) particles as well as more uniformly depositing this scintillator mixture on the substrate. Characterization of these screens is necessary to determine not only the effectiveness of the screen fabrication method, but to also measure their imaging performance.

Previously, borated scintillator screens were characterized with neutron imaging methods [8]. However, this method was limited by the fact that all screens must stay in the same location. The use of XCT in screen characterization may be useful when screens move between acquisitions or when neutron beamtime is limited and not all measurements on screens can be made.

In this work, the proof of principle for correlating XCT and neutron imaging was demonstrated. Future work may include a more rigorous application of this technique to characterize screen detection efficiency as well as comparing this method with previously utilized characterization methodologies. Subsurface defects and thin deposition layers, not optically observable could be detected with XCT. For the screen imaged (D2–03), a scintillator thickness of ~ 150 μm provided the maximum light output as thinner scintillator regions could not fully absorb the neutrons to produce photons while the thicker scintillator regions self-shielded the photons, causing them to be absorbed within the scintillator before they could escape and reach the detector.

While these results indicate boron-screen fabrication methods are advancing, further process development is ongoing with the goal of producing commercially viable screens that can offer improved spatial resolution and light output. Spatial resolution could be improved primarily by utilizing the smallest converter and phosphor particles, and secondarily by refining factors such as the substrate surface finish, coating surface conditions, and reducing pockets of trapped air or unescaped volatile solvents. Screen uniformity can be enhanced with new coating methods, the study of which are already underway. Better light output may be achieved by optimizing the homogeneity of converter and phosphor particle size and the ratio between converter atoms to phosphor molecules.

5 Conclusions

Neutron and X-ray imaging were successfully used together to characterize prototype neutron scintillator screens. These results suggest that the most recent iteration of borated scintillator screens appear more uniform than previous generations with the new coating method improving surface chemistry and scintillator material homogeneity. Additional fabrication methodology enhancements are underway to further improve screen quality with that goal of producing high-quality homogeneous boron-based scintillator screens.

Acknowledgements. The authors would like to thank the reactor operations staff at BNC. Screens were fabricated by DMI/Reading Imaging [16] and fabrication was funded through the INL Laboratory Directed Research & Development (LDRD) Program under DOE Idaho Operations Office Contract DE-AC07-05ID14517, LDRD Project ID #20A44-200FP. The image processing and registration work, as well as the X-ray computed tomography method development, were supported by the U.S. Department of Energy, Office of Nuclear Energy under DOE Idaho Operations Office Contract DE-AC07- 051D14517 as part of a Nuclear Science User Facilities project.

References

1. Katagiri, M., et al.: Scintillation materials for neutron imaging detectors. Nucl. Instrum. Methods Phys. Res., Sect. A **529**, 274–279 (2004). https://doi.org/10.1016/j.nima.2004.04.165

2. Miller, S.R., et al.: High resolution thermal neutron imaging with 10Boron/CsI:Tl scintillator screen. IEEE Trans. Nucl. Sci. **9499**, 1–5 (2020). https://doi.org/10.1109/tns.2020.3006741

3. Saha, S., et al.: Synthesis and characterization of borate glasses for thermal neutron scintillation and imaging. Radiat. Meas. **134**, 106319 (2020). https://doi.org/10.1016/j.radmeas.2020.106319

4. Chuirazzi, W., Craft, A., Schillinger, B., Cool, S., Tengattini, A.: Boron-based neutron scintillator screens for neutron imaging. J. Imaging **6** (2020). https://doi.org/10.3390/jimaging6110124

5. Schillinger, B., Chuirazzi, W., Craft, A., Cool, S., Tengattini, A.: Performance of borated scintillator screens for high-resolution neutron imaging. J. Radioanal. Nucl. Chem. **331**, 5287–5295 (2022). https://doi.org/10.1007/s10967-022-08477-w

6. Schillinger, B., et al.: New measurements on borated neutron imaging screens at Budapest Neutron Centre (BNC). J. Phys: Conf. Ser. **2605**, 012009 (2023). https://doi.org/10.1088/1742-6596/2605/1/012009

7. Nakamura, T., et al.: Development of a ZnS/10B2O3 scintillator with low- afterglow phosphor. In: International Workshop on Neutron Optics and Detectors 528, pp. 1–8 (2014). https://doi.org/10.1088/1742-6596/528/1/012043

8. Chuirazzi, W., Craft, A.: Measuring thickness-dependent relative light yield and detection efficiency of scintillator screens. J. Imaging **6**, 56 (2020). https://doi.org/10.3390/jimaging6070056

9. Kis, Z., Szentmiklósi, L., Belgya, T.: NIPS-NORMA station - A combined facility for neutron-based nondestructive element analysis and imaging at the Budapest Neutron Centre. Nucl. Instrum. Methods Phys. Res., Sect. A **779**, 116–123 (2015). https://doi.org/10.1016/j.nima.2015.01.047

10. Kis, Z.: The redesigned neutron imaging facility, NORMA at BNC, Budapest. Review of Scientific Instruments **95** (2024). https://doi.org/10.1063/5.0208844

11. Chuirazzi, W., Kancharla, R.R., Morankar, S.: Laboratory-based Micro-X-ray computed tomography of energy materials at Idaho National laboratory. JOM (2024). https://doi.org/10.1007/s11837-024-06970-z

12. Chuirazzi, W.C., Cool, S., Craft, A.E.: Quantifying neutron scintillator screens with X-ray computed tomography. Nuclear Instruments and Methods and Methods in Physics Research Section A (2024). https://doi.org/10.1016/j.nima.2024.169248

13. Gonzalez, R.C., Woods, R.E., Eddins, S.L.: (Prentice Hall, 2003)

14. Maurer, C.R., Qi, R., Raghavan, V.: A linear time algorithm for computing exact Euclidean distance transforms of binary images in arbitrary dimensions. IEEE Trans. Pattern Anal. Mach. Intell. **25**, 265–270 (2003). https://doi.org/10.1109/TPAMI.2003.1177156

15. Burger, W., Burge, M.J.: Digital Image Processing: An Algorithmic Introduction. Springer Nature (2022)

16. DMI/Reading Imaging (DMIReadingImaging@gmail.com)

Visualization of Liquid Water Behavior in a Polymer Electrolyte Fuel Cell Under High-Temperature Operation Using a Neutron Radiography

Yuto Obayashi[1]([envelope]), Hideki Murakawa[1], Katsufumi Nakagawa[1], Katsumi Sugimoto[1], Hitoshi Asano[1], Yuto Shirase[2], Christopher Schreiber[2], Solomon Wakolo[2], Eric Dzramado[2], Junji Inukai[2], Daisuke Ito[3], Naoya Odaira[3], and Yasushi Saito[3]

[1] Graduate School of Engineering, Kobe University, 1-1 Rokkodai, Nada, Kobe 657-8501, Japan
{236t313t,240t336t}@stu.kobe-u.ac.jp, {murakawa,sugimoto,
asano}@mech.kobe-u.ac.jp
[2] University of Yamanashi, 4-4-37, Takeda, Kofu 400-8510, Yamanashi, Japan
{g22dte02,g22tg002,g21dte03,g23ta018,jinukai}@yamanashi.ac.jp
[3] Institute for Integrated Radiation and Nuclear Science, Kyoto University, 2 Asashiro-Nishi,
Kumatori-Cho, Sennan-Gun, Osaka 590-0494, Japan
{ito.daisuke.5a,odaira.naoya.4e,saito.yasushi.8r}@kyoto-u.ac.jp

Abstract. Recently, research has been conducted on the installation of polymer electrolyte fuel cells (PEFCs) in heavy-duty vehicles. The PEFC stacks become large, as well as the heat produced during operation; a higher operating temperature is necessary to increase the power generation density and to reduce the size/weight of radiators for cooling the PEFCs. To increase the operating temperature, the operating pressure in the PEFCs must be controlled to avoid dryout of the proton exchange membrane (PEM) and flooding in the PEFCs. Therefore, it is crucial to evaluate the water transport in PEFCs under these conditions. In this study, visualization of the water distribution in a PEFC at an operating temperature of 105 °C using neutron radiography was performed by changing the operating pressure. Two-dimensional water distribution was measured every 30 s during the operation. When the operating pressure was atmospheric pressure, the current density could not be increased more than 0.6 A/cm^2 due to the PEM dryout. However, the operatable current density could be increased to 1.0 A/cm^2 at 250 kPaG, and water accumulation was clearly confirmed in the flow channels. To analyze the time variation of area-average water thickness, it was confirmed that the lowering of the cell performance was caused by liquid plug formation.

Keywords: PEFC · High Temperature Operation · Neutron Radiography

© The Author(s) 2026
A. E. Craft and H. Z. Bilheux (Eds.): WCNR 2024, SPPHY 348, pp. 180–187, 2026.
https://doi.org/10.1007/978-3-032-15003-5_21

1 Introduction

To operate polymer electrolyte fuel cells (PEFCs) stably, it is important to ensure gas supply to the catalyst layer (CL) through the gas diffusion layer (GDL), while the proton exchange membrane (PEM) must be kept at an appropriate humidity. For the operation of PEFCs, the supplied gases are often humidified to prevent the PEM from dryout. In contrast, water is generated at the cathode reaction site as an electrochemical reaction. If liquid water accumulates in the GDL or CL owing to supersaturation, it may inhibit gas diffusion as flooding. Therefore, water management is a critical issue for PEFC performance. Several investigations on water management have been conducted concerning water movement inside the PEM, water flooding in the GDL and CL, and water plugging in the channel [1–3]. It has been shown that the accumulation of liquid water under the ribs causes a decrease in performance. These findings will serve as guidelines for future cell development.

Recently, research has been conducted on the installation of PEFCs in heavy-duty vehicles. The PEFC stacks become large, as well as the heat produced during operation; a higher operating temperature is necessary to increase the power generation density and to reduce the size/weight of radiators for cooling the PEFCs. For this purpose, research has been undertaken to operate PEFCs at higher temperature, up to 120 °C, to increase the power generation density and reduce the size/weight of radiators for cooling the PEFCs, which are typically operated at 80–90 °C [1, 5, 6]. To operate the PEFCs at temperatures higher than 100 °C, controlling the back pressure is required to increase the operating pressure to increase the boiling point of the water to avoid PEM dryout. Because liquid water transport depends on the power output, operating pressure, and temperature in the PEFC, it is important to clarify the relationship between the liquid water and cell performance during the operation. However, the effect of liquid water in PEFC on the power generation performance under high-temperature conditions has not been well understood.

The purpose of this study is to evaluate the relationship between transient water distributions and cell performance under high-temperature conditions. Visualization of the water distribution in the PEFC under operation using neutron radiography was performed at an operating temperature of 105 °C.

2 Experimental Setup and Method

The PEFC was heated using ceramic heaters attached to the outside wall of the aluminum separators to maintain the cell temperature at 105 °C. The cell temperature was measured by using a thermocouple inserted into a small hole in the separator. The neutron beam was irradiated in a direction perpendicular to the electrode plane, and two-dimensional water distributions in the parallel direction of the PEM were obtained.

A schematic of the visualization PEFC is shown in Fig. 1. A proton exchange membrane was sandwiched between the gas diffusion layers and flow channels. The PEFC is a single cell having ten straight-parallel flow channels with a cross-sectional area of 1×1 mm^2. The channel length was 30 mm. The width and depth of the ribs were both 1 mm. Nafion® NRE-211 was used as the PEM with a thickness of approximately

20 μm having catalyst layers on both the anode and cathode sides. The active area was 20×20 mm^2. The GDL was made of carbon paper (SGL Carbon) with a thickness of 235 μm on both the anode and cathode sides. A microporous layer was attached to the GDL surface. The porosity of the GDL was approximately 40%. The cell was placed vertically. The hydrogen in the anode flowed vertically upward, and the air in the cathode flowed vertically downward to easily exhaust liquid water.

Visualization of the water distribution in the PEFC was carried out using a neutron radiography system at B4 port in the Kyoto University research reactor at a thermal power of 1 MW. Neutron radiographs on the scintillation converter were captured using a CMOS camera (Zyla 4.2, Oxford Instruments) with an array of 1024×1024 pixels, and the pixel size was 26.6 μm/pixel. The exposure time for obtaining the image was set to 30 s. L/D was set to 100. Experiment was conducted by increasing the current density in steps. Radiography measurements started 1 min before the current density was changed and continued for 20 min. Figure 2 shows the original visualized image without liquid water and during power generation. By performing image processing on these images, the two-dimensional distributions of the liquid water thickness, t_L, were calculated using the mass attenuation coefficient of water published by NIST [3]. The details of the image processing are described in the literature [4].

The experimental conditions are presented in Table 1. The cell temperature was set at 105 °C. The bubbler temperature for the humidification was set to 74.6 °C resulting in the inlet relative humidity of air and hydrogen being 34.1%. Gas flow rates were set at 200 ml/min defined at 25 °C and atmospheric pressure. Gas utilization is based on a current density of 1.0 A/cm^2. The back pressure was set at 0 and 250 kPa in gauge pressure, kPaG.

Table 1. Experimental condition.

Cell temperature, T_{cell} [°C]	105	Bubbler temperature, $T_{bubbler}$ [°C]	74.6
H$_2$ flow rate, Q_{H2} [cc/min]	200	Temperature of gas supply pipe, T_{pipe} [°C]	125
Air flow rate, Q_{Air} [cc/min]	200	Back pressure, P_{back} [kPaG]	0, 250
H$_2$ utilization [%]	15	Inlet relative humidity for H$_2$ and air RH_{in} [%]	34.1
Air utilization [%]	33		

3 Results and Discussion

3.1 Effect of Back Pressure on Cell Voltage

Figure 3 shows the time series change in the cell voltage as the current density increases. Figure 3 (a) shows the results for $P_{back} = 0$ kPaG, and Fig. 3 (b) shows the results for $P_{back} = 250$ kPaG. For $P_{back} = 0$ kPaG, the cell voltage is almost constant at $i = 0.4$ A/cm^2. However, the cell voltage suddenly dropped to 0.25 V at 57 min at $i = 0.6$ A/cm^2,

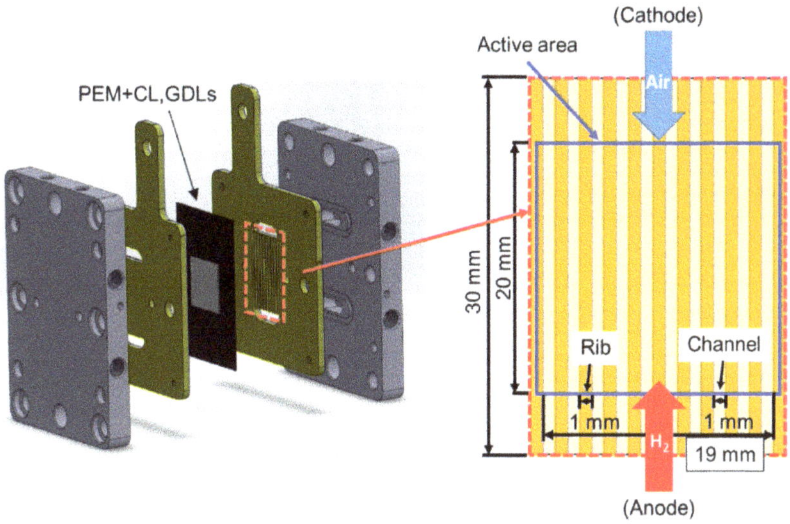

Fig. 1. Schematic diagram of the detail configuration of the tested PEFC.

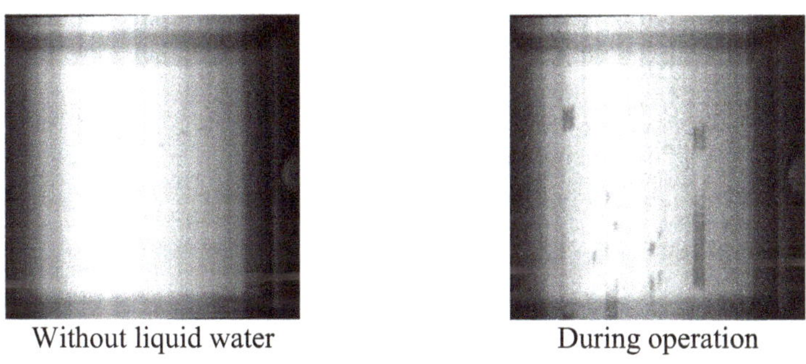

Fig. 2. Original visualized images.

and the power generation stopped. At this time, liquid water was not clearly confirmed in the cell, but it was inferred that the dryout prevented an increase in the current density. On the other hand, for $P_{back} = 250$ kPaG, the cell voltage was stable up to $i = 0.8$ A/cm^2. However, a decrease in the cell voltage with fluctuations was confirmed at $i = 1.0$ A/cm^2. Figure 4 shows the i-V curve at $P_{back} = 0$ kPaG and $P_{back} = 250$ kPaG. The plots show the average values at each current density and the error bars show the maximum and minimum values. For all current densities, the cell performance at $P_{back} = 250$ kPaG was superior to that at $P_{back} = 0$ kPaG. This is because the increased back pressure prevents the membrane from dryout.

3.2 Relationship Between Voltage Fluctuation and Water Thickness

Figure 5 shows the voltage variation with time at $P_{back} = 250$ kPaG and $i = 1.0$ A/cm^2. The two-dimensional water distributions at different times are shown in Fig. 6. The color scale represents the water thickness along the neutron beam. The edges of the ribs are highlighted by the white lines. The numbers indicate the flow channels. The red line represents the active area. The observed times for Figs. 6(a)–(c) correspond to the times in Fig. 5, respectively. After 82 min, the current density was increased to 1.0 A/cm^2. Small water droplets in tens to hundreds of micrometers were already seen mainly under the ribs. At 87 min, as indicated by the red circle, liquid water accumulated near the outlet of Channel 2. At 89 min, immediately after a large drop in the cell voltage from 0.38 to 0.35 V, the water droplets with the thickness close to the height of the flow channel (1 mm) grew in Channels 8 and spread across the entire cross-section of the channel, whereas the water in Channel 2 was seen to move to the outlet. Such liquid accumulations should significantly lower or even block the gas flow in the channels. Furthermore, at 98.5 min, on the upstream of the channels, accumulations of liquid water were observed as indicated by the dashed red circles, owing to supersaturation. At 101 min, Channel 8 was completely plugged. Interestingly, the locations of water accumulation in the flow channels were almost the same, probably because of the minute structure differences. From Fig. 6(a) to (c), the amount of water in the cell kept larger and the cell voltage became lower. The fluctuation of the voltage can be explained by the movement of liquid water in the cell to the outlet.

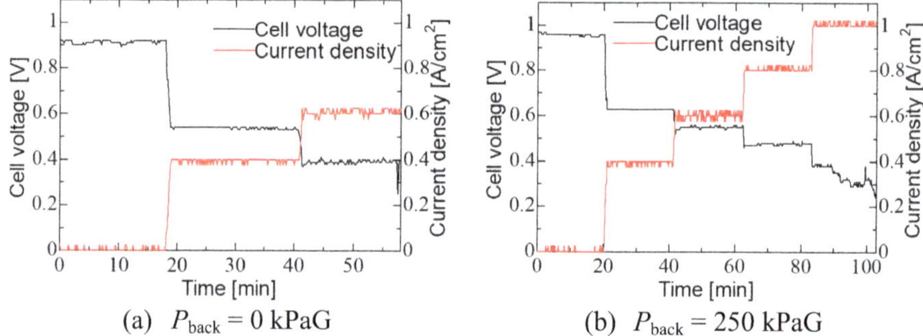

(a) $P_{back} = 0$ kPaG   (b) $P_{back} = 250$ kPaG

Fig. 3. Time-series of cell voltage variation with increasing current density.

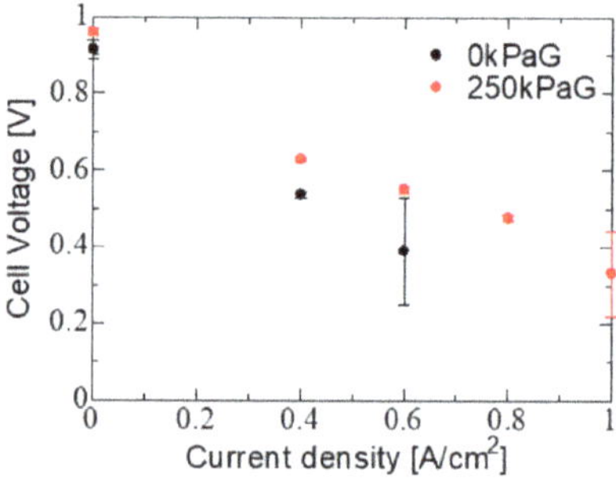

Fig. 4. Current density-cell voltage (i-V) curve ($P_{back} = 0$ kPaG, $P_{back} = 250$ kPaG).

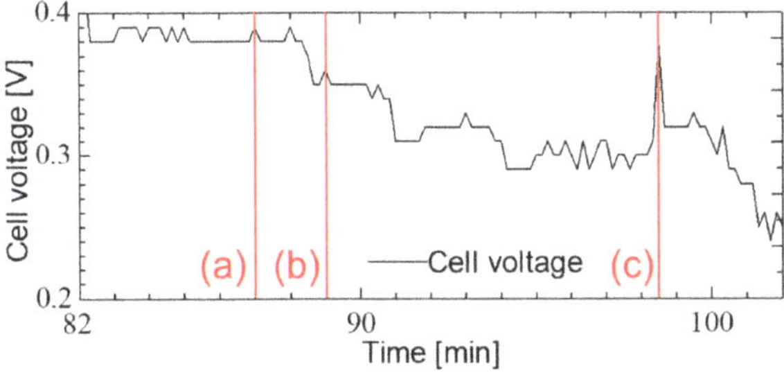

Fig. 5. Time-series of the cell voltage ($i = 1.0$ A/cm^2, $P_{back} = 250$ kPaG).

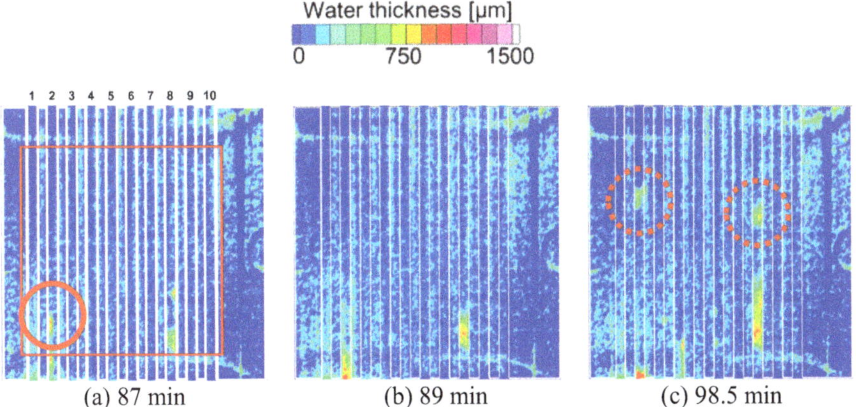

(a) 87 min   (b) 89 min   (c) 98.5 min

Fig. 6. Two-dimensional water distributions.

3.3   Effect of Liquid Water Plug in Flow Channels Downstream

The change in local area-average water thickness was calculated to analyze in detail. The calculation areas for (1)–(6) are shown in Fig. 7. The average water thickness was calculated over an area of 0.32 mm × 2.67 mm. Areas (2) and (5) are of channels, and Areas (1), (3), (4), and (6) are of the ribs on both sides. These six areas were selected to evaluate how the formation of a liquid plug downstream affects the formation of the liquid upstream. A moving average was applied to the time variation results. Figure 8 shows the results of the time variation of water thickness. The times in (a)–(c) correspond to the measurement times shown in Figs. 5 and 6. Immediately after the change in the current density, the water thickness increased proportionally. The difference between the channel and ribs and between the channels is small. The solid lines show the reference water accumulation rate calculated based on the electrochemical reaction, assuming that all generated water accumulated as liquid water. It was confirmed that the water thickness increased at a constant rate between 82 and 84 min and agreed well with the reference line. This trend was observed regardless of the channel and rib. Thus, the water generated by the electrochemical reaction mainly accumulates in the GDL during this period because a non-equilibrium phenomenon causes local supersaturation, resulting in liquid accumulation in the GDL. After 84 min, the water thickness remained nearly constant at approximately 100 μm. The thickness is the result of balancing the generated amount with the amount exhausted from the area. At 95 min, the water thickness in Area (5) started to increase. The rate of increase was the same as that of the reference rate. As described in Sect. 3.2, the formation of liquid water was observed downstream of Area (5) at 87 min and grew into a liquid plug at 89 min. Therefore, an increase occurred after the blockage of the gas flow. The gas supply was interrupted by a liquid plug downstream of the channel, causing local supersaturation. As a result, a rapid increase in the water thickness was thought to have occurred downstream of the channel.

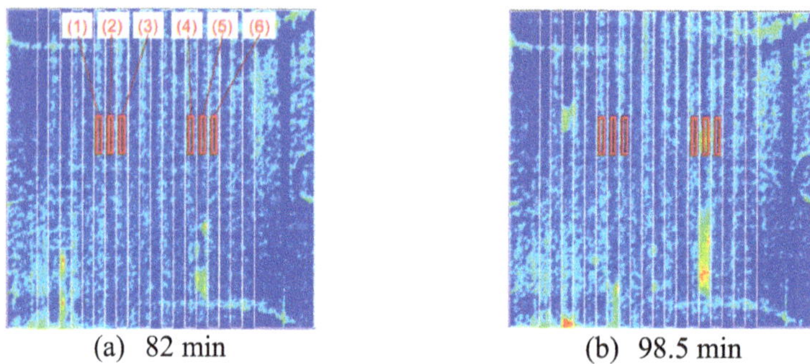

(a)   82 min   (b)   98.5 min

Fig. 7. Evaluation area of local water thickness.

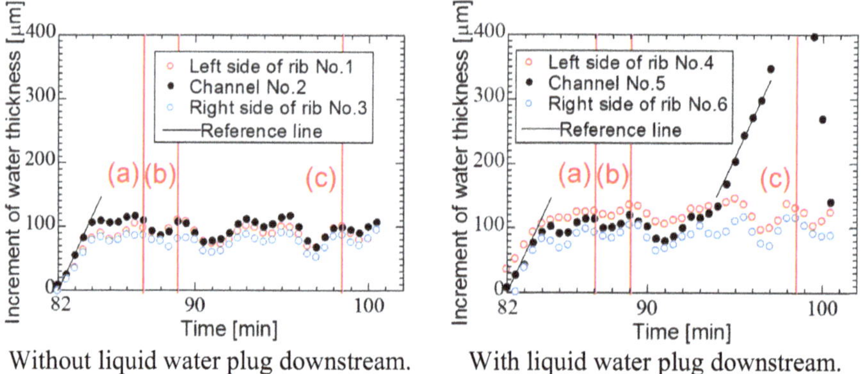

Without liquid water plug downstream.   With liquid water plug downstream.

Fig. 8. Time variation of water thickness.

4 Summary

Visualization of the water distribution in the PEFC under operation using neutron radiography was performed. The interaction between water distribution and fuel cell performance at 105 °C was evaluated by changing the operating pressure. When the operating pressure was atmospheric pressure, the current density could not be increased more than 0.6 A/cm^2 mainly due to the PEM dryout. However, the operatable current density could be increased to 1.0 A/cm^2 at 250 kPaG, and water accumulation was clearly confirmed in the flow channels. Owing to water accumulation, a decrease in cell performance was observed at $i = 1.0$ A/cm^2. To analyze the time variation of area-average water thickness, it was confirmed that the degradation of the cell performance was caused by liquid plug formation. In particular, the liquid plug formation downstream affected the liquid water generation upstream because the gas flow was blocked.

Acknowledgement. This work has been carried out in part under the Visiting Researchers Program of Kyoto University Institute for Integrated Radiation and Nuclear Science and ECCEED_GDL project of NEDO, Japan.

References

1. Baz, F.B., Elzohary, R.M., Osman, S., Marzouk, S.A., Ahmed, M.: A review of water management methods in proton exchange membrane fuel cells. J. Energy Conversion Manage. **302**, 118–150 (2024)
2. Nasu, M., et al.: Neutron imaging of generated water inside polymer electrolyte fuel cell using newly-developed gas diffusion layer with gas flow channels during power generation. J. Power. Sources **530**, 231–251 (2022)
3. J.H. Hubbell, S.M. Seltzer: Tables of X-Ray Mass Attenuation Coefficients and Mass Energy-Absorption Coefficients from 1 keV to 20 MeV for Elements Z = 1 to 92 and 48 Additional Substances of Dosimetric Interest. NISTIR 5632 (2004)
4. H. Murakawa, K. Sugimoto, N. Kitamura, H. Asano, N. Takenaka, Y. Saito: Visualization and Measurement of Water Distribution in Through-Plane Direction of Polymer Electrolyte Fuel Cell during Start-Up by Using Neutron Radiography. J. Flow Control, Measurement & Visualization 3, 122–133 (2015)
5. Kakinuma, K., Uchida, M., Taniguchi, H., Asakawa, T., Miyao, T., Aoki, Y., Iiyama, A.: The Possibility of Intermediate–Temperature (120 °C)–Operated Polymer Electrolyte Fuel Cells using Perfluorosulfonic Acid Polymer Membranes. J. Electrochemical Society 169 (2022)
6. Zhang, J., Tang, Y., Song, C., Cheng, X., Zhang, J., Wang, H.: PEM fuel cells operated at 0 % relative humidity in the temperature range of 23–120oC. J. Electrochimica Acta **52**, 5095–5101 (2007)

The Performance of Neutron Sensitive Scintillators for Imaging Purposes

E. Lehmann[1(✉)] and B. Walfort[2]

[1] Laboratory for Neutron Scattering and Imaging, Paul Scherrer Institut, Forschungsstrasse 111, CH-5232 Villigen, Switzerland
eberhard.lehmann@psi.ch
[2] RC Tritec AG/LumiNova AG, CH, CH-9053 Teufen, Switzerland

Abstract. Scintillation materials, excited by capturing of neutrons are the keys to obtain "shadow images" from observed materials as the basis of neutron imaging technologies like neutron tomography, real-time-imaging or grating neutron interferometry. Although the most promising material combinations have been known for a long time already, there is potential for further improvements in the selection of material combinations, manufacturing processes and adaptation to the sensor devices.

We present the currently available range of commercially available products and give an outlook to further potential materials, including their application fields.

Keywords: neutron detection · scintillator materials · spatial resolution · detection efficiency · afterglow · epithermal neutrons

1 Introduction

In the process of the transition of neutron imaging from film-based methods to digital ones (middle of the 90ies of last century), the neutron detection by scintillation received an essential role. Although the use of scintillators in detection systems for particle physics applications was already quite common, the production and application of plates with thin layers of the neutron sensitive materials with useful light output was not really established. Only a few companies provided such screens as a by-product of the more common X-ray imaging systems. There was no flexibility and interest of these companies to provide better and more efficient screens because the market is quite limited.

It was the initiative of Paul Scherrer Institute (PSI) to approach a small-sized company in Switzerland with the interest to develop and manufacture best suitable scintillator screens for neutron imaging applications, mainly used within camera-based detection systems. These detection systems have been proven to be the most flexible and best performing devices until now.

Two major lines of neutron sensitive scintillator materials have been found and developed towards commercial products: the combination of 6-LiF and ZnS – and Gadolinium oxysulfide, which is also used for X-ray imaging. However, the compounds, the mixture, the real enrichment and the production process is still a challenging procedure, defining

© The Author(s) 2026

A. E. Craft and H. Z. Bilheux (Eds.): WCNR 2024, SPPHY 348, pp. 188–199, 2026.
https://doi.org/10.1007/978-3-032-15003-5_22

the final detector performance. The creation of very thin material layers (1–5 μm) on Al plates with high homogeneity is still a problem to be solved. On the other hand, the application of other neutron absorber materials as converters is an ongoing challenge. Because only the best light output can be obtained if the neutron absorption is high, and the light excitation is stimulated effectively. These two part-processes needs to be optimized in the best way.

2   The Neutron Detection Process

Because neutrons as neutral particles carry no electrical charge and therefore cannot directly excite ionization processes, a neutron capture process is needed while the secondarily emitted radiation can perform light excitation in suitable materials. In most cases, the neutron scintillator detector consists of the neutron absorbing material and the scintillator material in a combined mixture in suitable ratios.

The process of neutron detection via light excitation is qualitatively described in Fig. 1.

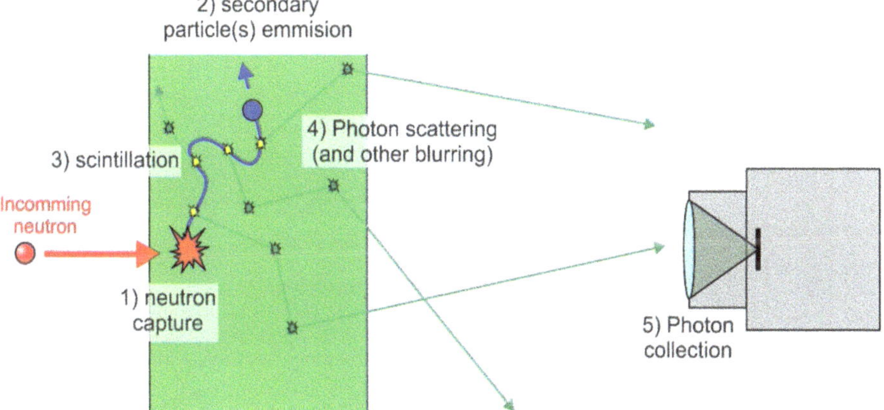

Fig. 1. The processes inside a composite scintillation layer of a neutron sensitive screen (picture: courtesy P. Boillat [1])

In this way, the grain size, the material layer thickness and also the optical transparency of the mixture play an important role to tune the most important parameters of the detection material. These are the intensity of the light emission, the spatial resolution, the signal/noise level and probably the afterglow properties. Because these parameters cannot be optimized at the same time in all respects and the demand for practical applications is not the same at all, specifically optimized material combinations have been found.

3   Tests with Discrete Material Combinations

Although the neutron absorption and the emission of secondary particles and radiation is well described in data libraries, the excitation process and the corresponding light emission is not yet been modelled accurately until now. Simulations of both parts of the process (absorption and emission, light excitation) still failed in the theoretical optimization procedures. Therefore, dedicated experimental verification is needed in the optimization process. In the search of the best combination of neutron absorbers and light emitters, ZnS has been used as one of the most shining scintillator materials.

All experiments were performed at the neutron imaging facility NEUTRA [2] at the Swiss spallation source SINQ, using the middle position and the MIDI detection system with the CCD camera ANDOR-icon-l [3].

As shown in Fig. 2, the camera setup was used for testing the combination of strong neutron absorbers (Au, In, Ho, Dy, Hf, Eu, Cd) with a layer of ZnS (200 um thick). There were considered two detector arrangements: The front- and the back-illumination as given in Fig. 3.

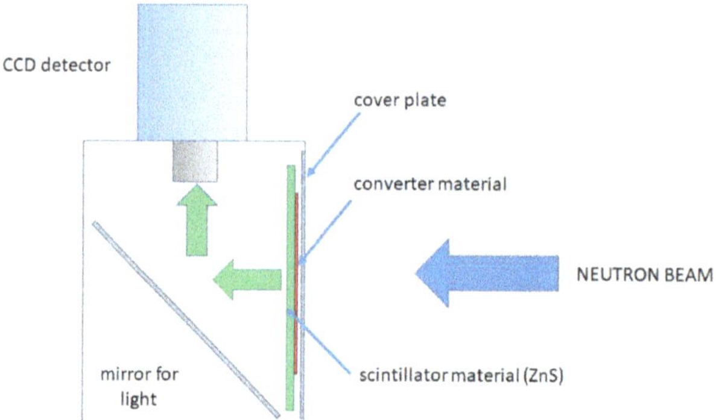

Fig. 2. Scheme of a camera based digital neutron imaging detector with separated converter/scintillator unit, used in this study.

As demonstrated by the data in Fig. 4, the back-illumination (BI) setup provides generally about the double number of countable photons, with deviations depending on the particular absorber material. The reason might be the self-absorption of the light coming back inside the ZnS screen and the better forward emission in the BI case.

All further investigations concerning the absorber materials were done by using the more effective BI – setup.

Concerning the light emission of the individual absorbers in metallic form, the data are given in Fig. 5. It has to be stated that all combinations with ZnS get a certain light output, however in different amounts. Most promising in this respect are In and Eu, considering the normalization to the absorption rate.

Unfortunately, there is no direct comparison to Li-6 possible due to its high chemical reactivity and the low availability of this enriched material. A combination is only possible as compounds like LiF.

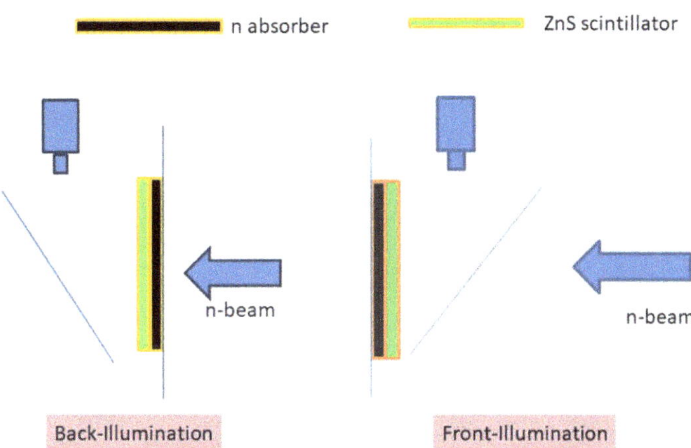

Fig. 3. Back- and front-illumination options in the study of heterogenous sample arrangements

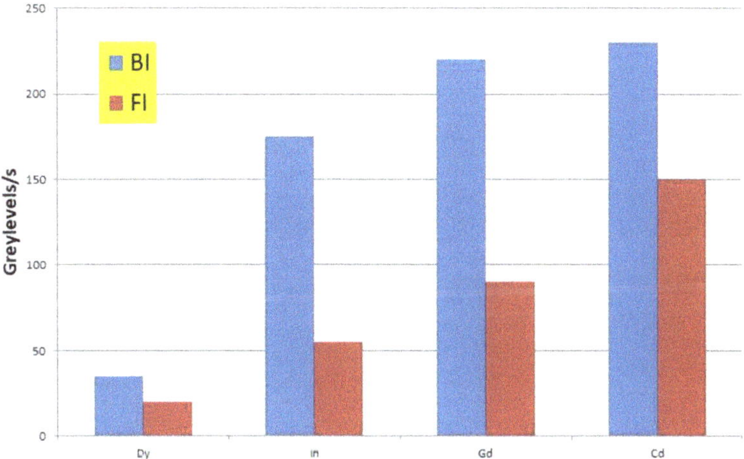

Fig. 4. Results of light efficiency measurements for the two cases of converter/scintillator arrangements as shown in Fig. 3; the values are not normalized and have to be considered as qualitative to show the effect of the detector setting (BI = back-illumination; FI = front-illumination)

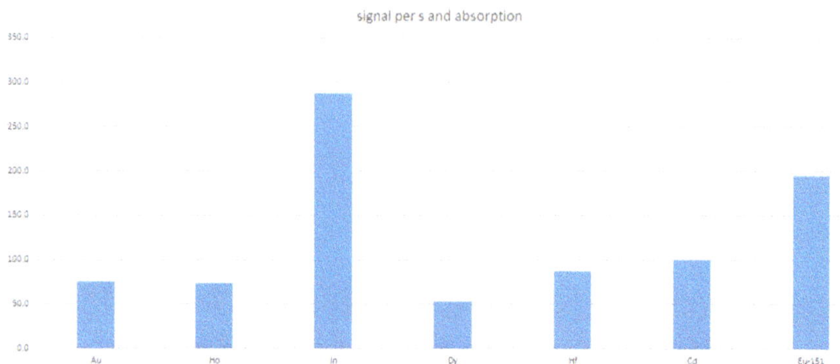

Results light excitation per absorbed neutrons

Element	signal per s and absorption
Au	75
Ho	75
In	288
Dy	52
Hf	88
Cd	100
Eu-151	195

Fig. 5. Comparison of the light emittance of different materials in contact with a 150 um ZnS layer, normalized to the absorption rate of the converter

4 Tests with Homogeneous Material Mixtures

Not all previously used materials are available as powders and mixable compounds. Therefore, only oxides of B-10, In, Ho, Eu and Hf were taken into account and measured in comparison to 6-LiF in the same mixing ratio to ZnS.

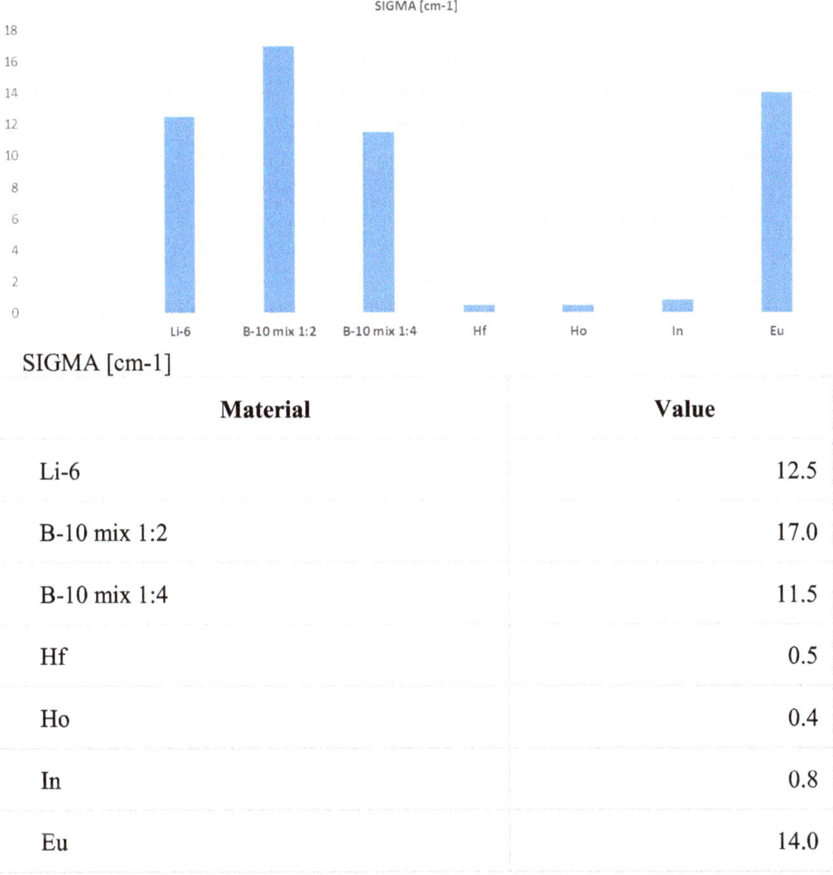

SIGMA [cm-1]

Material	Value
Li-6	12.5
B-10 mix 1:2	17.0
B-10 mix 1:4	11.5
Hf	0.5
Ho	0.4
In	0.8
Eu	14.0

Fig. 6. Absorption rate for different absorbers (as oxides) in homogenous mixture with ZnS in the same ratio

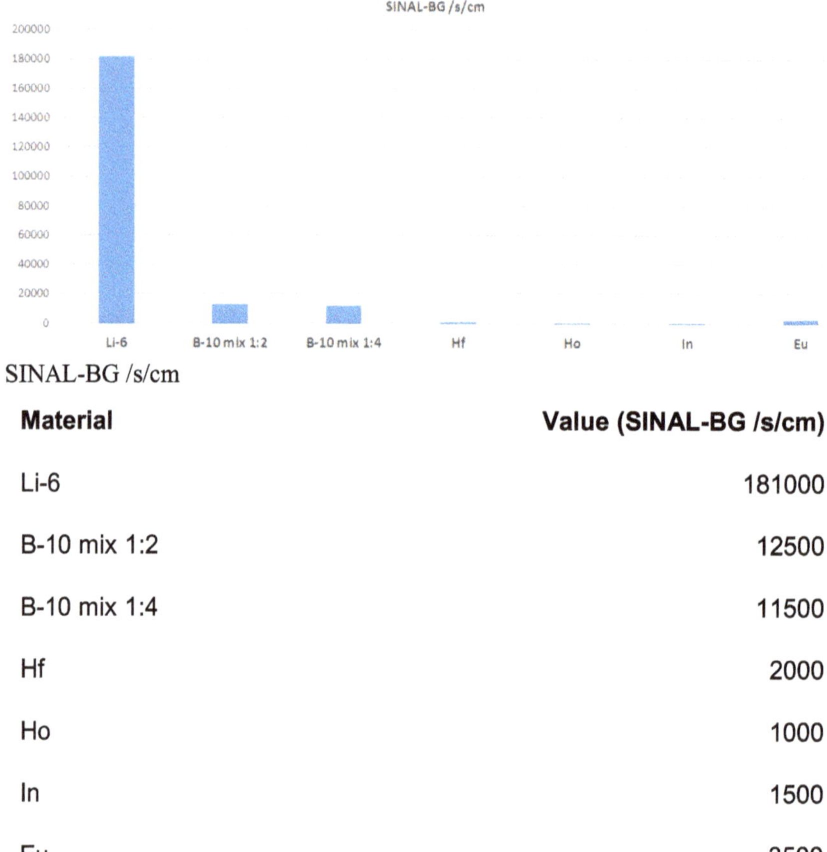

SINAL-BG /s/cm

Material	Value (SINAL-BG /s/cm)
Li-6	181000
B-10 mix 1:2	12500
B-10 mix 1:4	11500
Hf	2000
Ho	1000
In	1500
Eu	3500

Fig. 7. Light emission of the mixtures with oxides of the neutron absorbers and ZnS

As the results in Figs. 6 and 7 show, Eu_2O_3 and $B\text{-}10_2O_3$ have similar absorption rates (attenuation coefficients $[cm^{-1}]$) as 6-LiF. However, in the light emission they cannot compete by far with the common combination 6-LiF/ZnS.

5   Tests with Alternative Scintillation Materials

As Li-6 has been found to be best in the combination with ZnS regarding the excitation and light output, it was investigated if other scintillation materials than ZnS could be useful and superior. Table 1 gives the data of the used materials.

Table 1. Scintillator materials in combination with 6-LiF as neutron converter (not for Gd samples); mixtures as mass-per-cent

Absorber	Phosphor	mixture	thickness [mm]
6-LiF	ZnS:Cu	1 to 2	0.1
6-LiF	Y3(Al, Ga)5O12:Tb	1 to 2	0.1
6-LiF	Y2SiO3:Ce	1 to 2	0.1
6-LiF	Lu2SiO5:Ce	1 to 2	0.1
6-LiF	ZnO:Ga	1 to 2	0.1
6-LiF	Y3Al5O12:Ce	1 to 2	0.1
no extra	Gd2O2S:Tb		0.1
no extra	Gd2O2S:Pr,Ce,F		0.1

We investigated the absorption rate (Fig. 8) and the light output (Fig. 9) under identical conditions with the camera setup described above.

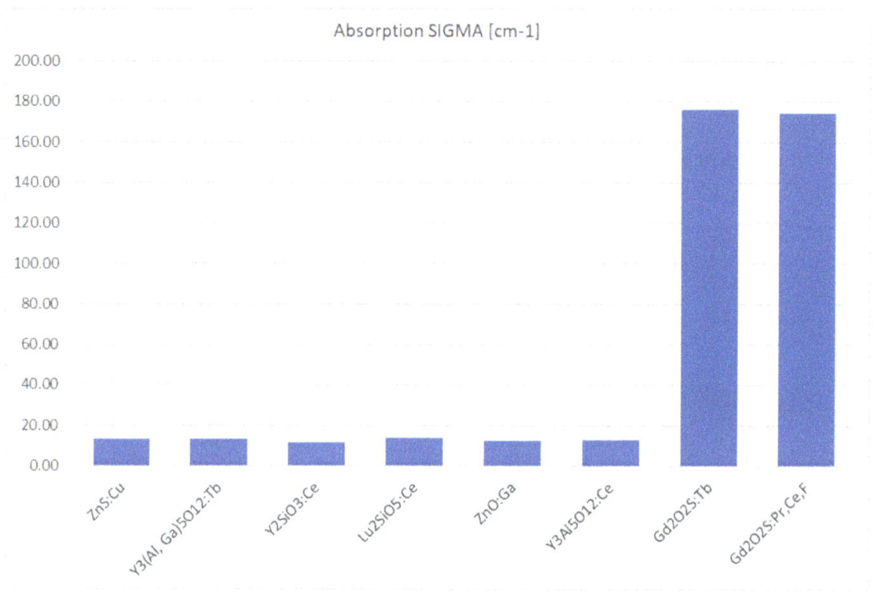

Absorption SIGMA [cm-1]

Material	Value [cm-1]
ZnS:Cu	14.00
Y3(Al, Ga)5O12:Tb	14.00
Y2SiO3:Ce	12.00
Material	**Value [cm-1]**
Lu2SiO5:Ce	14.00
ZnO:Ga	12.00
Y3Al5O12:Ce	12.00
Gd2O2S:Tb	177.00
Gd2O2S:Pr,Ce,F	177.00

Fig. 8. Capture rate as attenuation coefficient (in cm^{-1}) for all mixtures with Li-6, as described in Table 1

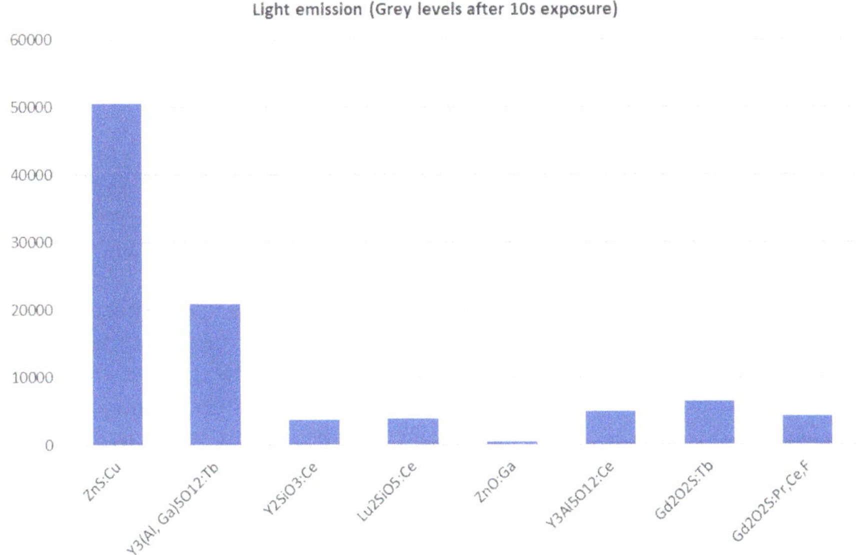

Light emission (Grey levels after 10s exposure)

Material	Light Emission (Grey levels after 10s exposure)
ZnS:Cu	50000
Y3(Al, Ga)5O12:Tb	22000
Y2SiO3:Ce	4000
Lu2SiO5:Ce	4000
Material	**Light Emission (Grey levels after 10s exposure)**
ZnO:Ga	800
Y3Al5O12:Ce	6000
Gd2O2S:Tb	6800
Gd2O2S:Pr,Ce,F	5800

Fig. 9. Light emission of the screens under identical conditions after 10 s exposure in the thermal beam at the NEUTRA facility.

Because the neutron absorber 6-Li is for all scintillator material combination with it the same, the absorption rate in Fig. 6 is identical. Only the Gd containing material have an about ten-time higher absorption rate.

On the other hand, the Gd containing scintillators are less shining as visible with the data in Fig. 9. And – the 6-Li based combinations are (again) dominated by ZnS.

6 Optimization Regarding the Applications

It depends on the particular application in neutron imaging experiments, which scintillation detector is the most useful one. Considering the parameters of light output, spatial resolution, statistical accuracy (given by the signal/noise ratio) and afterglow behavior has to be found.

In respect to very fast counting, the afterglow of ZnS has been found to be an issue. The replacement by ZnO:Zn in combination with 6-LiF has been found a good substitution option, even if its light output is only half of ZnS [6]. The applications domain will be "event-mode counting", time-of-flight applications in energy selective mode and very fast acquisition on very strong neutron sources. All other "standard" applications are not much affected by the afterglow behavior.

Although a good performance of currently used detector screens can be mentioned, there is still potential for improvements. The main parameters in this respect are: spatial resolution, light emission, signal/noise and afterglow behavior. As a GadoX screen with enriched Gd-157 converter [4] has shown its superior performance regarding spatial resolution in use of the NeutronMicroscope [5], the price for this technology is an issue. Cheaper and larger screens might be produced with nano-sized materials, still under development.

References

1. Boillat, P., et al.: Who made the noise. Opt. Express **32**(8), 14471–14489 (2024). https://doi.org/10.1364/OE.511939
2. Lehmann, E., Vontobel, P., Wiezel, L.: Properties of the neutron radiography facility NEUTRA. Nondestruct. Test. Eval. **16**(2–6) (2001)
3. https://andor.oxinst.com/products/ikon-xl-and-ikon-large-ccd-series/ikon-l-936
4. Trtik, P., Lehmann, E.: Isotopically enriched gadolinium-157 oxysulfide scintillator screens for the high-resolution neutron imaging. Nucl. Instrum. Methods Phys. Res. Sect. A **788**, 67–70Gd-157 (2015)
5. Trtik, P., et al.: The Neutron Microscope Project. Phys. Procedia **69**, 169–176 (2015)
6. Mann, S.E., Schooneveld, E.M., Rhodes, N.J., Liu, D., Sykora, G.J.: Position sensitive ZnO:Zn neutron detector – a high count rate alternative to ZnS:Ag scintillation detectors. Nucl. Inst. Methods Phys. Res. A **1057**, 168716 (2023)
7. Losko, A.S., Han, Y., Schillinger, B., et al.: New perspectives for neutron imaging through advanced event-mode data acquisition. Sci. Rep. **11**, 21360 (2021). https://doi.org/10.1038/s41598-021-00822-5

Review on Digital Neutron Imaging

Eberhard H. Lehmann[1](✉) and Burkhard Schillinger[2]

[1] Laboratory for Neutron Scattering and Imaging, Paul Scherrer Institute, Würenlingen, Switzerland
`eberhard.lehmann@psi.ch`
[2] Heinz Maier-Leibnitz-Zentrum (FRM II), Technische Universität München, Munich, Germany

Abstract. The transition from analogue, film-based detection methods to digital ones in the middle of the last decade of last century can be evaluated as a real "revolution" from the present point of view. We describe the main features of the detection processes in the neutron imaging and analyze which are the critical components for the different applications. In the outlook, we focus on the energy-selection by means of time-of-flight approaches and the options for single-event imaging with the suitable detection systems.

Keywords: neutron imaging · digital detectors · camera performance · scintillators screens · spatial resolution · frame rate · event mode

1 History and Starting Point

1.1 The Film Methods

Neutron radiography has been performed just after the availability of suitable neutron beams for imaging purposes [1, 2]. The combination of a converter layer of a neutron absorbing material with an X-ray sensitive film provided first images, which already showed the major power of neutron imaging: sensitivity for light elements, high penetration of heavy elements.

Over the last half of the 20th century, both the beam performance and the film methods have been improved and dedicated applications and their procedures were certified as "standard techniques". For special applications, these technological solutions are still valid and practiced until today, e.g. for critical components in space programs.

However, film methods have several disadvantages. Despite of the high spatial resolution over a wide field-of-view (FOV), the dynamic range is limited, and the detector response is non-linear. The information, frozen in the film cannot be post-processed and the data cannot be transferred easily. Even a digitalization with a suitable scanner device does not help to overcome the mentioned limitations much.

In particular, the acquisition of time series nend the investigations of sample motion has been impossible with film methods.

© The Author(s) 2026
A. E. Craft and H. Z. Bilheux (Eds.): WCNR 2024, SPPHY 348, pp. 200–205, 2026.
https://doi.org/10.1007/978-3-032-15003-5_23

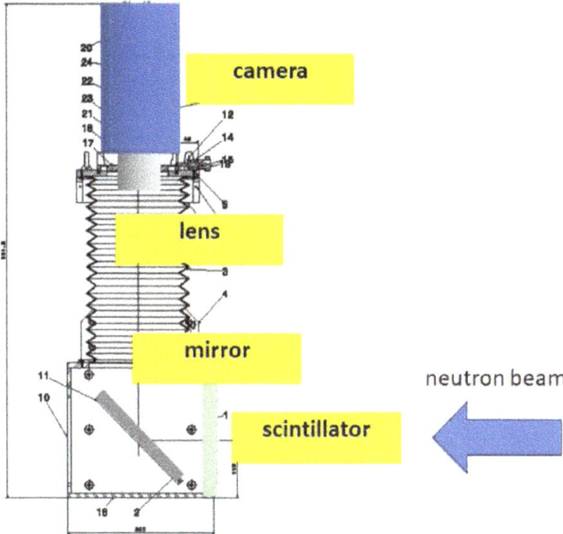

Fig. 1. Generic setup of a digital neutron imaging device, mostly camera based

1.2 Digital Systems

In order to overcome the before mentioned limitations, attempts with camera-based systems were started in the 90ies of the last century [3, 4]. The basic principle, as shown in Fig. 1, is valid until today: neutrons from the source are captured in an absorbing flat layer, which excite a scintillation process, then detected by a camera system.

The two critical components, the scintillator screens and the suitable camera system are both under further development and improvement in relation to the particular neutron imaging application.

Other detector features are the extremely light-tight cover of the system, a mirror which brings the camera out of the direct beam (risk of damage by neutrons and gammas) and an optical system that captures much of the emitted light.

Next to the camera approach, there have been developed other detection systems with the aim to overcome the analogue film method: semi-conducting flat panels with amorphous silicon as sensors [5] and imaging plates with pixelated readout scanners [6]. All techniques need a converter material (Gd, Li-6) for the excitation process before digitization.

2 State of the Art Today

Neutron imaging is dominated today by the camera solution. This is given by the high flexibility in the particular setup regarding FOV and spatial resolution. It is indicated in Fig. 2 with the five detector options, available at PSI. In this way, FOVs from only 1 cm up to 40 cm can be covered, when the lens system is tuned accordingly.

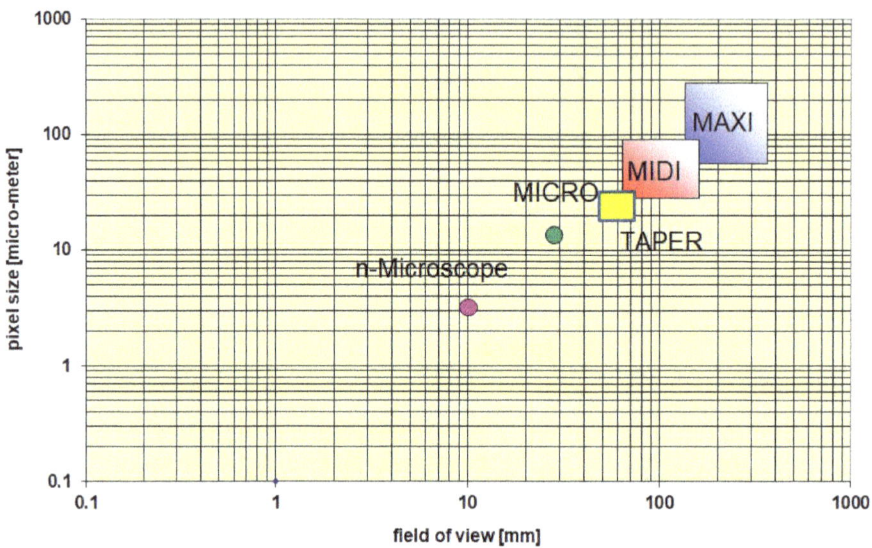

Fig. 2. Application range of camera-based neutron detection systems

The system with the highest spatial resolution was developed within the "neutron microscope" project by PSI [7]. Next to the direct view onto the scintillator screen, fiber-optical taper systems can "magnify" the FOV from the initial imaging process by a factor of 4 … 8 [8]. Many recent systems employ so-called 90 degree Heliflex lenses that had originally been developed for X-ray intensifiers. These lenses incorporate a 90 degree mirror system and project a 15 mm or 25 mm proximity-focused FoV to infinity, i.e. they produce a parallel beam that does not lose intensity over distance. The camera uses another inifinity-focused lens to look into the output of the Helifelx lens; in first order of approximation, the distance between the two lenses does not change the FoV or intensity of the projection [9]. With the appropriate lenses, this enables for a 1:1 projection with the pixel size of the camera equal to the effective pixel size of the detector.

All camera solutions are based on commercial systems, either with CCD or CMOS technologies. Their extremely high sensitivity and high quantum efficiency, coupled with a wide dynamic range and high linearity can only be achieved with cooling far below 0 °C, typically −20 °C …−60 °C, for a CCD, a little less for modern CMOS cameras.

An important advantage of the camera-based technology is its suitability for neutron tomography, where many projections of the object under investigation are taken over the full or partial angle range during rotation around the vertical axis. These projection data can be used for the reconstruction of the sample's volume with the option for slices at arbitrary positions and the segmentation of components within. Based on the improved computation power, neutron computed tomography is now a standard technology at most of the imaging facilities world-wide. The improvement of the data volumes in the tomography process is indicated in Fig. 3, mainly given by the pixel number of the detection system.

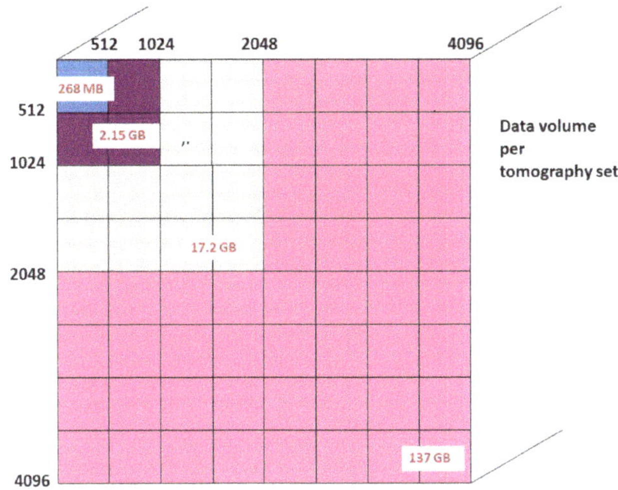

Fig. 3. Increase of the data volume for neutron tomography during the extension of the pixel numbers and the chip size of camera-based detectors

3   Further Developments

Specialized camera systems have been used coupled with an image intensifier for high-speed imaging (which mostly results in poor neutron statistics due to low incoming neutron flux) [10] or stroboscopic imaging with cooled camera for long-time exposure coupled with a gated intensifier that opens up at identical time windows of a periodic process, and integrated many such time windows into one image of good neutron statistics [11].

Apart from digital camera systems, several pixelated systems have been developed and practically applied. On first order, the micro-channel-plate (MCP) based device, initiated by A. Tremsin, needs to be mentioned. The MCP sensor contains B-10 as absorber/converter and the initiated and magnified electron cloud is detected and readout by a Mcdipix/Timepix chip [12]. Although the detector area is limited to only 28 mm, the advantage of the Multi-Channel Plate (MCP) system is its extremely high readout speed (100 ps). In this way, it is well suited for energy-selective measurements in the Time-of-Flight (TOF) principle, either at pulsed sources or by means of chopper systems. In the near future, systems with larger FOV will be made available.

Recently, an event-mode technology has been introduced and tested successfully [13]. It is based on an ultrafast detection and readout system with 1.5 ns (TimePix 3), which "sees" individual photons from neutron captures events and also gamma caused scintillator excitations, which can be distinguished due to their size and time characteristics. For a standard high speed camera and intensifier, the disadvantage of the method is the huge required storage capacity for image series and the software tools for producing the "valid" neutron image. In the newest versions, this is overcome by real-time event processing that stores only photon events with their location and time stamp, but not whole image frames. Due to the quick acquisition, the system is well suited for energy

= time dependent imaging applications. Both the storage capacities and the computation power in the future will help to make this technology more common and user-friendly.

Another key component for digital neutron imaging is the scintillator screen. Although the first such systems already used the combination of ZnS and ^{6}LiF, there was a lot of further development on this system and for other ones. This has been in particular the flexible, very homogeneous scintillator layer but also the best suited ratio of the composition in addition to the well selected ZnS supplier of this material, also the doping of ZnS that determines the color of emitted light which should be a dated to the best sensitivity range of cameras or image intensifiers. Furthermore, Gadolinium-oxysulfide has been found as most suitable for high spatial resolution applications, when very thin layers (few ten um) are used. The very best resolution has been achieved with GadoX using the the enriched isotope ^{157}Gd. The disadvantage of Gd-based screens is their low light output compared to ^{6}LiF + ZnS screens which introduces considerable photon noise into the pictures due to low photon statistics per neutron event. New screens are under development that use ^{10}B instead of ^{6}Li as a converter material. With a neutron cross section five times as high as that of ^{6}Li, and a light output a fifth of that of ^{6}Li + ZnS, These screens produce images of about the same brightness as 6LiF + ZnS, but five times higher neutron statistics. The detection efficiency is lower than that of Gadox, but the light output so much better that the brightness is again much higher than for Gadox, while achieving the same spatial resolution as thinned Gadox screens. [14]. Despite of the high prize, resolution on the order of a few micrometers have been verified for both Gadox and ^{10}B [15]. Further approaches in the scintillator technology are the use of nano-structured components, the use of alternative absorbers like Europium and to build screens most suitable for the application in the epithermal range of the neutron spectrum, based on Indium.

4 Conclusions and Outlook

The best possible use of neutrons at the imaging beamlines requires efficient and well-defined detectors, what is provided by digital systems. Their high performance regarding frame rate, spatial resolution, dynamic range and high statistical accuracy cannot easily be topped. However, there a some more needs for improvements, in particular for applications at the upcoming new pulsed sources with features for even higher energy resolutions and the imaging at Bragg edges [16].

Beside the hardware development there is a strong need for data treatment, image processing and automated data analyses. Some of the applications are manpower relevant because sophisticated data handling needs to be done hands-on, e.g. tomography visualization.

References

1. Kallmann, H., Kuhn, E.: Photographic detection of slowly moved neutrons, United States Patent 2,186,757 (1940)
2. Peter, O.: Neutronendurchleuchtungm Z. Naturforsch. A 1 557-9 (1946)

3. Lanza, R.C., et al.: An RFQ accelerator-based neutron radiography and tomography system. In: Proc. 5th WCNR, Berlin (1996)
4. Schillinger, B., et al.: 3D neutron computed tomography: requirements and applications. Physica B **276–278**, 59–62 (2000)
5. Lehmann, E., Vontobel, P.: The use of amorphous Silicon flat panels as detector in neutron imaging. Appl. Radiat. Isot. **61**(4), 567–571 (2004)
6. Takahashi, K., et al.: Imaging performance of imaging plate neutron detectors. Nucl. Instrum. Methods Phys. Res., Sect. A **377**(1), 119–122 (1996)
7. Trtik, P., et al.: Improving the Spatial Resolution of Neutron Imaging at Paul Scherrer Institut – the neutron microscope project. Phys. Procedia **69**, 169–176 (2015)
8. Morgano, M., et al.: Unlocking high spatial resolution in neutron imaging through an add-on fibre optics taper. Opt. Express **26**(2), 1809–1816 (2018). https://doi.org/10.1364/OE.26.001809
9. Tengattini, A., et al.: NeXT-Grenoble, the neutron and X-ray tomograph in Grenoble. NIM A **968**(11), 163939 (2020)
10. Tremsin, A.S., et al.: Detection efficiency, spatial and timing resolution of thermal and cold neutron counting MCP detectors. Nucl. Instruments Methods Phys. Res. Sect. A Page 10/11, Assoc. Equip. https://doi.org/10.1016/j.nima.2009.01.041 (2009)
11. Lani, C., Zboray, R.: Development of a high frame rate neutron imaging method for two-phase flows. NIM A **954**(21), 161707 (2020)
12. Schillinger, B. et al.: A study of oil lubrication in a rotating engine using stroboscopic neutron imaging. Physica B: Condensed Matter. **385–386**, Part 2, 921–923 (2006)
13. Losko, A.S., et al.: New perspectives for neutron imaging through advanced event-mode data acquisition. Sci. Rep. **11**, 21360 (2021)
14. Chuirazzi, W., et al.: Boron-based neutron scintillator screens for neutron imaging. J. Imaging **6**(11), 124 (2020). https://doi.org/10.3390/jimaging6110124
15. Trtik, P., Lehmann, E.: Isotopically-enriched gadolinium-157 oxysulfide scintillator screens for the high-resolution neutron imaging. Nucl. Instrum. Methods Phys. Res., Sect. A **788**(11), 67–70 (2015)
16. Malamud, F. et al.: Bragg edge imaging characterization of multi-material laser powder-bed fusion specimens, J. Phys.: Conf. Ser. 2605 012030. https://doi.org/10.1088/1742-6596/2605/1/012030

Fast Hyperspectral Reconstruction for Neutron Computed Tomography Using Subspace Extraction

Mohammad Samin Nur Chowdhury[1], Diyu Yang[2], Shimin Tang[3(✉)], Singanallur V. Venkatakrishnan[4], Andrew W. Needham[5], Hassina Z. Bilheux[3], Gregery T. Buzzard[6], and Charles A. Bouman[1]

[1] ECE, Purdue University, West Lafayette, IN 47907, USA
{chowdh31,bouman}@purdue.edu
[2] Apple Inc., Cupertino, CA 95014, USA
yang1467@purdue.edu
[3] Neutron Scattering Division, ORNL, Oak Ridge, TN 37830, USA
{tangs,bilheuxhn}@ornl.gov
[4] Electrical and Engineering Infrastructure Division, ORNL, Oak Ridge, TN 37831, USA
venkatakrisv@ornl.gov
[5] NASA Goddard Space Flight Center, Greenbelt, MD 20771, USA
andrew.w.needham@nasa.gov
[6] Department of Mathematics, Purdue University, West Lafayette, IN 47907, USA
buzzard@purdue.edu

Abstract. Hyperspectral neutron computed tomography enables 3D non-destructive imaging of the spectral characteristics of materials. In traditional hyperspectral reconstruction, the data for each neutron wavelength bin is reconstructed separately. This per-bin reconstruction is extremely time-consuming due to the typically large number of wavelength bins. Furthermore, these reconstructions may suffer from severe artifacts due to the low signal-to-noise ratio in each wavelength bin.

We present a novel fast hyperspectral reconstruction algorithm for computationally efficient and accurate reconstruction of hyperspectral neutron data. Our algorithm uses a subspace extraction procedure that transforms hyperspectral data into low-dimensional data within an intermediate subspace. This step effectively reduces data dimensionality and spectral noise. High-quality reconstructions are then performed within this low-dimensional subspace. Finally, the algorithm expands the subspace reconstructions into hyperspectral reconstructions. We apply our algorithm to measured neutron data and demonstrate that it reduces computation and improves reconstruction quality compared to the conventional approach.

Keywords: neutron computed tomography · hyperspectral imaging · hyperspectral reconstruction · non-negative matrix factorization

© The Author(s) 2026
A. E. Craft and H. Z. Bilheux (Eds.): WCNR 2024, SPPHY 348, pp. 206–215, 2026.
https://doi.org/10.1007/978-3-032-15003-5_24

1   Introduction

Neutron computed tomography (nCT) enables volumetric reconstruction of a sample from radiographs recorded by exposing the sample to a neutron source at multiple angles. nCT has been used to identify foreign materials or contamination inside an object [1], to study the effect of environmental factors on material properties, and to monitor the aging of polymers or other materials [2]. Importantly, nCT provides information that is complementary to traditional X-ray CT because neutrons interact directly with the nucleus rather than the electron cloud [3].

Hyperspectral neutron computed tomography (HSnCT) is a more advanced technique in which a pulsed neutron source illuminates a sample, and a time-of-flight detector measures the projection images across a range of wavelengths - potentially of the order of a few thousand. Using HSnCT, it is possible to analyze material characteristics like crystallographic phases [4] and isotopic compositions [5].

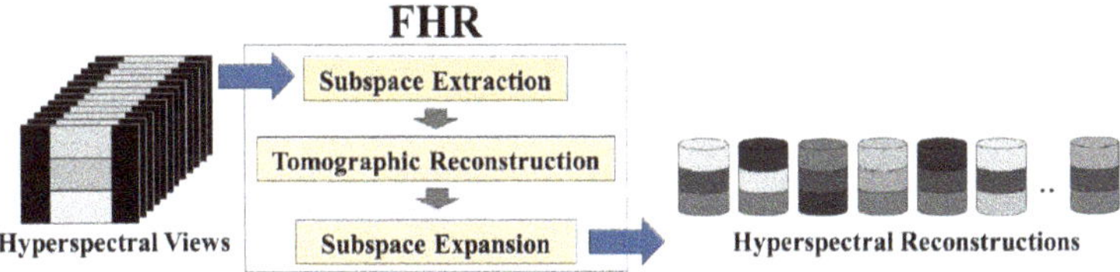

Fig. 1. Overview of the fast hyperspectral reconstruction (FHR) algorithm. FHR reduces the spectral dimension during subspace extraction and restores it during subspace expansion. In between, FHR performs tomographic reconstruction to transition from the sinogram domain to the spatial domain.

HSnCT reconstruction is challenging since data from potentially thousands of neutron wavelength bins must be reconstructed into individual 3D volumes. Direct hyperspectral reconstruction (DHR) [6] is the most established solution to this problem. In DHR, data from each wavelength bin is reconstructed separately, typically using filtered back projection (FBP) due to its computational speed. However, even with FBP, DHR is still highly time-consuming since it may require thousands of 3D tomographic reconstructions. Moreover, producing high-quality reconstructions with FBP requires a large number of projections (also referred to

This manuscript has been authored by UT-Battelle, LLC, under contract DE-AC05-00OR22725 with the US Department of Energy (DOE). The US government retains and the publisher, by accepting the article for publication, acknowledges that the US government retains a nonexclusive, paid-up, irrevocable, worldwide license to publish or reproduce the published form of this manuscript, or allow others to do so, for US government purposes. DOE will provide public access to these results of federally sponsored research in accordance with the DOE Public Access Plan (http://energy.gov/downloads/doe-public-access-plan).

as views) with high signal-to-noise ratios (SNR), which are extremely difficult to acquire due to time and resource constraints. Consequently, DHR reconstructions using FBP often suffer from severe noise and reconstruction artifacts. While DHR can alternatively use an advanced algorithm like model-based iterative reconstruction (MBIR) [7] to improve quality, the computation time would be too high to make it practically implementable.

In this paper, we present a fast hyperspectral reconstruction (FHR) algorithm for effective and efficient reconstruction of HSnCT data. As shown in Fig. 1, the first step of FHR is a subspace extraction procedure [8] that transforms the high-dimensional hyperspectral measurements into low-dimensional subspace projection views. FHR then performs reconstruction within this low-dimensional subspace. Finally, FHR expands the subspace reconstructions into hyperspectral reconstructions.

The use of subspace extraction in our algorithm serves two purposes:

- Substantially improves SNR in the reconstructions by approximating the data within a low-dimensional subspace.
- Significantly accelerates computation by reducing the required number of tomographic reconstructions.

We apply our algorithm to measured neutron data collected at the Oak Ridge National Laboratory Spallation Neutron Source SNAP beamline and demonstrate that it is substantially faster and yields better results compared to the baseline DHR method.

2 Hyperspectral Imaging System

Figure 2 illustrates a standard hyperspectral neutron imaging system that is used to collect wavelength-resolved hyperspectral data from multiple orientations of the sample at a pulsed neutron source [9]. The pulse of neutrons passes through the sample and is detected by a 2D time-of-flight (ToF) imaging array [10]. The

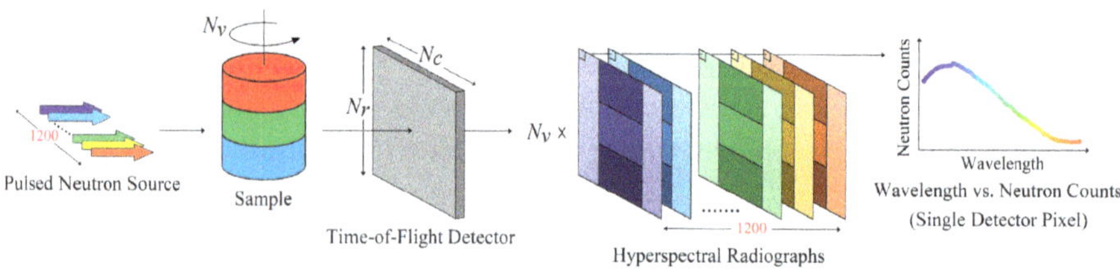

Fig. 2. Illustration of a hyperspectral imaging setup at a spallation neutron source. The arrows on the left represent a pulsed neutron source, which generates a beam of neutrons across a range of wavelengths. The beam travels through the sample to a $N_r \times N_c$ time-of-flight (TOF) detector, which produces wavelength-resolved radiographs for $N_k = 1200$ bins. The right plot shows neutron counts across the wavelength bins for a single detector pixel. Such data are collected for N_v orientations of the sample.

ToF detector counts the number of neutrons at each pixel and for each time interval bin. These time interval bins then correspond to each neutron's velocity or wavelength. The specific relationship between the neutron time of flight, Δt, and the wavelength, λ, is given by

$$\lambda = \frac{h}{m_n} \frac{\Delta t}{L} \ , \tag{1}$$

where h is Planck's constant, m_n is the neutron mass, and L is the distance between the source and the detector.

We now introduce the following notation to describe the relevant quantities:

- N_r is the number of detector rows
- N_c is the number of detector columns
- N_k is the number of wavelength bins
- N_v is the number of tomographic views

The output of the ToF detector is a hyperspectral neutron radiograph in the form of a $N_r \times N_c \times N_k$ array. In our experiment, we used $N_r = N_c = 512$ and $N_k = 1200$. So, a single hyperspectral radiograph in our experiment contains approximately 300 megapixels.

In order to perform a tomographic scan, the object is rotated to N_v orientations [11,12] and at each orientation, a hyperspectral radiograph, $y_{v,r,c,k}$, is measured where v, r, c, k represent the discrete view, row, column, and wavelength indices. In addition, a single hyperspectral radiograph is measured with the object removed, which we denote by $y^o_{r,c,k}$. From this, we can compute the normalized hyperspectral projection views p as

$$p_{v,r,c,k} = - \log \left(\frac{y_{v,r,c,k}}{y^o_{r,c,k}} \right) . \tag{2}$$

We note that in practice, various corrections must be made in the calculation of (2) to account for effects like scatter, detector bias, and detector drift [5].

3   Fast Hyperspectral Reconstruction (FHR)

Figure 3 illustrates the three steps in FHR consisting of subspace extraction, tomographic reconstruction, and subspace expansion. FHR reduces the spectral dimension during subspace extraction, then performs volumetric reconstruction in this lower dimensional space, and finally restores the spectral dimension using subspace expansion. In more colorful terms, the algorithm is "dehydrating" the data in the sinogram domain and then "rehydrating" in the reconstruction domain.

We use the following notation in the remaining sections:

- N_s is the dimension of the subspace
- $N_p = N_v \times N_r \times N_c$ is the number of measurements for each wavelength bin

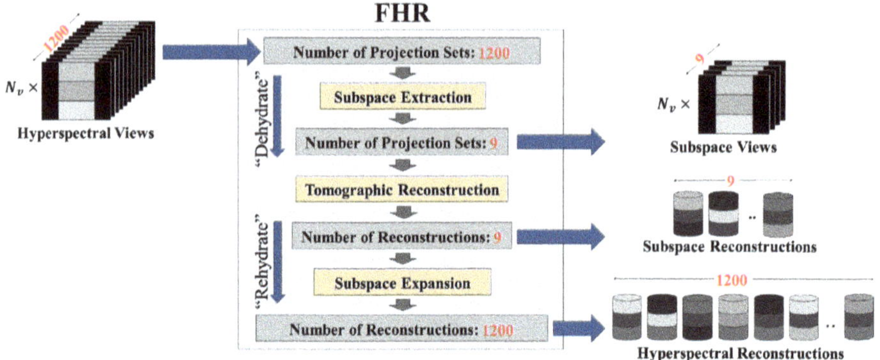

Fig. 3. Illustration of the FHR algorithm. FHR first transforms the $N_k = 1200$ dimensional hyperspectral views into $N_s = 9$ dimensional subspace views. Then, it performs tomographic reconstruction to produce 9 subspace reconstructions. Finally, it expands the 9 subspace reconstructions into 1200 hyperspectral reconstructions.

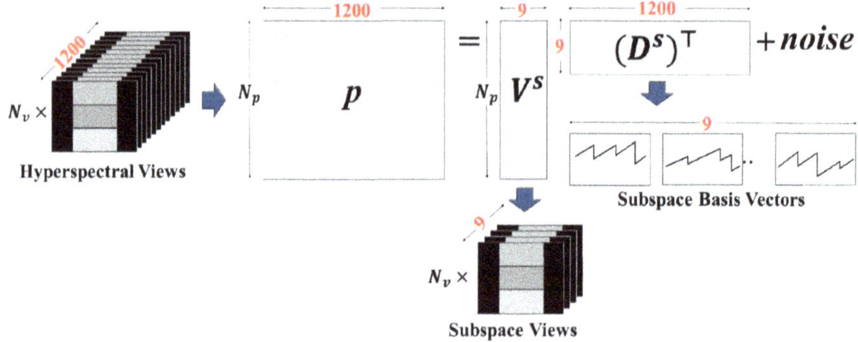

Fig. 4. Subspace extraction: FHR employs NMF to decompose the $N_k = 1200$ dimensional hyperspectral views (p) into $N_s = 9$ dimensional subspace views (V^s) and corresponding subspace basis vectors (D^s). This approach effectively reduces data dimensions and also eliminates significant spectral noise.

- $N_x = N_r \times N_c \times N_c$ is the number of voxels for each wavelength bin

Subspace Extraction:

Figure 4 illustrates the subspace extraction step of FHR in which the high-dimensional hyperspectral views $p \in \mathbb{R}^{N_p \times N_k}$ are decomposed into low-dimensional subspace views $V^s \in \mathbb{R}^{N_p \times N_s}$ and corresponding subspace basis vectors $D^s \in \mathbb{R}^{N_k \times N_s}$. The subspace dimension N_s is chosen such that $N_m < N_s << N_k$, where N_m is the number of materials present in the sample. The decomposition can be obtained by solving the non-negative matrix factorization (NMF) [13] problem.

Since $N_s << N_k$, operations in the subspace domain are much faster than in the hyperspectral domain. Also, empirically, the residual difference from the decomposition $\epsilon = p - V^s(D^s)^\top$ is primarily spectral noise. This greatly reduces the noise in the final reconstructions.

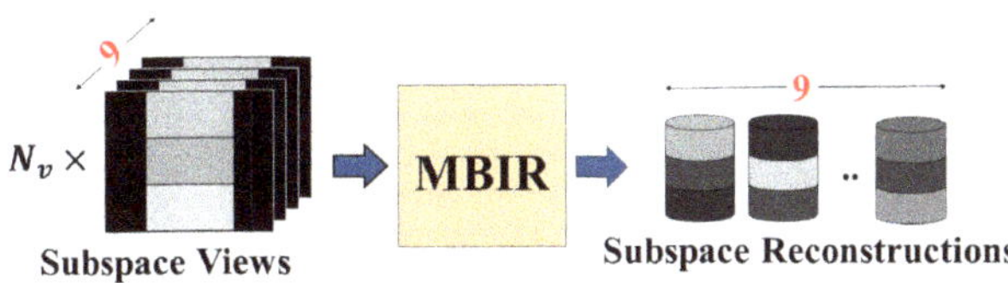

Fig. 5. Tomographic reconstruction: FHR computes $N_s = 9$ reconstructions from the extracted 9 sets of subspace views using MBIR.

Tomographic Reconstruction:

Figure 5 illustrates the reconstruction step of FHR in which N_s reconstructions within the subspace $x^s \in \mathbb{R}^{N_x \times N_s}$ are computed using MBIR [7]. MBIR is well known to produce superior reconstruction when dealing with sparse and low SNR measurements - hence is the ideal choice for HSnCT [14]. In order to perform MBIR, FHR uses the recently released svMBIR Python package [15].

We note that MBIR reconstruction tends to be much slower than FBP reconstruction, but since MBIR enables sparse view reconstruction, it can effectively reduce the data acquisition time for HSnCT. Also, the increased reconstruction time for MBIR is of much less concern since FHR only requires the reconstruction of N_s volumes rather than the N_k 3D volumes required for DHR.

Subspace Expansion:

Figure 6 illustrates the subspace expansion step of FHR in which the algorithm expands the subspace reconstructions into hyperspectral reconstructions using D^s. Since D^s maps each voxel from subspace to hyperspectral coordinates, the resulting hyperspectral reconstruction is of size $x^h \in \mathbb{R}^{N_x \times N_k}$.

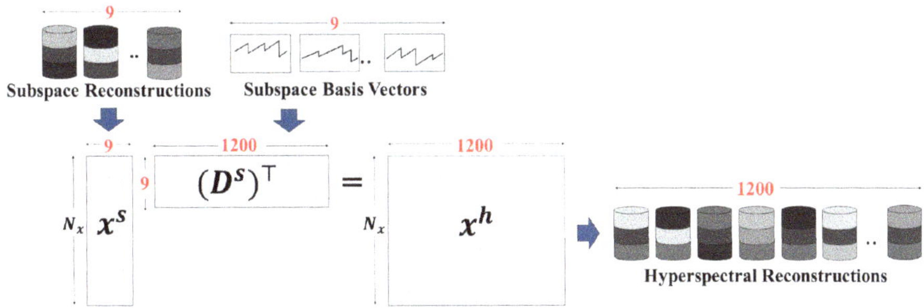

Fig. 6. Subspace expansion: FHR expands the $N_s = 9$ subspace reconstructions (x^s) to $N_k = 1200$ hyperspectral reconstructions (x^h) using the subspace basis vectors (D^s).

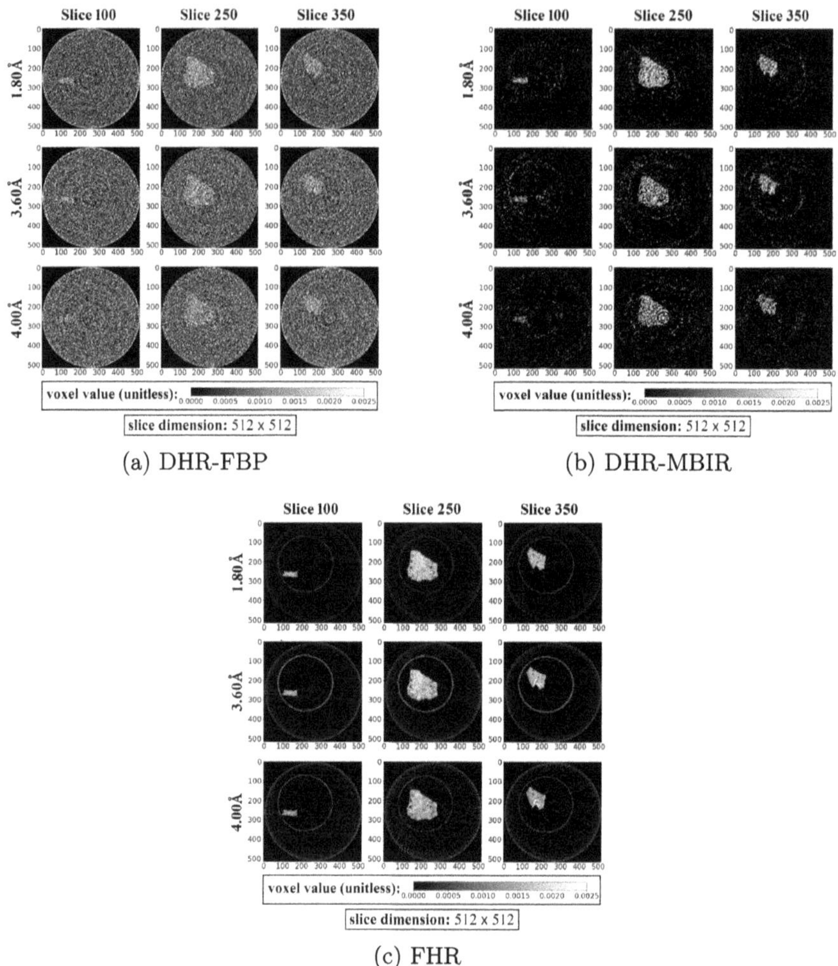

(a) DHR-FBP

(b) DHR-MBIR

(c) FHR

Fig. 7. Hyperspectral reconstruction of a moon-rock sample: (a) using the baseline DHR method with FBP, (b) using the baseline DHR method with MBIR, and (c) using the proposed FHR algorithm. The images shown are three different slices from the 3D volumes at three different wavelength bins. While DHR-MBIR reconstructions exhibit significantly fewer artifacts compared to DHR-FBP, they are still impacted by noise due to the low SNR per wavelength bin. In contrast, the proposed FHR dramatically improves the reconstruction quality by suppressing noise while preserving edge details.

4   Results

Below, we present results for hyperspectral reconstruction of a Moon rock sample from Apollo Mission 14, measured at the Spallation Neutron Source at Oak Ridge National Laboratory. HSnCT data was collected with $N_v = 53$ views, $N_r = N_c = 512$, and $N_k = 1200$ hyperspectral wavelength bins. So, the resulting

3D volumes have dimensions $512 \times 512 \times 512$. We heuristically selected $N_s = 9$ as the dimension of the subspace.

Figure 7 presents hyperspectral reconstructions for the measured data using DHR-FBP, DHR-MBIR, and the proposed FHR algorithm. Notice that while DHR-MBIR reconstructions demonstrate superior quality compared to DHR-FBP, they still contain significantly more noise and artifacts than FHR reconstructions. This is because of the initial subspace extraction in FHR, which removed the spectral noise orthogonal to the $N_s = 9$ dimensional space. Table 1 shows a quantitative performance comparison among DHR-FBP, DHR-MBIR, and FHR. For this dataset, FHR resulted in an SNR approximately 32 dB higher than DHR-FBP and 8.5 dB higher than DHR-MBIR. Moreover, FHR was over 10 times faster than DHR-FBP and over 100 times faster than DHR-MBIR, highlighting the computational efficiency of our approach.

Table 1. Quantitative performance comparison among DHR-FBP, DHR-MBIR, and the proposed FHR algorithm.

Algorithm	SNR	Computation Time
DHR-FBP	-6.18 dB	557.03 min
DHR-MBIR	17.40 dB	6181.84 min
FHR	25.94 dB	50.87 min

5   Conclusion

We present a fast hyperspectral reconstruction (FHR) algorithm for HSnCT that produces higher-quality reconstructions with dramatically less computation when compared to a conventional direct HSnCT reconstruction method. The intermediate subspace extraction introduced in FHR decreases data dimensionality by a factor of up to 100 and reduces spectral noise. In our experiments, this resulted in a decrease in overall computational time by factors of 10 and 100 compared to DHR-FBP and DHR-MBIR, respectively. We also observed an increase in reconstruction SNR by approximately 32 dB over DHR-FBP and 8.5 dB over DHR-MBIR.

Acknowledgment. C. Bouman was partially supported by the Showalter Trust. This research used resources at the Spallation Neutron Source, a DOE Office of Science User Facility operated by the Oak Ridge National Laboratory. The beam time was allocated to the Spallation Neutrons and Pressure Diffractometer (SNAP) instrument on proposal number IPTS-25265.

References

1. Kim, K.H., Klann, R.T., Raju, B.B.: Fast neutron radiography for composite materials evaluation and testing. Nucl. Instrum. Methods Phys. Res., Sect. A **422**(1–3), 929–932 (1999)
2. Chin, J., et al.: Temperature and humidity aging of poly(p-phenylene-2,6-benzobisoxazole) fibers: Chemical and physical characterization. Polym. Degrad. Stab. **92**(7), 1234–1246 (2007)
3. Vlassenbroeck, J., et al.: A comparative and critical study of X-ray CT and neutron CT as non-destructive material evaluation techniques. Geological Soc. London, Special Public. **271**(1), 277–285 (2007)
4. Woracek, R., et al.: 3D mapping of crystallographic phase distribution using energy-selective neutron tomography. Adv. Mater. **26**(24), 4069–4073 (2014)
5. Balke, T., Long, A. M., Vogel, S. C., Wohlberg, B., Bouman, C. A.: Hyperspectral neutron CT with material decomposition. In: 2021 IEEE International Conference on Image Processing (ICIP), pp. 3482-3486, IEEE, (2021, September)
6. Daugherty, M. C., et al.: Assessment of dose-reduction strategies in wavelength-selective neutron tomography. SN Computer Science, 4(5), 586 (2023)
7. Bouman, C. A.: Foundations of computational imaging: a model-based approach. Society for Industrial and Applied Mathematics (2022)
8. Chowdhury, M. S. N., et al.: Autonomous polycrystalline material decomposition for hyperspectral neutron tomography. In: 2023 IEEE International Conference on Image Processing (ICIP), pp. 1280-1284, IEEE (2023, October)
9. Nelson, R. O., et al.: Neutron imaging at LANSCE-From cold to ultrafast. J. Imaging, 4(2), 45 (2018)
10. Tremsin, A. S., Vallerga, J. V., McPhate, J. B., Siegmund, O. H., Raffanti, R.: High resolution photon counting with MCP-timepix quad parallel readout operating at > 1 KHz frame rates. IEEE Transactions on Nuclear Science, 60(2), 578-585 (2012)
11. Yang, D., et al.: An edge alignment-based orientation selection method for neutron tomography. In: ICASSP 2023-2023 IEEE International Conference on Acoustics, Speech and Signal Processing (ICASSP), pp. 1-5, IEEE (2023)
12. Tang, S., et al.: A machine learning decision criterion for reducing scan time for hyperspectral neutron computed tomography systems. Scientific Reports, 14(1), 15171 (2024)
13. Pauca, V.P., Piper, J., Plemmons, R.J.: Nonnegative matrix factorization for spectral data analysis. Linear Algebra Appl. **416**(1), 29–47 (2006)
14. Venkatakrishnan, S., et al.: Improved acquisition and reconstruction for wavelength-resolved neutron tomography. J. Imaging **7**(1), 10 (2021)
15. SVMBIR development team: super-voxel model based iterative reconstruction (SVMBIR). Software Library (2020). https://github.com/cabouman/svmbir

Preliminary Design Study on Neutron Imaging Facilities in a New Research Reactor in Japan

Daisuke Ito[1]([⊠]) [iD], Naoya Odaira[1] [iD], Yasushi Saito[1] [iD], Takenao Shinohara[2] [iD], and Yoshiaki Kiyanagi[3] [iD]

[1] Institute for Integrated Radiation and Nuclear Science, Kyoto University, Kumatori 590-0494, Osaka, Japan
ito.daisuke.5a@kyoto-u.ac.jp
[2] J-PARC Center, Japan Atomic Energy Agency, Tokai 319-1184, Ibaraki, Japan
[3] Hokkaido University, Sapporo, Hokkaido 060-0808, Japan

Abstract. A new research reactor is planned to be constructed in Japan, and a neutron imaging facility is considered one of the priority facilities on the site. The specifications for the construction of the facility have now been discussed. Installing two separate instruments using thermal and cold neutrons is desirable for performing a wide range of application and development research. Furthermore, they should be the most advanced facilities in the world. In this paper, we summarize the current status of the specifications required to perform state-of-the-art neutron imaging research in each instrument and establish the basis for instrument design.

Keywords: Neutron imaging · Research reactor · Thermal neutron · Cold neutron · Imaging facility

1 Introduction

The new research reactor is now planned to be constructed at the fast breeder reactor Monju site, which the Japanese government had decided to decommission [1]. It aims to become a core center for research and development in the new nuclear field and human resource development in Japan. The site will include state-of-the-art facilities.

The neutron imaging facility is one of the most essential parts of the site. The main neutron energies used in neutron imaging are thermal and cold neutrons. Thermal neutrons generally have a higher transmission characteristic than cold neutrons. On the other hand, cold neutrons are characterized by their ability to produce higher contrast images than thermal neutrons. Since each measurement has different characteristics, installing independent experimental apparatuses for thermal and cold neutrons in the new research reactor is desirable.

The new research reactor will be constructed in Fukui Prefecture, Japan's most active dinosaur fossil research site. Various analyses have been conducted on fossils using X-rays or other methods [2]; however, the use of neutrons could lead to the clarification of the existence of organic matter as well as 3-D structure data. The application of neutron

© The Author(s) 2026
A. E. Craft and H. Z. Bilheux (Eds.): WCNR 2024, SPPHY 348, pp. 216–221, 2026.
https://doi.org/10.1007/978-3-032-15003-5_25

imaging is expected to result in discoveries of fossils excavated in Fukui. On the other hand, the feasibility of applying neutron imaging to study the varved sediments in Lake Suigetsu [3], located in Fukui Prefecture, is being investigated. Clarifying geomagnetic features by magnetic imaging is considered a key to unraveling the history of the Earth. With these applications, neutron imaging can demonstrate a significant role in the local contribution of the new research reactor, which is one of its objectives.

To conduct characteristic research such as high-speed behavior measurement, high spatial resolution observation, and magnetic imaging, including new studies, it is necessary to develop the most advanced imaging equipment in the world. In this paper, we summarize the essential specifications of the thermal and cold neutron imaging facilities currently under consideration and report on the current status toward establishing a foundation for the equipment design.

2   Preliminary Design of Imaging Facilities

2.1   Thermal Neutron Imaging Facility

The thermal neutron imaging facility has been selected as one of the priority facilities (small angle scattering, powder diffraction and reflectometry) in the new research reactor. It features high-speed dynamic imaging, 3-dimensional fast computed tomography, and in operando measurement using actual equipment. Therefore, experimental equipment using high-intensity neutron beams is required. Figure 1 shows the schematics of the thermal neutron imaging facility under consideration. The neutron beam is extracted from the beam tube installed in the D_2O tank of the research reactor. Main beam shutters are placed on the reactor shielding wall, and the beam size is adjusted by the collimator drum installed just behind the shutter. Then, the neutron beam is guided to the irradiation room using the beam collimating duct. To allow for high-speed imaging with short exposure time, the neutron beam must be extracted close to the reactor core to obtain high thermal neutron flux. As for neutron flux, 10^9 n/sec/cm^2 should be achieved, which is higher than the neutron flux of 10^8 n/sec/cm^2 of the thermal neutron radiography facility (TNRF) in JRR-3 [4]. For beam collimation L/D, it is desirable to be able to select from low L/D (100 or less) for high-speed imaging to high L/D (1000 or more) for high spatial resolution measurement of several hundred micrometers or less (L/D = 176@TNRF, JRR-3). Therefore, an extended space is required in the axial direction, and L/D will be 1000 at the downstream sample position when the collimator hole size is 10 mm in diameter. It is also important to have enough space in the irradiation position to allow measurement of relatively large industrial products and easy access to and carrying of samples. There are two sample stages which are placed in different beam directions. The upstream stage is relatively small and for high-intensity beam experiments. The downstream stage is for typical neutron imaging and tomography, and it can be used to move large samples. The flight tube is installed between two stages. Detectors should be available for high-sensitivity imaging and high-speed imaging.

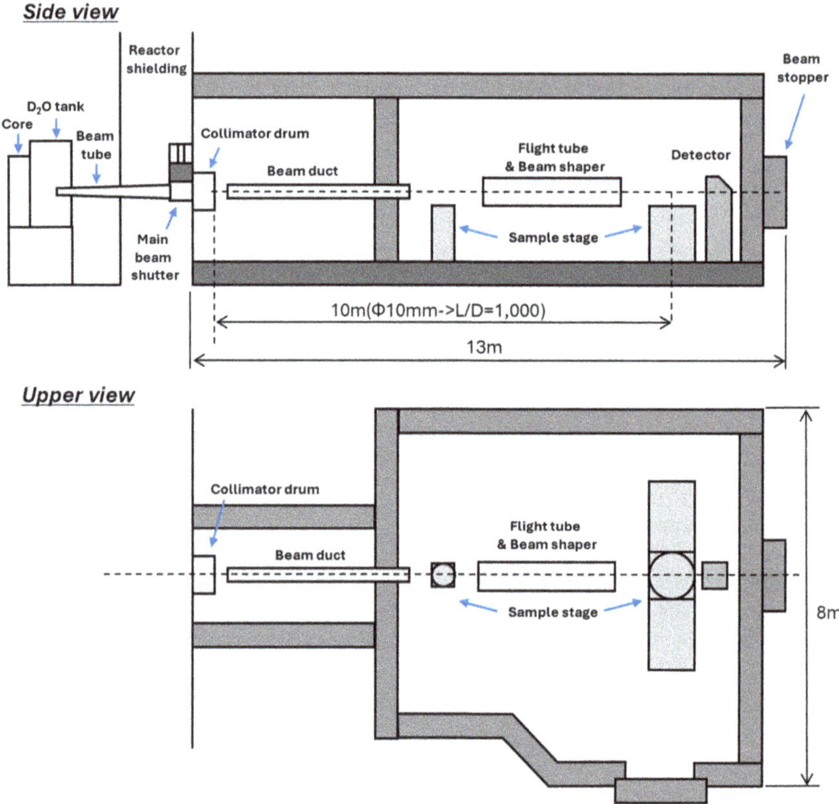

Fig. 1. Schematic diagram of the thermal neutron imaging facility.

2.2 Cold Neutron Imaging Facility

The cold neutron imaging facility is also essential for industrial applications and a wide range of scientific applications, such as materials science. It features high-contrast, high spatial resolution for precise measurement and innovative imaging using interferometry or polarization analysis. For this reason, it is desirable to have a large area along the beam direction to allow the installation of neutron optical devices for beam control and to make the experimental setup flexible. Figure 2 shows the schematics of the cold neutron imaging facility in the new research reactor. Cold neutrons from the cold neutron source are delivered through the neutron guide. The diffuser is placed at the end of the guide to smear the intensity deviation in the beam. The time-of-flight analysis option is introduced by installing a disk chopper. A velocity selector and a double crystal monochromator are also planned to be equipped for cases where monochromatic neutrons are required, such as Bragg edge imaging. In the irradiation room, a flight tube, which is segmented to allow different flight path lengths so as to install various devices on the beam axis, is placed to reduce intensity loss due to air scattering. As for devices, a grating interferometer and polarization analysis equipment are currently expected; however, flexible arrangements of other devices will be possible, depending on future development. The sample stage and detector can be moved along the beam direction according to the experimental conditions. The field of view should be about 200 mm square at the downstream position. The L/D will be variable between 200 and 2000 by changing the size of the aperture using an aperture wheel. Neutron flux should be at least more than 10^8 n/sec/cm^2 under minimum L/D conditions. However, since the higher flux is better, the neutron flux of the order of 10^9 n/sec/cm^2 is targeted to be achieved by optimizing the shape of the neutron guide to focus the beam. The peak wavelength of 0.3~0.4 nm is desirable. The irradiation room requires a space of at least 8 m in the beam direction to install optical instruments and 2 m higher than the beam height to use sample environments like cryogenic refrigerators or magnets. In addition, motorized stages for sample mounting and movable optical benches to switch optical devices will be equipped to allow flexible experimental arrangement.

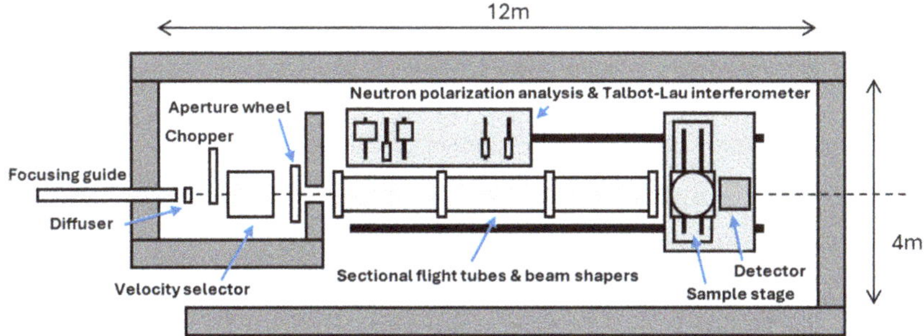

Fig. 2. Schematic diagram of the cold neutron imaging facility.

3 Current Requirements for Imaging Facilities

The requirements for the two facilities described above are summarized in Table 1. Both imaging facilities require high neutron flux. However, neutron flux is determined by the design around the reactor core. So, we need to work with the reactor design division to obtain higher fluxes. Field of view and L/D should be selectable for the required measurement. In addition, the imaging system, which is expected to improve further resolution and sensitivity in the future, and peripheral equipment will be provided for characteristic measurements at each instrument.

Table 1. Requirements for new imaging facilities.

	Thermal neutron imaging facility	Cold neutron imaging facility
Required neutron flux	10^9 n/sec/cm^2 or more	10^8 n/sec/cm^2 or more
Required Field of View	300 mm square at the rear sample position	Less than 10 mm~200 mm square
Required L/D	100~1000	200~2000

(*continued*)

Table 1. (continued)

	Thermal neutron imaging facility	Cold neutron imaging facility
Spatial resolution	50 μm~500 μm	few μm~100 μm
Imaging system	- Imaging system with high-sensitivity camera - High-speed imaging system	- Normal imaging system - High-resolution imaging system
Other required devices	- Pin-hole collimator - Flight tube and beam shaper - Sample stages - Small crane - X-ray tube and detector	- Monochromatic device (velocity selector/Chopper) - Variable aperture, beam shaper, and focusing supermirror - Interferometer, polarized analysis device - Sample stage, CT stage, sample changer

4 Conclusions

The new neutron imaging facility is critical because it will make a significant contribution to the development of the neutron imaging field. In Japan, the start of RADEN utilization at J-PARC has enabled advanced imaging. However, the number of research reactors is decreasing, and there has been a demand for a new facility capable of conducting neutron imaging. The new research reactor constructed at the Monju site could solve such problems. Currently, requirements and specifications for thermal and cold neutron instruments are being considered, and research and development must continue to complete the world's most advanced instruments in 10–15 years.

References

1. New research reactor promotion office homepage. https://www.jaea.go.jp/04/nrr/en. Accessed 30 Sep 2024
2. Kawabe, S., Hattori, S.: Complex neurovascular system in the dentary of Tyrannosaurus. Hist. Biol. **34**(7), 1137–1145 (2022)
3. Hyodo, M., et al.: Intermittent non-axial dipolar-field dominance of twin Laschamp excursions. Commun. Earth Environ. **3**, 79 (2022)
4. Kurita, K., et al.: Introduction to neutron radiography facilities at the Japan research reactor-3. J. Phys. Conf. Ser. **2605**, 012005 (2023)

Experimental MeV Neutron Imaging at RADEN with Monte Carlo Simulations

Naoya Odaira[1]([✉]) [iD], Yusuke Tsuchikawa[2], Daisuke Ito[1] [iD], Yasushi Saito[1],
Takenao Shinohara[2] [iD], Yoshiaki Kiyanagi[3,4] [iD], Saerom Kwon[5] [iD], Kentaro Ochiai[5],
and Satoshi Sato[5]

[1] Institute for Integrated Radiation and Nuclear Science, Kyoto University, Asashiro-nishi,
2-1010, Kumatori-cho, Sen-nan-gun, Osaka 590-0494, Japan
`odaira.naoya.4e@kyoto-u.ac.jp`
[2] Japan Atomic Energy Agency, Shirakata 2-4, Tokai-mura, Naka-gun, Ibaraki 319-1195, Japan
[3] Japan Neutron Optics Inc., Takeshima-cho 20-5, Gamagori-gun, Aichi 443-0031, Japan
[4] Emeritus Prof. Hokkaido University, Kita 13, Nishi 8, Kita-ku, Sapporo 060-8628, Japan
[5] National Institutes for Quantum Science and Technology, Obuchi-Omotedate 2-166,
Rokkasho-mura, Kamikita-gun, Aomori 039-3212, Japan

Abstract. Neutron imaging has been widely used for internal inspections of products and for research and developments in mechanical and chemical engineering, architecture, and agriculture. Nevertheless, the industrial products requiring internal inspection may exceed the penetration capability of thermal neutron imaging. In such cases, MeV neutron imaging would be an optimal candidate for observing their interior because it may usually have higher penetration capability. In this study, an experimental study of the MeV neutron imaging at RADEN in J-PARC MLF was conducted using some representative metallic and hydrogenous materials. Additionally, Monte Carlo simulations were performed using PHITS to complement the experimental data. The transmissions in the MeV neutron imaging exhibited an ideal exponential decrease with the object thickness for any materials used in the experiment, which enables us to yield an accurate quantification of the thickness distribution of the object from the transmission images. The deposition energy evaluations in the scintillators obtained from the numerical simulations confirmed less impact of the scattered and moderated neutrons on the plastic scintillator, which is agreed with the transmission dependence on the material thickness by the experiments and simulations. As a result, it was confirmed that the MeV neutron imaging at RADEN has a significant advantage on the accurate quantification of a thick object including metallic and hydrogenous materials.

Keywords: MeV neutron imaging · Epithermal neutron imaging · Monte Carlo simulation · Thickness dependence of transmission

1 Introduction

The thermal neutron imaging technique has been extensively studied and developed, and it has many applications in various fields, such as non-destructive inspections of metallic materials or research and developments in mechanical engineering, architecture,

© The Author(s) 2026
A. E. Craft and H. Z. Bilheux (Eds.): WCNR 2024, SPPHY 348, pp. 222–234, 2026.
https://doi.org/10.1007/978-3-032-15003-5_26

and agriculture. Nevertheless, thicknesses of the industrial products requiring internal inspection may exceed the penetration capability of thermal neutron imaging. In such cases, MeV neutron imaging would be an optimal candidate for observing their interior due to the high penetration capability of high-energy neutrons. Therefore, various developments and experiments were performed [1–5]. High energy X-ray also has a high penetration capability, and thus it may be used for similar applications. The main difference between them is the sensitivity to the low-Z materials such as water or oil. For example, the detection of explosives [6] and the observation of two-phase flows in nuclear fuel bundles [7, 8] can be performed by using high-energy neutron imaging, whereas these may be difficult to observe by using high-energy X-ray imaging.

Although many studies of MeV neutron imaging were conducted, its capability and characteristics have not fully been described yet because the materials and thicknesses of the samples, the neutron energies, the detectors, and the methodologies to evaluate its performance have been different for each researcher and facility.

A Monte Carlo simulation code for particle transportation is a powerful tool to simulate neutron imaging. PHITS is an open-source Monte Carlo simulation code developed by JAEA [9] and is used in a wide variety of fields. Thus, PHITS can simulate MeV neutron imaging, which capability and characteristics can be understood for any condition. Also, reasonable detailed information such as neutron energy spectra can be obtained and complement the neutron transmission images since these images have limited information.

The imaging instrument RADEN at MLF (Materials and Life science experimental Facility) in J-PARC (Japan Proton Accelerator Research Complex) [10] serves us a wide energy range of neutrons from cold to MeV neutrons. MeV neutron imaging has scarcely been performed in RADEN though the facility is capable of it. Thus, the MeV neutron imaging may expand the capability of the facility.

Therefore, the purpose of this study is an investigation of the characteristics of MeV neutron imaging at RADEN based on the experiment. However, a neutron transmission image has limited information, i.e. a two-dimensional neutron intensity distribution which is also called a radiograph. Thus, a numerical simulation by PHITS on the MeV neutron imaging was also performed to complement the experimental data. The characteristics of the epithermal neutron imaging was also investigated simultaneously at the same beam line. The obtained data will be helpful to understand the characteristics of and to emphasize the advantages of the MeV neutron imaging.

2 Experiments

2.1 Experimental Method

Experiments were performed on RADEN [10]. In this experiment, the L/D ratio was set as 250. The neutron beam was filtered by cadmium ($t = 1.0$ mm) and lead ($t = 75$ mm) to eliminate the thermal neutrons and to reduce the intensity of the gamma rays in the beam, respectively. The expected neutron spectra in RADEN under unfiltered and filtered conditions were simulated by PHITS and they are illustrated in Fig. 1. The input neutron energy spectrum for the present numerical simulation at 50 cm from the moderator surface [11] and the expected filtered neutron spectrum at the detector position were indicated simultaneously. The neutron spectrum obtained by the numerical simulation has been experimentally characterized by Harada at BL-10 [12].

A simultaneous MeV and epithermal neutron imaging experiment was performed to obtain transmission images at different energy ranges of neutrons. Two scintillators were utilized in this experiment: a PP/ZnS (mixed with polypropylene and ZnS) scintillator ($t = 2.5$ mm) for MeV neutron detection and a ^{6}LiF/ZnS scintillator ($t = 0.1$ mm) for epithermal neutron detection (both manufactured by RC TRITEC). To obtain images of each scintillator in a single measurement, they were placed side by side, as shown in Fig. 2. CCD camera iKon-L 936 (Andor) was used for the image acquisitions.

Simple rectangular blocks of 5 cm × 5 cm × t (1 ~ 5 cm) were used as samples in the present experiment. The blocks were made of SS400 (Fe > 99.8%), C1020 (Cu > 99.96%), A5052 (Al-2.8%Mg alloy with less than 1% of impurities), and acrylic (PMMA: polymethyl methacrylate). In addition to the abovementioned blocks, a distilled water step wedge with thicknesses ranging from 1 to 5.6 cm sealed in an aluminum container was utilized to obtain water transmission images. The distance between the back of the sample and the scintillator was 5.6 cm. The exposure time to obtain a single image was two minutes. Five images were taken for each sample through the present experiment. The net counts (the gray values of the images after subtracting the dark current) were approximately 340 and 2300 for the MeV neutron images (obtained by the PP/ZnS scintillator) and the epithermal neutron images (obtained by the ^{6}LiF/ZnS scintillator), respectively. The MeV neutron images were much darker than the epithermal neutron images.

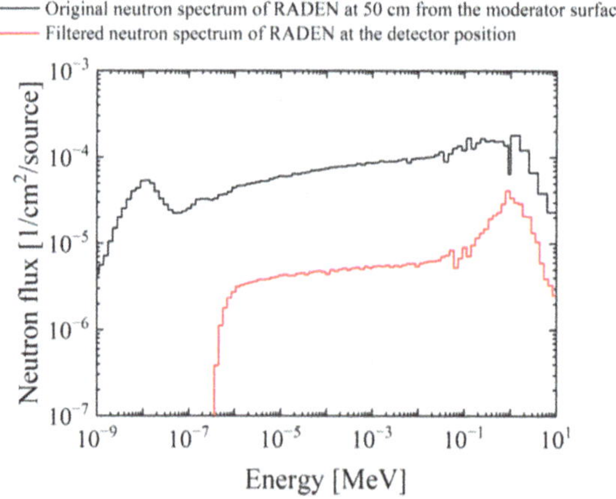

Fig. 1. Estimated neutron energy spectra at RADEN [11]

2.2 Experimentally Obtained Transmission Image

Figure 3 shows the experimentally obtained transmission image of the 5 cm thick iron block (SS400) as a typical example. The transmission image on the left is the epithermal neutron image, while the right is the MeV neutron image. As shown in the figure, the MeV

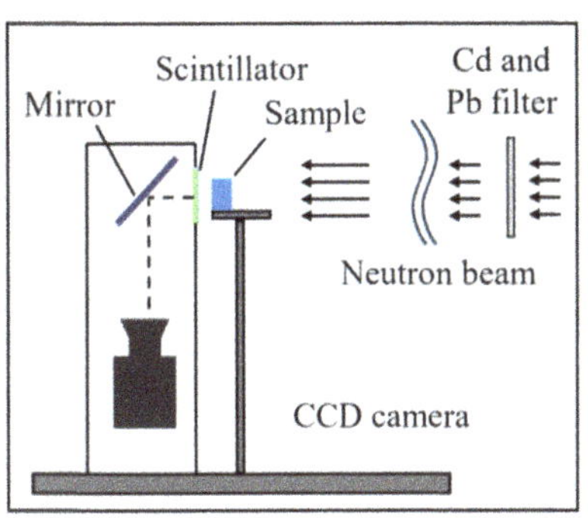
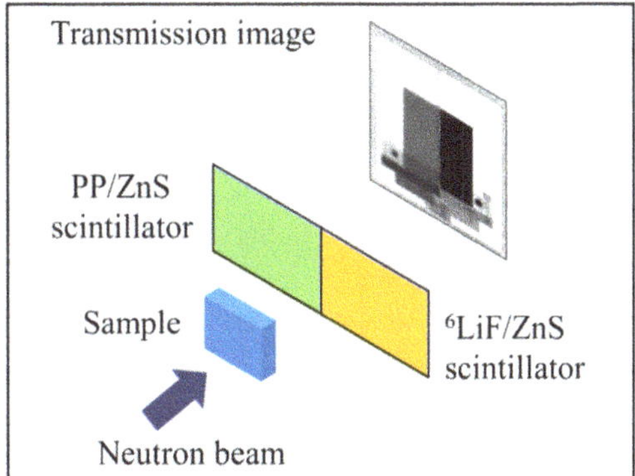

Fig. 2. Schematic of scintillator-camera system used in the present experiment

neutron image was relatively blurred in comparison to the epithermal neutron image. However, the transmission in the MeV neutron image is significantly higher than that in the epithermal neutron image. This difference represents the remarkable penetration capability of the MeV neutrons.

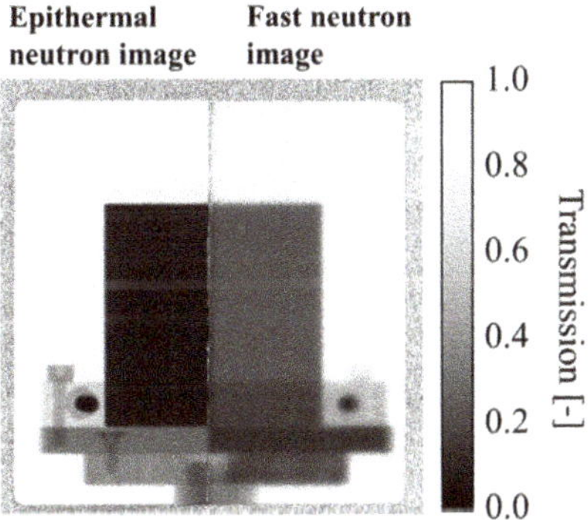

Fig. 3. The epithermal and MeV neutron images of Fe (SS400) block ($t = 5$ cm)

2.3 Resolution Assessment

The knife-edge function originally designed for evaluating a scintillator response, as shown in Eq. (1) is only applicable to fully absorbing materials such as cadmium for thermal neutrons, as introduced by Harms [13].

$$S = \frac{1}{2} + \frac{1}{\pi}\tan^{-1}(\lambda x) \tag{1}$$

S is the signal of the edge, λ is the resolution parameter, and x represents the position of each signal. S in Eq. (1) takes zero to one which corresponds to the ideal situation as mentioned above. In the present case, the iron block is not a fully absorbing material for MeV and epithermal neutrons. However, most of the MeV and epithermal neutrons could be scattered in the case of the 5 cm thick iron block. In order to estimate the resolution parameter, it may be possible to apply the knife-edge function to the edges of the iron block image if an appropriate scaling factor is included in the equation. In addition, the edge center in Eq. (1) is defined as zero. Including the shift of the edge position in the knife-edge function is valuable to fit the present experimental data. Accordingly, the resolution parameter λ was evaluated using the following modified knife-edge function as shown in Eq. (2).

$$S = T + A[\frac{1}{2} + \frac{1}{\pi}\tan^{-1}\{\lambda(x - x_0)\}] \tag{2}$$

T is the transmission of the material, A is the scale parameter corresponding to $1/T$, x_0 represents the position of the specific position of the edge, respectively. Least-squares fits were performed on the MeV and epithermal neutron images of the 5 cm thick iron block. The fitting results and the location of the edges are indicated in Fig. 4. The obtained resolution parameters λ and the full-width at half-maximum (FWHM) values derived from the line spread function with the obtained resolution parameter λ are presented in Table 1. FWHM can be calculated by Eq. (3)

$$FWHM = 1/\lambda \tag{3}$$

The resolution parameter λ for the MeV neutron image is significantly smaller than the epithermal neutron image, indicating decrease of the resolution. This can be attributed to the thicknesses of the scintillators which were $t = 2.5$ mm and $t = 0.1$ mm for the PP/ZnS and the ^{6}LiF/ZnS scintillators, respectively. It is important to note that the obtained resolution parameters include the blurring effects induced by the scattered neutrons in the sample and the beam divergence which can be estimated as approximately 0.22 mm ($= 56$ mm / 250). When considering the geometric blurring in this experiment and the thickness of the scintillator, it exhibits that the resolution performance may be further enhanced. This may be induced by the scattered neutrons at the edge. On the other hand, Lehman shows the MeV neutron image obtained by a layer-type scintillator which consists of 3 mm polyethylene plate with 0.06 mm ZnS photoluminescence layer, and the result shows higher resolution than the image obtained by a mixed-type scintillator [14]. Hence, the resolution of the MeV neutron image may be improved if a layer-type scintillator is used.

3   Numerical Simulations

3.1   Simulation Method

Numerical simulations were performed using PHITS Ver. 3.34 (Particle and Heavy Ion Transport code System) developed by Japan Atomic Energy Agency, which was a general-purpose Monte Carlo particle transport simulation code [9] with the nuclear data library JENDL-5 [15]. The geometry in the simulations is illustrated in Fig. 5. The expected neutron spectrum at RADEN was utilized in the present simulation (see Fig. 1).

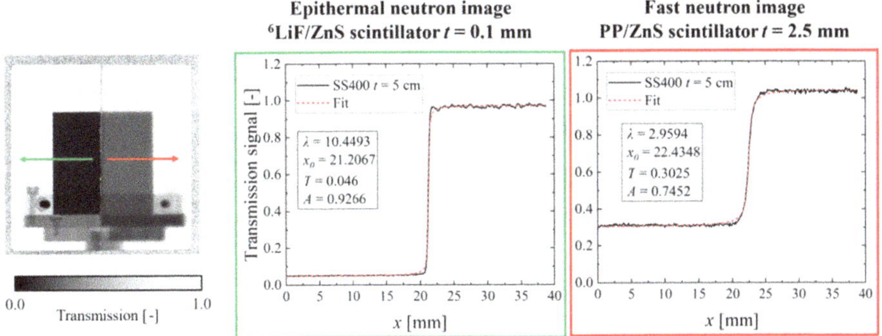

Fig. 4. Experimental and knife-edge function fitting lines obtained from the epithermal neutron image by the ^{6}LiF/ZnS scintillator (left) and the MeV neutron image by the PP/ZnS scintillator (right).

Table 1. Obtained resolution parameters.

Condition	λ [mm^{-1}]	FWHM [mm]
MeV neutron image obtained by PP/ZnS scintillator $t = 2.5$ mm (performed at RADEN)	2.95	0.68
Epithermal neutron image obtained by ^{6}LiF/ZnS scintillator $t = 0.1$ mm (performed at RADEN)	10.45	0.19

The distance between the back of the sample and the scintillator was 5.6 cm, which was the same situation as the experiment. The materials used in the experiment were simplified as follows: pure Fe for SS400, pure Cu for C1020, Al-2.8Mg alloy for A5052, $C_5O_2H_8$ for acrylic, and H_2O for distilled water. The deposition energy distributions on the scintillators which may correspond to the images obtained in the experiments were acquired by PHITS. The transmissions for each condition were evaluated from the deposition energy distributions using Eq. (4).

$$T = E_{dep,tr}/E_{dep,in} \tag{4}$$

T is the transmission and $E_{dep,in}$ and $E_{dep,tr}$ are the deposition energies on the scintillators at each position for the incident (open beam) and the sample transmitted neutron beam, respectively. The simulated transmission image for 5 cm thick iron block as a typical example is shown in Fig. 6.

3.2   Efficiency of the Scintillators

The PP/ZnS and ^{6}LiF/ZnS scintillators used in the experiments were modeled in the numerical simulations. These scintillator efficiencies were estimated by a dedicated numerical simulation. In this simulation, the neutron beam was directed onto the scintillators with a uniform probability distribution between 0.01 eV and 10 MeV. The

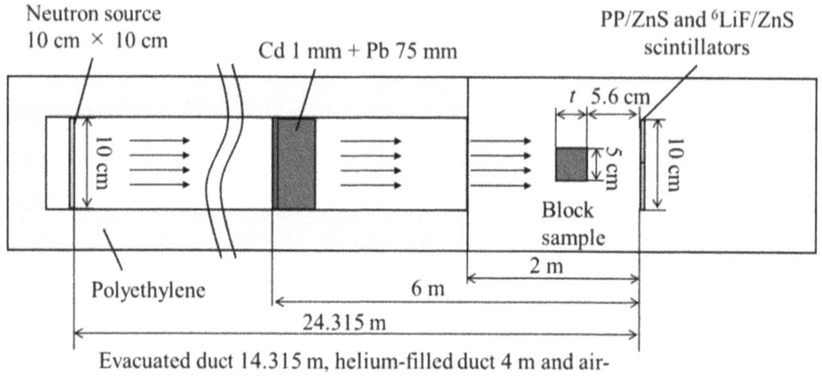

Fig. 5. Geometry in the numerical simulations

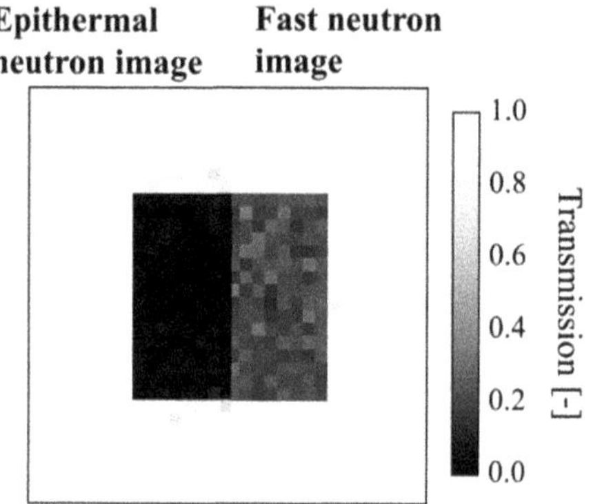

Fig. 6. Simulated transmission image of 5 cm thick iron block

deposition energies on the two scintillators were then obtained, which might represent the efficiency of the neutron-to-light conversion by the scintillators. The main deposition mechanism on the scintillators would be by the recoil protons in the case of the PP/ZnS scintillator, and by the alpha and triton particles emitted induced by the ^{6}Li absorption reaction in the case of the ^{6}LiF/ZnS scintillator. Figure 7 shows the deposition energies on the scintillators as a function of the incident neutron energy. The neutron energy in the figure has been determined from the time-of-flight (TOF) analysis. Figure 7 shows that the ^{6}LiF/ZnS scintillator is sensitive to low-energy neutrons, and the conversion efficiency increases with decreasing incident neutron energy. The energy of the scattered neutrons is typically lower than that of the incident neutrons, and thus the ^{6}LiF/ZnS scintillator can detect more scattered neutrons than the incident neutrons. On the other hand, the deposition energy distribution on the PP/ZnS scintillator shows that it is reactive to high-energy neutrons and the conversion efficiency increases with increasing the incident neutron energy. Thus, the PP/ZnS scintillator would detect scattered neutrons with a lower efficiency than the incident neutrons.

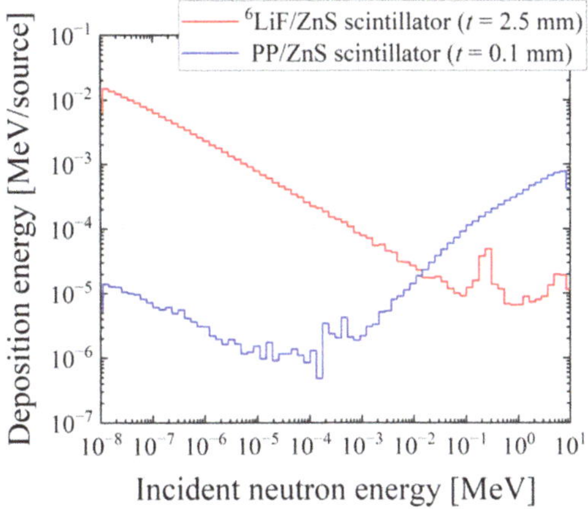

Fig. 7. Deposition energy in the PP/ZnS and ^{6}LiF/ZnS scintillators as a function of incident neutron energy.

4 Results

4.1 Comparison of Transmission in Each Image

Spatially averaged transmissions were derived from both the experimental and simulated images. Figure 8 shows the comparisons of the transmissions between the experiments and the numerical simulations for each material. Figure 8(a) to (c) show the transmissions of metallic materials, while (d) and (e) show the transmissions of hydrogenous materials. The relative standard deviation of the spatially averaged transmission was expected to be less than 10^{-6} for the experiment and the relative statistical error would be less than 0.02 for the numerical simulations. Although the statistical error in the numerical simulation was relatively high compared to the experiment data, the error bars in the graphs were difficult to see and thus these were not indicated in the figures.

The transmissions obtained from the experimental and simulated images are reasonably agreed in all cases. The transmissions of Fe and Cu show similar trends. Both results exhibit considerably elevated penetration capabilities in MeV neutron imaging as we expected. If 0.2 is set as a requirement to visualize interior structure, 6 cm of Fe or 5 cm of Cu can be visualized. However, the difference in the Fe transmissions between epithermal (0.2 at about 2.0 cm thick) and MeV neutron imaging (0.2 at about 6 cm thick) seems to be larger than that of the Cu transmissions, which in turn makes the advantage to perform MeV neutron imaging of Fe larger than that of Cu. On the other hand, the transmissions of Al-2.8 Mg alloy show higher penetration capability in epithermal and thermal neutrons than in MeV neutrons, as can be seen from the dependence of the Al cross-section on the neutron energy.

The transmissions of the hydrogenous materials, i.e., PMMA and water, in the epithermal neutron images deviate from the ideal exponential decrease. This is obviously due to the neutron scattering and moderation in the sample. Consequently, the actual transmission, defined as the number of transmitted neutrons that do not react with the sample, is predicted to be significantly lower than the data presented in the figure.

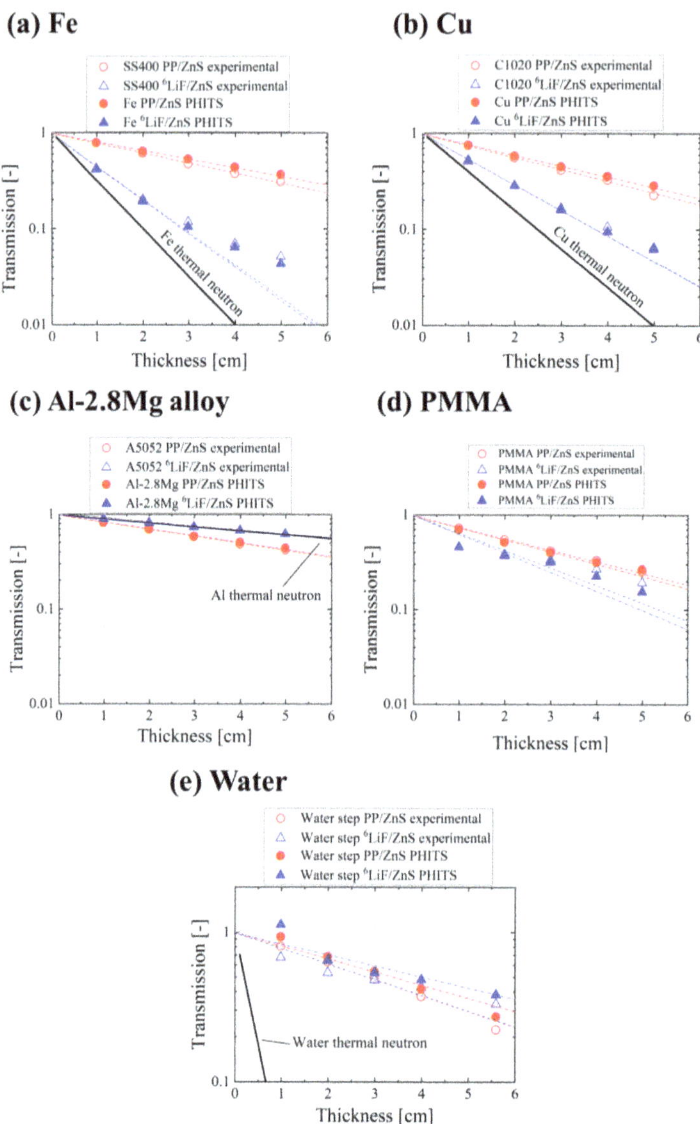

Fig. 8. Spatially averaged transmissions for each material in the MeV and epithermal neutron imaging obtained from both the experiments and the numerical simulations together with each thermal neutron transmission [16]

This result shows the quantitative evaluations of these thicknesses from the epithermal neutron images can be challenging. On the other hand, the transmissions of hydrogenous materials in the MeV neutron images are reasonably consistent with ideal exponential decreases. It may indicate that the scattered neutrons have less impact on the MeV neutron imaging. Consequently, the quantitative evaluations in the MeV neutron images would be more straightforward than those in the epithermal neutron images.

4.2 Energy Spectrum of the Neutrons

The transmissions evaluated from the PHITS simulations exhibited reasonable agreement with the experimentally obtained transmissions in all cases, including the MeV and epithermal neutron imaging. The neutron energy spectra and other information obtained from the PHITS simulations may be in reasonable agreement with the experimental situation.

The neutron energy spectra of the neutron beams incident on the scintillators in each condition obtained from the PHITS simulations were indicated in Fig. 9 together with the area of interest in the simulation model which was defined as the projected area of the sample on the scintillators. The open beam spectrum corresponds to the incident beam without any sample, which was also illustrated in Fig. 1. The remaining lines show transmitted neutron spectra through the sample.

The transmitted neutron energy spectra of hydrogenous materials, i.e. PMMA and water, show a significant decrease in the neutron fluxes below 1 MeV, which may be attributed to the energy dependence of the scattering cross-section of hydrogen [15]. Furthermore, these spectra indicate an augmentation in thermal neutrons which clearly represents the moderation of neutrons. The transmitted neutron energy spectrum of pure Fe is relatively similar to that of Al-2.8Mg alloy or the open beam above 0.1 MeV. However, the neutrons lower than 0.1 MeV seem to be significantly attenuated by Fe, which clearly shows the dependence of the Fe cross-section on the neutron energy. This behavior indicates that Fe provides significant change on the neutron energy spectrum which is a hardening. On the other hand, the spectral change by Cu seems to be smaller than the Fe case. This can be attributed to the larger difference in the transmissions of Fe between the epithermal and MeV neutron imaging compared to the Cu transmissions. Interestingly, the neutron energy spectrum of Al-2.8 Mg alloy shows a larger decrease around 1 MeV neutrons compared to the neutrons lower than 10 keV. It can be said that Al-2.8Mg alloy softens the neutron beam.

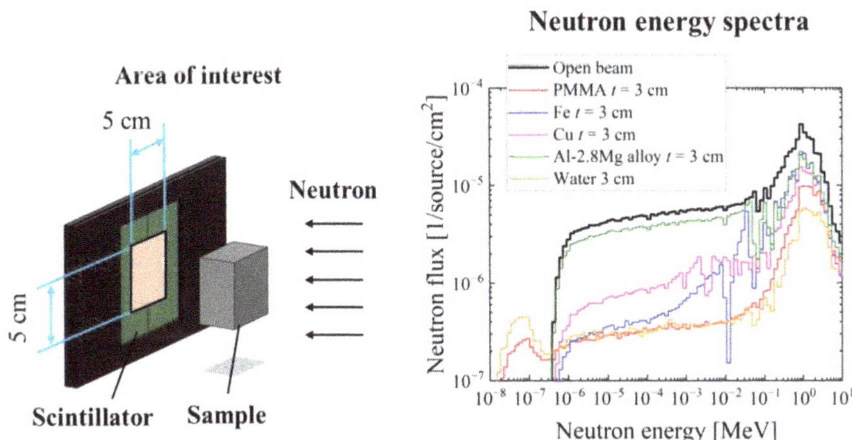

Fig. 9. Transmission neutron energy spectra for each condition on the area of interest obtained from the numerical simulations

4.3 Energy Dependent Contribution of the Energy Deposition to the Images

As mentioned above, the photoluminescence intensity of the scintillators may depend on the energy deposited on them. In general, each scintillator has a different sensitivity to a given incident neutron energy, which is difficult to evaluate in the present experiment. On the other hand, PHITS can track the dose variation with the TOF of the neutrons which can be converted to the incident neutron energy to the sample. The absolute deposition energies on the scintillators are difficult to compare with the experimental results, while the deposition fractions DF may provide insight into the contributions to the images. Thus, DF has been calculated according to the following Eq. (5).

$$DF\,(E_{in})\, =\, E_{dep}(E_{in})\,/\,\int E_{dep}(E_{in})dE_{in} \tag{5}$$

DF may represent the contributions to the transmission images obtained by each scintillator and E_{dep} is the deposition energy in each scintillator. Note that DF and E_{dep} depend on the incident neutron energy E_{in} towards the samples which is not the neutron energy toward the scintillators unfortunately because of a difficulty to obtain it in the numerical simulation.

The results, as shown in Fig. 10, indicate that the epithermal neutron images obtained by the ^{6}LiF/ZnS scintillator exhibit a relatively broad energy region for the absorptions of neutrons. The neutrons transmitted through the water and the PMMA show different trend compared to the open beam condition. This can be considered as the result of the enhancement of the influence of the scattered and moderated neutrons. Consequently, the epithermal neutron images in the case of hydrogenous materials might be strongly affected by the scattered neutrons. The deposition fractions in the MeV neutron images obtained by the PP/ZnS scintillator show sharper peaks around 1 MeV than those for the ^{6}LiF/ZnS scintillator cases. It can be said that the MeV neutron images are mostly contributed by 100 keV to 10 MeV of neutrons and it seems that the neutrons around 1 MeV are the dominant contributors to the images. The deposition fractions in the sample-transmitted MeV neutron images remain similar to that for the open beam. This may prove that the MeV neutron images were not significantly influenced by the scattered neutrons for all materials used in the present study.

5 Conclusion

This study investigated the characteristics of the MeV neutron imaging at RADEN. The MeV neutron imaging experiment was performed with simple rectangular blocks and water in an aluminum container by using a simple scintillator-camera system. The obtained Fe transmission trend exhibited that there is a large advantage to perform the MeV neutron imaging for a thick iron sample because of the significant elevation of penetration capability compared to the epithermal neutron imaging. Even a 6 cm thick Fe may be visualized by the MeV neutron imaging while 2 cm thick Fe may be able to characterize by the epithermal neutron imaging if assuming 0.2 transmission. On the other hand, the result of Al-2.8 Mg alloy exhibited opposite trend to Fe and Cu case which is lower transmission in the case of the MeV neutron imaging than the epithermal neutron imaging. The transmissions of hydrogenous materials in the epithermal neutron imaging exhibited a build-up by scattered and moderated neutrons. On the contrary, these transmissions in the MeV neutron imaging exhibited nearly ideal exponential decreases.

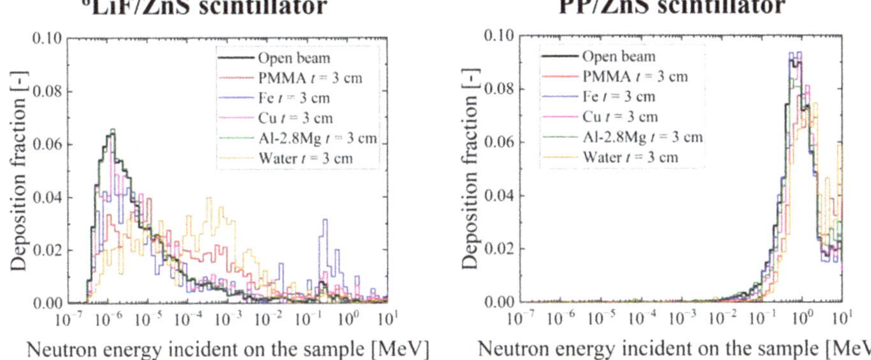

Fig. 10. Contributions of the deposition energy to the transmission images in the scintillators as a function of the incident neutron energy obtained from the numerical simulations

The applicability of PHITS was confirmed by a comparison of the spatially averaged transmissions with the experimental data. The transmissions between the experimental and simulation results exhibited a reasonable agreement in all cases including the MeV and epithermal neutron imaging at RADEN.

The obtained neutron spectrum by numerical simulation under the open beam condition exhibited relatively high neutron flux around 1 MeV, which was preferable to perform MeV neutron imaging. The transmitted neutron spectra clearly indicate: a moderation of neutrons through the hydrogenous materials, a beam hardening through Fe, and a beam softening through Al-2.8Mg alloy. This distributions of the spectral change of the neutron energy might be similar situation to the experiment. The trend of the deposition energy on the scintillators under the open beam condition exhibited a relatively broad peak around 1 eV, where 0.5 to 100 eV of neutrons will react with the ⁶LiF/ZnS scintillator, and a sharper peak around 1 MeV for the PP/ZnS scintillator. The trend of deposition energy on the ⁶LiF/ZnS scintillator was significantly changed by hydrogenous materials. On the other hand, the trend for the PP/ZnS scintillator was the same for all the samples including the open beam condition. From these results and the conversion efficiency dependence of the scintillators, the MeV neutron images using the PP/ZnS scintillator at RADEN was less influenced by scattered and moderated neutrons than the epithermal neutron image using the ⁶LiF/ZnS scintillator. This characteristic is a significant advantage to perform quantitative analysis of a thick object including metallic and hydrogenous materials by the MeV neutron imaging.

Acknowledgement. This work has been funded by National Institute for Quantum Science and Technology. The neutron experiment at the Materials and Life Science Experimental Facility of the J-PARC was performed under a user program (Proposal No. 2023B0224).

References

1. Yoshii, K., Miya, K.: A study on the mechanism model of luminescence in fast neutron television converter. Nucl. Instrum. Methods Phys. Res., Sect. A **346**(1–2), 253–258 (1994)

2. Mor, I., Vartsky, D., Bar, D., et al.: High spatial resolution fast-neutron imaging detectors for pulsed fast-neutron transmission spectroscopy. J. Instrum. **4**, 05016 (2009)
3. Yang, W., Bin, T., Heyoung, H., et al.: The study of zinc Sulphide scintillator for fast neutron radiography. Phys. Procedia **43**, 205–215 (2013)
4. Makowska, M., Walfort, B., Zeller, A., Grunzweig, C., Bucherl, T.: Performance of the commercial PP/ZnS:Cu and PP/ZnS:Ag scintillation screens for fast neutron imaging. J. Imag. **3**(4), 60 (2017)
5. McCall, K.M. Sakhatskyi, K., Lehmann, E, et al.: Fast neutron imaging with semiconductor nanocrystal scintillators, ACS nano **4**, 14686−14697 (2020)
6. Mor, I., Dangendorf, V., Reginatto, M., et al.: Reconstruction of material elemental composition using fast neutron resonance radiography. Phys. Procedia **69**, 304–313 (2015)
7. Takenaka, N., Asano, H., Fujii, T., et al.: Application of fast neutron radiography to three-dimensional visualization of steady two-phase flow in a rod bundle. Nucl. Instrum. Methods Phys. Res., Sect. A **424**, 73–76 (1999)
8. Zboray, R., Dangendorf, V., Mor, I., et al.: Time-resolved fast-neutron radiography of air-water two-phase flows in a rectangular channel by an improved detection system. Rev. Sci. Instrum. **86**, 075103 (2015)
9. Sato, T., Iwamoto, Y., Hashimoto, S., et al.: Recent improvements of the Particle and Heavy Ion Transport code System - PHITS version 3.33. J. Nucl. Sci. Technol. **61**, 127–135 (2024)
10. Shinohara, T., Kai, T., Oikawa, K., et al.: The energy-resolved neutron imaging system. RADEN, Re. Sci. Instrum. **91**(4), 043302 (2020)
11. https://j-parc.jp/researcher/MatLife/en/instrumentation/ns3.html
12. Harada, M., Teshigawara, M., Ooi, M., et al.: Experimental characterization of high-energy component in extracted pulsed neutrons at the J-PARC spallation neutron source. Nucl. Instrum. Methods Phys. Res., Sect. A **1000**, 165252 (2021)
13. Harms, A.A., Wyman, D.R.: Mathermatics and Physics of Neutron Radiography, D. Reidel Publishing Company (1986)
14. Lehman, E.H., Mannes, D., Strobl, M., et al.: Improvement in the spatial resolution for imaging with fast neutrons. Nucl. Instrum. Methods Phys. Res., Sect. A **988**, 164809 (2021)
15. Shibata, K., Iwamoto, O., Nakagawa, T.: JENDL-4.0: a new library for nuclear science and engineering. J. Nucl. Sci. Technol. **48**(1), 1–30 (2011)
16. Hardt, P., Röttger, H.: Neutron Radiography Handbook. Springer, Dordrecht (1981). https://doi.org/10.1007/978-94-009-8567-4

Condensation of the Vapor in Sessile Water Droplets and Open-Ended Water-Filled Capillaries Revealed by Neutron Radiography

Hyeonjun An[1], Youngtak Koo[1], Jiyong Cheon[1], Pavel Trtik[2], and Joonwoo Jeong[1(✉)]

[1] Department of Physics, Ulsan National Institute of Science and Technology, Ulsan, Republic of Korea
`jjeong@unist.ac.kr`
[2] Laboratory for Neutron Scattering and Imaging, Paul Scherrer Institut, Villigen PSI 5232, Switzerland

Abstract. Evaporation and condensation occur simultaneously at interfaces. Even when a sessile water droplet on a substrate evaporates to disappear eventually, the droplet absorbs water vapor from the surrounding air of a finite humidity. Exploiting a significant contrast in neutron scattering cross section between hydrogen and its isotope, deuterium, neutron radiography can quantify how much water vapor is absorbed into a sessile D_2O droplet. Although water vapor enters the droplet only via the air-droplet interface, absorbed water exists almost uniformly over the droplet, regardless of various sizes and contact angles we investigate. This spatial homogeneity also exists in the pendant droplet; absorbed H_2O is lighter than D_2O. To minimize flow-induced mixing, *e.g.*, by capillary or Marangoni flow, we observe water vapor absorption into an open-ended capillary filled with D_2O. With no sample-wide mixing, the distribution of absorbed water indeed exhibits non-uniform distribution. These experiments demonstrate the usefulness of neutron radiography in studying mass transport processes.

Keywords: sessile droplet · pendant droplet · evaporation · condensation · neutron radiography · mass transport

1 Introduction

Spatiotemporal visualization and quantification of material composition are paramount in elucidating various material transport problems. Fluorescence tagging of molecules of interest is an archetypal example, although not all systems allow optical observation [1]. For instance, if a system is opaque or refracts light with a curved interface, such as the air-liquid interface of a spherical droplet, one cannot see inside optically. Other non-optical methods, including electron

A. E. Craft and H. Z. Bilheux (Eds.): WCNR 2024, SPPHY 348, pp. 235–243, 2026.
https://doi.org/10.1007/978-3-032-15003-5_27

microscopy, may require particular environments or sample preparation, such as vacuum or microtoming, preventing *in situ* observation. Choosing a proper observation method providing enough spatiotemporal resolution and contrast between the components of interest and background is essential in studying material transport.

Neutron radiography is an irreplaceable tool for this purpose because of its unique interactions with materials. The neutral particles interact with nuclei, with relatively low scattering cross-section to metals compared to electromagnetic waves such as X-rays, also discriminating isotopes [2,3]. Neutron radiography study of fuel cells manifests this capability that neutron visualizes the distribution of H_2O inside the cell made of metal [4,5]. Neutron radiography has also been employed to study material transport problems associated with liquid matter far from equilibrium, offering insights into the distribution of diverse constituents within complex systems. In the specific case of a sessile D_2O droplet surrounded by ambient air, neutron imaging techniques enable quantitative monitoring of H_2O absorption and subsequent evaporative processes [2].

This work revisits the problem of D_2O absorbing water vapor from ambient air, revealing the roles of flow mixing and gravity in the spatiotemporal distribution of absorbed H_2O. Our observation of homogeneous H distribution, despite the water uptake only via the air-D_2O interfaces of sessile and pendant droplets having various sizes and contact angles, is consistent with the previous work, hinting at mixing by droplet-wide flow. Open-ended capillary experiments minimizing the flow-induced mixing exhibit the gravity-dependent H distribution and support our hypothesis. These findings demonstrate neutron radiography's capability for *in situ* characterization of liquid matter at nonequilibrium and make us look forward to its application to a variety of material transport problems.

2   Results and Discussion

2.1   Absorbed Water Is Almost Uniformly Distributed in Sessile Water Droplets

We measure the neutron transmittance through a sessile D_2O droplet exposed to ambient air and find a uniform distribution of absorbed H_2O within the droplet. As the D_2O evaporates, it simultaneously absorbs H_2O from the surrounding atmosphere, leading to a progressive enrichment of H_2O within the droplet. Neutron radiography provides quantitative evidence of this dynamic exchange, revealing that the overall neutron transmittance and the droplet volume decrease over time. Assuming the droplet's axial symmetry, we calculate the neutron path length at each pixel [2] and convert the transmittance at each pixel into an effective attenuation coefficient, as shown in Fig. 1. After applying the inverse Abel transform, this map of the effective attenuation coefficient can generate a H-composition map in three dimensions, of which the horizontally averaged values are shown in Fig. 2. Because of the H-D exchange reaction being coupled with the diffusion of H_2O within the droplet, what we actually measure is the spatiotemporal distribution of the mole fraction of H, rather than the local mole fraction of H_2O. The local mole fraction of H means the number of H divided by the sum of numbers of H and D, *i.e.*, [H]/([H]+[D]). Figures 1 and 2 both indicate that the absorbed H exists homogeneously, despite that H_2O enters via the air-droplet interface. This observation is consistent with the observation in J. K. Im et al. [2]. There is a caveat that we cannot detect any spatial inhomogeneity at the timescale shorter than the image acquisition time, *e.g.*, typically 5 s in these experiments.

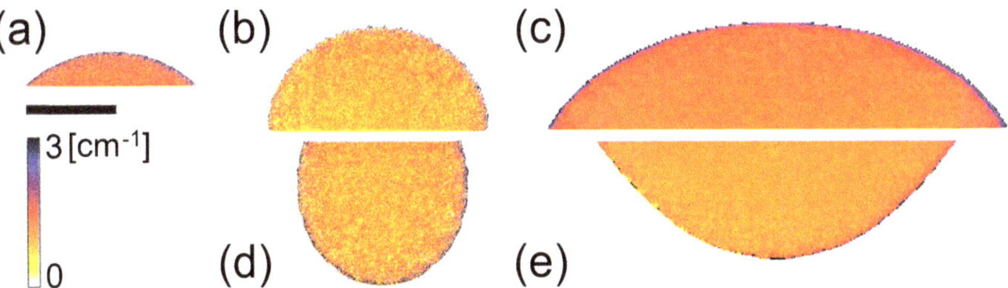

Fig. 1. Neutron attenuation coefficient map of D_2O droplets at 10 min after their exposure to ambient air. (a-c) sessile droplets of different volumes: 8, 30, and 100 μl. (d and e) pendant droplets of different volumes: 30, and 100 μl. Droplets are placed on glass substrates except (b and d) on PTFE substrates with higher contact angles. The scale bar represents 2 mm, and the color code indicates the attenuation coefficient value.

Our systematic variation of droplet geometry does not affect the homogeneity of the H distribution. First, as shown in Fig. 1(a-c), we vary the droplet volume from 8 to 100 μl and find droplet size had no discernible impact on the internal homogeneity of the H/D mixture. There is a caveat to the interpretation of the attenuation coefficients in the vicinity of the interface, *e.g.*, the boundaries of Fig. 1(c) and the top region of Fig. 2: partial volume effects, edge effects [6] and uncertainties in the edge detection contribute to erroneous numbers at the boundaries. Note also that the higher contact angle of the droplet on its substrate in Fig. 1(b) does not affect the spatial distribution.

Droplets' orientation to the gravitational field neither matters in the distribution. Since the density of H_2O is approximately 10% smaller than D_2O's, we alter the droplet orientation relative to the gravitational field, *i.e.*, from a sessile droplet to a pendant one, and assess potential buoyancy effects. As shown in Fig. 1(d) and (e), H is distributed homogeneously within droplets. These findings hint at a mixing mechanism that traverses the whole droplet as large as the one in Fig. 1(c) and counteracts any density-induced stratification at a long time-scale, *i.e.*, longer than the acquisition time of the radiography. Moreover, because the spatial homogeneity is retained from the initial observation of millimeter-scale droplets shown in Fig. 2, we dispute slow diffusion as the origin of mixing. Furthermore, considering the negligible difference in the surface tension of D_2O and H_2O, we exclude the solutal Marangoni flow from possible origins of mixing, but the capillary flow [7].

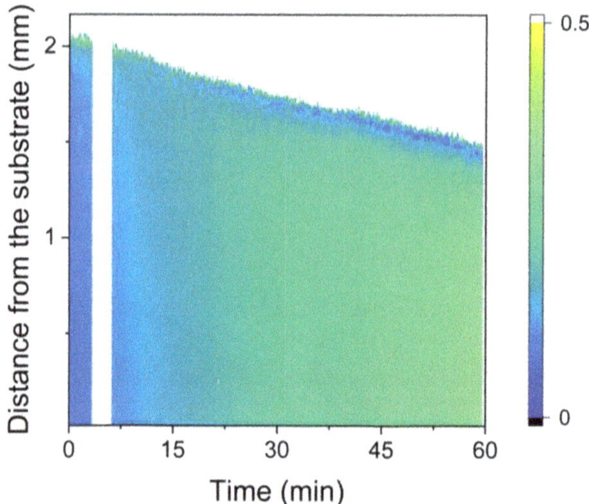

Fig. 2. H mole fraction map of a D_2O droplet of 30 μl on a glass substrate. The plot shows horizontally averaged values of $[H]/([H]+[D])$ according to the time after the exposure to ambient air. Note that the data region around 5 min is missing because the neutron beam was off, and the mole fractions at the top regions are considerably affected by boundary effects, giving slightly erroneous numbers.

2.2   Gravity Affects the Distribution of Absorbed Water in Open-Ended Capillaries

We design an open-ended capillary experiment to minimize the flow mixing and observe the expected inhomogeneous H distribution by gravity. A rectangular capillary filled with D_2O has only one end open, allowing the influx of H_2O but preventing the sample-wide flow [8]. As shown in Fig. 3, neutron radiography maps the effective attenuation coefficient spatiotemporally, according to the location of the open end, *i.e.*, the directions of density gradient and gravity. As in the droplet experiments, the evaporation of D_2O and condensation of H_2O cooccur. However, unlike the droplet case, gravity significantly impacts the H distribution within the capillary. As shown in Fig. 3, whether the open end faced upward or downward relative to the gravity direction results in distinct attenuation coefficient maps.

Figure 4 represents the full spatiotemporal map of $[H]/([H]+[D])$. In the case of the 'sessile' capillary with the open end at the top, H is concentrated near the open end and shows a vertical gradient because of its diffusion towards the bottom. In contrast, the 'pendant' capillary exhibits spatially uniform but time-varying H distribution, where gravitational convection facilitates homogenization. The overall H concentration increases over time, as shown in the bottom panel of Fig. 3. However, the slight increase in the mole fraction is not apparent in Fig. 4 because of the color map shared with the 'sessile' capillary. These find-

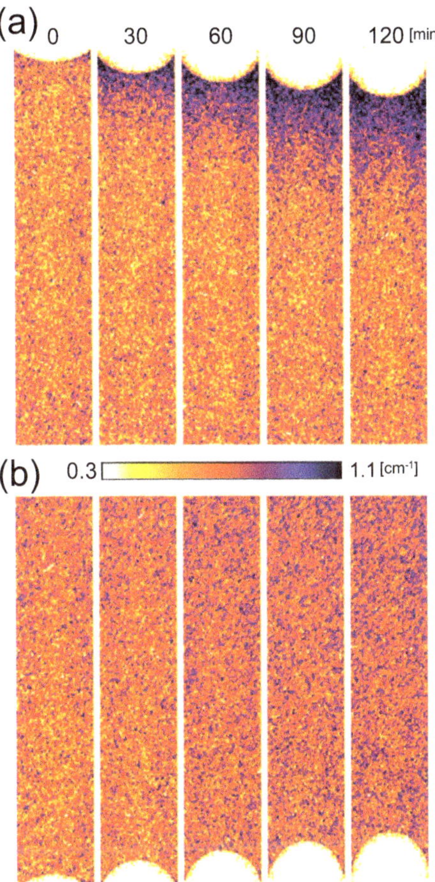

Fig. 3. Neutron attenuation coefficient maps of D_2O inside open-ended capillaries according to the time after the exposure to ambient air. The capillary dimensions are 4 mm wide, 2 mm deep, and 20 mm long. (a) The 'sessile' capillary with the open end located at the top. (b) The 'pendant' capillary with the open end located at the bottom.

ings emphasize the importance of understanding the intricate interplay among diffusion, advection, flow, and confinement in mass transport phenomena.

3   Materials and Methods

3.1   Sample Preparation

We purchased D_2O (99.9 atom % D isotopic purity) from Sigma Aldrich and used it as is. For both sessile and pendant configurations, we manually placed a droplet of a designated volume on a substrate and ensured its axial symmetry. To vary the contact angle, we employed glass and polytetrafluoroethylene (PTFE)

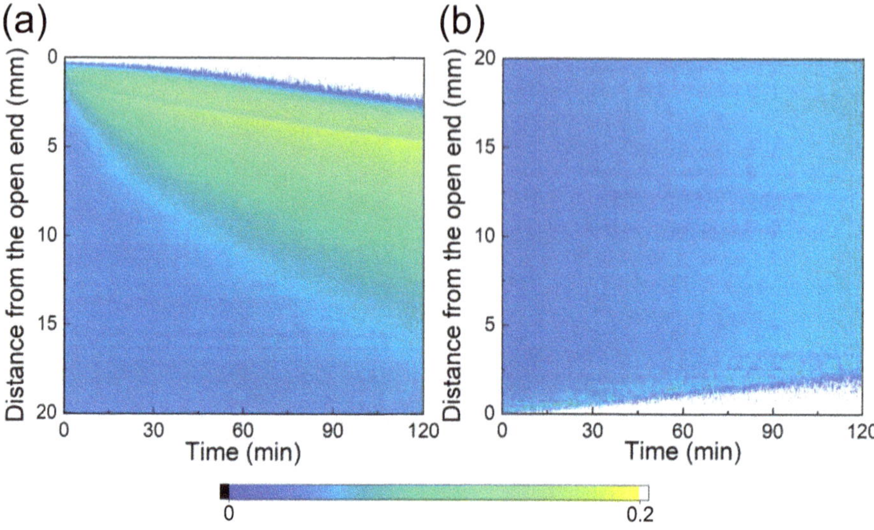

Fig. 4. H mole fraction maps of D_2O confined in (a) the sessile capillary and (b) the pendant capillary, according to the time after the exposure to ambient air. The plot shows horizontally averaged values of $[H]/([H]+[D])$ according to the time after the exposure to ambient air.

as substrates. Before taking the first neutron image, the droplet and substrate were contained in an environment chamber purged with dry nitrogen. When the remotely controlled chamber was lifted, the droplet started to be exposed to ambient air with monitored temperature, $26.5 \pm 0.5°C$, and relative humidity ranging from 29 to 35%.

In the capillary experiment, we injected D_2O into a rectangular quartz capillary (Friedrich & Dimmock Inc.) via one sealed end and filled it up to the open end. The capillary's neutron path length and width were 2 mm and 4 mm, respectively, and the length was longer than 20 mm.

3.2   Neutron Radiography

For the neutron radiography, we used the ICON [9] and NEUTRA [10] beamlines of the SINQ at Paul Scherrer Institut, Switzerland. Table 1 shows the information about the beamline and imaging setup of each radiographic data presented. For all measurements, we set the sample-to-detector distance as close as possible: approximately 5 mm in typical setups. The droplets and capillaries were monitored for 60 and 120 min, respectively. In addition to a typical flat-field and dark-current correction of the images, we employed the blackbody correction [11,12] to mitigate artifacts arising from systematic biases due to scattering components. These artifacts typically manifested as increased transmission signals, particularly in the center of bulk samples, leading to a bias towards lower computed attenuation coefficients.

Table 1. Beamline and imaging setup information of each radiographic data presented.

Data	Beamline	Field of view (mm^2)	Exposure time (sec)	Pixel size (μm)
Figure 1a	ICON	5.5 × 5.5	60.0	2.7
Figure 1b, 1d, and 2	ICON	37.1 × 37.1	5.0	18.1
Figure 1c and 1e	ICON	69.2 × 69.2	3.0	33.8
Figure 3 and 4	NEUTRA	66.1 × 66.1	20.0	32.3

A processed transmittance map was converted into an effective attenuation coefficient map with known pathlengths at each pixel, *e.g.*, 2.00 mm in the capillary case. For droplets of axial symmetries, the inverse Abel transform further mapped this into a three-dimensional map. Utilizing the measured attenuation coefficients of pure D_2O and H_2O, we eventually estimated [H]/([H]+[D]) at each voxel. Note that we used different methods to determine the attenuation coefficients of the pure samples in the different beamlines. In the NEUTRA experiments, the attenuation coefficients of neat D_2O and H_2O were measured using the average of 27 images of each specimen contained in a sealed aluminum cuvette with a known pathlength of 2.00 mm, *i.e.*, the capillary pathlength: D_2O: 0.514 ± 0.046 cm^{-1}, H_2O: 3.59 ± 0.06 cm^{-1}. These values are similar to the reported ones [13]. In the ICON experiments, we estimated the attenuation coefficients by fitting the transmittance profiles of droplet specimens [2]: D_2O: 0.598 ± 0.018 cm^{-1}, H_2O: 4.36 ± 0.05 cm^{-1}, which are comparable to the reported values [2].

4   Conclusion

This study leverages neutron radiography to investigate the mass transport dynamics of D_2O exposed to ambient air *in situ*. In the droplet experiments, regardless of the droplet sizes, contact angle, and gravitational direction, the absorbed H_2O is uniformly distributed within the droplet, at the timescale longer than the image acquisition time. In contrast, the capillary system exhibits gravity-induced spatial inhomogeneity, implying that flow-induced mixing is responsible for the homogeneity in the droplets. These observations underscore not only the critical role of the system's geometry and orientation in transport phenomena but also neutron radiography's unique capability to map materials spatiotemporal compositions *in situ*. Future research should systematically explore a wider range of geometries and environmental conditions to delineate further the relative contributions of diffusion and advection in mass transport phenomena involving multiple components of scientific and industrial applications, *e.g.*, binary liquids with water and alcohol.

Acknowledgment. The authors gratefully acknowledge the financial support from the National Research Foundation Korea through NRF-2021R1A2C101116312. This work was also supported by 2023 Research Fund (1.230077.01) of Ulsan National Institute of Science and Technology (UNIST). This work is based on experiments performed at the Swiss spallation neutron source SINQ, Paul Scherrer Institute, Villigen, Switzerland (beamtime proposal nos. 20212820 and 20222552). This paper has been submitted for review on 30th September, 2024, and accepted for publication on 17th December, 2024

References

1. Wagner, D., et al.: Neutron, fluorescence, and optical imaging: an in situ combination of complementary techniques. Rev. Sci. Instrum. **86**, 093703 (2015). https://doi.org/10.1063/1.4930068
2. Im, J., et al.: High-resolution neutron imaging reveals kinetics of water vapor uptake into a sessile water droplet. Matter **4**(6), 2083–2096 (2021). https://doi.org/10.1016/j.matt.2021.04.013
3. Fumey, B., et al.: Enhanced gas-liquid absorption through natural convection studied by neutron imaging. Int. J. Heat Mass Transf. **182**, 121967 (2022). https://doi.org/10.1016/j.ijheatmasstransfer.2021.121967
4. Boillat, P., et al.: In situ observation of the water distribution across a PEFC using high resolution neutron radiography. Electrochem. Commun. **10**, 546–550 (2008). https://doi.org/10.1016/j.elecom.2008.01.018
5. Satija, R., et al.: In situ neutron imaging technique for evaluation of water management systems in operating PEM fuel cells. J. Power Sources **129**, 238–245 (2004). https://doi.org/10.1016/j.jpowsour.2003.11.068
6. Strobl, M., et al.: High-resolution investigations of edge effects in neutron imaging. Nucl. Instrum. Methods Phys. Res., Sect. A **604**, 640–645 (2009). https://doi.org/10.1016/j.nima.2009.03.075
7. Jones, G., Ray, W.A.: The surface tension of deuterium oxide and of its mixtures with water. J. Chem. Phys. **5**, 505–508 (1937). https://doi.org/10.1063/1.1750067
8. Chen, A., et al.: Marangoni convection analysis during ethanol natural evaporation in a capillary tube. Microfluid. Nanofluid. **26**, 98 (2022). https://doi.org/10.1007/s10404-022-02628-x
9. Kaestner, A.P., et al.: The ICON beamline–A facility for cold neutron imaging at SINQ. Nucl. Instrum. Methods Phys. Res., Sect. A **659**(1), 387–393 (2011). https://doi.org/10.1016/j.nima.2011.08.022
10. Lehmann, E.H., Vontobel, P., Wiezel, L.: Properties of the radiography facility NEUTRA at SINQ and its potential for use as European reference facility. Nondestructive Testing Eval. **16**(2–6), 191–202 (2001). https://doi.org/10.1080/10589750108953075
11. Boillat, P., et al.: Chasing quantitative biases in neutron imaging with scintillator-camera detectors: a practical method with black body grids. Opt. Express **26**, 15769 (2018). https://doi.org/10.1364/oe.26.015769
12. Carminati, C., et al.: Implementation and assessment of the black body bias correction in quantitative neutron imaging. PLoS ONE **14**, e0210300 (2019). https://doi.org/10.1371/journal.pone.0210300
13. Lehmann, E.H., et al.: The inspection of tube fitting systems in use for drinking water supply by means of neutron imaging methods. European Conference Nondestructive Testing (2006)

Introduction of Camera System for CNRF at JRR-3

Keisuke Kurita[1]($\boxtimes$) (iD), Hiroshi Iikura[1], Isao Harayama[1], Yusuke Tsuchikawa[2], Satoshi Morooka[1], Tetsuya Kai[2], Takenao Shinohara[2], Naoya Odaira[3], Daisuke Ito[3], Yasushi Saito[3], and Masahito Matsubayashi[1]

[1] Materials Sciences Research Center, Japan Atomic Energy Agency (JAEA), 2-4 Shirakata, Tokai, Naka, Ibaraki 319-1195, Japan
`kurita.keisuke@jaea.go.jp`
[2] J-PARC Center, Japan Atomic Energy Agency (JAEA), 2-4 Shirakata, Tokai, Naka, Ibaraki 319-1195, Japan
[3] Institute for Integrated Radiation and Nuclear Science, Kyoto University, 2-1010 Asashironishi, Kumatori, Sennan, Osaka 590-0494, Japan

Abstract. CNRF is a neutron radiography facility that uses cold neutrons installed in the JRR-3 beam hall. After JRR-3 resumed operations in 2021 following a 10-year shutdown, only an imaging system using imaging plates (IPs) was available at the CNRF. Although high-resolution images could be acquired, the system was unsuitable for continuous imaging required for computed tomography (CT) scans and dynamic imaging. Additionally, performing quantitative analysis from these images was difficult. Thus, we have introduced a new imaging system with a light-tight box, scintillator, and CMOS camera. We demonstrated the performance of the system using a new Gd test pattern. The developed camera system successfully obtained the transmission images of the test pattern. Additionally, the spatial resolution of the system was estimated to be approximately 60 μm. In contrast, CNRF was not suitable for CT imaging etc. owing to its poor L/D value, and the neutron beam must be enhanced in future research.

Keywords: Neutron Radiography · Cold Neutron · Neutron Radiography Facility

1 Introduction

The CNRF is a neutron radiography facility using cold neutron beams installed at the C2–3–3–1 port of the beam hall of JRR-3 [1]. Neutrons from the JRR-3 reactor are converted into cold neutrons (~5 meV or ~4 Å) by the cold neutron source (CNS) using liquid hydrogen as a moderator. These cold neutrons are transported to the beam hall through three neutron guides (C1 to C3). In addition to the CNRF, the C2 beamline houses the triple-axis spectrometer LTAS [2], the neutron reflectometer SUIREN [3], neutron spin echo NSE [4], and the pulsed neutron instrument CHOP. A neutron bender downstream from SUIREN splits the cold neutron beam into three beam-lines (0°, 10°,

A. E. Craft and H. Z. Bilheux (Eds.): WCNR 2024, SPPHY 348, pp. 244–251, 2026.
https://doi.org/10.1007/978-3-032-15003-5_28

and 20° reflection angles) [5]. The CNRF and CHOP are installed on the 20° beamline. Compared to the thermal neutron radiography facility (TNRF) in the JRR-3 reactor room, the CNRF has a much smaller imaging room. The neutron flux and beam size at CNRF are also an order of magnitude smaller than at TNRF, making it unsuitable for imaging large samples and dynamic imaging. However, cold neutrons can produce high-contrast images, making CNRF ideal for thin-film samples. When JRR-3 resumed operations in 2021 after a 10-year shutdown, only an imaging system using imaging plates (IPs) was available at CNRF [1]. While high-resolution images could be obtained, the system was not suitable for continuous imaging, such as CT scans or dynamic imaging, and quantitative analysis was challenging. A previous system using a scintillator and SIT camera is no longer available, leading to the introduction of a new imaging system with a light-tight box, scintillator, and CMOS camera. A performance demonstration was conducted using a newly created Gd test pattern. This paper discusses the current state of the CNRF, the camera system, the Gd test pattern, and the demonstration results.

2 CNRF Specifications

2.1 CNRF Beam Specifications and Imaging Chamber Information

Figure 1 shows a schematic of the CNRF imaging chamber, which measures 1150 mm (W) × 1500 mm (H) × 800 mm (D). A lithium fluoride and lead beam shutter is installed upstream. The chamber has a rotary-stage, linear-stage, and Z-stage for imaging samples, with a linear-stage for the light-tight box downstream. A hole in the downstream wall allows the neutron beam to pass through to the CHOP facility, blocked by a beam stopper when CNRF is in use. The sample position is about 4 m from the bender. The CNRF beam size is approximately 20 mm (W) × 30 mm (H), with a neutron flux of 1.7×10^7 n/cm^2/s [1]. Tamura et al. reported a flux of 1.58×10^7 n/cm^2/s at the C2-3-3 port with a 4.8 Å peak [5]. The slight flux increase is likely due to the replacement of part of the upstream conduit with a supermirror conduit.

Until recently, only imaging plates (IPs) were available for the CNRF imaging system. The IPs were BAS-ND 2025 and developed by Fuji Photo Film CO., LTD. The IPs used for neutron imaging were scanned using an imaging reader (BAS-2500, Fuji Film Corp.). The spatial resolution of the CNRF using the IP system was approximately 80 μm [1].

2.2 Camera System Installed at CNRF

The TNRF installed at the JRR-3 reactor room uses a cooled CCD camera (iKon-L 936, Andor Ltd.) as its main camera system for capturing still images. However, this camera is relatively large, making it unsuitable for CNRF, which has a small imaging room. Therefore, the CNRF introduced a CMOS camera (ORCA Flash 4.0 v3, Hamamatsu Photonics Co., Ltd.), which is smaller and lighter than the one used at the TNRF. The specifications of this camera are given in Table 1. The camera has an imaging size of 2048 × 2048 pixels, pixel size of 6.5 μm × 6.5 μm, and 16-bit analog-to-digital (A/D) converter. The maximum exposure time is 10 s. However, in cases of experiments where

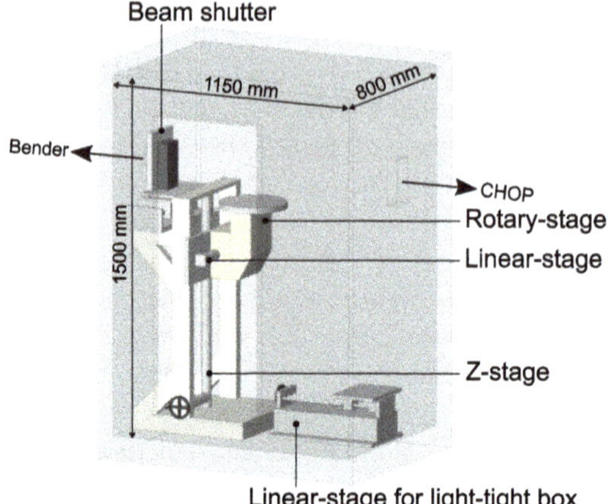

Fig. 1. A schematic diagram of the CNRF imaging chamber.

10 s is insufficient exposure time, multiple images are accumulated. The camera was installed in the light-tight box shown in Fig. 2. The size of the light-tight box was 250 mm (W) × 530 mm (H) × 350 mm (D), and the basic components were made of aluminum. A 74 mm extension was attached to this light-tight box and a scintillator screen was mounted at its end. The emission from the scintillator was bent 90° downwards by a mirror mounted on the neutron beam axis, which was then detected by the camera. The camera was mounted on a Z-stage and could be moved up and down by about 60 mm. The light-tight box was mounted on a frame stand so that the center of the scintillator was approximately at the center of the beam. The stand was mounted on a linear-stage for the light-tight box, which could be moved 300 mm along the beam axis, making it easy to adjust the distance to samples.

Table 1. Specifications of the CMOS camera introduced to the CNRF.

Effective no. of pixels	2048 (H) × 2048 (V)
Cell size	6.5 μm × 6.5 μm
A/D converter	16 bit

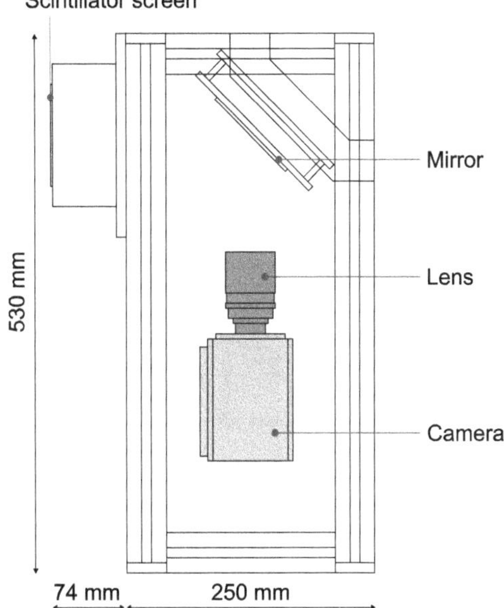

Fig. 2. A schematic diagram of the CNRF light-tight box.

3 Demonstration of CNRF Camera System

3.1 Gd Test Pattern Target

A new Gd test pattern was fabricated for the CNRF camera system demonstration. An overview of the fabricated Gd test pattern is shown in Fig. 3. This pattern was produced with reference to the Gd test pattern made by Segawa et al. [6]. Gd was deposited on a 1 mm thick quartz glass plate as a lattice pattern, a JRR-3 letter pattern, and four additional test patterns. The deposition thickness was 3 μm. The lattice pattern consisted of lines 1 mm wide, spaced 10 mm apart, with an overall size of 41 mm $\times$ 41 mm. The following four test patterns were produced within this grid pattern:

(a) **Vertical and horizontal line-pairs:** Groups of lines with decreasing widths of 0.5, 0.4, 0.3, 0.2, and 0.1 mm, each with five cycles.
(b) **Siemens star:** 10 mm diameter with 16 spokes. The minimum distance between spokes is 0.1 mm, and the maximum is 1 mm. There is a 0.1 mm groove for every 0.2 mm narrowing in width.
(c) **Line pairs for spatial resolution verification** (vertical, horizontal, and 45°): These line pairs have three cycles with widths of 0.3, 0.2, and 0.1 mm.
(d) **Vertical and horizontal line pairs for finer spatial resolution verification:** Ten line pairs decreasing from 0.1 mm to 0.01 mm in 0.01 mm increments. The line pairs have a period of five cycles.

The glass plate with these patterns was fixed to an aluminum frame, with an overall size of 55 mm $\times$ 55 mm $\times$ 5 mm.

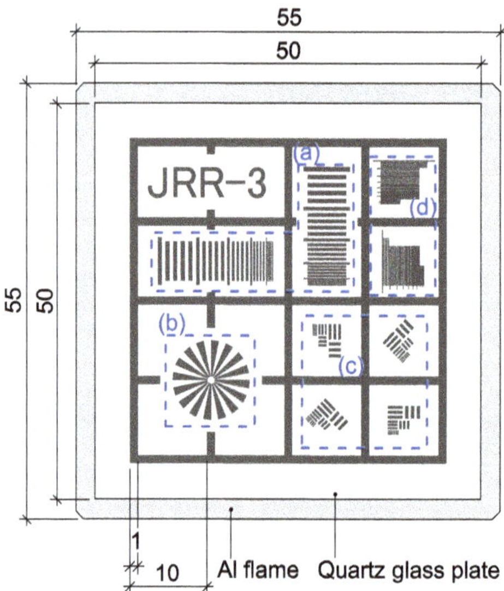

Fig. 3. Design patterns for the test targets (sizes in units of mm).

3.2 Demonstration

A demonstration of the newly introduced camera system in the CNRF was conducted using the fabricated test pattern. A 105 mm Micro-Nikkor lens (Nikon) was attached to the CMOS camera, and a 50-μm thick ^{6}LiF/ZnS scintillator was used as a detector. The field of view in this setup was approximately 35 mm $\times$ 35 mm. The exposure time was 10 s, and five images were taken per imaging target. The center pixel values of the five images were extracted using ImageJ software (ver. 1.50i, public domain software developed by NIH). The test pattern was imaged under two conditions: in close contact with the scintillator and at a distance of 1 cm from it. The acquired images were converted to transmission images for evaluation.

Figure 4 shows a direct beam image from CNRF and the vertical and horizontal line profiles of the image. Although the beam and image centers are offset, the beam distribution was successfully imaged. Brightness variations and fine lines in the image are likely caused by the structure of the CNRF neutron beam and scratches on the scintillator. These artifacts disappear in the neutron transmission images and are not currently considered problematic.

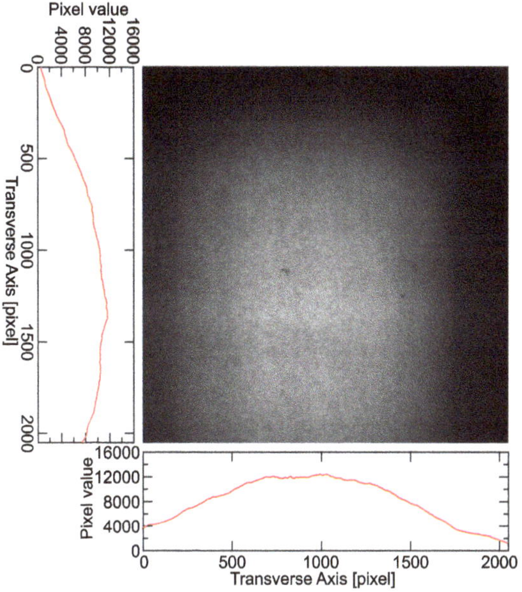

Fig. 4. Design patterns for the test targets (sizes in units of mm).

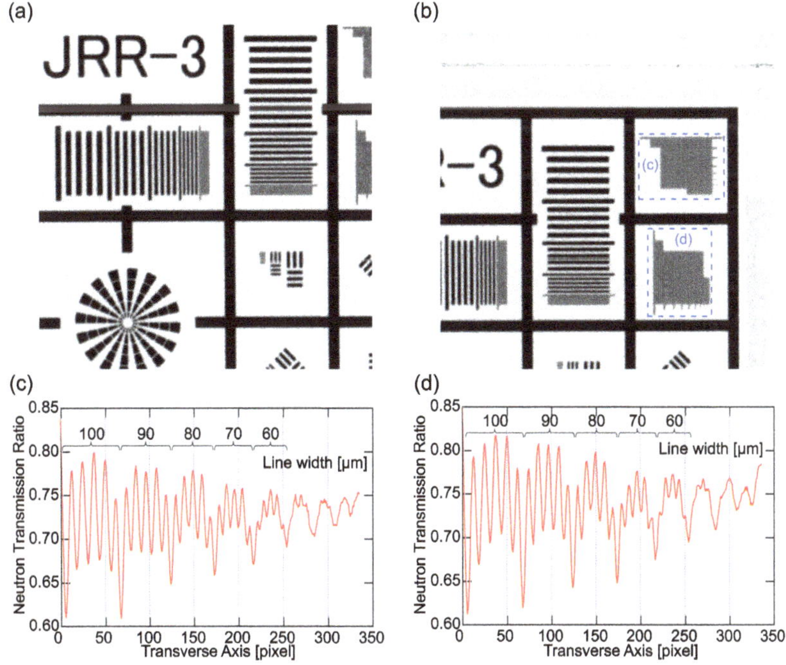

Fig. 5. (a, b) transmission images of Gd test pattern and the line profile of (c) vertical and (d) horizontal line-pairs from 0.1 to 0.01 mm.

Figure 5(a) shows transmission images of the Gd test pattern when placed in close contact with the scintillator. Figure 5(a) and 5(b) capture different parts of the test pattern. In Fig. 5(a), the 0.5–0.1 mm line pairs, Siemens star, and 0.3–0.1 mm line pairs were imaged. In Fig. 5(b), the 0.1 to 0.01 mm line pairs were imaged. From Fig. 5(a), the 0.1 mm line pairs were distinguishable in all patterns. In Fig. 5(b), the line profiles for the 0.1 to 0.01 mm line pairs (shown in areas (c) and (d)) were obtained to examine them in more detail. Figure 5(c) and 5(d) show the line profiles for vertical and horizontal line pairs, with the values indicating line widths. From these profiles, the spatial resolution of the CNRF camera system was approximately 60 μm in both directions.

(a)   (b)

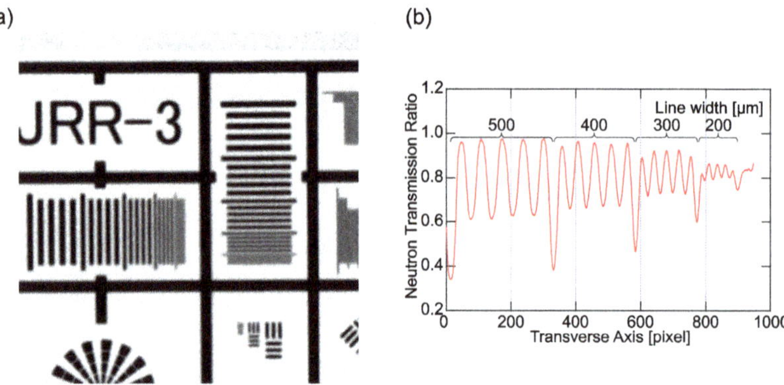

Fig. 6. (a) transmission images of the Gd test pattern and (b) line profile of the horizontal line pairs from 0.5 to 0.1 mm when the distance between the pattern and scintillator is 1 cm.

Figure 6(a) and (b) shows the transmission image and line profile of the horizontal line pairs (0.5 to 0.1 mm) when the test pattern was 1 cm away from the scintillator, respectively. The image appears more blurred compared to Fig. 5, and the 0.1 mm line pairs are no longer distinguishable. The spatial resolution was estimated to be approximately 200 μm in this case. Based on the transmission images, the L/D value, which indicates the parallelism of the neutron beam, was determined to be approximately 71. This value is similar to the estimate by Hayashida et al. [7] but less than half the L/D value of the TNRF [1]. Equipment with a low L/D value is unsuitable for CT imaging, where sample rotation is required. The actual beam structure in the CNRF is expected to be more complex, as the neutron beam is supplied by the bender. Improving the neutron beam will be a future challenge for CT imaging in the CNRF.

4   Conclusion

The new camera system was introduced at CNRF, and a demonstration was performed using a fabricated Gd test pattern. The system successfully captured transmission images of the test pattern, with an estimated spatial resolution of approximately 60 μm. Due to the low L/D value at CNRF, achieving good spatial resolution requires the scintillator and sample to be in close contact, making CT imaging challenging. Compared to various neutron radiography facilities worldwide [8–10], the L/D value is rather low, which limits the experiments that can be performed at CNRF. Future efforts will focus on improving the neutron beam to broaden the system's applications.

Acknowledgements. This work was supported by JSPS KAKENHI Grant Number 23H00237.

References

1. Kurita, K., et al.: Introduction to neutron radiography facilities at the japan research reactor-3. J. Phys. Conf. Ser. **2605**, 012005 (2023)
2. Metoki, N., Shibata, K., Matsuura, M., Kitazawa, H., Suzuki, H.S.: Hyperfine splitting and nuclear spin polarization in $NdPd_5Al_2$ and $Nd_3Pd_{20}Ge_6$. J. Phys. Soc. Jpn. **91**, 054710 (2022)
3. Yamazaki, D., et al.: Polarized neutron reflectometer SUIREN at JRR-3. Physica B **404**, 2557–2560 (2009)
4. Funama, F., Tasaki, S., Hino, M., Oda, T., Endo, H.: Double-focusing geometry for phase correction in neutron resonance spin-echo spectroscopy. Nucl. Instrum. Methods Phys. Res., Sect. A **1010**, 165480 (2021)
5. Tamura, I., et al.: Development of a compact vertical splitting system for the cold neutron beam at JRR-3. J. Phys. Conf. Ser. **528**, 012012 (2014)
6. Segawa, M., et al.: Spatial resolution test targets made of gadolinium and gold for conventional and resonance neutron imaging. In: JPS Conference Proceedings, vol. 22, p. 011028 (2018)
7. Hayashida, H., Segawa, M., Yasuda, R., Iikura, H., Sakai, T., Matsubayashi, M.: Development of multi-pinhole collimator for large imaging area with high spatial resolution. Nucl. Instrum. Methods Phys. Res., Sect. A **605**, 77–80 (2009)
8. Tengattini, A., Lenoir, N., Andò, E., Viggiani, G.: Neutron imaging for geomechanics: a review. Geomech. Energy Environ. **27**, 100206 (2021)
9. Schulz, M., Schillinger, B.: ANTARES: Cold neutron radiography and tomography facility. J. Large-Scale Res. Facil. **1**, A17 (2015)
10. Hussey, D.S., et al.: A new cold neutron imaging instrument at NIST. Phys. Procedia **69**, 48–54 (2015)

Crystallographic Phase Transformations and Corresponding Temperature Distributions During GTAW of Supermartensitic Stainless Steel Visualized by Neutron Bragg-Edge Imaging

Axel Griesche[1]([✉]), Tobias Mente[1], Alessandro Tengattini[2], Stefano Dal Pont[3], and Nikolay Kardjilov[4]

[1] Federal Institute for Materials Research and Testing (BAM), Berlin, Germany
axel.griesche@bam.de
[2] Institut Laue-Langevin (ILL), Grenoble, France
[3] Université Grenoble-Alpes, Grenoble, France
[4] Helmholtz-Zentrum Berlin Für Materialien Und Energie (HZB), Berlin, Germany

Abstract. We investigated the phase transformations during butt-welding of supermartensitic steel plates with help of Neutron Bragg-Edge Imaging (NBEI). Gas tungsten arc welding (GTAW) was used with a motorized torch allowing for automated weldments. The austenitization in the heat affected zone (HAZ) could be clearly visualized at $\lambda = 3.95$ Å, a wavelength smaller than the Bragg edge wavelengths of both austenite and martensite phases. The re-transformation into the martensitic phase during cooling was clearly detected. However, we observed an unexpected additional change in transmission at $\lambda = 4.4$ Å, a wavelength larger than the wavelengths of the Bragg edges of both the martensitic and austenitic phases. We attribute this change to the temperature dependence of coherent scattering at a crystal lattice. The observed two-dimensional attenuation map corresponds well with a temperature distribution modelling by software macros in ANSYS. Here, the absolute temperature values could be achieved by calibrating the modelled attenuation with help of a thermocouple placed at the steel plate. This allows in return for a direct two-dimensional temperature reading based on the relation between the neutron attenuation caused by the inelastic neutron scattering and the sample temperature. The method was extended further to 3D mapping of temperature distribution in bulky steel samples.

Keywords: Bragg Edge Imaging · Gas Tungsten Arc Welding · Steel · Inelastic Neutron Scattering

1 Introduction

The neutron imaging technique is a unique experimental method for non-destructive investigation of the microstructure of materials. Metals can be easily radiographed, even larger engineering components with wall thicknesses in the cm-range. Neutron imaging

© The Author(s) 2026
A. E. Craft and H. Z. Bilheux (Eds.): WCNR 2024, SPPHY 348, pp. 252–258, 2026.
https://doi.org/10.1007/978-3-032-15003-5_29

can be applied to tackle a broad range of problems in material's research as well as metallurgical or fundamental science aspects [1, 2]. For special applications, e. g. for visualizing the transformation of a crystal lattice as a function of time and temperature, monochromatic imaging is a very useful tool. It allows, e.g., to image the phase transition of γ-austenite to α'-martensite during free cooling of a heat-treated martensitic steel sample [3]. The transmitted intensity is significantly reduced during cooling below the martensite start temperature M_s since the emerging martensitic phase has a $\sim 10\%$ higher attenuation than the austenitic phase at the used wavelength of $\lambda = 3.95$ Å (see Fig. 2), resulting in a large image contrast. In the wavelength range beyond the last Bragg-edge from about 4.1 Å upwards, the coherent scattering contribution is no more present. This is due to the larger wavelengths than the lattice constant of the steel material. Energy-selective neutron imaging in this wavelength region, where the thermal motion of atoms causes significant temperature-dependent changes of the neutron transmission, shows temperature dependent contrast. In this work, we measured the two-dimensional neutron transmission through a vertically positioned steel plate sample during remelting of its top side by tungsten inert gas welding and extended this technique to 3D mapping of the temperature distribution in bulky steel samples. Three-dimensional FE modelling as a function of time was used to simulate the welding process where the obtained data were compared to the experimentally evaluated results.

2 Experiment

In a welding experiment at the former CONRAD-2 beamline of the BER II reactor at HZB, the phase transformations and the 2D temperature distribution during the remelting process were visualized [4]. In Gas Tungsten Arc Welding (GTAW), a voltage between sample and tungsten electrode allows establishing a welding arc that gives a high energy input in the sample. The melt region is protected against oxidization by an argon stream through the nozzle around the electrode. The local heat input in the sample can be controlled by the welding current and the travelling speed of the torch. Welding currents of 40 A, 60 A and 80 A and feed rates of 5 cm/s and 10 cm/s were used. The welding apparatus was a Castolin CastoTIG 1611 DC. A calibrated type K thermocouple covered by Inconel™ sheathing was positioned in a bore hole inside the sample side 2 mm below the welded surface and 20 mm behind the starting point of welding. The used set-up is shown in a photograph in Fig. 1.

The material was a low transition temperature (LTT) martensitic steel. The temperature range for the martensite-to-austenite transition starts around 80 °C and reaches up to 250 °C. The chemical composition of the low carbon (0.054 wt.-%) steel was Cr-Ni-Mo-Mn-Si 10.3-10.2-0.28-0.9-0.55 wt.-%. The sample dimensions were LxWxH 80x5x15 mm.

The neutron transmission images were recorded by a digital detector based on a sCMOS camera Andor NEO (6.5 µm, 2560 x 2160 pixels), which is focused through a lens on a scintillating screen. The further optical path is turned by a mirror to prevent radiation damage of the sensor. The effective pixel size was 54 µm. The scintillator is based on ^{6}Li:ZnS active material with a thickness of 200 µm. A double-crystal monochromator was used to select the neutron wavelength in the range between 2 Å and 6 Å with a

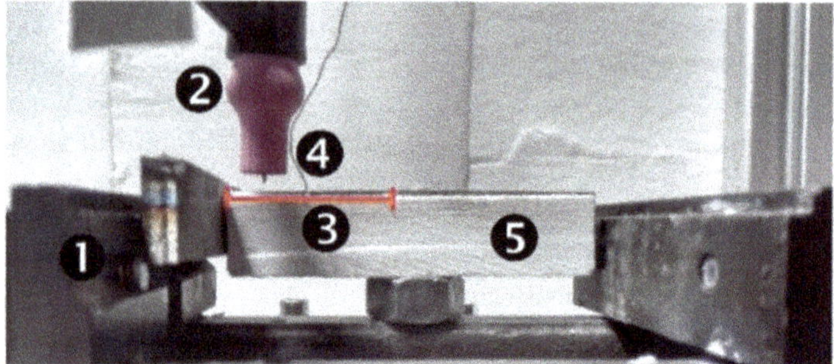

Fig. 1. Photograph of the experimental set-up. A vice (1) clamps the sample (3), which is thermally isolated by alumina felt (5) to prolong the cooling phase and to help to gain more images. A thermocouple (4) sticks in a bore hole close to the weldment. The torch (2) is set alight and travels from the side to the mid of the vertically mounted 5 mm thick steel sheet and images at different wavelengths are acquired. The viewing direction in the picture is in flight direction of the neutrons.

resolution of $\Delta\lambda/\lambda \approx 3\%$ [5]. The exposure time for detecting the monochromatic images was set to 2 s per image for a L/D of 167.

3 Results and Discussion

In the heat affected zone (HAZ), underneath the travelling welding torch, an image contrast in the sample is visible for monochromatic light at wavelengths smaller than those of the Bragg edges – 3.95 Å. This contrast is generated by the local austenitization (in the HAZ) of the martensitic weld metal with subsequent reconversion. As shown in Fig. 2a, the difference in attenuation between austenitic and martensitic phases is about $\sim 12\%$ that gives the clear image contrast at the used wavelength of $\lambda = 3.95$ Å.

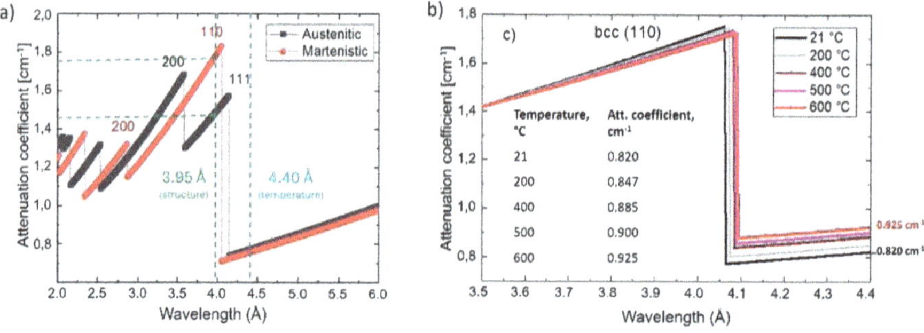

Fig. 2. a) Theoretical attenuation curves for both austenitic and martensitic crystallographic lattice types at room temperature. b) Calculated neutron total cross sections as a function of wavelength and temperature [7, 9].

The effect of this contrast variation can be seen in Fig. 3a, where the radiography of the sample at 32 s after the start of the welding is shown. The image was postprocessed by a median filter of 3x3 pixels followed by Gaussian blur filter with a window of 7 pixels for improving the signal-to-noise ratio of the transmission map. The intensity profile along the marked red arrow shows the rise in the transmission signal due to the austenitization in the heat affected zone (HAZ).

In addition to this, welding was repeated with the same image and welding parameters at $\lambda = 4.4$ Å, a wavelength larger than those of the Bragg edges. The obtained contrast in the HAZ, see Fig. 3b, shows a drop of the transmission (red curve), which can be related to the rapid increase of the temperature.

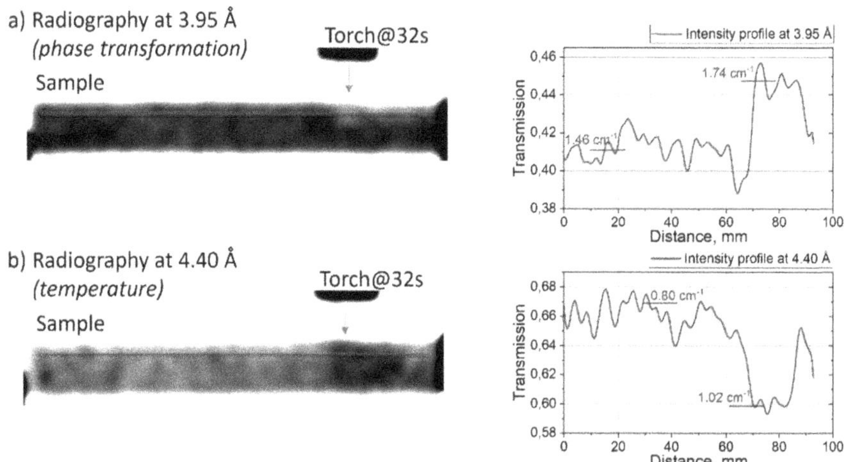

Fig. 3. a) Monochromatic radiography at $\lambda = 3.95$ Å (32 s after the welding started). The plot on the right is showing the intensity profile along the red arrow in the radiography image on the left. b) Monochromatic radiography at $\lambda = 4.4$ Å (32 s after the welding started). The plot on the right is showing the intensity profile along the red arrow in the radiography image on the left.

This finding can be explained by the temperature dependence of the inelastic neutron scattering as shown in refs. [6, 7]. Therefore, the contrast is not due to the change of the crystallographic structure but due to scattering of the neutrons induced by the thermal motion of the lattice.

The effect of temperature on the total cross section for neutrons is shown in Fig. 2b. The data were computed using the program *nxs* [8].

The results from Fig. 2b were used for correlation of the neutron transmission at 4.4 Å to the local temperature of the steel material. This correlation was used to convert the transmission variations in the measured neutron images to the temperature distribution in the steel samples as shown in [9]. The obtained results were compared with attenuation coefficient maps calculated from the series of radiography images taken at the neutron wavelength of 3.95 Å. The maps show the location of the temperature driven martensitic-austenitic phase transformation during the welding which takes place at temperatures higher than 700 °C. The comparison of the two types of maps shows clear relation between temperature distribution and austenitic phase transformation areas in the martensitic sample.

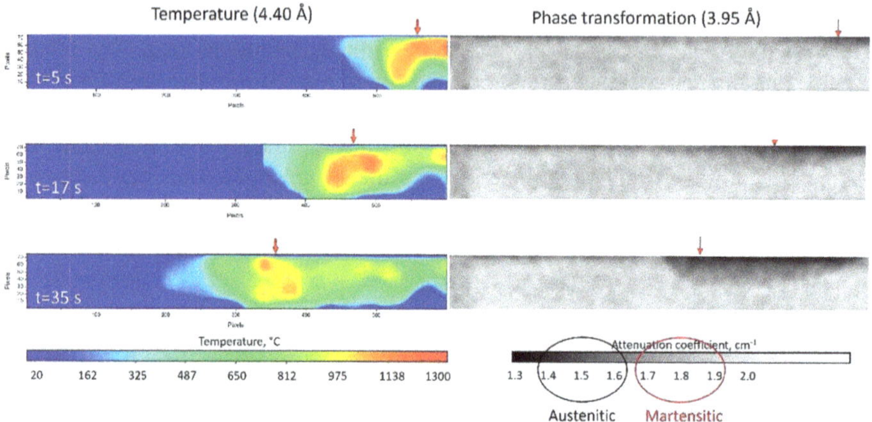

Fig. 4. Comparison between temperature distribution during welding (GTAW) of LTT steel plates [9] and attenuation coefficient maps calculated from the radiography series of images taken at wavelength of 3.95 Å.

To obtain more data regarding the influence of the inelastic neutron scattering on the image contrast, we performed neutron tomography experiments at the NeXT instrument at the ILL in Grenoble. In a furnace with radiation heating from above, a stack of two ferritic steel cylinders with 2 cm diameter and 4 cm length were heated up to a fixed temperature, which was kept constant during tomographic imaging. A fiber insulation prevented thermal convection to stabilize the thermal gradient along the vertical axis during imaging.

Fig. 5. Fotos of the gradient furnace used for the tomography experiments. Left: View inside the ceramic tube with steel cylinder samples and ceramic fiber isolation. Right: Ceramic tube on turntable with heating head mounted on top.

For the experiment, a monochromatic neutron beam with a wavelength of 4.4 Å was set by the double crystal monochromator of the NeXT instrument. The imaging para-meters were 3 s exposure time and 300 projections per 360°. The pixel size is 200 μm. Figure 6d shows a cut through the reconstructed 3D model of the tomography of the two cylinders. With help of the data from Fig. 2b and the procedure shown in [9], the gray values are converted into temperature values as shown in the scale. A NiCr-Ni thermocouple attached to the lower cylinder (Fig. 6) allowed for a single point calibration of the absolute temperature values.

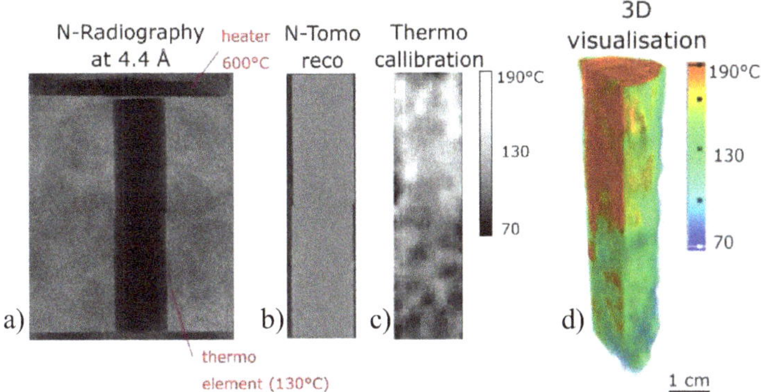

Fig. 6. a) Single radiography of the stack of two steel cylinders at neutron wavelength of 4.40 Å. b) Tomographic slice of the sample representing the distribution of attenuation coefficients. c) Temperature distribution map after applied calibration. d) Cut through the 3D reconstruction of the stack of two steel cylinders with the temperature color code.

Clearly visible is that the upper cylinder, which is closer to the heater, has a higher inner temperature than the lower one. Obviously, the thermal conductivity via the front surface towards the lower one is limited.The skin surface of the upper cylinder is colder than the interior probably due to the heat loss by radiation.

4 Summary

For the first time, we could show that the temperature dependence of coherent scattering at a crystal lattice can influence the image formation in Neutron Bragg Edge Imaging (NBEI) for wavelengths larger than those of the Bragg edges. The temperature dependence of attenuation in this wavelength range allows the calculation of temperature from image grey values or vice versa. The experimental findings are supported by independent results gained from temperature field finite element modelling of the welding process. Thus, NBEI allows principally the quantitative temperature field visualization in crystalline materials.

References

1. Kardjilov, N., Manke, I., Hilger, A., Strobl, M., Banhart, J.: Mater. Today **14**(6), 248–256 (2011)

2. Griesche, A., Große, M., Schillinger, B.: Neutron imaging. In: Fritzsche, H., Huot, J., Fruchart, D. (eds.) Neutron scattering and other nuclear techniques for hydrogen in materials, pp. 193–225. Springer International Publishing, Cham (2016). https://doi.org/10.1007/978-3-319-227 92-4_7

3. Dabah, E., et al.: J. Mater. Sci., 1–7 (2016)

4. Griesche, A., Pfretzschner, B., Taparli, U.A., Kardjilov, N.: Appl. Sci. **11**(22), 1–9 (2021)

5. Kardjilov, N., et al.: J. Appl. Crystallogr. **49**(1), 195–202 (2016)

6. Sato, H., et al.: Sci. Rep. **13**, 688 (2023). https://doi.org/10.1038/s41598-023-27857-0

7. Al-Falahat, A.M., et al.: J. Appl. Crystallogr. **55**, 919–928 (2022). https://doi.org/10.1107/S16 00576722006549

8. Boin, M.: J. Appl. Crystallogr. **45**(3), 603–607 (2012). https://doi.org/10.1107/S00218898120 16056

9. Jamro, R., et al.: J. Phys. Conf. Ser. **2605**(1) (2023)

Neutron Tomography and the Virtual World of Palaeontology

Joseph J. Bevitt[✉]

Australian Centre for Neutron Scattering, Australian Nuclear Science and Technology Organisation (ANSTO), Lucas Heights, NSW 2234, Australia
joseph.bevitt@ansto.gov.au

Abstract. Neutron tomography is experiencing a rapid uptake by palaeontologists for the digital excavation of specimens not achievable using mechanical means or X-ray tomography. Interest in neutrons is driven by their ability to penetrate iron-rich minerals, and their high sensitivity to hydrogen, which acts as a natural contrast agent to enhance the visibility of select fossils and preserved soft-tissue structures. The applications of neutron imaging in palaeontology extend beyond the description of extinct species, and can inform conservation efforts, authenticity investigations and contribute to museum exhibitions. Neutron-assisted discoveries are making headlines, and discoveries such as the first evidence that crocodiles ate dinosaurs, new pterosaur species, neurosensory diversity in early reptiles, the oldest amniote skin and world's oldest heart illustrate how neutron tomography is leading a revolution in palaeontology.

Keywords: Palaeontology · fossils · evolutionary studies · morphology · anatomy · soft-tissues · museums

1 Introduction

Evolutionary palaeontology is the study of how species evolve, the changes in their characteristics, relationships, and diversity over time. Physical evidence is obtained by examining species that exist today, and those that have gone extinct, via the fossil record. The need to physically extract fossils from their surrounding rock using mechanical or chemical methods to record their morphology is a major hindrance to the field. These procedures are generally conducted blind, without an exact prior knowledge of the number or location of fossils within the matrix, and do not retain a full record of the spatial relationships between elements, or amongst organisms. Mineralised and non-mineralised soft-tissue and contextual environmental information present in the host rock are also lost during preparation, while internal morphologies are accessible only through destructive sampling. The physical extraction of fossils from rock results in delayed deterioration, posing challenges for the preservation of specimens for future generations.

A. E. Craft and H. Z. Bilheux (Eds.): WCNR 2024, SPPHY 348, pp. 259–267, 2026.
https://doi.org/10.1007/978-3-032-15003-5_30

X-ray radiography and computed tomography (CT) have revolutionised palaeontology, with CT becoming a critical non-destructive tool for the characterisation of bone surfaces and internal features via digital 3D models. While CT works well in many geological settings, it does not work well for dense rock samples, the imaging of low-density fossils in a higher density matrix, or where the chemical composition or material density of fossils and their matrices are similar.

2 Background

Neutron tomography (NT) provides a non-destructive alternative to CT for the imaging of fossilised material, based on the different interaction of neutrons with matter compared to X-rays. The potential complementarity of neutrons to X-rays in palaeontology varies depending on material composition, mode of fossilisation (diagenesis and taphonomy) and structural features of interest. The most frequent issue encountered with the use of CT in palaeontology is the presence of iron-rich minerals, which strongly attenuate X-rays. Authigenic pyrite forms when active, localised sulfate reduction in iron-rich pore waters results in a strong concentration gradient, which confines pyrite precipitation to decaying organic matter [1]. For this reason, pyrite is often found both in direct contact with bones, and within bone cavities. Aragonitic fossils are susceptible to pyritization, and in exceptional circumstances, it preserves non-biomineralised tissues. In the latter situation, the pyrite acts as a convenient natural contrast agent, yielding exceptional CT visualisation of soft-tissue remains. In the former situations, pyrite results in beam hardening artefacts that obscure fossil remains. Increasing the X-ray beam energy sufficiently to penetrate pyrite generally causes a reduction in contrast between the matrix and the embedded fossils. As pyrite serves as an indicator for potentially preserved organic remains, such specimens are ideally suited to NT, owing to the low neutron attenuation by pyrite, and high sensitivity to hydrogen and potential organic remains.

Until recently, the utilisation of NT by palaeontologists and natural history museums has been limited due to a lack of awareness of the technique, concerns over long-term neutron activation, the unclear benefits over CT, and limited accessibility to neutrons. A strategic focus on the application of NT to palaeontology at the Australian Nuclear Science and Technology Organisation (ANSTO), combined with the high availability of neutrons, extensive consultation with museums and a careful assessment of the risks involved in working with fossil specimens has resulted in the remarkable uptake of NT by the palaeontological community.

3 The First Fossil Samples

Engagement with palaeontologists at our facility began in 2014, soon after the construction and commissioning of Dingo, the thermal-neutron radiography/tomography/imaging instrument at ANSTO [2], a facility funded via the Australian Government's National Collaborative Research Infrastructure Strategy program. The Australian Age of Dinosaurs Museum of Natural History (AAOD) supplied a group of specimens to test the applicability of NT for the investigation of material from two localities that yielded poor results with medical CT: dinosaur-bearing ironstone from the

Winton formation, and marine nodules from Ilfracombe, Queensland. In contrast to the medical CT which failed to reveal the contents of the nodules, the sensitivity of neutrons to hydrogen-rich chitin achieved a clear visualisation of the marine arthropods, such as the carapace of a 112-million-year-old fossil crab, *Torynomma quadrata*, Fig. 1.

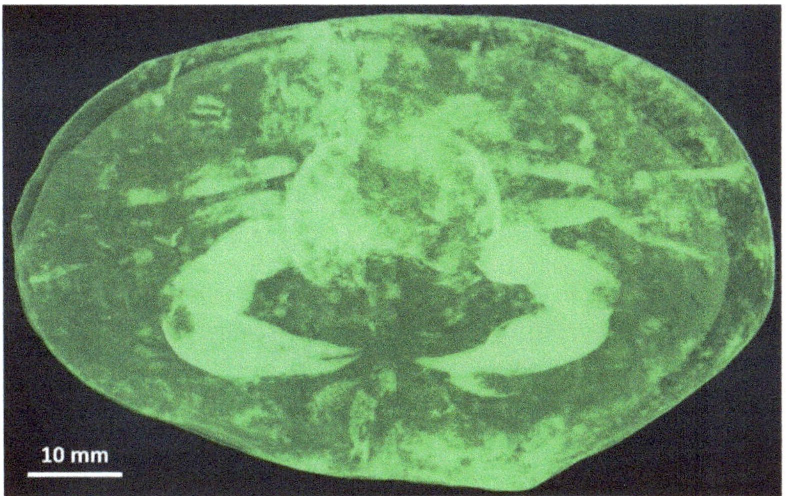

Fig. 1. Digital reconstruction of Ilfracombe crab *Torynomma quadrata.*

Digital reconstructions of the fossilized crabs and crayfish revealed the orientation of the specimens within the nodules, enabling the assessment of their completeness, suitability for exhibition and providing a guide to their physical preparation; factors that greatly reduce the cost of labour involved with the production of museum specimens. Scans of the fist-sized, dense ironstone samples were also successful, achieving sufficient penetration and contrast between sediments, included plant and vertebrate fossils. Concerns regarding long-term neutron activation of were alleviated with daily monitoring of the specimens post-irradiation. Radioactivity of the marine nodules decayed to natural levels within 2 weeks, and the ironstone within 5 weeks of irradiation. The specimens were radiologically cleared and returned to the museum within 6 weeks of receipt, paving the way for the next specimens and many fruitful collaborations.

4 *Confractosuchus*, the Broken Dinosaur Killer

While using heavy machinery to clear away the overburden in search of preserved sauropod dinosaur remains, staff from the Australian Age of Dinosaurs Museum of Natural History (AAOD) observed a large, shattered ironstone concretion with exposed bone elements. The rubble was gathered, CT scans attempted and failed, and two years of effort spent reassembling the three-dimensional jigsaw puzzle, identified as a Cretaceous crocodilian (AODF0890). Following the successful feasibility NT scans, unidentified parts of specimen were scanned with the aim of determining their positions relative to the assembled concretion, and to guide potential preparation of the delicate bones.

One block suspected of containing a crocodilian coracoid was imaged, resulting in the serendipitous discovery of numerous non-crocodilian bones. These included the femur of a small ornithopod dinosaur, identified by the presence of a pendant fourth trochanter on the femur. One end of the femur had been sheared off, likely due to oral processing, and a bite mark was evident in the shaft. The collection of dinosaur bones was encircled by bones and skin of the crocodilian, suggesting that NT had enabled visualization of the abdominal contents of the crocodilian. Further evidence was required to determine whether these ornithopod remains were directly associated with the crocodilian; a process involving 5 years of concerted effort.

NT scans of other parts of the concretion achieved mixed results due to internal volumes of neutron-attenuating clay. The Imaging and Medical Beamline at the Australian Synchrotron was employed for these blocks, using the highest-energy pink-beam configuration available with a peak X-ray intensity of 260 keV and voxel size of 200 μm. At this resolution, bone-matrix boundaries were unclear, but the data were sufficient to guide preparation of the largest blocks encasing the skull and the overall reduction in mass of others. Following this process, a second set of higher resolution NT and synchrotron CT scans were acquired, and the datasets combined to yield a complete digital excavation of the holotype of a near complete ~2.5 m long crocodilian, *Confractosuchus* ('broken crocodile'), referring to the shattered concretion, and *sauroktonos* ('lizard killer'), referring to the abdominal contents, Fig. 2. The combination of scans permitted visualisation of the abdominal cavity, revealing a remarkable insight into the diet of *Confractosuchus*, and the first evidence that crocodiles ate dinosaurs; the partially digested remains of a juvenile ornithopod dinosaur were found concentrated in the anterior part of the abdominal cavity in the vicinity of the pectoral girdle and bound unambiguously between the axial skeleton and osteoderms of the presumed gastral shield.

This project is the first instance in which a combination of NT and synchrotron-CT were used to enable the manual preparation and holistic analysis of a natural history museum specimen. The 3D digital reconstruction is now permanently exhibited alongside the partially prepared crocodilian fossil and a 3D print of the stomach contents at the Australian Age of Dinosaurs Museum, Queensland, Australia.

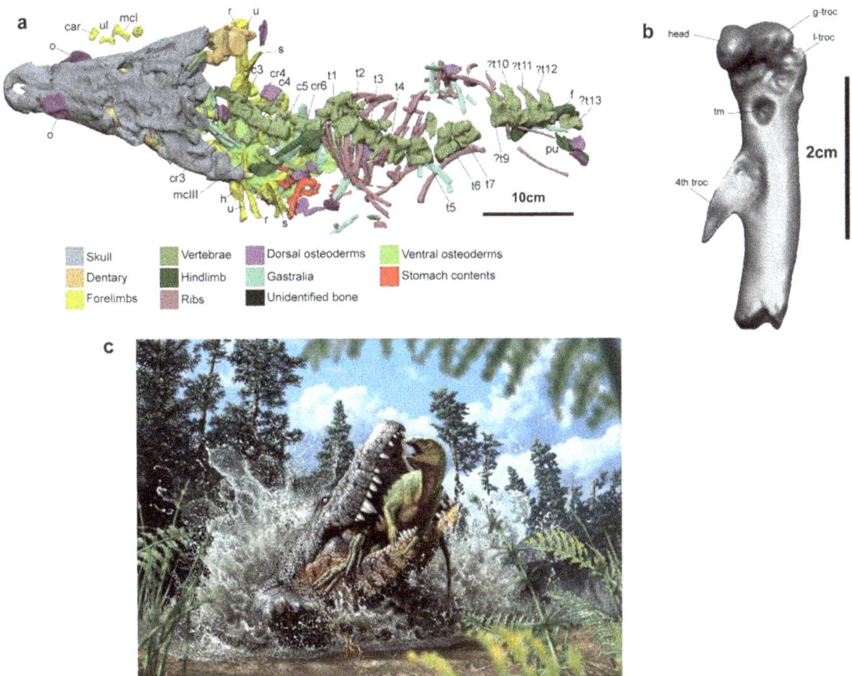

Fig. 2. a) Digital reconstruction of *Confractosuchus sauroktonos* in dorsal aspect with abdominal ornithopod remains shown in red. Abbreviations: c, coracoid; car, carpal; c(no.), cervical vertebrae (number); cr3, cervical rib 3; f, femur; h, humerus; mcI, metacarpal 1; mcIII metacarpal III; o, osteoderm; pu, pubis; r, radius; s, scapula; t(no.), thoracic vertebrae (number); u, ulna; ul, ulnare; b) the ingested ornithopod left femur with tooth mark. Abbreviations: g-troc, greater trochanter; l-troc, lesser trochanter; tm, tooth mark; head, femur head; 4th-troc, fourth trochanter. Source: [3]; c) life restoration of the crocodilian eating the dinosaur. Credit: Julius Csotonyi.

5 The Permian Fossils of Richards Spur

The Richards Spur Dolese Brothers Quarry, near Fort Sill, Oklahoma preserves the most diverse assemblage of early Permian terrestrial tetrapods, with more than 40 recognised species found within calcareous claystone and mudstone fissure fills of a karst system eroded out of the Ordovician Arbuckle Limestone. The unique feature of the fossils from this locality is the superb quality of bone preservation, undistorted by geological processes. Fossils in some fills are coloured white and found in a white, mostly calcite matrix, while in other fills the fossils are stained black due to the infusion of oil-seep hydrocarbons from the underlying Devonian age Woodford Shale [4], and encrusted by diagenetic pyrite formed under anoxic conditions. CT analysis of unprepared white specimens is challenging due to the low X-ray contrast between bone and calcite, and the visualisation of black bones is hampered by the presence of highly attenuating pyrite. With neutrons, little to no contrast is achieved between white bone and matrix, and the inherent higher spatial resolution of synchrotron CT is the preferred method for imaging this material.

The infusion of hydrogenous kerogen into the black-stained bones acts as a natural stain, enhancing NT contrast between bone and matrix. Further, pyrite and calcite attenuate thermal-neutrons to the same extent, rendering pyrite invisible in NT. The high penetrating ability of neutrons through this material enables the imaging of dense assemblages of bone in blocks up to 20 cm in diameter, corresponding to the maximum field of view of Dingo. The extreme contrast between bone and calcite matrix reduces the time required to yield high-quality NT data, enabling the efficient inspection of bulk samples, and high-resolution, low-noise scans on individual specimens. NT has enabled the first detailed anatomical study of neurological diversity [5–7] Fig. 3, ontogenesis [8], and mandibular-dental pathologies [9] of fauna from this locality, along with the oldest-known evidence of amniote skin [4], providing a unique resource for the study of evolutionary transformations that occurred with the development of a fully terrestrial lifestyle among vertebrates.

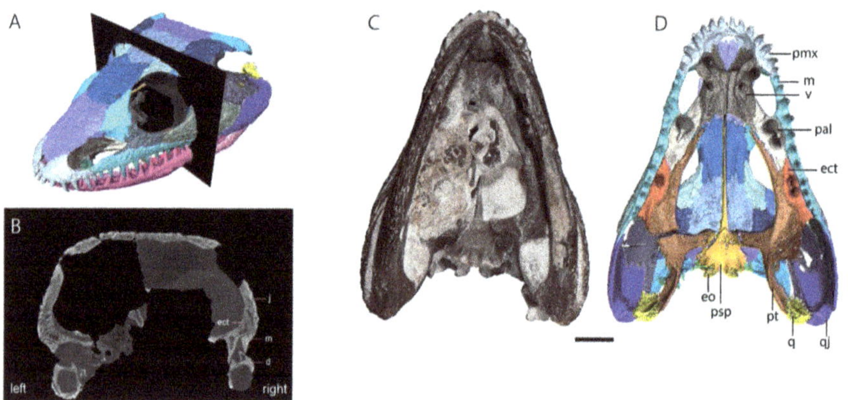

Fig. 3. OMNH 79318, cf. *Acheloma*, entire specimen in a) anterolateral profile, b) transverse slice through NT reconstruction showing exceptional contrast and sutural relationships of the ectopterygoid, jugal and maxilla; ventral profile. c) photograph; d) NT digital segmentation of the skull roof and palate. Scale bar, 1 cm. Source: Robert Reisz and [8].

6 Plants and Fish

The sensitivity of neutrons to hydrogenous organic content is particularly relevant to the visualisation of morphology and internal anatomy of carbonised fossil plant material, material that only appears as low-density pockets in equivalent CT scans. In one study, the high attenuation of neutrons by fossil resin yielded direct evidence for fire-adaptive traits including resin canals and reservoirs in the Cretaceous polar conifer *Protodammara*. Fire-adaptive strategies are common in plants living in fire-prone areas today, and provide evidence for widespread polar wildfires during the mid-Cretaceous hothouse (ca. 99-90 Ma) [10].

The application of NT to silicified fossils, those in which organic or inorganic matter is replaced by silica, generally fails due to low neutron attenuation and low contrast. This is the case for most silicified plant and fish specimens. An exception are the opalised fossil remains, including Cretaceous dinosaurs, mammals and pine cones from the Lightning Ridge fossil locality, which are currently under investigation. Opalisation preserves internal details in fossils when silica solution seeps into the organic material before it decomposes, replacing the organics with silica and preserving the fine internal structures in bone or plant. Internal details are not preserved in 'jelly mould' fossils, where the silica solution fills the cavity left by an organism that has rotted away.

The Gogo Formation in Western Australia is a Konservat-Lagerstätte that preserves extraordinary fossils of a Late Devonian reef community. Fish and arthropod fossils from this locality have been found within limestone concretions with unparalleled biomineralized, three-dimensional, and undistorted digestive systems, umbilical structures, nerves and muscle fibres that retain features at the cellular level. These structures have been visualised directly via the acetic acid preparation technique, which dissolves the surrounding limestone while conserving the delicate phosphatised soft-tissues and bone. Once prepared in this manner, some specimens become so extremely brittle that movement from their museum casings for further study becomes impossible. For some time, non-destructive inspection and reconstruction of the musculature within this fauna has been achieved with high-resolution CT at the European Synchrotron (ESRF) [11].

The application of NT to specimens from the Gogo Formation is challenging due to the material's high neutron attenuation, and low contrast between the phosphate-rich limestone and phosphatised fossils. In a study of a unique specimen of a 380-million-year-old arthrodire, a form of armoured fish, published in *Science*, NT uncovered the only known example of a three-dimensionally fossilized heart and cardiac vessels from this lineage, and the oldest heart known [12]. NT showed that the heart was more complex than expected, two-chambered and S-shaped like the heart of many modern animals, including humans. The heart was also positioned more forward than in earlier fish; an evolutionary step that eventually made space for the evolution of lungs, Fig. 4.

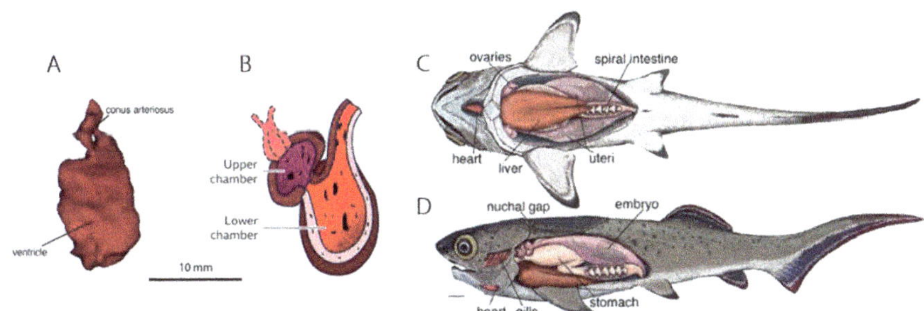

Fig. 4. a) Digital reconstruction from neutron tomography, and b) model of the Gogo fish's two-chambered heart and cardiac vessels. Internal arthrodire anatomy in c) lateral view and b) ventral view. Scale bar, 1 cm. Source: Kate Trinajstic and [12].

7 Conclusions and Future Work

Neutron tomography has proven to be an exceptional tool for the non-destructive analysis of fossils, accounting for 25% of Dingo instrument usage at ANSTO. The ability to penetrate iron-rich minerals and achieve contrast based on the presence of hydrogen enables the digital excavation of soft-tissue remains, stomach contents and organisms that are not accessible via manual extraction, nor X-ray tomography. By visualizing organic consolidants and adhesives, NT can inform conservation efforts and identify fraud in palaeontology. Challenges do remain for successful NT of large and highly attenuating fossils. A new sample stage is to be installed in late 2026, enabling the imaging of bulkier specimens up to 100 cm tall. Preliminary experiments with cadmium-filtered epithermal neutrons on Dingo are promising, with the benefit of reduced radioactivation of specimens. Ongoing programs of work with museums continue to materialise opportunities of integrating NT-based visuals and life reconstructions into exhibitions. The uptake of neutron tomography by palaeontologists has demonstrably increased public awareness of neutron-based science and the outcomes serve to inspire the next generation of scientists.

References

1. Farrell, Ú.C.: Pyritization of soft tissues in the fossil record: an overview. Paleontol. Soc. Papers **20**, 35–58 (2014)
2. Garbe, U., et al.: A new neutron radiography / tomography/imaging station DINGO at OPAL. Phys. Procedia **69**, 27–32 (2015)
3. White, M.A., et al.: Abdominal contents reveal cretaceous crocodyliforms ate dinosaurs. Gondwana Res. **106**, 281–302 (2022)
4. Mooney, E.D., et al.: Paleozoic cave system preserves oldest-known evidence of amniote skin. Current Biol. **34**(2), 417–426, e4 (2024)
5. Bazzana, K.D., et al.: Endocasts of the basal sauropsid Captorhinus reveal unexpected neurological diversity in early reptiles. Anat. Rec. **306**(3), 552–563 (2023)
6. Bazzana-Adams, K.D., et al.: Neurosensory anatomy and function in Seymouria. J. Morphol. **284**(5), e21577 (2023)
7. Bazzana, K.D., et al.: Neurosensory anatomy of Varanopidae and its implications for early synapsid evolution. J. Anat. **240**(5), 833–849 (2022)
8. Gee, B.M., Bevitt, J.J., Reisz, R.R.: A juvenile specimen of the trematopid Acheloma from Richards Spur, Oklahoma and challenges of trematopid ontogeny. Front. Earth Sci. **7**, 38 (2019)
9. Mooney, E.D., et al.: An intriguing new diapsid reptile with evidence of mandibulo-dental pathology from the early Permian of Oklahoma revealed by neutron tomography. PLoS ONE **17**(11), e0276772 (2022)
10. Mays, C., Cantrill, D.J., Bevitt, J.J.: Polar wildfires and conifer serotiny during the cretaceous global hothouse. Geology **45**(12), 1119–1122 (2017)
11. Trinajstic, K., et al.: Fossil musculature of the most primitive jawed vertebrates. Science **341**(6142), 160–164 (2013)
12. Trinajstic, K., et al.: Exceptional preservation of organs in devonian placoderms from the Gogo lagerstätte. Science **377**(6612), 1311–1314 (2022)

Enhancement of Spatial Resolution in High-speed Neutron Imaging Using a Multi-slit Collimator

Yasushi Saito[1]($\boxtimes$) , Daisuke Ito[1], Naoya Odaira[1] , Isao Harayama[2], and Keisuke Kurita[2]

[1] Institute for Integrated Radiation and Nuclear Science, Kyoto University, Asashiro-Nishi, Kumatori 2-1010590-0494, Sennan, Osaka, Japan
`saito.yasushi.8r@kyoto-u.ac.jp`
[2] Japan Atomic Energy Agency, Shirakata 2-4, Tokai, Naka 319-1195, Ibaraki, Japan

Abstract. A multi-slit collimator system, which is effective for enlarging imaging area with high spatial resolution, has been demonstrated at the Thermal Neutron Radiograph Facility at JRR-3 in JAEA. The developed multi-slit is made of a cadmium plate, with 1mm-wide slits arranged at 2 mm intervals. The neutron beams emitted from each slit spread radially depending on the collimation of the incident neutrons, so the distance between the converter multi-slits was adjusted to optimize it for the neutron beam used. The effectiveness of the multi-slit was confirmed using the standard indicator for the J-PARC, and the beam collimation and image defects caused by the slits were demonstrated. In addition, a multi-pinhole system was also demonstrated by using double multi-slit collimators.

Keywords: Neutron Radiography · Multi-slit Collimator · Spatial Resolution · Non-destructive Testing · Image Enhancement

1 Introduction

In the field of neutron imaging, the challenge of simultaneously achieving high spatial and temporal resolution is significant, particularly for dynamic imaging applications. While using a pinhole collimator can improve beam collimation, it simultaneously results in a reduction of the neutron flux, affecting the required exposure time or the imaging speed for dynamic imaging. The beam collimation, L/D is directly related to the image blurring. The image blurring, d, in transmission imaging can generally be expressed by the following equation.

$$d = \frac{L_s D}{L} \tag{1}$$

where L_s is the distance between object and detector, D is the aperture size and L is the distance between aperture and object. The image blurring, d should be possibly small for sharp image. The L_s is limited by the experimental conditions and L is restricted by a

A. E. Craft and H. Z. Bilheux (Eds.): WCNR 2024, SPPHY 348, pp. 268–275, 2026.
https://doi.org/10.1007/978-3-032-15003-5_31

facility space. Therefore, use of a pinhole aperture is effective for high collimator ratio. In the Neutron imaging instrument installed at the National Institute of Standards and Technology, a pinhole aperture is used, achieving a high collimator ratio [1]. Use of the pinhole aperture results in smaller imaging area and smaller neutron flux is restricted by the original aperture size, the distance between the aperture and the pinhole. Hayashida et al. [2] developed a multi-pinhole collimator to improve the spatial resolution and to expand the effective imaging area. They conducted a multi-pinhole collimator experiment at the Cold Neutron Radiography Facility (CNRF) at JRR-3, where they achieved 450 beam collimations using 70 original beam collimations. Bingham et al. [3] has developed a coded source imaging (CSI) system and tested at the High Flux Isotope Reactor (HFIR) CG1D beamline at Oak Ridge National Laboratory (ORNL).

However, high-speed imaging cannot be performed by using such a multi-pinhole collimator or coded source due to its smaller neutron flux. By using the multi-pinhole, it is possible to improve the spatial resolution in the two-dimensional directions. However, it would be possible to detect a curved line, whose normal vectors are not limited to horizontal or vertical directions, by improving the spatial resolution only in the one-dimensional direction by using a multi-slit collimator. By using a multi-slit collimator, which improve the beam collimation in only one direction, the attenuation of neutron flux can be minimized.

In this paper, we demonstrate the improvement of beam collimation in either the horizontal or vertical direction using a single multi-slit collimator, and we also discuss the results of using two orthogonal multi-slit collimators to achieve a multi-pinhole effect.

2 Experimental Setup

2.1 Design of Multi-slit Collimator

Figure 1 shows the schematic of the developed multi-slit collimator. The developed multi-slit is made of a 0.5 mm thick cadmium plate, with 1 mm-wide slits arranged at 2 mm intervals as shown in Fig. 1(b). The slit area is 150 mm × 150 mm, having 50 slits in the horizontal direction, as shown in Fig. 1(a). Based on the geometric aperture ratio, the neutron flux emitted from the multi-slit is calculated to be almost 33 percent of the incident neutron flux.

2.2 Setup of Multi-slit Collimator

Figure 2 shows a typical setup of the multi-slit collimator. The distance between the imaging plane and the multi-slit, L_s, is adjusted by using a one-dimensional motorized traverser. To realize a collimation ratio of 600, L_s is set at around 600 mm in this study. The TNRF installed at the JRR-3 reactor room uses a cooled CCD camera (iKon-L936, Andor Ltd.). The distance between the sample and the imaging plane, ls, is kept at 150 mm. Figure 3 shows the J-PARC indicator [5], which is used as the test sample.

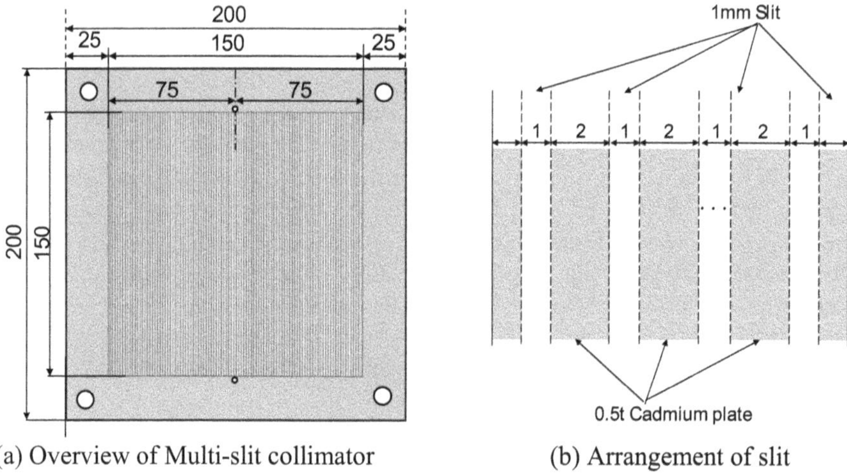

(a) Overview of Multi-slit collimator   (b) Arrangement of slit

Fig. 1. Schematic of Multi-slit collimator (a) Overview, (b) arrangement of each slit.

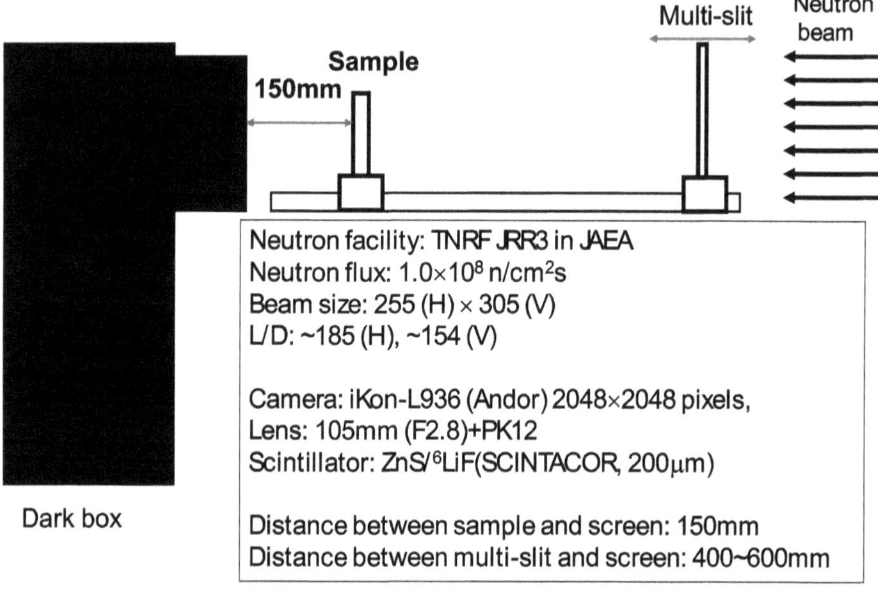

Fig. 2. Setup of Multi-slit collimator.

2.3 Adjustment of Single Multi-slit Collimator

Figure 4 shows the typical setup of the multi-slit collimator. To prevent adjacent neutron beams from overlapping. The optimal distance, L_m, is 575 for vertical setup 475 for horizontal setup, respectively. As shown in Table 1 [4], the initial L/D of TNRF takes different value for each direction. The difference in the optimal distance depends of the initial L/D.

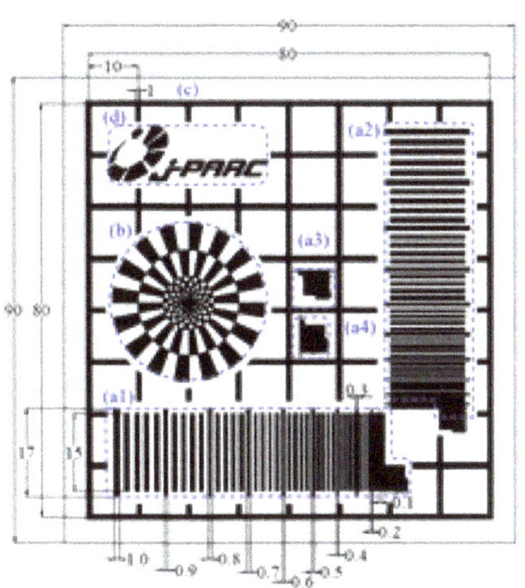

(a) Detailed design of J-PARC indicator.   (b) Typical Neutron image

Fig. 3. Setup of Multi-slit collimator.

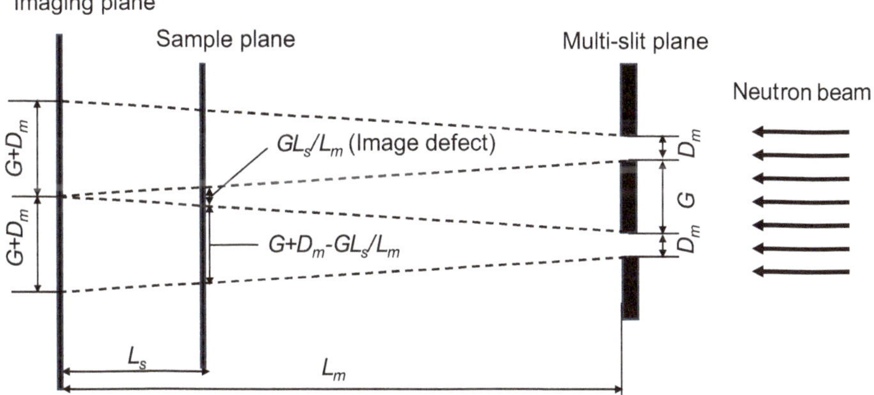

Fig. 4. Optimal setup of Multi-slit collimator.

Table 1. Specifications of TNRF.[4]

Neutron flux	1.0×10^8 n/cm^{2}s
Beam size	255(Horizontal) $\times$ 305(Vertical)
L/D	~185 (Horizontal) ~154 (Vertical)

As shown in Fig. 4, It was found that avoiding beam overlap on the detector surface leads to image defects. In this figure, D_m (=1mm) is the slit width, G (=2 mm) is the gap between slits, L_m is the distance between the detector and the multi-slit, and L_s (=150 mm) is the distance between the detector and the sample plane, respectively. The image defect can be estimated as GL_s/L_m and the local magnification rate as $(G + D_m)/(G + D_m\text{-}GL_s/L_m)$ as indicated in Fig. 4.

2.4　Adjustment of Double Multi-slit Collimator

Figure 5 shows the final setup for double multi-slit collimator (or multi-pinhole). The optimal distance between the detector and the multi-slit was obtained separately for horizonal and vertical directions. Using the double multi-slit collimators, the neutron flux emitted through two multi-slit collimators can be calculated to be 10.9% of the incident neutron flux. The resultant beam collimations are 475 in the horizonal, 575 in the vertical directions, respectively.

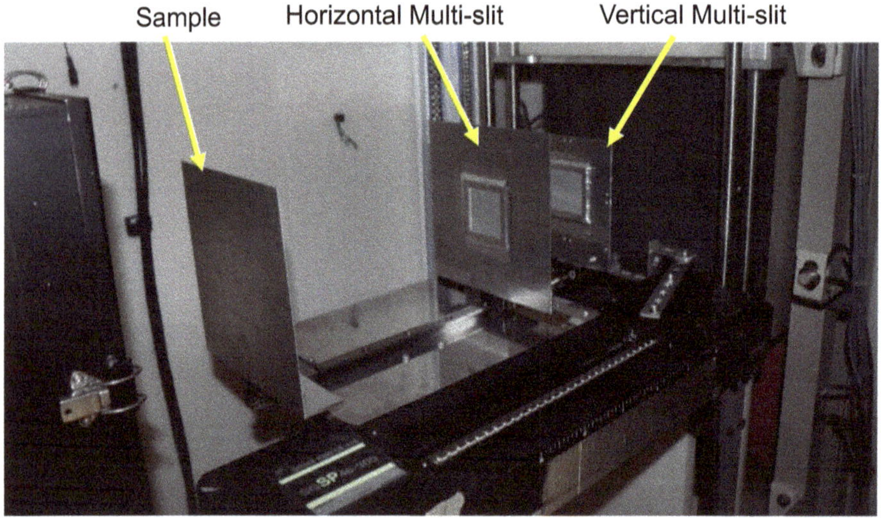

Fig. 5.　Setup of Double Multi-slit collimator.

3　Results

Figure 6 show the comparison of the transmission images with and without the vertical multi-slit collimator. Spatial resolution enhancement can be observed by using the multi-slit collimator. However, some severe image distortion can be found, which is generated by the local magnification and the image defect.

The local magnification can be calculated as:

$$(G + D_m)/(G + D_m - GL_s/L_m) = (2 + 1)/(2 + 1 - 2 * 150/575) = 1.21$$

Therefore, on the detector plane, the sample will be magnified 1.2 times in the perpendicular direction to the slit direction. At the same time, about 20% of the image on the sample plane cannot detected. Such local magnification and image defect may cause the image distortion by using the multi-slit collimator.

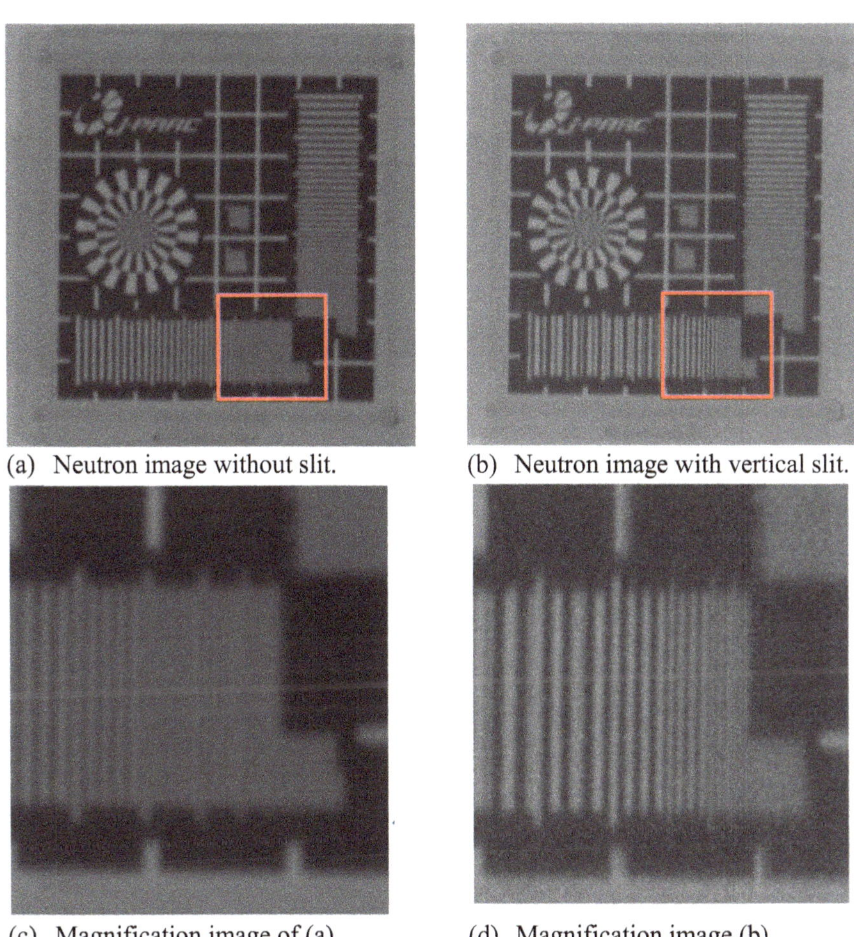

(a) Neutron image without slit.

(b) Neutron image with vertical slit.

(c) Magnification image of (a)

(d) Magnification image (b)

Fig. 6. Comparison of neutron transmission images with/without vertical slit.

Figure 7 shows the neutron images of the vertical multi-slit collimator and the processed J-PARC indicator with the boundary lines between the slits. From these images, the local magnification and the image defect can be clearly observed. The red line in this figure denotes the boundary lines between the slits and the area of the image defect on the sample plane.

Figure 8 shows the comparison of the indicator images, with and without the multi-slit collimator, where the red lines denote the corresponding boundary lines between the slit. As shown in these figures, the curved structures in the central region can be clearly detected by the multi-slit collimators.

(a) Neutron image of Multi-slit

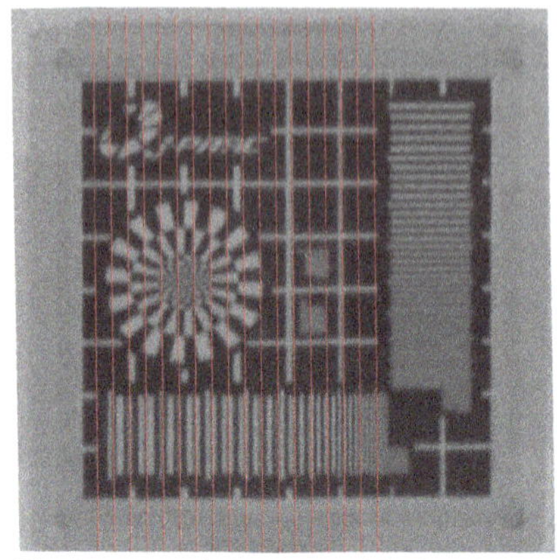

(b) Processed image with boudary line between the slits.

Fig. 7. Comparison of neutron images of (a) vertical multi-slit and (b) processed image with boundary lines between the slits.

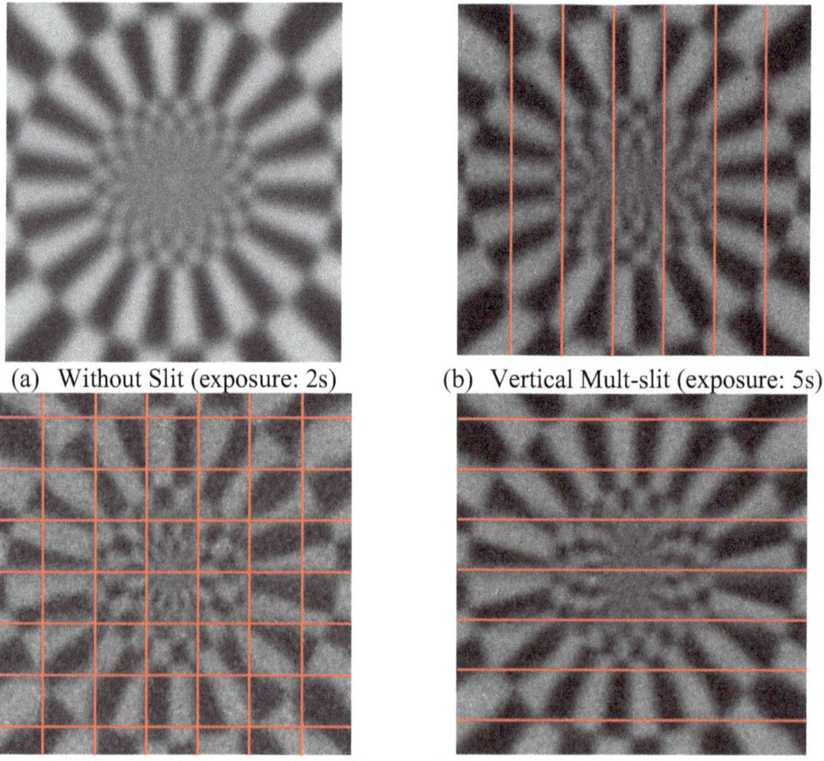

(a) Without Slit (exposure: 2s)

(b) Vertical Mult-slit (exposure: 5s)

(c) With double slit (exposure: 15s)

(d) Horizontal Multi-slit(exposure: 5s)

Fig. 8. Neutron images (a) without slit, (b) with vertical multi-slit, (c), with double slit, and (d) with horizontal multi-slit.

4 Conclusions

To minimize the reduction in neutron flux and achieve a balance between spatial resolution and time resolution, we developed a multi-slit collimator system and conducted experiments at the TNRF of JRR-3. As a result, we demonstrated that the L/D ratio improved to 575 in the horizontal direction and 475 in the vertical direction. However, it was also revealed that avoiding beam overlap resulted in partial image loss and image enlargement in certain areas. Such image loss and enlargement should be corrected with an appropriate kernel.

Acknowledgement. This work at the JRR-3 of the JAEA was performed under a user program. The authors would like to express their gratitude to Mr. Masaya Kanayama of Institute for Integrated Radiation and Nuclear Science, Kyoto University, for his cooperation in the development of the multi-slit collimator system.

References

1. Hussey, D.S., Jacobson, D.L., Arif, M., Huffman, P.R., Williams, R.E., Collk, J.C.: Nucl. Instr. and Meth. A **542**, 9–15 (2005)
2. Hayashida, H., et al.: Nucl. Inst. Methods Phys. Res. A **605**, 77–80 (2009)
3. Bingham, P., et al.: Phys. Procedia **69**, 218–226 (2015)
4. Matsubayashi, M., et al.: Nucl. Technol. **132**, 309–324 (2000)
5. Segawa, M., et al.: Proc. Int. Conf. Neutron Optics (NOP2017) **22**, 011028 (2018)

Characterization of Dual-Mode Scintillators for Fast Neutron and X-ray Radiography

Nicholas Anastasi[1(✉)] [ORCID], Stuart Miller[2] [ORCID], Pijush Bhattacharya[1] [ORCID],
Tawan Jamdee[1] [ORCID], Edgar van Loef[1], Guillermo Velasco[1], Charles Sosa[1] [ORCID],
Paul Rose[3] [ORCID], Bipin Singh[1] [ORCID], Lakshmi Soundara Pandian[1] [ORCID],
and Vivek V. Nagarkar[1] [ORCID]

[1] Radiation Monitoring Devices, Inc., Watertown, MA 02472, USA
{nanastasi,pbhattacharya,tjamdee,evanloef,gvelasco,csosa,bsingh,
lspandian,vnagarkar}@rmdinc.com
[2] Nevada National Security Sites, 232 Energy Way North, Las Vegas, NV 89030, USA
millersr@nv.doe.gov
[3] Oak Ridge National Laboratory, 1 Bethel Valley Road, Oak Ridge, TN 37830, USA
rosepb@ornl.gov

Abstract. X-rays and neutrons serve as sources for various radiographic applications, and scintillators for use with each are well-established. While high-energy X-ray radiography is useful for applications involving high-Z materials, this technique is limited for probing objects with low-density composition. Alternatively, fast neutron radiography can be used for detecting low-Z compositions without significant attenuation by dense materials. By combining fast neutrons and X-rays in a single radiographic application, the complementary information provided by each source can be used to produce images with more information about a subject. Currently, there are no commercially available detectors capable of high-resolution imaging with both fast neutrons and X-rays. Under sponsorship through the DOE/NNSA's Office of Defense Nuclear Nonproliferation, RMD is developing a high-resolution radiography detector for both high-energy X-ray and fast neutron imaging. The envisioned detector maximizes detection efficiency for fast neutrons and high-energy X-rays, while also producing high-contrast radiographs with sub-mm spatial resolution. Multiple scintillator formats for use in a dual-mode detector have been developed including 1) Embedded ZnS converters, 2) Layered CsI converters, 3) Tin-loaded Organic Glass Scintillators, and 4) Pixelated Organic Glass Scintillators. These scintillators have produced X-ray radiographs with spatial resolutions below 100 μm and fast neutron radiographs with spatial resolutions as low as 550 μm. Here we report on imaging data collected with these dual-mode scintillators using separate fast neutron and X-ray sources.

Keywords: Fast neutron radiography · organic glass scintillators · layered and composite scintillators

A. E. Craft and H. Z. Bilheux (Eds.): WCNR 2024, SPPHY 348, pp. 276–289, 2026.
https://doi.org/10.1007/978-3-032-15003-5_32

1 Introduction

X-rays and neutrons serve as sources for various radiographic applications, and scintillators for detecting each are widely used and commercially available. The electromagnetic interactions of X-rays with atoms' electron clouds means that the probability of absorption is directly related to a material's effective atomic number Z. This makes X-rays useful for detecting high-density materials but limited in their detection of low-Z materials [1]. In contrast, neutron interactions consist primarily of scattering by atomic nuclei, making them useful for the detection of low-Z materials such as those rich in hydrogen, carbon, nitrogen, and oxygen. More specifically, fast neutrons ($>$ 1 MeV) are useful in detecting hydrogenous materials such as water, plastics, and organics while penetrating deeper into high-Z materials than low-energy neutrons. By combining fast neutrons and X-rays for a single radiographic application, the complementary information provided by each source can be utilized to produce images with more information about an object [2]. Due to the high penetration depths of fast neutrons and high-energy X-rays, this remains true even when the object being imaged is heavily shielded by dense materials. Currently, there are no commercially available detectors capable of high-resolution imaging with both fast neutrons and X-rays.

To address the need for a dual-mode radiography detector, RMD is developing scintillator formats capable of detecting both fast neutrons and high-energy X-rays. These scintillators can be integrated into various electronic readout platforms to form a portable, high-resolution, combined fast neutron/X-ray detector for radiography. The envisioned detector can capture both fast neutron and MeV X-ray radiographs with sub-millimeter spatial resolution in short time frames. To this end, four types of dual-mode scintillators have been developed at RMD: 1) Embedded ZnS converters, 2) Layered CsI converters, 3) Tin-loaded Organic Glass Scintillators, and 4) Pixelated Organic Glass Scintillators. The radiographic performance of these scintillators has been evaluated using multiple fast neutron and X-ray sources ranging from 1–14.1 MeV neutrons and 70 kVp – 7.5 MeV X-rays.

While many X-ray sources are capable of generating high fluxes (10^8–10^{10} photons/s/cm^2 for standard medical X-ray tubes [3]), the fluxes offered by most available fast neutron sources are orders of magnitude less. Additionally, fast neutrons are significantly more difficult to detect than X-rays due to their low interaction probabilities with most matter [4]. Another concern in the context of transmission radiography is source geometry; the collimation of incoming radiation from most fast neutron sources is less than that of X-ray generators [5]. These factors make fast neutron imaging generally more challenging than imaging using X-rays. Thus, much of the development for a dual-mode scintillator and detector is based on the demonstration of radiography with fast neutrons.

To form an imaging detector, scintillators can be coupled to various readout platforms such as large-area a:Si:H flat panels, lens-coupled CMOS/CCD cameras, SiPM arrays, event-mode cameras [6], or other readout platforms. Each type of scintillator/readout configuration offers its own advantages and limitations, and each configuration may find use in particular applications requiring dual-mode imaging with X-ray and fast neutrons. Homeland security, cargo inspection, nondestructive evaluation, material science, and biological imaging are just some of many fields that stand to benefit from a dual-mode imager [7].

2 Dual-Mode Scintillators

Layered CsI:Tl: CsI:Tl is well-established as a high-resolution, bright scintillator for X-ray radiography. Through deposition of columnar CsI:Tl onto plastic substrates (Fig. 1a), this material can be used as a scintillator for detecting both fast neutrons and X-rays. In this configuration, fast neutrons interact with protons in the plastic layer, which recoil a short distance to excite the CsI:Tl and produce scintillation. Fast neutron detection efficiency is thus limited by the recoil distance of protons in the plastic layer. Figure 2 shows the GEANT4 [8] simulations of probabilities for various recoil proton outcomes for incident DD and DT neutrons as a function of converter layer thickness. The "Protons crossed into scint" trace shows the probability that a recoil proton is generated and crosses into the scintillating layer. This probability plateaus near 100 μm for DD (2.5 MeV) neutrons and near 2000 μm for DT (14.1 MeV) neutrons. This suggests that fast neutron detection efficiency is maximized for converter layers of 0.1 mm thickness for DD neutrons and 2 mm thickness for DT neutrons. On the other hand, detection efficiency for X-rays is determined by the CsI:Tl layer thickness. Though increased thickness can improve total X-ray absorption, thinner CsI:Tl layers can produce fast neutron/X-rays radiographs with high spatial resolution.

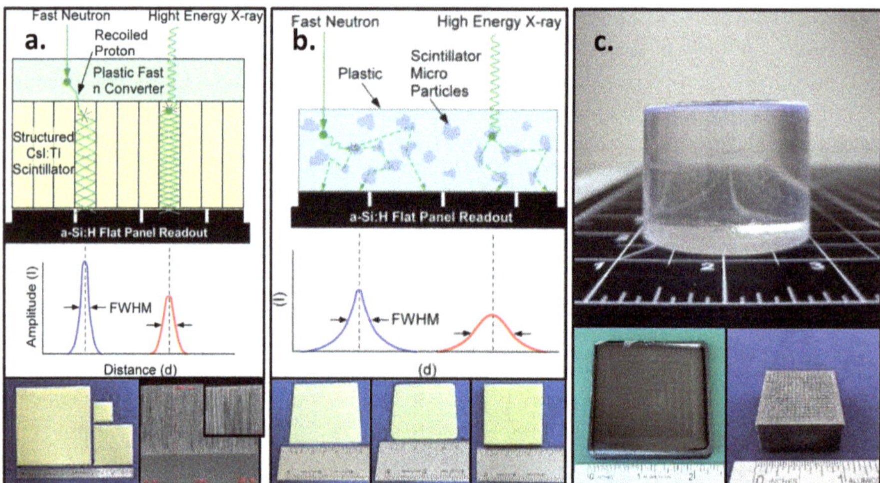

Fig. 1. **a.** Schematic of RMD's Layered CsI:Tl converter. Photographs and SEM micrographs of 700 μm-thick columnar CsI:Tl deposited on 1.6 mm polystyrene are shown below. **b.** Schematic of RMD's Embedded ZnS:Cu converter with 5×5 cm^2 sample photographs. **c.** Photographs of 50 mm-thick Sn-OGS cylinder, 6 mm-thick Sn-OGS imaging plate, and OGS cast into a $1'' \times 1'' \times 0.5''$ tungsten matrix with 1 mm^2 pixels.

Embedded ZnS:Cu: Embedded ZnS:Cu scintillators (Fig. 1b) consist of ZnS:Cu microparticles suspended in a polymer. The polymer bulk is responsible for interacting with fast neutrons through proton recoil and transferring energy to neighboring ZnS:Cu particles. Embedded ZnS converters produce scintillation light both from indirect fast neutron excitation and direct X-ray conversion. These scintillator structures are durable and non-hygroscopic. For the best radiographic performance, two main parameters must be optimized: scintillator thickness, and ZnS:Cu concentration. Testing of various thicknesses and compositions shows optimal thickness between 2–5 mm and optimal ZnS:Cu concentration between 30–50 wt%. Maximum neutron absorption at these parameters is estimated at ~20% for 2.45 MeV DD neutrons and ~5% for 14.1 MeV DT neutrons.

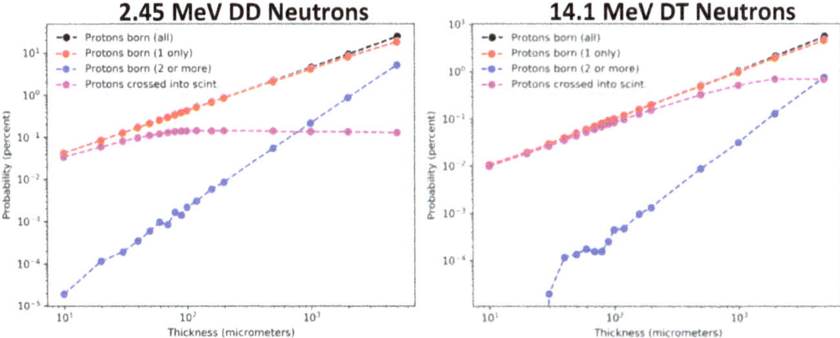

Fig. 2. GEANT4 simulations of recoil protons in a layered CsI:Tl with HDPE converter.

Tin-Loaded and Pixelated Organic Glass Scintillator: Organic Glass Scintillator (OGS) is an intrinsic fast neutron scintillator with impressive properties including up to 19,000 photons/MeV light yield, FOM > 3 for neutron/gamma discrimination, and decay time < 2 ns [9]. Tin-loading of OGS (Sn-OGS) increases the detection efficiency of high-energy X-rays, making for a scintillator capable of combined fast neutron/X-ray radiography. Sn-OGS samples of various thickness have been fabricated at RMD (Fig. 1c) and tested in the context of high-resolution combined imaging. One of the advantageous properties of (Sn-) OGS is high transparency to its own scintillation light [10]. This allows for a scaleup in scintillator thickness, and thereby detection efficiency, without introducing self-absorption. To use thick OGS while maintaining high spatial resolution in radiographic applications, OGS is cast into metallic matrix structures. This effectively pixelates the scintillator such that a given spatial resolution can be achieved irrespective of thickness. Thus far, RMD has successfully cast OGS (no Sn-loading) into $1'' \times 1'' \times 0.5''$ metal matrices with 1 mm^2 pixel area (Fig. 1c).

3 Experimental Methods

3.1 X-ray and Fast Neutron Sources

Table 1. X-ray & Fast Neutron sources used in this work.

Site	Source	Type	Energy	Source Dist./Fluence
RMD	Gendex GX-1000	Dental X-ray	70 kVp	45 cm
RMD	GE XRS-3	Pulse X-ray	370 kVp	45 cm
ORNL	JME PXB 7.5	Betatron	7.5 MeV	160 cm
UML-RR	Accelerator	Fast neutron	1–2 MeV	2×10^6 n/s
ORNL	DT Generator	Portable fast neutron	14.1 MeV	3×10^8 n/s
Adelphi	DT Generator	Portable fast neutron	14.1 MeV	1×10^8 n/s

Radiographic evaluations were performed using in-house sources at RMD as well as at numerous outside facilities. At RMD, a Gendex GX-1000 dental radiography X-ray source and a Golden Engineering XRS-3 370 kVp pulsed X-ray generator were used for characterization of scintillators. High-energy 7.5 MeV X-rays from the JME PXB 7.5 betatron were used at Oak Ridge National Lab (ORNL) alongside a Thermo Scientific miniGen 14.1 MeV DT neutron generator. Additional fast neutron imaging data was collected at the University of Massachusetts, Lowell Research Reactor (UML-RR) and Adelphi. The details and specifications of the X-ray and neutron sources used are given in Table 1.

3.2 Data Acquisition

Several image acquisition parameters are important for generating the highest quality radiographs. These include source-to-detector distance, exposure time/frame rate, total acquisition time, source flux, and L/D ratio. L/D is the ratio of the distance between the beam's entrance aperture and the image plane (L) to the smallest diameter of beam collimation (D), and it is a key factor in determining neutron image resolution. For each detector configuration and source used, these parameters were adjusted to optimize imaging performance. While most of these parameters were varied between different imaging setups, some data acquisition principles remained constant. For fast neutron image acquisition, detectors were placed at a distance from the source in such a way to maximize flux while maintaining acceptable L/D. Exposure times were set between 1–10 s, and a series of image frames, along with background and flat-field sets, were acquired. Total image acquisition times are given by:

$$t_{acquisition} = t_{exposure} \times n_{frames} \tag{1}$$

The readout platforms tested and used to evaluate scintillator performance include a lens-coupled sCMOS camera and a lens-coupled EMCCD camera. The specifications for

Table 2. Detector specifications for combined fast neutron/X-ray imaging experiments.

Detector	Andor Marana	Andor iXon
Type	sCMOS	EMCCD
Sensor size	22.5 mm × 22.5 mm	13.3 mm × 13.3 mm
Resolution	2048 × 2048 pixels	1024 × 1024 pixels
Pixel size	11 μm × 11 μm	13 μm × 13 μm
Field-of-View	7 cm × 7 cm	122 mm × 122 mm
Effective pixel size	34 μm × 34 μm	119 μm × 119 μm

these readouts are given in Table 2. Lens-coupled cameras were housed in a light-tight box and aligned perpendicular to the scintillator such that readout electronics could sit outside of the beam path and be shielded from the source. In this configuration, a mirror placed at 45° directs scintillation light to the camera as shown in Fig. 3. For fast neutron imaging, detectors were shielded with lead bricks to minimize any potential contribution to signal caused by gammas emitted from the fast neutron source.

3.3 Image Processing

Many sources contribute to noise during image acquisition. These include dark current, read noise, and direct hits to the sensor by incoming radiation. Dark current and read noise manifest as a persistent background signal, and direct hits can appear in an image frame as saturated pixels. Removal of this noise from image data is critical for improving the signal-to-noise ratio (SNR) of a radiograph. In this work, image data was processed by taking the median pixel value for a series of images. This removes all saturated pixel values caused by direct hits and significantly improves SNR by combining the statistics generated by a large collection of image frames. Persistent signal caused by dark current and read noise is mitigated through background subtraction, in which a set of background images acquired in the absence of a radiation source is subtracted from a collection of image data. Background subtraction can be done pre- or post-processing by subtracting a single background frame from each subsequent image frame or by subtracting the median of a set of background images from the median of a set of image frames.

Additional image artifacts can result from non-uniformity and defects in the scintillator. These defects appear as constant features in all images but can be removed through flat-field correction. In flat-field correction, a set of images is acquired of only the scintillator under exposure to radiation. The median of this set of flat-field images is taken to produce a reference flat-field image. Background subtraction is applied, and signal images are normalized by dividing each pixel value by that of the reference image and rescaling. This correction is described by the following equation:

$$I_{FFC} = \frac{I - B}{F - B} \tag{2}$$

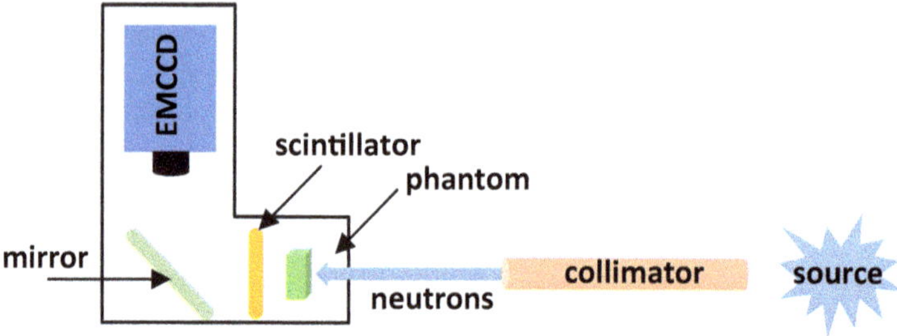

Fig. 3. EMCCD detector configuration for fast neutrons/X-ray imaging.

where I_{FFC} is the flat-field corrected image, I is the signal image, B is the background image, and F is the flat-field image. This process removes persistent artifacts caused by the scintillator and can improve image quality.

4 Results

4.1 X-ray Imaging

Layered CsI:Tl: Micro-columnar CsI:Tl is deposited on a non-reflective polymer substrate to form a scintillator screen capable of producing high-resolution radiographs using both X-rays and fast neutrons. The performance of these screens was tested at RMD using 70 kVp X-rays and an sCMOS camera. Spatial resolution was characterized under X-rays by imaging a line-pair phantom and by measuring the Modulation Transfer Function (MTF). The MTF of layered CsI:Tl films was measured using X-ray images of a tungsten slit phantom. A 10-micron wide tungsten slit phantom was imaged under 70 kVp X-rays using a layered CsI:Tl converter. A series of 4 slit images, 4 flat images (scintillator with no phantom), and 4 dark images were taken and used to generate a final flat-field corrected slit image. A line profile was then taken across the slit image to generate a Line Spread Function (LSF). Taking the magnitude of the Fourier Transform of the LSF yields the MTF of the screen. A 350 µm-thick layered CsI:Tl converter fabricated at RMD shows 5.5 lp/mm at 10% MTF (Fig. 4). This corresponds to a spatial resolution of 90 µm.

Embedded ZnS:Cu: X-ray imaging data using embedded ZnS:Cu scintillators has not yet been collected as part of this work. However, ZnS screens have been demonstrated as effective X-ray imaging scintillators [11–13]. Thus, given the fast neutron imaging performance of these screens (see Sect. 4.2), it is reasonable to assume that the same image quality could be reproduced under X-ray irradiation.

Sn-OGS: Tin-loaded Organic Glass Scintillator (Sn-OGS) was tested under the same X-ray imaging configuration at RMD mentioned above. Tests were performed for multiple thicknesses of Sn-OGS from 3 mm up to 25 mm. The Contrast Transfer Function (CTF) was measured for a 3 mm-thick, 50 mm diameter sample of Sn-OGS. CTF was measured by taking line profiles across a flat-field corrected image (as described above) of a line-pair phantom at various line spacings. The line profile measures the pixel intensities

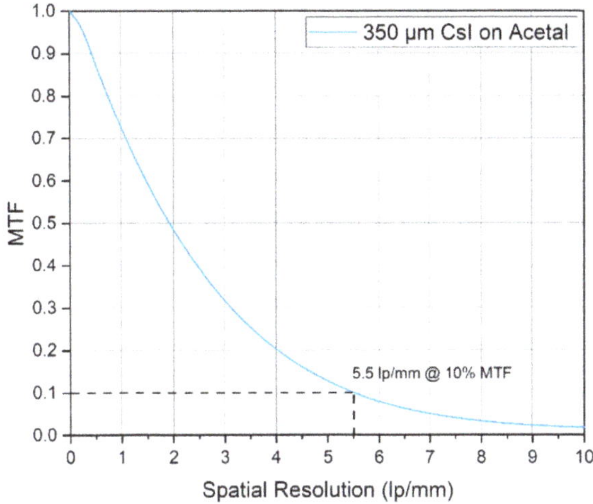

Fig. 4. MTF of 350 µm-thick CsI:Tl layer deposited on 1.6 mm-thick polypropylene converter shows 5.5 lp/mm at 10% MTF, corresponding to an intrinsic spatial resolution below 90 µm.

along a given line in an image. The average amplitude of oscillations in pixel intensities along a line profile are normalized to the total estimated dynamic range of the radiograph. This estimates image contrast (C) at a given spatial frequency according to the following equation:

$$C = \frac{P - V}{B - D} \tag{3}$$

where P (peak) is the average maximum value of an oscillation, V (valley) is the average minimum value of an oscillation, D (dark) is the average pixel value of a selected area in the image covered by a solid section of the line-pair phantom, and B is the maximum achievable pixel intensity in the image. The 3 mm-thick Sn-OGS sample showed 16.5% CTF at 7 lp/mm (70 µm feature size) (Fig. 5). The MTF of 6 mm- and 25 mm-thick square Sn-OGS slabs show 6.2 lp/mm (80 µm feature size) and 2.8 lp/mm (180 µm feature size) at 10% MTF respectively (Fig. 6).

Pixelated OGS. Additional X-ray imaging was performed using pixelated Organic Glass Scintillator (pOGS). Here, OGS was cast into additively manufactured metal matrix structures of dimensions 1″ x 1″ x 0.5″ with 1 mm² pixels. Each pixel is separated by a 300 µm matrix wall thickness. This scintillator structure serves to confine scintillation light to an area defined by the matrix pixels, maintaining a fixed spatial resolution for arbitrary scintillator thickness. Hence, detection efficiency can be increased while maintaining radiographic image sharpness. An additional benefit of using high-Z metals is enhanced absorption efficiency of high-energy X-rays. Tungsten and stainless-steel pOGS matrices were lens-coupled to an sCMOS camera and imaged under X-rays (Fig. 7). The tungsten pOGS matrix produced a high-contrast radiograph of a steel nut phantom with exposure to 70 kVp X-rays, and using a 370 kVp portable X-ray source, the stainless steel pOGS matrix resolved a 2 mm slit from a steel resolution phantom.

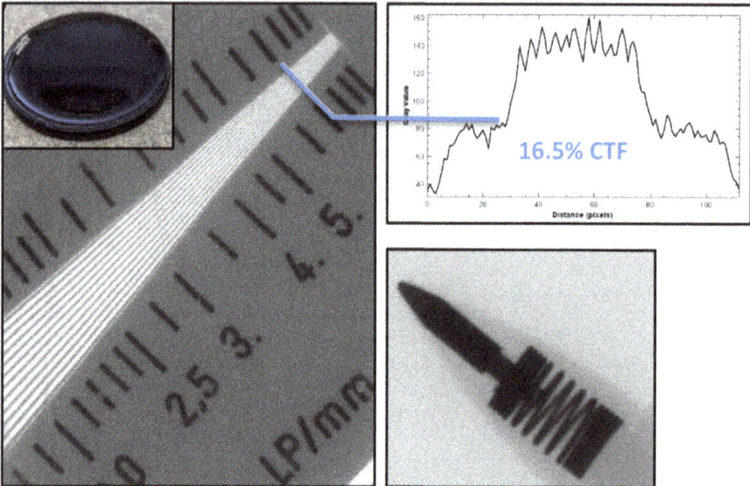

Fig. 5. 70 kVp X-ray image of a line-pair phantom and ballpoint pen using 3 mm-thick Sn-OGS. This sample shows 16.5% contrast at 7 lp/mm (70 μm feature size).

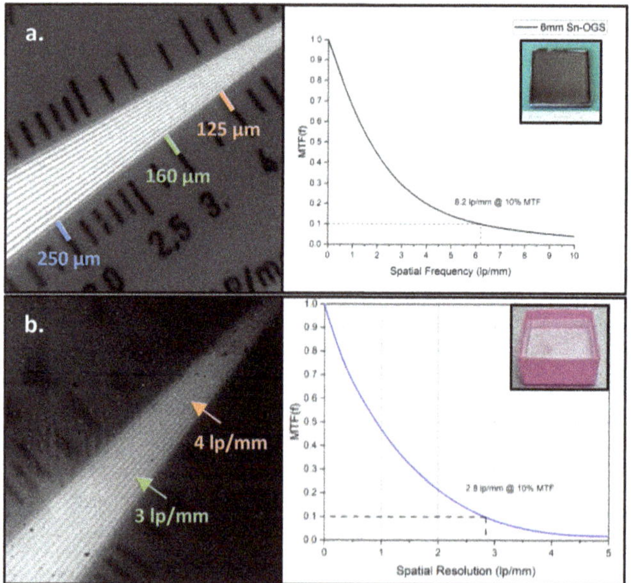

Fig. 6. a. 70 kVp X-ray image of a line-pair phantom using 6 mm-thick Sn-OGS. The MTF of this sample (right) shows an intrinsic resolution of 6.2 lp/mm (80 μm). **b.** 70 kVp X-ray image of a line-pair phantom using 25 mm-thick Sn-OGS. The MTF of this sample (right) shows an intrinsic resolution of 2.8 lp/mm (180 μm).

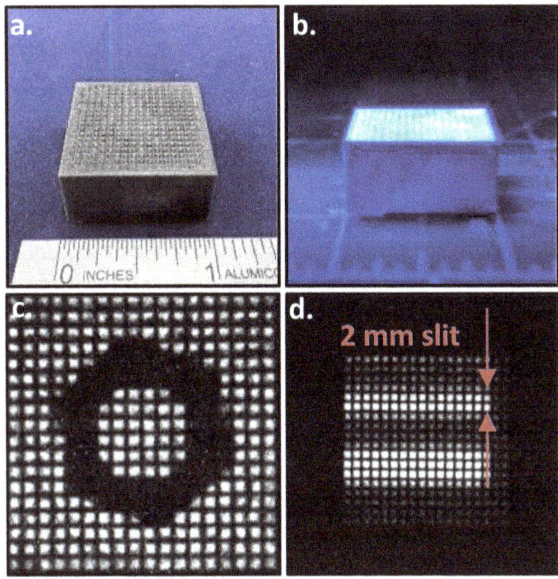

Fig. 7. Photos under ambient (**a**) and UV (**b**) light of OGS cast into $1'' \times 1'' \times 0.5''$ tungsten matrix with 1 mm² pixels. 70 kVp X-ray image of a nut using tungsten pOGS (**c**) and 370 kVp X-ray image of steel slit phantom using stainless steel pOGS (**d**).

4.2 Fast Neutron Imaging

Embedded ZnS:Cu: Fast neutron imaging was performed with Embedded ZnS:Cu converters at Adelphi using their DT neutron generator. Here, a $2''$-thick stainless-steel phantom with 7.5 mm holes was imaged under exposure to 14.1 MeV neutrons using embedded ZnS:Cu converters lens-coupled to the iXon EMCCD detector. Images were acquired at various time scales from 20 s to 10 min (Fig. 8).

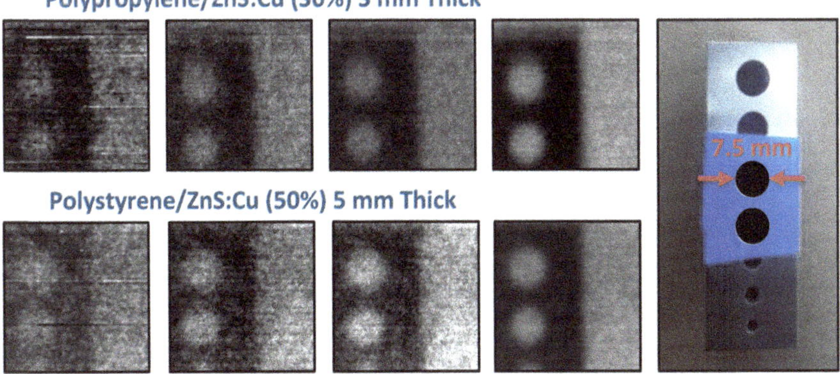

Fig. 8. Images of a $2''$-thick stainless-steel phantom with 7.5 mm holes using embedded ZnS:Cu. Exposure from 14.1 MeV fast neutrons from DT generator at Adelphi.

Layered CsI:Tl: Dual-mode imaging was performed using Layered CsI:Tl converters at Oak Ridge National Lab. Here, scintillators were lens-coupled to an EMCCD detector and exposed (separately) to both 7.5 MeV X-rays and 14.1 MeV neutrons. Images were acquired of a 3 cm-thick tungsten line-pair phantom. This detector configuration resolved feature sizes as small as 350 μm under X-rays and 550 μm under fast neutrons. Additional dual mode imaging was performed using this detector configuration on a toy phantom with both plastic and metal parts. Complementary information about the composition of this phantom is seen in the neutron and X-ray radiographs (Fig. 9). Something to consider is that much of the blurring seen at higher spatial frequencies in the fast neutron radiograph of the line-pair phantom (Fig. 9c) is caused by a combination of finite spot size of the neutron source, imperfect source-to-detector alignment, and the non-negligible thickness of the imaging phantom. These factors contribute to shadowing seen in the resulting radiograph, so the minimum resolvable feature sizes of this resolution phantom can be seen as underestimating the intrinsic spatial resolution of the detector.

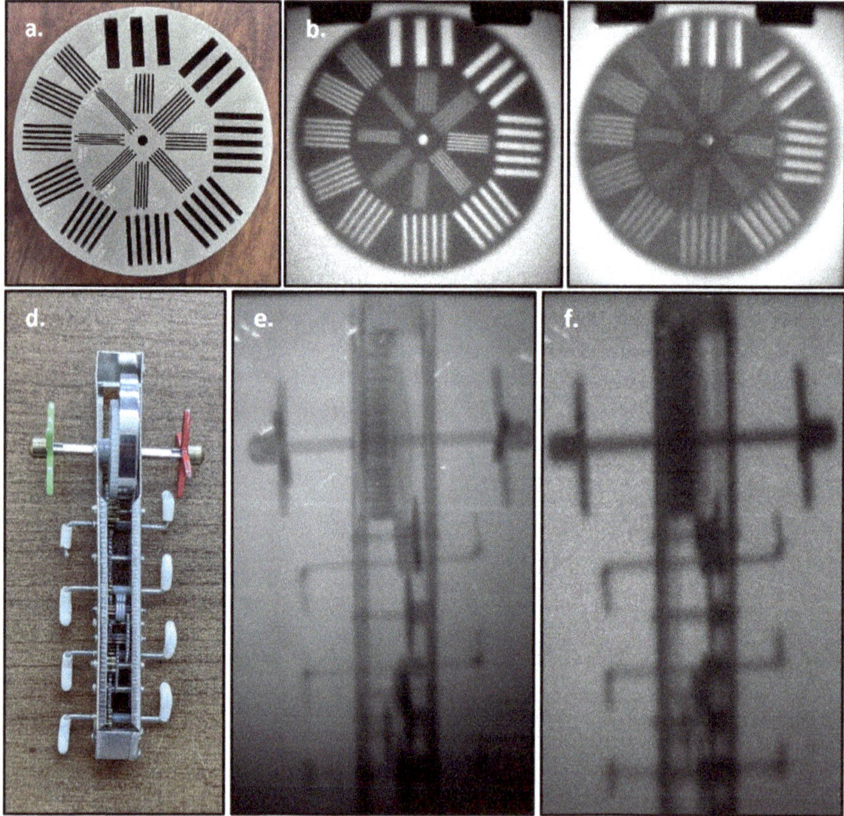

Fig. 9. 3 cm-thick tungsten resolution phantom with slit sizes ranging from 0.3 – 4 mm (**a**) imaged under 7.5 MeV X-rays from a betatron (**b**) and 14.1 MeV neutrons from a portable DT neutron generator (**c**). Images were acquired using 350 μm-thick layered CsI:Tl/acetal (0.8 mm) lens-coupled to an EMCCD. Wind-up toy (**d**) imaged with the same source and detector configurations with X-rays (**e**) and fast neutrons (**f**).

Sn-OGS: Sn-OGS was also tested for fast neutron imaging at UML-RR using their 5.5 MV Pulsed van-de-Graaff accelerator. A 3 mm-thick sample of Sn-OGS was lens-coupled to an EMCCD camera and used to acquire images of a 1″-thick tungsten cylinder with an 8 mm diameter hole. Using this setup, a fast neutron image was acquired in just 10 min of exposure time. This same sample and phantom were imaged under 70 kVp X-ray at RMD using a lens-coupled sCMOS camera (Fig. 10), demonstrating the dual-mode imaging capabilities of Sn-OGS.

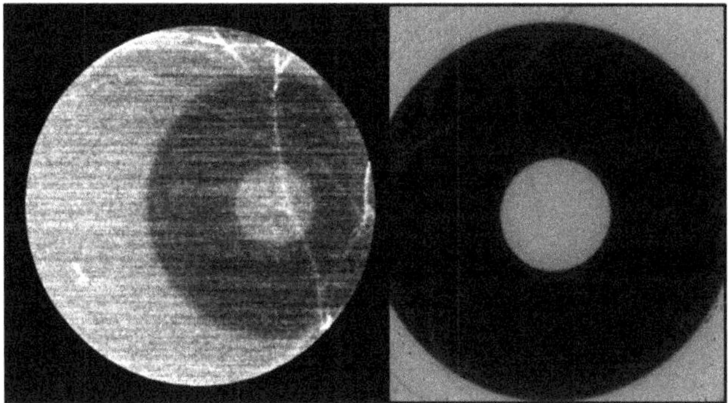

Fig. 10. 10-min exposure fast neutron image (2 MeV, 5×10^5 n/cm²/s flux) and X-ray image of the same tungsten cylinder using 3 mm-thick Sn-OGS lens-coupled to EMCCD.

5 Summary and Conclusion

The results obtained through X-ray and fast neutron imaging using dual-mode scintillators developed at RMD demonstrate significant progress towards a high-resolution, imaging detector capable of dual-mode radiography. Each scintillator and readout configuration presents unique advantages, which can be tailored to specific applications based on the required imaging modality, spatial resolution, and detection efficiency.

Layered CsI:Tl scintillator proved highly effective for X-ray imaging, achieving sub-90 μm spatial resolution for a 350 μm-thick scintillator. This performance highlights the suitability of CsI:Tl for applications where high spatial resolution is critical. While fast neutron detection efficiency is limited by proton recoil distances, the results from ORNL confirmed that fast neutron imaging with sub-millimeter resolution was achievable using a portable DT neutron generator. **Embedded ZnS:Cu scintillators** also performed well for fast neutron imaging, showing adequate image quality within 10-min time frames under exposure to 14.1 MeV DT neutrons. The performance of embedded ZnS:Cu converters could be further optimized for fast neutron and X-ray imaging by adjusting the thickness and composition of the scintillator/polymer matrix. **Tin-loaded organic glass scintillators (Sn-OGS)** showed great promise for dual-mode imaging. The high fast neutron detection efficiency of OGS, coupled with enhanced X-ray detection via tin loading, enabled fast neutron radiography in short time frames and

X-ray imaging with resolutions of 70–180 μm for scintillator thicknesses between 3–25 mm. Additionally, **pixelated OGS** structures present excellent potential for improving neutron and X-ray detection without sacrificing spatial resolution.

The results presented herein show that these scintillators, coupled with appropriate readout platforms, can produce sub-mm resolution radiographs for both X-rays and fast neutrons. The potential flexibility in choosing between various readout platforms provides the opportunity to optimize the detector based on specific application needs, balancing resolution, sensitivity, and acquisition time. Applications in homeland security, materials identification, nondestructive testing, and other fields stand to benefit greatly from a high-resolution dual-mode imaging system. Continued scintillator development, along with optimization of detector configurations, will further enhance dual-mode imaging performance. Future work will focus on improving neutron detection efficiency, scaling up detector size, and refining pixelated scintillator structures to achieve even higher resolution and faster image acquisition times.

Acknowledgements. A. Rogers–This work is supported by DOE R&D Small Business Innovation Research program under contract DE-SC0020942.

References

1. Licata, M.: The potential of real-time, fast neutron and γ radiography for the characterization of low-mass, solid-phase media. Nucl. Instr. Methods Phys. Res. Sect. A: Accelerators Spectromet. Detect. Assoc. Equip. **954**, 161706 (2020)
2. Kumar, R., et al.: Gamma in addition to neutron tomography (GIANT) at the NECTAR instrument. Sci. Rep. **13**(1) (2023)
3. Bartzsch, S., Oelfke, U.: Line focus x-ray tubes—a new concept to produce high brilliance x-rays. Phys. Med. Biol. **62**(22) (2017)
4. Bergaoui, K.: Design, testing and optimization of a neutron radiography system based on a Deuterium-Deuterium (D–D) neutron generator. J. Radioanal. Nucl. Chem. **299**, 41–51 (2013)
5. Li, J.J.: An estimation method of the spatial resolution for magnifying fast neutron radiography. AIP Adv. **12**(5), 055117 (2022)
6. Losko, A.S., et al.: New perspectives for neutron imaging through advanced event-mode data acquisition. Sci. Rep. **11**(1), 21360 (2021)
7. Liu, Y.: Comparison of neutron and high-energy X-ray dual-beam radiography for air cargo inspection. Appl. Radiat. Isot. **66**(4), 463–473 (2008)
8. Agostinelli, S.: Geant4—a simulation toolkit. Nucl. Instrum. Methods Phys. Res., Sect. A **506**(3), 250–303 (2003)
9. Carlson, J.S.: Melt-cast organic glasses as high-efficiency fast neutron scintillators. Nucl. Instrum. Methods Phys. Res., Sect. A **832**, 152–157 (2016)
10. Shirwadkar, U., et al.: Development of multi-mode organic glass scintillators for nuclear physics applications. In: 2021 IEEE Nuclear Science Symposium and Medical Imaging Conference (NSS/MIC), Piscataway, NJ, USA, pp. 1–4 (2021)
11. Kandarakis, I., et al.: Evaluation of ZnS: Cu phosphor as X-ray to light converter under mammographic conditions. Radiat. Meas. **39**(3), 263–275 (2005)
12. Li, X., et al.: High-efficiency ZnS: Cu+, Al3+ scintillator for X-ray detection in a non-darkroom environment. Inorgan. Chem. **62**(20), 7914–7920 (2023)

13. Saleh, M.: Evaluation of undoped ZnS single crystal materials for x-ray imaging applications. In: SPIE, 2017, Window and Dome Technologies and Materials XV, vol. 10179 (2017)

Overview of MARS: The Multimodal Advanced Radiography Station at the High-Flux Isotope Reactor

James Torres[1] ⓘ, Hassina Z. Bilheux[1] ⓘ, Jean-Christophe Bilheux[1] ⓘ,
Chen Zhang[2] ⓘ, Lowell Crow[3] ⓘ, Erik Stringfellow[1], Ian Turnbull[3],
Violet Freudenberg[3], Roger Hobbs[1], Mike Harrington[3], Thomas Huegle[3] ⓘ,
Erik Iverson[3] ⓘ, Rohit Srivastava[1], and Yuxuan Zhang[1(✉)] ⓘ

[1] Neutron Scattering Division, Oak Ridge National Laboratory, Oak Ridge, TN 37830, USA
zhangy6@ornl.gov
[2] Computer Science and Mathematics Division, Oak Ridge National Laboratory, Oak Ridge, TN 37830, USA
[3] Neutron Technologies Division, Oak Ridge National Laboratory, Oak Ridge, TN 37830, USA

Abstract. Neutron imaging is a powerful tool used for inspecting internal structures and studying macroscale particle dynamics. This work is performed at the Multimodal Advanced Radiography Station (MARS), which is the cold-neutron imaging beamline at the High-Flux Isotope Reactor at Oak Ridge National Laboratory. Herein is a summary of recent upgrades to instrumentation, software, and methods, driven by needs from the user community. Planned upgrades within the next decade are included that will significantly enhance instrument capability and performance.

Keywords: Neutron imaging · radiography · tomography · reactor

1 Introduction

Neutron imaging non-destructively reveals material internal structure and composition with broad application and complementarity to x-ray imaging. The Multimodal Advanced Radiography Station (MARS) is the neutron imaging facility located on Cold Guide 1D (CG-1D) of the High Flux Isotope Reactor (HFIR) at Oak Ridge National Laboratory (ORNL). Since our last update [1], there have been major upgrades to hardware, software, and imaging modalities to meet the needs of the user community. Herein, we summarize the community, layout, performance, capabilities, and future of MARS. Because MARS is accessed through a competitive scientific review process, we include some guidance on proposing feasible and realistic neutron imaging experiments. Finally,

© The Author(s) 2026
A. E. Craft and H. Z. Bilheux (Eds.): WCNR 2024, SPPHY 348, pp. 290–298, 2026.
https://doi.org/10.1007/978-3-032-15003-5_33

MARS is expected to relocate to a new location in the HFIR cold guide hall as part of the upcoming beryllium reflector replacement maintenance activity. As such, we summarize the anticipated changes as well as MARS's role within the ORNL source optimization for neutron imaging.

2   History of the Facility

CG-1D was first commissioned in 2009 as a development beamline where imaging prototyping followed shortly thereafter [2]. Based on a high community demand for imaging capabilities at HFIR, CG-1D was converted to a dedicated imaging beamline and entered the ORNL neutron scattering user program in 2011. The vibrant scientific imaging program at CG-1D eventually led to the recent construction of VENUS (BL-10) [3] at the Spallation Neutron Source, which has entered commissioning in July 2024, and will be followed by general user access in 2025. Since its inception, CG-1D (now MARS) has grown to be a world-class and versatile instrument serving a growing user community.

2.1   Science and Community

Imaging serves one of the most diverse user communities in neutron-based research, reaching all natural and engineering sciences as well as drawing users from industry. The community for MARS is represented in Fig. 1 where a significant fraction of experiments focuses on either materials science or energy materials. Recently, there has been increased attention on advanced manufacturing, water sorption studies in biological tissues and structural materials *in situ*, diffusion dynamics measurements in batteries *in operando*, and thermophysical property measurements of nuclear materials.

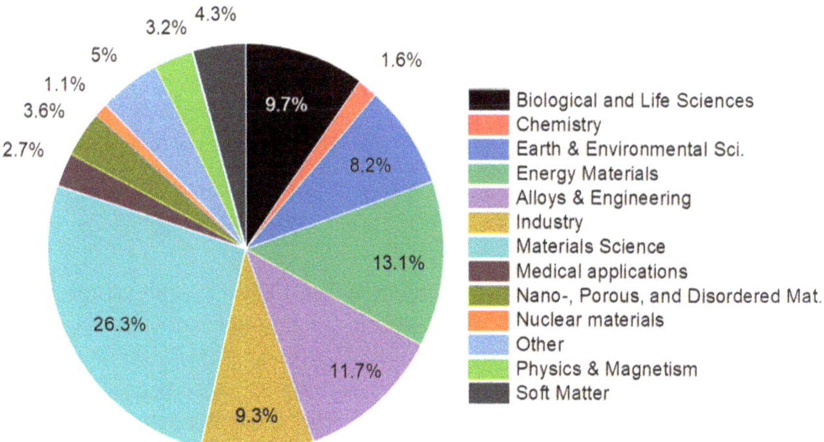

Fig. 1. Representative research areas served by MARS/CG-1D from 2013 to 2023.

2.2 Instrument Layout and Performance

MARS uses the conventional pinhole camera geometry with the layout described in our previous report [1]. A two-channel, curved guide with a combined rectangular 150 mm tall by 25.4 mm wide cross section is used to guide neutrons from the liquid hydrogen cold source to a shielded optics box in the HFIR cold-guide hall. The upper 50 mm part of the beam is captured by two monochromators that direct mono-energetic neutrons to the neighboring CG-1A development beamline and CG-1B crystal alignment station. CG-1C is currently shielded and not used. The bottom ~ 90 mm part of the guide is available for MARS. The beam is further constrained by a 5 mm thick boron nitride plate with a ~ 29 mm diameter opening at the center of the bottom section. The masked beam is then incident on a diffuser, which is an aluminum container filled with either graphite (C) with 2 to 15 μm particle size or aluminum oxide (Al_2O_3) with 50 nm particle size powders, a few millimeters thick each. These are designed to diffuse strong guide reflections. Various pinhole apertures are used to define the L/D, where L is the aperture-to-detector distance (typically ~ 6.5 m), and D is the pinhole aperture diameter. The L/D ratio therefore defines the collimation of the instrument where a higher L/D enables higher spatial resolution but at the cost of decreased flux. A list of instrument performance specifications and options is given in Table 1.

Table 1. Summary of instrument specifications and options.

Parameter	Value			Detail
Wavelength range [Å]	1 to 9 Å			Peak at 2.6 Å
Pinhole diameters (D) [mm], number flux for each [10^8 n cm^{-2} s^{-1}] and L/D ratios	Pinhole 3.3 4.1 8.2 11.0 16.0	Flux 0.4 0.6 2.4 4.5 9.1	L/D 219 176 88 66 45	Aperture assembly is a 2-layer boron nitride and cadmium sheet with knife-edge pinholes. Flux was measured at the optics box exit using a low-efficiency beam monitor calibrated using dysprosium foil activation.
Sample positions	Variable			Typically, L ~1.0 m and ~6.5 m
Max beam size	10×10 cm^2			At 6.5 m position
L/D ranges	60–300 400–2000			At L ~ 1.0 m position At L ~ 6.5 m position

Figure 2 shows a photo of the instrument (as of 2024), which extends from the upstream optics box to the downstream detector system and beam stop. Spanning 4.5 m of the neutron flight path are two helium-filled steel tubes with boron nitride and cadmium beam scrapers used to minimize air scattering and absorb stray neutrons. At the exit of the last tube is a boron nitride slit system. The area between the flight tubes and detectors accommodates the wide range of samples and sample environments. Finally, the detector system was recently upgraded to a gantry that supports up to four detectors to provide flexible and rapid changes between high spatial and high temporal resolution setups, as illustrated in Table 2. An optional beam monitor is also available.

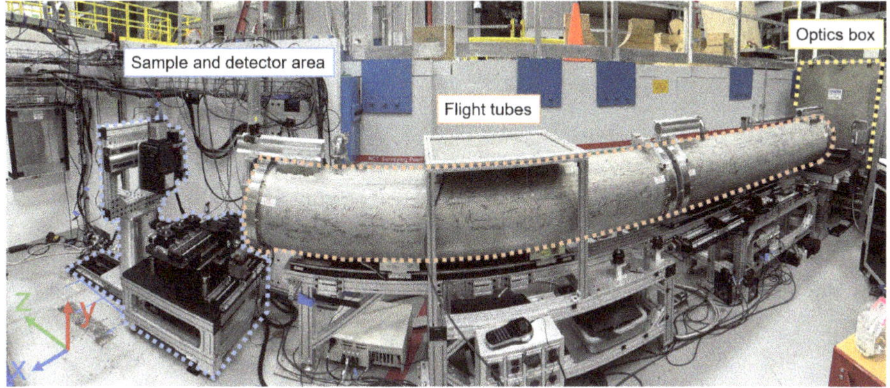

Fig. 2. Panoramic photo of MARS with the overall instrument layout and sample area.

Table 2. Typical image acquisition parameters for each detector system.

#	Detector system	Max FOV (cm^2)	Pixel size (µm)	Best spatial res. (µm)	Acquisition time (s)	Max frame rate @16-bit
1	Highest res., small FOV	3.6 × 2.4	3.8	10–15	900	1
2	High-res., good FOV	5.0 × 4.8	7.6	20–25	300	1
3	Large FOV, good res	9.0 × 9.0	16	~50	30–90	1
4	High-speed, coarse res	5.6 × 5.6	27	~100	–	74
5	Large view, coarse res	9.0 × 9.0	43	~100	–	24

2.3 Capabilities

Imaging Modes

The following modes are currently supported (*in commissioning):

- Radiography
- Tomography
- Monochromatic imaging
- Spectroscopic imaging
- Polarized imaging
- Laminography*
- Neutron grating interferometry (nGI)

Both symmetric and asymmetric nGI options are available with scannable correlation length ranges between ~40 and ~3400 nm; scans of this entire range require a combination of symmetric and asymmetric nGI. Also, nGI measurements are either qualitative or quantitative depending on whether using white beam or monochromated beam, respectively.

Software Tools

The beamline control system is built with the Experimental Physics and Industrial Controls System (EPICS) [4] and controlled via Control System Studio [5]. Data files contain input/output control metadata such as camera details and acquisition parameters, motor positions, and sample environment process variables (PVs) – all used for archiving, analyzing, and auditing measurements. In addition, users can request integration of PVs from their own equipment. Users access their data on the ORNL analysis cluster [6].

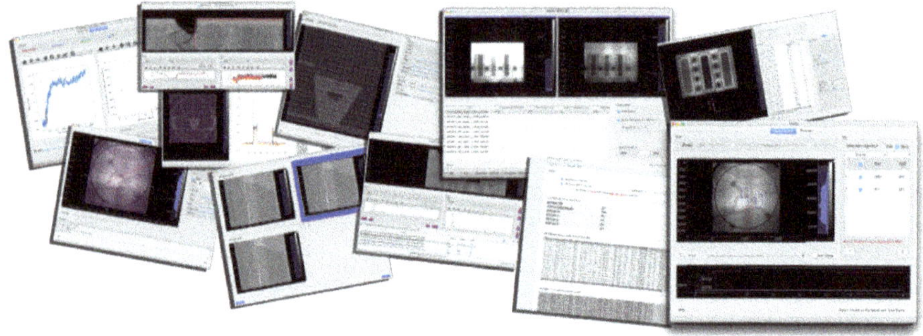

Fig. 3. Jupyter notebooks for neutron imaging.

Data preprocessing, reduction, analysis, and some visualizations are performed using Python-based Jupyter notebooks [7] [8] (Fig. 3). This framework is straightforward (top-to-bottom execution), interactive (widgets), and flexible, such that notebooks can be customized by instrument staff or users for their applications. Some advanced tasks include image co-registration, wavefront dynamics, and radial profiles. In addition to the notebooks, a few standalone tools are also available for more complex analysis such as grating interferometry, complex computed tomography, and laminography. For many routine computed-tomography measurements such as image reconstructions, workflows have been automated, starting shortly after measurement completion. At the conclusion of an experiment, users should expect to depart with reduced and analyzable data as well as knowledge of all available tools. Any additional tools can be developed upon reasonable request. Additionally, our development group maintains an open-source software organization on GitHub [9], providing access to various software tools and resources used for our neutron imaging. A selected subset of the software is also published on our anaconda channel [10]. Finally, advanced visualization like image segmentation is performed with commercial software (e.g., *Amira/Avizo* [11] or *Dragonfly* [12]), which users have access to after their experiment.

Sample Environment

Modern beamlines are increasingly viewed as miniature or *ad hoc* laboratories, multiplexed with different probes (e.g., x-ray imaging [13] [14]) and sample environments.

Furthermore, imaging beamlines like MARS must reconfigure significantly between experiments to tailor to user needs. The following list includes all systems that affect the sample condition, chosen with the guiding philosophy of maximum versatility.

- X-Y-Z sample platform with loading capacity from 75 kg to 200 kg.
- Various sized rotation, translation, and tilt stages as well as goniometers.
- Wide array of optomechanical equipment to customize sample setups.
- Six-channel *Biologic* VSP-300 potentiostat.
- *Conviron* Adaptis A1000 plant growth chamber.
- McHugh-type pressure cell [15]: up to ~ 2 kbar of CO_2, He, N_2, and Ar with heating option up to ~ 80 °C using cartridge heaters.
- Amteco Hot-Rail 1500 °C air furnace.
- Vertical (5–8 T) and horizontal (5–11 T) magnet systems.
- Vacuum (30–1600 °C) and controlled atmosphere (30–1500 °C) furnaces.
- Cryostats (1.7–300 K) and closed-cycle refrigerators (4–300 K).

3   Proposing a Feasible Experiment

To access MARS, prospective users must propose experiments through a rigorous scientific review process. There are online guides with writing tips, templates, experiment preparations, and more [16] [17]. Engaging instrument staff early in the proposal development process is strongly encouraged to increase the likelihood of a feasible and ultimately successful experiment. As an example, attenuation-based imaging requires several considerations in terms of spatial and temporal resolutions (see Table 2), adequate transmission ($\geq 20\%$), sufficient contrast (~5%), and accommodation for custom equipment. Figure 4 illustrates the total approximate image acquisition time for attenuation-based CT scans using each detector system and various sample diameters. An online tool was developed to estimate transmission [17]. Finally, if experiment feasibility is questionable or signal/contrast needs to be tested, then there exist proof-of-principle proposals (up to 1 day of beam time) for such tests that can later support a full proposal.

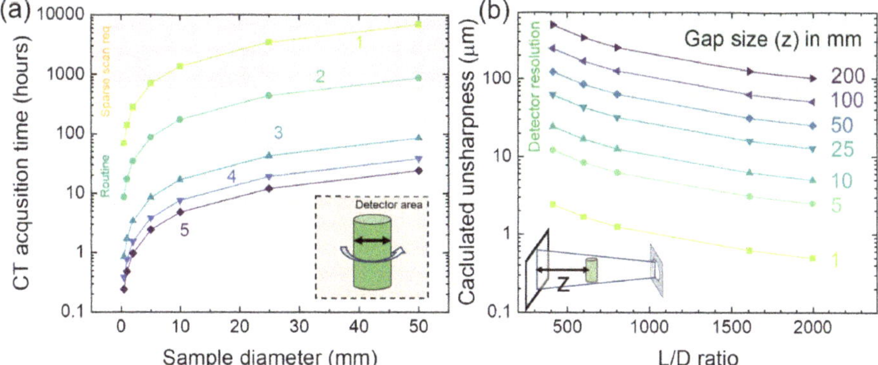

Fig. 4. (a) Total approximate CT acquisition time for each camera system (see Table 2) and various sample diameters (see inset). The number of projections is based on the Nyquist sampling theorem [18]. (b) Effects of sample-to-detector gap size (z) in millimeters on spatial resolution, i.e., calculated "un-sharpness," as a function of *L/D* ratio.

4 Future of MARS

4.1 Relocation in the Guide Hall

Every 20 years or so, the permanent beryllium reflector in HFIR must be replaced. During the next replacement, ca. 2028–2032, all instruments will go offline, and a new six-guide system will be installed in the cold guide hall. The future MANTA triple-axis spectrometer is planned to be built at the CG-1 location (NB-6 after replacement). Therefore, MARS will relocate between the SANS tanks onto guide NB-4 as illustrated in Fig. 5. The scientific requirements have been established, and the conceptual design is under development. The new configuration will feature a larger experiment area and longer maximum flight path, thus offering high operational flexibility to support the broad user community. Most notably, the dedicated guide is designed for imaging, resulting in higher flux and more uniform intensity profile at the guide exit (see Fig. 5). Note that these are concept designs with simulated performance that is subject to change during optimization and construction.

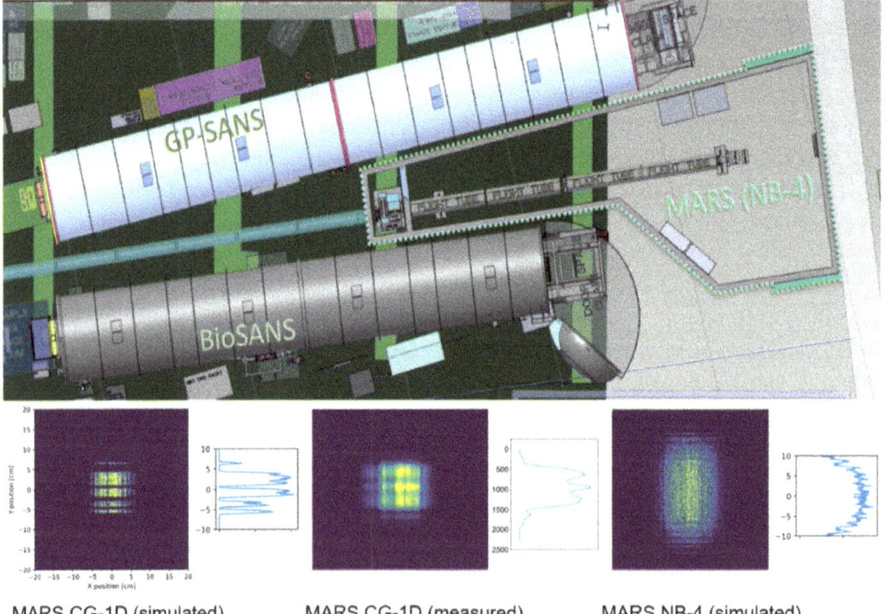

Fig. 5. (a) Future location of MARS on the NB-4 guide. (b) Comparison between current and anticipated beam intensity profiles.

4.2 Planned Upgrades

Below are several upgrades in various stages of planning and implementation:

- Artificial intelligence/machine learning tools for image acquisition and analysis
- Compact, simultaneous x-ray imaging system

- (Continuous) detector developments and optics upgrades for high spatial and temporal resolutions
- *Dedicated* polarization setup (currently loaned from sample environment group)
- Environmental chamber (vacuum, inert gas, humidity) for air-sensitive samples
- Robotic sample changer
- Various mechanical load frames
- Velocity selector and upgraded monochromator

4.3   ORNL Imaging Suite

Optimization of neutron imaging beamlines is based on the source intrinsic characteristics. ORNL is currently steward to two neutron sources: HFIR and Spallation Neutron Source's First Target Station (SNS-FTS) optimized for sharp pulses of thermal neutrons. The latter is home to VENUS – the versatile neutron imaging instrument, which will take a leadership role in energy-resolved imaging, namely strain mapping via Bragg-edge imaging (BEI) and isotope mapping via resonance imaging using thermal and epithermal neutrons. By 2032, MARS will be optimized for tomographic, time-resolved, and parametric studies using the highest integrated cold flux of neutrons across the ORNL facilities. Finally, the future Second Target Station (STS) will offer the highest peak brightness of cold neutrons, complementing both MARS and VENUS. The CUPI^{2}D neutron imaging concept instrument at STS [19] is designed to lead time-dependent, wavelength-resolved measurements like simultaneous BEI and nGI in the study of multiscale structures *in situ*.

Acknowledgements. The authors would like to thank the support teams at ORNL that ensure safe and reliable operation of MARS and the user program. This research used resources at the High Flux Isotope Reactor and Spallation Neutron Source, DOE Office of Science User Facilities operated by the Oak Ridge National Laboratory. The beam time was allocated to MARS on proposal numbers IPTS-23768.1, IPTS-26647.1, and IPTS-26032.1.

References

1. Santodonato, L., et al.: The CG-1D neutron imaging beamline at the Oak Ridge National Laboratory High Flux Isotope Reactor. In: 10 World Conference on Neutron Radiography (2015)
2. Crow, L., et al.: The CG1 instrument development test station at the high flux isotope reactor. Nucl. Inst. Methods Phys. Res. A **634**, S71–S74 (2011)
3. Bilheux, H.Z., et al.: The VENUS imaging beamline at the Spallation Neutron Source: layout, expected performance and status of the construction project. In: 9th International Topical Meeting on Neutron Radiography (ITMNR-9) (2023)
4. Kraimer, M.R.: EPICS: input/output controller (IOC) application developer's guide. EPICS release 3.12 specific documentation.[Experimental Physics and Industrial Control System]. Argonne National Laboratory (2022)
5. CSS landing page. https://www.controlsystemstudio.org. Accessed 30 Sept 2024
6. Analysis homepage. https://analysis.sns.gov/. Accessed 30 Sept 2024
7. Bilheux, J.C., et al.: Neutron imaging software tools at the Oak Ridge National Laboratory. In: Journal of Physics: Conference Series, vol. 2605 (2023)

8. Neutron Imaging website. https://neutronimaging.ornl.gov/. Accessed 30 Sept 2024
9. GitHub repository for neutron imaging. https://github.com/ornlneutronimaging. Accessed 30 Sept 2024
10. Anaconda website for neutron imaging. https://anaconda.org/neutronimaging. Accessed 30 Sept 2024
11. Stalling, D., Westerhoff, M., Hege, H.-C.: Amira: a highly interactive system for visual data analysis. In: The Visualization Handbook, vol. 38, pp. 749–7672005
12. Dragonfly 2022.2 [Computer software]. Comet Technologies Canada Inc., Montreal, Canada; software. https://www.theobjects.com/dragonfly
13. Tengattini, A., et al.: NeXT-Grenoble, the Neutron and X-ray tomograph in Grenoble. Nucl. Instruments Methods Phys. Res. A: Accelerators Spectromet. Detectors Assoc. Equip. **968** (2020)
14. LaManna, J.M., Hussey, D.S., Baltic, E., Jacobson, D.L.: Neutron and X-ray Tomography (NeXT) system for simultaneous, dual modality tomography. Rev. Sci. Instruments **88**, 113701 (2017)
15. Teixeira, S.C.M., Leao, J.B., Gagnon, C., McHugh, M.A.: High pressure cell for Bio-SANS studies under sub-zero temperatures or heat denaturing conditions. J. Neutron Res. **20**, 11–21 (2018)
16. ORNL User homepage. https://neutrons.ornl.gov/users. Accessed 30 Sept 2024
17. Neutron transmission calculator. https://neuit.ornl.gov/transmission. Accessed 30 Sept 2024
18. Kak, A.C., Slaney, M.: Principles of Computerized Tomographic Imaging. Society for Industrial and Applied Mathematics (2001)
19. Brugger, A., et al.: The complex, unique, and powerful imaging instrument for dynamics (CUPI2D) at the spallation neutron source. Rev. Sci. Instruments **94** (2023)

Friction Stir Additive Manufacturing: Neutron Interferometry and Bragg Edge Imaging

Saber Nemati[1], Huan Ding[1], Kyungmin Ham[2], Shengmin Guo[1], James Torres[3], Anton Tremsin[4], and Leslie G Butler[5(✉)]

[1] Department of Mechanical Engineering, Louisiana State University, Baton Rouge 70803, LA, USA
{mnemat2,hding3,sguo2}@lsu.edu
[2] CAMD, Louisiana State University, Baton Rouge 70803, LA, USA
kham1@lsu.edu
[3] Neutron Sciences, Oak Ridge National Laboratory, Oak Ridge 37831, TN, USA
torresjr@ornl.gov
[4] Space Sciences Laboratory, University of California at Berkeley, Berkel 94720, CA, USA
astr@berkeley.edu
[5] Department of Chemistry, Louisiana State University, Baton Rouge 70803, LA, USA
lbutler@lsu.edu

Abstract. The layered structure of aluminum 6061 cross-sections was studied with several neutron imaging methods: transmission, grating interferometry (dark field), and Bragg-edge imaging. The multilayered samples were prepared by additive friction-stir deposition (AFSD) of aluminum 6061 with 2.54 mm print layer thickness. The AFSD aluminum printing was done in air; no evidence of aluminum hydroxide impurities was found in the neutron transmission images. The neutron dark-field data, however, showed a distinct layering structure that was affected by post-build annealing. These measurements used Talbot-Lau interferometry at a cold-neutron beamline without a monochromator and fitted, in favorable cases, to a single scattering model. Anisotropic scattering in the 6061 build plate was noted and measured. The neutron Bragg-Edge images showed crystal grain alignment in patterns correlated with the AFSD tool path.

Keywords: additive friction stir deposition · neutron interferometry · Bragg edge imaging

1 Introduction

Additive friction stir deposition (AFSD) is a purely additive layer build-up manufacturing process that offers unique speeds and build sizes on the scale of tens

---

S. Nemati, H. Ding, K. Ham, S. Guo, J. Torres, A. Tremsin and L. G. Butler—This Authors contribute equal to this work.

A. E. Craft and H. Z. Bilheux (Eds.): WCNR 2024, SPPHY 348, pp. 299–307, 2026.
https://doi.org/10.1007/978-3-032-15003-5_34

of kilograms per hour [1]. Combined with its scalable process, ability to fabricate in air, and crack-free products, AFSD is a promising solution for manufacturing large complex metal components for aircraft, naval vessels, automobiles, and other various machinery.

Multilayer deposition via AFSD creates a visibly layered structure in the fabricated object. The focus of our research is whether the interleaved aluminum layers contain a dispersion of reaction products, such as oxide and hydroxide phases that are generated due to the processing of AFSD in ambient air [2]. Contamination of this type has the potential to initiate cracks or promote crack propagation [2,3], thus degrading mechanical properties and lowering the useful life of the component. To address this concern, our approach in the present work is a full-volume inspection of AFSD test specimens in search of contamination using neutron imaging techniques, leveraging the unique interaction of neutrons with atomic nuclei, including high penetration through thick aluminum specimens.

As will be discussed in the following, conventional radiographic imaging based on neutron transmission was shown to be weakly sensitive to the layered structure. However, a relatively new technique, neutron grating interferometry (nGI), with sensitivity to small-angle scattering from mesoscopic particles [4–6], clearly revealed alternating layers of low and high dark-field contrast attributed to a heterogenous microstructure. Finally, the diffraction contrast provided by Bragg-edge neutron imaging (BEI) [7–10] revealed highly localized regions of Al grains with a preferred orientation that inspired follow-up studies.

2   Experimental

An AFSD bar was printed out of aluminum alloy 6061 with a mass composition of 1.0% Mg, 0.61% Si, 0.31% Cu, and Al balance where the Al matrix contained Mg_2Si precipitates. A graphite lubricant was used in samples labeled 'G'. The print parameters were: 2.54 mm layer thickness, 400 RPM spindle speed, $\sim$152 mm/min spindle traverse speed, and $\sim$229 mm/min feedstock rate. The build plate was also Al6061 with a typical rolled microstructure consisting of $\sim$200-micron-long grains elongated along the rolling direction. Our experimental setup was inspired by previous neutron characterization work [11] that explored transient microstructures within friction-stir-welded Al6061. In our work, samples were cut from the printed bar in 20-mm-thick sections appropriate for neutron characterization. The G3T-labeled sample represents the as-built condition, and the G3B-labeled sample underwent annealing (O1 temper) consisting of a 45- m hold at 530 °C followed by cooling in still air.

Neutron interferometry was performed at the Multimodal Advanced Radiography Station (MARS) at Oak Ridge National Laboratory (ORNL) [12]. The fundamental principles as well as applications of the nGI technique can be found in the literature (e.g., Ref. [4,5,13,14]). In brief, the setup at MARS consisted of three gratings (source, phase, and analyzer) configured for symmetric Talbot-Lau. As detailed by Strobl [4], the autocorrelation length (ξ_{corr}) is a function of neutron wavelength (λ), sample-to-grating distance (L_s), and grating period (p):

$$\xi_{corr} = \frac{\lambda L_s}{p} \tag{1}$$

At MARS, white beam was used with a peak wavelength of $\lambda = 2.6$ Å and grating period of 56.5 μm on all three gratings. Transmitted neutrons behind the analyzer grating were converted to light using a 50-μm-thick ^{6}LiF-ZnS scintillator optically coupled to an Andor model DZ936 CCD camera in a light-tight box. The pixel size was 7.8 μm with an effective resolution of about 42 μm. The moiré interference pattern formed by the arrangement of gratings was modulated on the detector by translating the phase grating in a direction normal to the beam. In this way, each pixel records a sinusoidal change in intensity, which, when compared with and without a sample, can be analyzed to reveal neutron scattering and refraction effects from the sample. In this setup, the ξ_{corr} range of about 20–900 nm was scanned by changing L_s from 2 mm to 5 mm and then 5 mm to 195 mm in steps of 10 mm. The sinusoidal intensity pattern was analyzed on a per pixel basis using a vectorized least squares algorithm [15].

3 Results and Discussion

3.1 Neutron Grating Interferometry (nGI)

nGI analysis provided three image contrasts within the AFSD samples: transmission, refraction (phase contrast), and dark field (DF). The transmission and DF contrasts are shown in Figs. 1a and b, respectively, where the layered structure was weakly visible via transmission. Similarly, phase-contrast images (not shown) contained very little detail of the internal structure. On the other hand, AFSD layering was clearly observed in the DF images, especially at the longest ξ_{corr} of about 800 nm.

The process of converting DF layered structure into a numerical quantity went through several iterations. The process shown here used a mask set at the 50% threshold for pixel values (i.e., light or dark pixels) in the DF images. These distributions are shown in Fig. 1c and d. With this process, we find a well-layered structure in the as-built (G3T) sample a structure which is retained with annealing (G3B), though modified.

Strobl [4] described the scattering function for monodisperse spheres in neutron DF imaging, which is used herein as a first approximation for this image analysis:

$$DFI_{sphere} = \exp\left[\sigma t \left(\exp\left[-\frac{9}{8}\left(\frac{\xi_{corr}}{R}\right)\right] - 1\right)\right] \tag{2}$$

where t is sample thickness, R is the sphere diameter, and σ is a fitting parameter with units of inverse length. Figure 2 plots the DF intensity (DFI) versus autocorrelation length for both samples. With the assumption of monochromatic beam, this simple scattering function shows that there is a decrease in

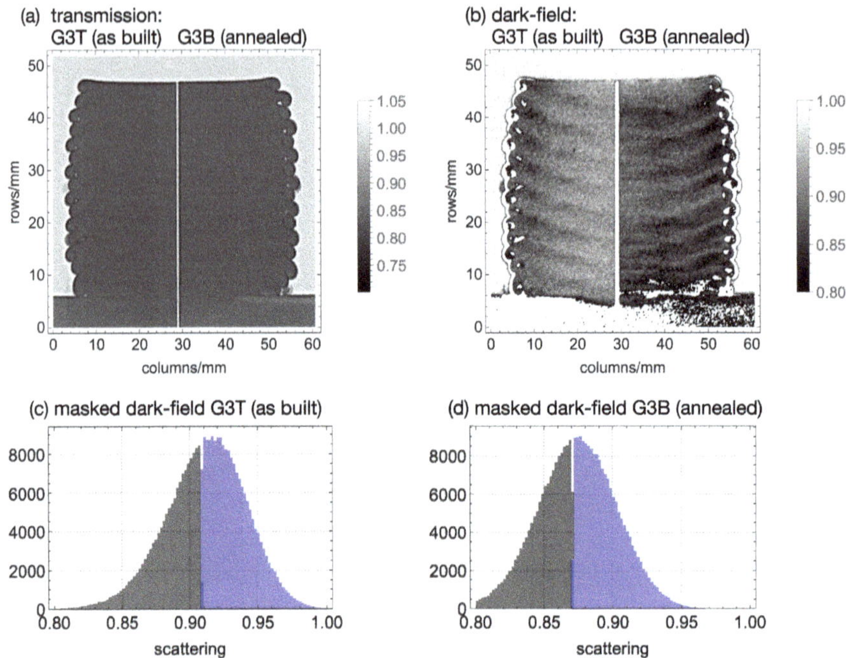

Fig. 1. Neutron transmission and dark-field (scattering) of AFSD samples as-built, G3T, and after annealing, G3B. The dark-field (DF) images are masked at the 50% threshold to separate the pixel intensities into two regions. Annealing shifts the mean values of both regions and narrows the distribution of the lefthand peak, the peak corresponding to larger scattering centers.

the DFI as a result of annealing, shown as a shift in the histogram in Figs. 1c and d. We attribute this shift to an increase in the size of the particles that contribute to beam scattering. We note, however, that proper application of Eq. 2 requires wavelength discrimination, which we plan to use in future measurements. Nevertheless, these studies using white beam provide relatively fast, qualitative inspections. We hypothesize that these particles are the contamination products of interest; however, a composition or microscopy analysis is required to confirm.

During nGI measurements, we observed DF contrast changed not only within the AFSD samples but in the base plate as well. We hypothesized that this contrast arises from crystallographic texture based on the *a priori* knowledge that the base plate is expected to have columnar texture oriented along the rolled direction during manufacturing. Furthermore, in a recent report, Kim *et al.* [16] demonstrate that for DF imaging in the presence of appreciable microstructural variation, the scattering cross section may exceed that of attenuation as the dominant contribution to image contrast. To explore this, three small cubes were cut from a spare base plate where the rolled direction was marked as shown in

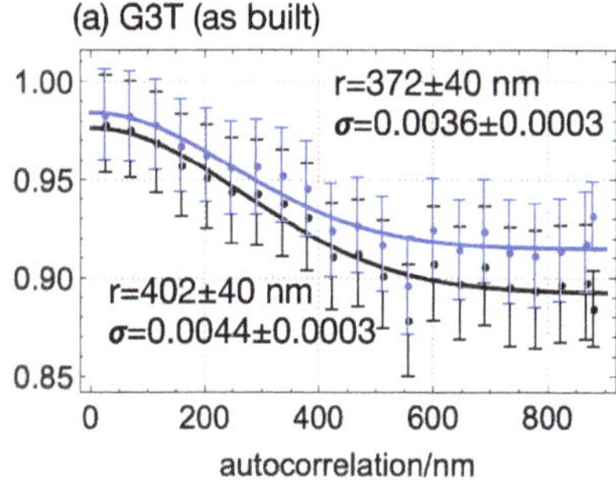

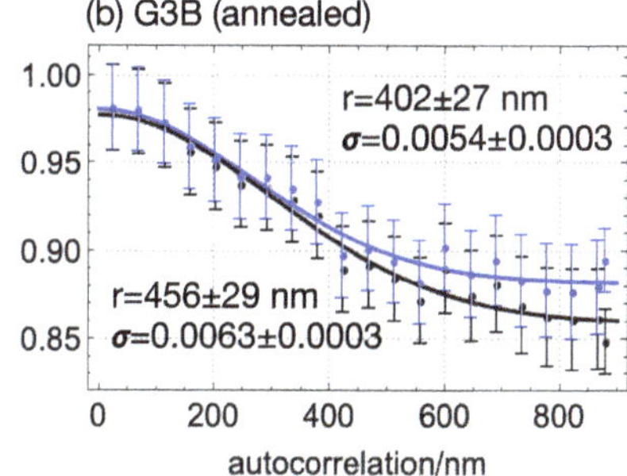

Fig. 2. Fits of dark-field (scattering) of the AFSD samples (a) as built, G3T, and (b) after annealing, G3B, to Eq. 1. The blue traces correspond to the lighter layers in Fig. 1(b) and the black traces to the darker layers.

Fig. 3a. Representative transmission and DF images are shown in Fig. 3b and c, respectively. Transmission contrast appears to be only sensitive to sample thickness, whereas DF images show high sensitivity when the rolled direction is parallel to the beam direction. Equation 2 was applied to both the cube-up and cube in beam direction samples and plotted in Fig. 3d and e, respectively. Given that dark field contrast was observed in both the samples as well as their build plates, a conclusion cannot yet be reached to whether the contrast arises from small-angle scattering from contamination particles or from texture in sub-micron grains.

3.2 Bragg-Edge Neutron Imaging (BEI)

The Bragg-edge imaging (BEI) technique provides spatially resolved diffraction contrast over large samples [17]. The same AFSD samples (G3T and G3B) were studied with BEI at the BL10 NOBORU cold-neutron time-of-flight beamline at J-PARC [18]. The wavelength range of interest was 2–4 Å, which spans the largest Bragg edges of Al. Figure 4 shows three images of the same samples taken with different neutron wavelengths, which clearly reveals not only the AFSD layering but also highly localized dark spots in alternating layers around a neutron wavelength of 3.98 Å. Because this wavelength approximately corresponds to twice the lattice spacing of Al (200) planes, following Bragg's law in transmission geometry, we attribute these regions to highly oriented Al domains. Furthermore, these structures were observed in all AFSD samples and appear to be correlated with the tool path. We make the proposition that this mesoscale crystal domain alignment is a controllable feature which may be usefully employed in applications in harsh environments. More work is planned to develop this concept.

We note here that a residual stress analysis (e.g., [8,9,19]) was attempted as a function of deposition layer depth using the HIDRA diffractometer [20] at ORNL.

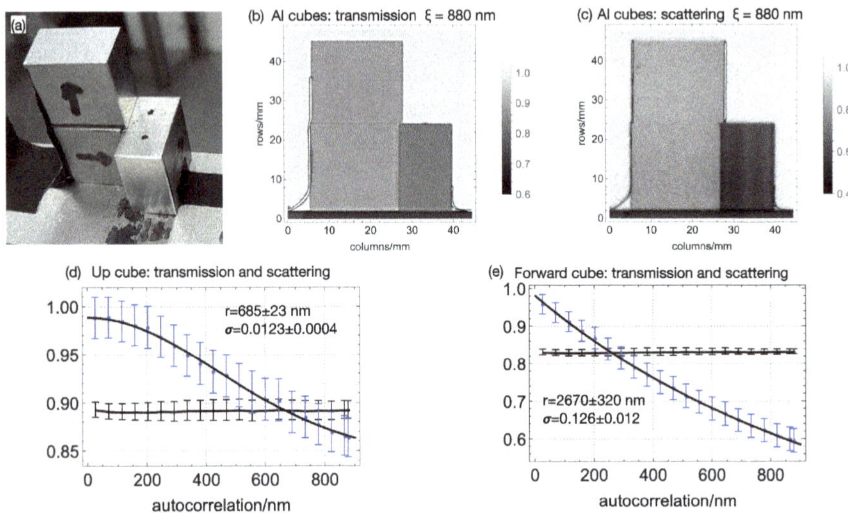

Fig. 3. (a) Three cubes cut from an Al6061 rolled plate with arrows parallel to the roll direction. (b) Neutron transmission showing similar neutron mass attenuation coefficients for all three cube orientations. (c) Neutron DFI showing enhanced neutron scattering for a cube with a roll direction parallel to the neutron beam. The neutron DF (scattering) signal (blue trace) evolves as the interferometer autocorrelation length increases. The black trace shows the neutron transmission. As more scatters of larger and larger sizes contribute, the value of the cumulative distribution function decreases from initially one to a limiting value of zero. The cube-up orientation (a,c), shows a dependence consistent with isolated single particle scattering sites while the cube oriented along the neutron beam direction (a,c) has a different scattering function. We note that the value of the interferometer autocorrelation length is assigned with the assumption of a monochromatic neutron beam with a wavelength of 2.6 Å.

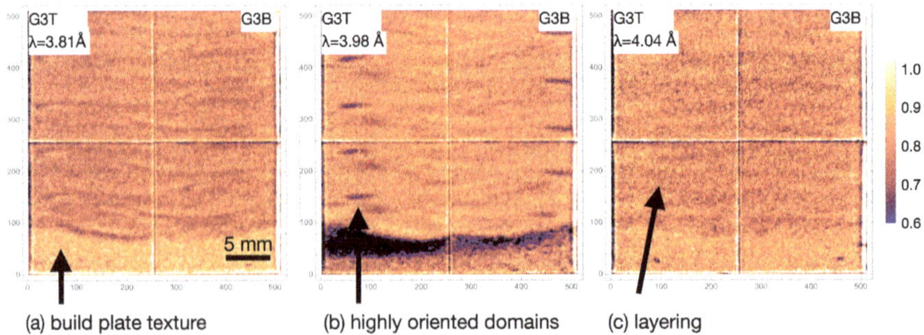

Fig. 4. Bragg edge images at three wavelengths, 3.81, 3.98, and 4.04 Å, show detail within the AFSD layers, including highly oriented domains with dimensions of several millimeters.

However, counting statistics within $\sim$2.5 mm thick layers were insufficient in this one-day feasibility study; a future, longer duration study will be considered.

4   Summary

Additive friction-stir deposition (AFSD), with the attractive features of high deposition rate and operating in air, is shown by neutron transmission imaging to have no detectable contamination products. However, the AFSD test bars show a layering structure in the neutron interferometry dark-field images hypothesized to arise from contamination. When the same parts were studied with neutron Bragg-edge imaging, the layering structure was also observed along with signs of crystal grain alignment. Because these grains were not visible using nGI, further work such as wavelength-resolved nGI and microanalysis is planned to determine the origin of nGI contrast. The effects of AFSD process variables such as deposition temperature and nature of the graphite lubricant will be assessed as well. The present work demonstrates the utility of the nGI and BEI neutron techniques for full volume inspection of structures in the mesoscale range (tens of nanometers to microns) that allow for more targeted microscopy studies. For these techniques to be combined and quantitative, specialized time-of-flight neutron imaging facilities are required such as VENUS beamline [21] at the Spallation Neutron Source as well as the proposed CUPI^{2}D beamline [22,23] at the future Second Target Station at ORNL.

Acknowledgements. This work is supported by the U.S. National Science Foundation under grant number OIA-1946231 and the Louisiana Board of Regents for the Louisiana Materials Design Alliance (LAMDA). A portion of this research used resources at the High Flux Isotope Reactor, a DOE Office of Science User Facility operated by the Oak Ridge National Laboratory. The beam time was allocated to MARS and HIDRA on proposal numbers IPTS-30890.1 and IPTS-31978.1.

Declarations. The authors declare no conflicts of interest in the performance of this research nor its presentation in this proceedings.

References

1. Shao, J., et al.: Additive friction stir deposition of metallic materials: Process, structure and properties. Mater. Des. **234**, 112356 (2023). https://doi.org/10.1016/j.matdes.2023.112356
2. Phillips, B.J., et al.: Effect of parallel deposition path and interface material flow on resulting microstructure and tensile behavior of al-mg-si alloy fabricated by additive friction stir deposition. J. Mater. Proc. Tech. **295**, 117169 (2021). https://doi.org/10.1016/j.jmatprotec.2021.117169
3. Tang, M., Pistorius, P.C.: Oxides, porosity and fatigue performance of alsi10mg parts produced by selective laser melting. Int. J. Fatigue **94**, 192–201 (2017). https://doi.org/10.1016/j.ijfatigue.2016.06.002
4. Strobl, M.: General solution for quantitative dark-field contrast imaging with grating interferometers. Sci. Rep. **4**(1), 7243 (2014). https://doi.org/10.1038/srep07243

5. Brooks, A.J., et al.: Intact, commercial lithium-polymer batteries: spatially resolved grating-based interferometry imaging, bragg edge imaging, and neutron diffraction. Appl. Sci. **12**(3), 1281 (2022). https://doi.org/10.3390/app12031281

6. Wolf, C.M., et al.: Simulation of neutron dark-field data for grating-based interferometers. J. Appl. Crystallogr. **57**(Pt 2), 403–412 (2024). https://doi.org/10.1107/S1600576724001201

7. Tremsin, A.S., et al.: Monitoring residual strain relaxation and preferred grain orientation of additively manufactured Inconel 625 by in-situ neutron imaging. Addit. Manuf. **46**, 102130 (2021). https://doi.org/10.1016/j.addma.2021.102130. Accessed 2024-09-29

8. Busi, M., et al.: Bragg edge tomography characterization of additively manufactured 316L steel. Phys. Rev. Mat. 6(5), 053602 (2022) https://doi.org/10.1103/PhysRevMaterials.6.053602

9. Su, Y., et al.: Residual stress relaxation by bending fatigue in induction-hardened gear studied by neutron Bragg edge transmission imaging and X-ray diffraction. Int. J. Fatigue **174**, 107729 (2023). https://doi.org/10.1016/j.ijfatigue.2023.107729

10. Gaudez, S., et al.:Evolution of texture and residual stresses in 2205 duplex stainless steel during laser powder bed fusion. Materials Design 251, 113658 (2025). https://doi.org/10.1016/j.matdes.2025.113658

11. Woo, W., Ungár, T., Feng, Z., Kenik, E., Clausen, B.: X-Ray and neutron diffraction measurements of dislocation density and subgrain size in a frictionstir-welded aluminum alloy. Metallurgical and Materials Transactions A 41(5), 1210–1216 (2010). https://doi.org/10.1007/s11661-009-9963-5

12. Torres, J., et al.: Overview of mars: the multimodal advanced radiography station at the high-flux isotope reactor. In: Proceedings of the 12th World Conference on Neutron Radiography - WCNR-12 (2025)

13. Pfeiffer, F., Grünzweig, C., Bunk, O., Frei, G., Lehmann, E., David, C.: Neutron Phase Imaging and Tomography. Phys. Rev. Lett. 96(21), 215505 (2006). https://doi.org/10.1103/PhysRevLett.96.215505

14. Strobl, M., et al.: Neutron dark-field tomography. Phys. Rev. Lett. 101(12), 123902 (2008). https://doi.org/10.1103/PhysRevLett.101.123902

15. Marathe, S.: Improved algorithm for processing grating-based phase contrast interferometry image sets. Rev. Sci. Instrum. **85**(1), 013704 (2014). https://doi.org/10.1063/1.4861199

16. Kim, Y., et al.: Application of neutron grating interferometry and tomography to the nineteenth century korean copper coins. Sci. Rep. **15**, 14848 (2025). https://doi.org/10.1038/s41598-025-99235-x

17. Kockelmann, W., Frei, G., Lehmann, E.H., Vontobel, P., Santisteban, J.R.:Energy-selective neutron transmission imaging at a pulsed source. Nucl. Instrum.Methods Phys. Res. Sect. A-Accel. Spectrom. Dect. Assoc. Equip. 578(2), 421–434 (2007). https://doi.org/10.1016/j.nima.2007.05.207

18. Oikawa, K., et al.: Design and application of noboru—neutron beam line for observation and research use at j-parc. Nucl. Inst. Methods Phys. Res. A **589**, 310–317 (2008). https://doi.org/10.1016/j.nima.2008.02.019

19. Tremsin, A.S., et al.:Calibration and optimization of Bragg edge analysis in energy-resolved neutron imaging experiments. Nucl. Instrum. Methods Phys. Res. Sect. A-Accel. Spectrom.Dect. Assoc. Equip. 1009, 165493 (2021). https://doi.org/10.1016/j.nima.2021.165493

20. Bunn, J.R., et al.: The high intensity diffractometer for residual stress analysis (HIDRA), a third generation residual stress mapping neutron diffractometer at

the high flux isotope reactor. Rev. Sci. Instrum. **589**, 035101 (2023). https://doi.org/10.1063/5.0122250

21. Bilheux, H.Z., et al.: The VENUS imaging beamline at the spallation neutron source: layout, expected performance and status of the construction project. J. Phys: Conf. Ser. **2605**, 012004 (2023). https://doi.org/10.1088/1742-6596/2605/1/012004

22. Brugger, A., et al.: The Complex, Unique, and Powerful Imaging Instrument for Dynamics, CUPI2D, at the Spallation Neutron Source (invited). Rev. Sci. Instrum. **94**(5), 051301 (2023). https://doi.org/10.1063/5.0131778

23. Bilheux, H.Z., Torres, J., Brugger, A., Lin, J.: Conceptual optics design of the CUPI2D beamline at the spallation neutron source second target station. In: Proceedings of the 12th World Conference on Neutron Radiography - WCNR-12(2025)

Neutron Imaging of Hydrogenous Gases in Porous-Carbon Materials Around Phase Transitions

Frederik Ossler[1]($\boxtimes$) , Charles E. A. Finney[2] , Louis J. Santodonato[3] , Jean-Christophe Bilheux[2] , Hassina Z. Bilheux[2] , Douglas P. Armitage[2], Rebecca A. Mills[2] , Yuxuan Zhang[2] , and Luke L. Daemen[3]

[1] Combustion Physics, Lund University, Lund, Sweden
frederik.ossler@fysik.lu.se
[2] Oak Ridge National Laboratory, Oak Ridge, TN 37831, USA
[3] Oak Ridge National Laboratory (Retired), Oak Ridge, TN 37831, USA

Abstract. Focus herein is on hydrogenous gases interacting with porous-carbon materials on activated biochar at pressures ranging from 0 to 100 bar and temperatures from room (ambient) to cryogenic. Neutron radiography was conducted using a custom-made cell system to simultaneously image pure gas and gas interacting with the porous carbon. From neutron attenuation, the cross sections of methane and ethane and number concentrations were determined for different pressures and temperatures and compared with corresponding reference data. Sorption properties of the porous materials were investigated *in situ* with methane, ethane, and propane.

Keywords: Hydrogen & processes · Image and data processing · New techniques & analysis

1 Introduction

The high scattering/attenuation cross sections of hydrogen compared to most other elements make neutron radiography (NR) and neutron computed tomography (NCT) powerful tools to study physical and chemical process of hydrogenous substances and hydrogen-infused materials. The techniques are especially useful when bulky materials and gases require containment in metal casings where extreme conditions of pressure and temperature are needed. These techniques can therefore play an almost unique role in the fields of energy, materials, food, and environmental research and technology. Porous carbons are interesting materials for energy storage and transportation [1]. Understanding the dynamics of phase transitions between liquid and gas is of interest for loading, storing and releasing fuel/energy from a porous medium. Based on a previous study on NR applied to methane interreacting with soot materials [2], this work focuses on describing NR experiments where hydrogenous gases are interacting with activated pinewood char, inside a metal cell at different thermodynamic conditions (pressure and temperature,

© The Author(s) 2026

A. E. Craft and H. Z. Bilheux (Eds.): WCNR 2024, SPPHY 348, pp. 308–316, 2026.
https://doi.org/10.1007/978-3-032-15003-5_35

p,T). The experiments were performed at the CG-1D IMAGING (now called MARS) beamline at the High Flux Isotope Reactor at Oak Ridge National Laboratory (ORNL) between 2016 and 2020. Some of this material has been presented as oral presentations at Carbon conferences between 2018 and 2022 [3, 4]. Here, we have extended the analysis and described parts and methodology that could be of interest for researchers on hydrogenous gases and porous carbon.

2 Hydrogenous Gases Interaction with Porous Materials

Based on preceding NR studies on CH_4 with soot [2], a new cell (in aluminum, yielding less activation from neutrons and with higher thermal conductivity) was designed and built to enable easier operation for *in-situ* measurements at different pressures (0–100 bar) and temperatures (cryogenic–room). Figure 1 describes the measurement apparatus.

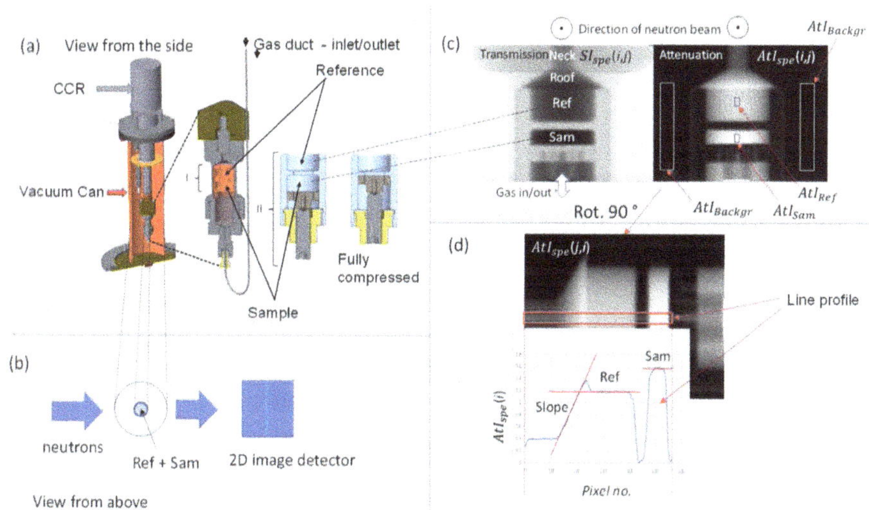

Fig. 1. (a) Main parts of the cell system showing the CCR and the vacuum can: (*I*) refers to the first version sample cell and (*II*) a second unified version developed into a clamp cell for the sample; (b) schematic of the setup (peak neutron wavelength of 2.6 Å; for more details regarding the neutron source and detector system, see [5]). (c) and (d) are assistant pictures for the data-evaluation schemes described in the text below.

The cell system depicted in Fig. 1a was connected to a temperature-controlled chiller (ARS DE-204 closed cycle refrigerator, CCR [6]) inside a vacuum chamber including a heat shield surrounding the cell system to prevent radiant heat losses from the cold stage of the CCR and the sample. The setup was installed in the beam path (see Fig. 1b). The chamber was then evacuated, the pressure and temperature of the specimen gas (here, CH_4) were set, and measurements started. Neutron radiographs were acquired with typical exposure times of 30 s with a camera system (see [2]) in continuous framing mode to capture dynamics. Image analysis was performed with ImageJ [7]. The field of view allowed simultaneous observation of the cell working areas and reference regions of interest (*ROIs*, Fig. 1c), to measure both the reference (*Ref*) and sample (*Sam*) gas/sorption dynamics.

The sample images (*SI*) were recorded, subtracted by the I_{DF} (dark field/no-beam image), and then normalized by $I_{OB} - I_{DF}$ (I_{OB}, open unobstructed beam image, i.e., without any obstruction between beam and detector) to produce a transmission image (see, e.g., Fig. 1c). The attenuation image (*AtI*) was produced by dividing the *SI* for vacuum (*SI*$_{vac}$) by *SI* with the specimen gas introduced (*SI*$_{spe}$) finalized by taking the natural logarithm ($\log_e$) of the ratio.

$$AtI_{spe}(i,j) = \log_e[SI_{vac}(i,j)/SI_{spe}(i,j)] \tag{1}$$

Assuming that the spatial distribution across the beam is maintained constant during the experiments one can as a first approximation exclude the normalization by $I_{OB} - I_{DF}$ when using Eq. (1), since it divides out. In Fig. 1c the left image shows the transmission image $SI_{spe}(i,j)$, where i and j are row and column indices, respectively, of the digital images, for $(p,T) = (57.97\,\text{bar}, 200\,\text{K})\,CH_4$ with pinewood (*pinus silvestris*) char (activated by CO_2 oxidization at 800 °C; BET specific surface of 1038 m^2/g) in *Sam*. Pinewood was chosen as an alternative sustainable char source and can reach high specific surface areas comparable with other commercial products [8]. The right image shows the attenuation image $AtI_{spe}(i,j)$. One observes that the average background has a finite offset that varies slightly from frame to frame. This is expected to be primarily caused by variations in beam intensity with time, e.g., and influence open beam and the specimen recording, since they are performed at different times, and can affect evaluation at very low signals for *Ref* in Eq. (1), when performing normalization with the vacuum image (cell system in vacuum) recorded earlier. Under the assumption that the intensity changes with the same factor for each pixel of the frames, an average value of the background, I_{BG} ($\sim -0.3\%$ of *Ref*-signal level before and $\sim -3\%$ after pressure release) can be determined for each specimen frame for the *ROIs* shown in Fig. 1c. It varies only weakly with distance from the cell and can be disregarded in most cases. The amount of specimen, e.g., CH_4 in Fig. 1, can in principle be calculated from AtI_{spe}, i.e., $\overline{AtI}_{spe}(i,j) - I_{BG}$, the average of the selected *ROIs* assuming Lambert–Beer law (*LBL*), when $AI_{spe,Ref}$ is low. Since the error from not considering the cylindrical shape is only $-0.4 - -0.3\%$ for d_{Ref} and -2% for d_{Neck}, LBL yields:

$$AtI_{spe} = \sigma_{CH4} \times n_{CH4} \times d \tag{2}$$

where σ_{CH4}, n_{CH4}, and d ($d = 1.27$ cm in II), are the microscopic attenuation cross section, number concentration of CH_4, and the *Ref* and *Sam* diameters (same), respectively. There is an additional 2% to consider from the miss/spatial gap between outer clamp cell wall and inner cell wall that contains specimen and contributes to the attenuation, but can be disregarded at this stage of level of demonstration. If we discard any deviations from LBL, one can define the storage capacity (SC, corresponding to "DF" in [2]) as the amount of specimen in *Sam* relative to that in *Ref* as:

$$SC = AtI_{spe,Sam}/AtI_{spe,Ref} = n_{CH4,Sam}/n_{CH4,Ref} \tag{3}$$

From $AtI_{spe,Sam}$ one can invert Eq. (2) to calculate n_{Sam}, which multiplied by the sample volume, V_{Sam}, and divided by the quantity of porous material can be calculated in, e.g., units of mmol CH_4/g absorber or g CH_4/g absorber.

2.1 Measurements of σ_{spe} for CH_4 and C_2H_6 as Non-Ideal Gases in *Ref*

Room temperature (RT, 296 K) measurements of neutron attenuation were performed on porous carbon (here, soot) in *Sam* and *Ref* for the cell system version *I*.

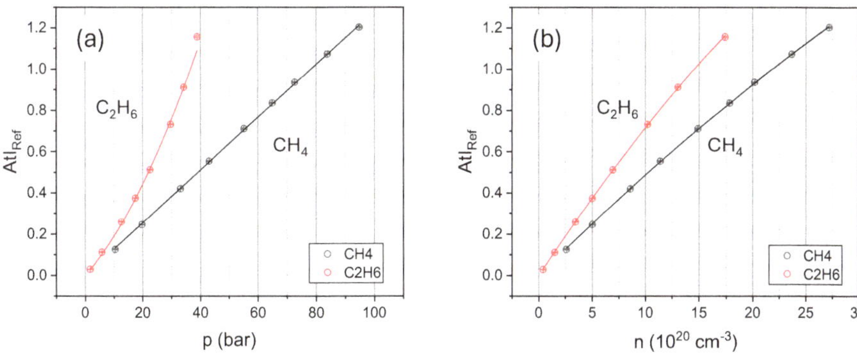

Fig. 2. (a) Attenuation of pure CH_4 and C_2H_6 in the reference cell as a function of pressure (p) and (b) number concentration. Curves show the corresponding regression fits of $AtI_{Ref} = 0 + B1 \cdot 10^{20} \cdot n + B2 \cdot (10^{20} \cdot n)^2$. Figure 2b can be used as a source for calibration curves of n vs AtI_{Ref}.

The measurements were performed at different pressures under static conditions with sequential stepdown of pressure by release from *ca* 100 bar for CH_4 and 40 bar for C_2H_6. In this setup the cell was not evacuated but washed with helium according to the description in [2]. The vacuum image was replaced by a helium image (4.3 bar in this case). The attenuation data were collected as a function of p then converted to n by using thermodynamic data from NIST REFPROP 9.1 (RP) [9]. Figure 2 shows plots with fits using a 2nd-order polynomial with intercept 0. The results of the fits are given in Table 1. From the B1, the attenuation cross section for H can be approximated by:

$$\sigma_{CH4} = \sigma_C + 4 \times \sigma_H; \sigma_{C2H6} = 2 \times \sigma_C + 6 \times \sigma_H \tag{4}$$

However, σ_H is expected to show sensitivity to wavelength and chemical bonding [10, 11], while σ_C considerably less so. The first choice is to put $\sigma_C = 5.556$ barn (b) (see [2] and references therein) while σ_H as an unknown in Eq. 4. σ_H then becomes 77 and 78 b for CH_4 and C_2H_6, respectively. σ_{CH4} calculated in the previous study was 300 b, which yields $\sigma_H = 74$ b. Due to experimental uncertainties an apparent attenuation cross section and concentration are measured when using Eq. (2).

The system was subjected to CH_4, C_2H_6, and C_3H_8 and investigated at different pressures and temperatures across the saturation–phase transition lines defined by RP. The sorption properties of the char were also analyzed. Figure 3 and Fig. 4 results from

Table 1. Fitting parameters and resulting cross sections [b] for CH_4 and C_2H_6

Gas	B1	Error	B2	Error	σ [b]	Error	σ_H [b]
CH_4	0.0513	2.57×10^{-4}	-2.53×10^{-4}	1.21×10^{-5}	315	± 1.6	77
C_2H_6	0.0781	4.92×10^{-4}	-6.43×10^{-4}	4.29×10^{-5}	480	± 3.0	78

operando NR monitoring of the dynamics of CH_4 and the char. The vacuum reference image was measured at 300 K, compared to 145–200 K for the sample measurements. This caused a vertical shift between the images, which however did not significantly affect the evaluation at the reference points measured (see Fig. 4f). The pressure and temperature data were recorded manually from display monitors, yielding a time resolution of the order of a frame time, 30 s. The experiments started with CH_4 at a supercritical condition $(p,T) = (54.57$ bar, 200 K$)$. To access different regions, pressure was released on one occasion. The (p,T) trace then moved well into the gas-phase region, and temperature was changed to reach the saturation curve again.

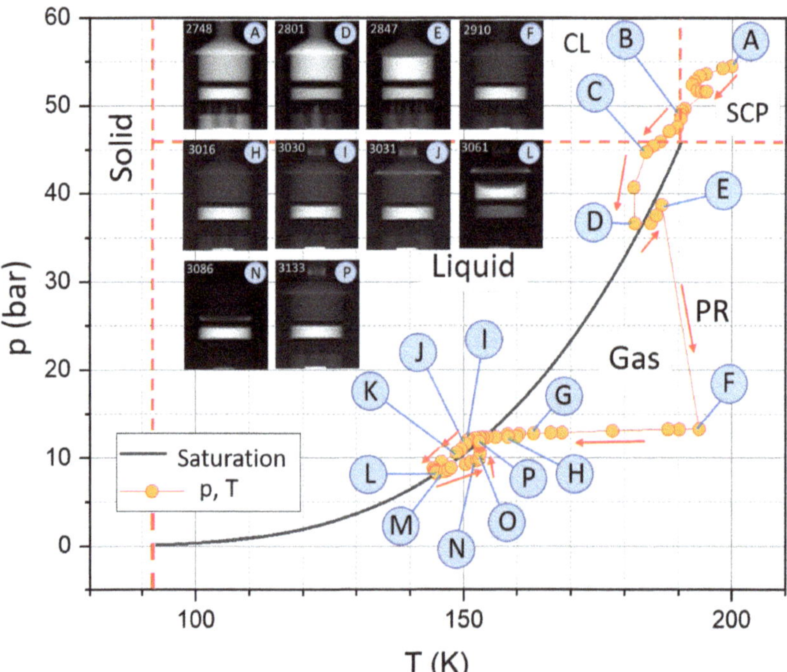

Fig. 3. Panel showing the saturation curve for CH_4 and the (p,T) measurement points as well as selected attenuation images. CL and SCP stand for compressible liquid and supercritical phase, respectively. PR stands for pressure release between E and F.

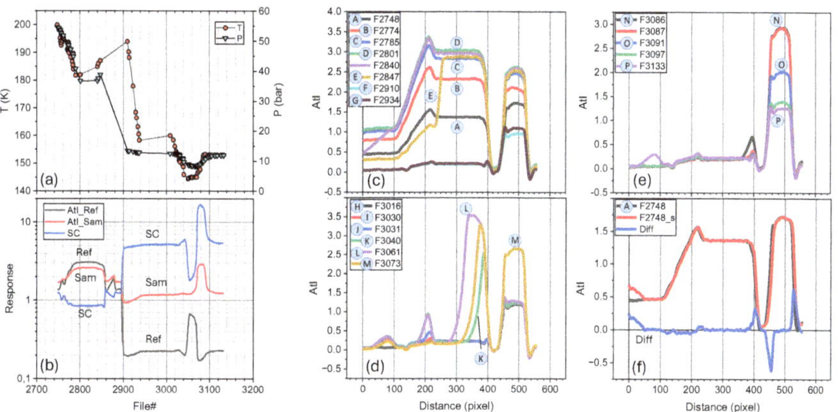

Fig. 4. (a) Measured (p,T) for different File#; (b) evaluated AtI_{Ref}, AtI_{Sam}, and SC (LBL) vs File# for the ROIs given in Fig. 1c; (c)–(e) profiles along the cell (see Fig. 1d) for selected frames. AtI_{Ref} and AtI_{Sam} are sampled at pixel distance 250 and 495, respectively; (f) comparison between non-shifted and shifted images accounting for cell/sample movement with temperature.

At the starting point of Fig. 3, $(p,T) = (54.57$ bar, 200 K), the concentrations n_{CH4} $= 4.05 \times 10^{21}$ cm^{-3} and 3.41×10^{21} cm^{-3} were obtained from solving the 2nd-order polynomial equation (2op) by Eq. 5 and by using the LBL as expressed in Eq. 2, with cross section given in Table 1, respectively.

$$n_{CH4,2op} = \frac{d_I}{d_{II}} \times \left\{ -\frac{B_1}{2B_2} - \left[\frac{1}{4} \times \left(\frac{B_1}{B_2} \right)^2 + \frac{AtI_{Ref}}{B_2} \right]^{\frac{1}{2}} \right\}; 0 < AtI_{Ref} < 2.6 \quad (5)$$

where d_I and d_{II} are the diameters valid for types I and II. Other symbols and values are from Table 1 specified above.

In comparison, with reference data from RP and also NIST WebBook [12], for frame 2748 ("F2748") one obtains $n_{CH4,F2748} = 4.23 \times 10^{21}$ cm^{-3}, which is considerably close to the value obtained via Eq. 5. These three types of number concentration data are referred to as "2op", "LBL" and "NWB", respectively, in the following text.

The correspondence with RP was considerably good for the first two frames, F2748 and F2774. However, when entering the liquid zone, there is a limited region with poor agreement which appeared, close to the transition line between liquid and gas. With the NIST WebBook tool (NWB) it was possible to obtain improved results using saturation data for CH$_4$, when the measured pressure data were used but the temperature was allowed to change to that given by the saturation curve.

Consequently, with respect to the measured temperature, the temperature increased for the liquid-phase and lowered for the vapor-phase region, respectively. Figure 5 shows the results from the comparison between data obtained from NWB, 2op, and LBL for different (p,T) measurement points along the experiment. The agreement between 2op evaluation and predicted values by NWB are considerably good. Data within the blue oval were out of range (could not be solved) for 2op.

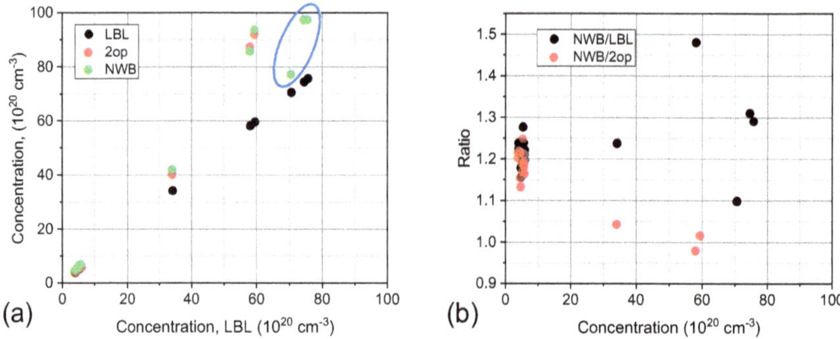

Fig. 5. (a) Comparison between number concentrations retrieved from measurements and LBL, 2op evaluation, and NWB data; (b) Ratio between number concentrations from NWB/LBL and NWB/2op.

As the system crossed the saturation line from liquid to gas phase, the liquid started to evaporate, and both phases were seen while close to the line. CH_4 pressure was then released to 13.3 bar, and the temperature was set to 160 K. Meanwhile, *Sam* (char) showed high CH_4 content, while low for *Ref* once p,T was well inside the gas-phase region. *SC* was close to 5. As the system later passed the saturation line back again liquid, condensation started at the surface of the wall of the *Neck* (see Fig. 1c), and it was seen only when on the transition line or in the liquid phase.

Good agreement between 2nd-order polynomial (2op)- and RP (NWB)-retrieved n_{CH4} were obtained also when the system once again had passed back into the gaseous region at 150 K. Concentrations obtained using LBL (cross section specified in Table 1) could be applied outside the applicable range of the 2nd-order solver (2op) for $AtI >$ 2.6. However, they are estimated to yield lower n_{CH4} values than expected from an exact solver: for Frame 2748 (54.57 bar, 200 K) the concentration was 84% of that of the 2nd-order fit. From AtI_{Sam} one obtains the absorption capacity (AC) ranging from ca 9 to 28 mmol CH_4 /g pine char ($m = 0.153$ g). SC in the case of LBL is also expected to yield a lower value than expected from a perfect solver.

Measurements were also performed with C_2H_6 (8–36 bar, 235–291 K) and C_3H_8 (1.2–8.8 bar, 270–296 K). We used the NWB tool vapor density from measured p and T from NIST saturation data. We obtained an agreement between the n_{C2H6} from the NWB calculations and those from the 2op C_2H_6 solver (using Eq. 5 with C_2H_6 data from Table 1). A first evaluation shows an agreement with a deviation of 3–11% between 2op and NWB for p,T data points that closely followed the liquid–gas transition line, on the liquid side. The images present vapor density in the *Ref* section and local condensation below the *Sam* section. This demonstrated the simultaneous presence of vapor and gas in the cell system that would be vapor saturated in C_2H_6.

In the case of C_3H_8, results between NIST-based number concentrations and those obtained from attenuation measurements agreed reasonably. Concentrations from AtI_{Ref} were obtained using the cross section by linear combination of cross sections according to the molecular formula for C_3H_8 analogous to Eq. 4. Tests included choice of σ_H corresponding to that of ethane, 78 and 82.50 b [2].

Preliminary results based on LBL evaluation show that for the char, the AC were around 9–28, 9–15, and 1–5 mmol/g for CH_4, C_2H_6, and C_3H_8, respectively. C_3H_8 showed a visible presence in the char in the final part of the measurements, during evacuation and vacuum at 293 K that corresponded to *ca* 1 mmol/g char. To acquire a higher accuracy in the retrieval of number concentrations, it would be interesting to evaluate the possibilities of using shorter wavelengths and monochromatic neutron radiation. Measurements could then be performed at pulsed accelerator sources to study attenuation and evaluate number concentrations for different wavelengths.

2.2 Summary/Conclusion

In this paper we have focused on characterizing the transition of methane and its interaction with activated pinewood char at pressures between 55 and 8 bar and temperatures between 200 and 145 K in a specially built measurement cell. We have been able to measure the microscopic attenuation cross sections for methane and ethane and demonstrated a method to retrieve number concentrations that were compared with NIST reference data, with considerably good results.

First tests were also made with ethane and propane. First data analysis included *quantitative* measurements on *number concentrations* that were compared with NIST reference data with encouraging results. The absorption capacity of activated pinewood char was given a first evaluation for methane, ethane, and propane. In the future, access to new bright neutron sources with adjustable and higher energies will increase the possibility to tune the attenuation cross section of H to achieve less saturation and facilitate and increase the accuracy of concentration retrieval by, e.g., LBL. This will be helpful for dynamics studies in the more complex gas–liquid sorption systems in the future, e.g., in heat pump and storage applications [13].

Acknowledgements. Frederik Ossler (FO) acknowledges the support by Oak Ridge National Laboratory and the U.S. Department of Energy (DOE) for access to experiments at HFIR. FO acknowledges the support by the Physics Department, Lund University, Generic Research for Optimized Energy Conversion Processes (GRECOP), project number 38913–2, supported by the Swedish Energy Agency, and funding from Horizon Europe – the Framework Programme for Research and Innovation (2021/2027) under grant agreement No 101115506. This research used resources at the High Flux Isotope Reactor and Spallation Neutron Source, DOE Office of Science User Facilities operated by the Oak Ridge National Laboratory. The beam time was allocated to CG-1D IMAGING on proposal numbers IPTS-15577.1 and IPTS-20748.1. FO thanks Indu Dhiman and Erik E. Stringfellow of ORNL for technical assistance with the experiments.

References

1. Memetova. A., et al.: Porous carbon-based material as a sustainable alternative for the storage of natural gas (methane) and biogas (biomethane): a review. Chem. Eng. J. **446**, 137373 (2022)
2. Ossler, F., Santodonato, L., Bilheux, H.: In-situ neutron imaging of hydrogenous fuels in combustion generated porous carbons under dynamic and steady state pressure conditions. Carbon **116**, 766–776 (2017)

3. Ossler, F., et al.: Neutron imaging and scattering based characterization of the interaction between hydrogenous substances and soot materials. In: Oral Presentation, Carbon 2018, Madrid, 1–6 July 2018 (2018)
4. Ossler, F., et al.: Characterization of soot and char adsorption of volatile organic compounds. In: Oral Presentation, Carbon 2022, London, 3–8 July 2022 (2022)
5. Santodonato, L., et al.: The CG-1D neutron imaging beamline at the Oak Ridge National Laboratory High Flux Isotope Reactor. Phys. Procedia **69**, 104–108 (2015)
6. https://www.arscryo.com/de-204. Accessed 15 Sept 2024
7. Schneider, C., Rasband, W., Eliceiri, K.: NIH Image to ImageJ: 25 years of image analysis. Nat. Methods **9**, 671–675 (2012)
8. Contescu, C.I., et al.: Activated carbons derived from high-temperature pyrolysis of lignocellulosic biomass. C **4**, 51 (2018)
9. Bell, I.: NIST Reference Fluid Thermodynamic and Transport Properties Database (REF-PROP) Version 9 - SRD 23, National Institute of Standards and Technology (2013). https://doi.org/10.18434/T4JS3C. Accessed 15 Sept 2024
10. Fritz, D., et al.: Total thermal neutron cross section measurements of hydrogen dense polymers from 0.0005–20eV. Ann. Nucl. Eng. **183**, 109651 (2023)
11. Rauch, H., Zeilinger, A.: Hydrogen transport studies using neutron radiography. Atomic En. Rev. **15**, 249–290 (1977)
12. Lemmon, E.W., Bell, I.H., Huber, M.L., McLinde, M.O.: Thermophysical properties of fluid systems. In: Lindström, P.J., Mallard, W.G. (eds.) Chemistry WebBook, NIST Standard Reference Data Base 69. National Institute of Standards and Technology (USA) 20889 (2010). https://doi.org/10.18434/T4D303
13. Palomba, V., Nowak, S., Dawoud, B., Frazzica, A.: Dynamic modelling of adsorption systems: a comprehensive calibrated dataset for heat pump and storage applications. J. Energy Storage **33**, 102148 (2021)

Commercial Neutron Imaging with an Accelerator-Based Neutron Source

Martin Wissink$^{(\boxtimes)}$ (iD)

Phoenix LLC, Fitchburg, WI 53711, USA
martinwissink@phoenixwi.com

Abstract. Neutron imaging is a powerful and complementary tool to other non-destructive testing (NDT) methods due to the high penetration of neutrons through dense, high atomic number materials and high contrast to certain low atomic number elements such as hydrogen. The primary NDT applications for neutron radiography are the detection of residual core material in single-crystal turbine blades and the inspection of mission-critical aerospace pyrotechnic devices. Historically, production-scale commercial neutron radiography has only been available at a handful of nuclear research reactors, which face challenges with limited availability, low scalability, high regulatory burden, and service interruptions due to both planned maintenance and unplanned outages. To address these challenges, Phoenix has designed and deployed an accelerator-based neutron source built from the ground up for the express purpose of neutron radiography. The Phoenix Neutron Imaging Center, operating from our facility in Fitchburg, Wisconsin, has been in commercial operation since 2020 and has served dozens of customers across aerospace, defense, and academia, imaging several hundred thousand components per year. This article will provide an overview of the facility, examples of typical NDT applications, and case studies for industrial neutron computed tomography and time-resolved neutron imaging. The development of this highly parallel and inherently scalable approach to neutron imaging has demonstrated that a privately funded neutron source is commercially viable, and that we can democratize access to neutrons and potentially significantly increase the user base by bringing neutron production to the open market.

Keywords: neutron imaging · nondestructive testing · neutron radiography

1 Introduction

Nondestructive testing (NDT) is a wide array of testing and analysis techniques that allow for examination of an object or material without impacting its future usefulness, and it is an essential part of many industries including aerospace, energy, infrastructure, manufacturing, and transportation. The practice of NDT is classified into several methods, which include well-known and heavily used techniques such as electromagnetic testing, magnetic particle testing, liquid penetrant testing, (x-ray) radiographic testing, and ultrasonic testing [1]. Neutron radiography (imaging) is a niche but nonetheless vital method in the aerospace and defense industries. The primary value of neutron imaging

© The Author(s) 2026
A. E. Craft and H. Z. Bilheux (Eds.): WCNR 2024, SPPHY 348, pp. 317–325, 2026.
https://doi.org/10.1007/978-3-032-15003-5_36

for NDT stems from its unique combination of high penetration through common aluminum, ferrous, and nickel-based alloys, relatively high contrast to certain light elements such as H, Li, B, and Cl, and incredibly high contrast to Gd, as illustrated in Fig. 1. Neutrons also have high penetration through dense refractory metals such as Ta and W and through x-ray shielding materials such as Pb and Bi, making them very complementary to the more common x-ray method.

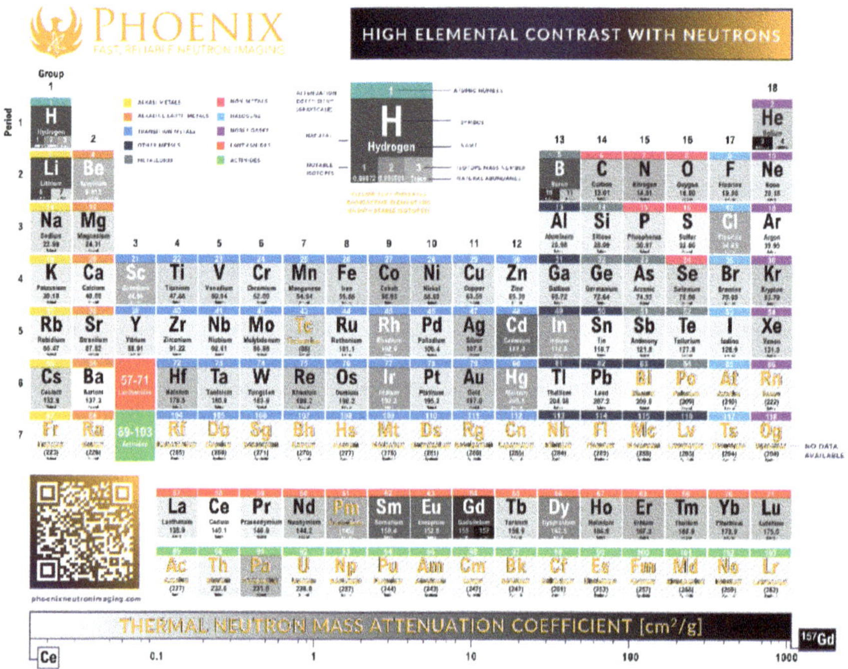

Fig. 1. Periodic table shaded in grayscale by logarithmically scaled thermal neutron mass attenuation coefficient illustrates elemental contrast spanning 5 orders of magnitude.

Historically, neutron imaging for NDT purposes has been performed at research reactors, the majority of which were built during the 1960s and 1970s. The global fleet peaked in 1975 and has been steadily decreasing since. Of the 843 research reactors recorded by the International Atomic Energy Agency (IAEA), there are 227 listed as operational, 22 under construction/planned, 75 temporary/extended/permanent shutdown, and 519 under decommissioning/decommissioned [2]. Of those operational, over 70% are more than 40 years old, and of those under construction, there is only 1 in the western hemisphere (RA-10, Argentina). Among the operational reactors, only a fraction have neutron imaging facilities; the IAEA lists 45 as of 2022 [3]. An even smaller fraction have the capability to perform neutron imaging for NDT at an industrially relevant scale, as the majority operate as academic research centers. As of this writing, there are only 3 in North America: the McMaster Nuclear Reactor (McMaster University, Ontario, CA, operated in partnership with Nray Services Inc.), the PULSTAR Reactor (North Carolina State University, North Carolina, USA, operated in partnership with Nray Services Inc.), and the McClellan Nuclear Research Center (University of California, Davis, California, USA).

This shortage of usable facilities presents a significant supply chain vulnerability. An unplanned outage or shutdown at just one reactor can irreparably impact an entire program, such as by causing a rocket to miss its launch window due to the inability to inspect mission critical components. To address this need, Phoenix has developed and built a privately financed, accelerator-based neutron source with a sole focus on performing neutron radiography commercially and at production scale.

The Phoenix Neutron Imaging Center, shown in Fig. 2, is a high neutron yield cyclotron-based Be(p,n) system featuring advanced target, moderator, and collimator designs to provide reliable and high flux beams. The system uses no tritium, uranium, or other radioactive fuel, greatly simplifying regulatory compliance and enabling higher energetics possession limits inside the building than is possible at nuclear reactors. The facility features two neutron sources: the left bunker shown in Fig. 2 is an unmoderated fast neutron source with a single collimator, which offers high penetration through thick components, while the right bunker contains a D_2O moderated thermal neutron source, which is used for production NDT radiography. The 10 thermal beam ports situated radially around the moderator operate in parallel, allowing for high throughput radiography. Each of the thermal beam ports offers a neutron flux of 3.8×10^5 n cm^{-2} s^{-1} at an L/D ratio of 70. Higher L/D ratios are available using boron aperture inserts, with a corresponding decrease in neutron flux.

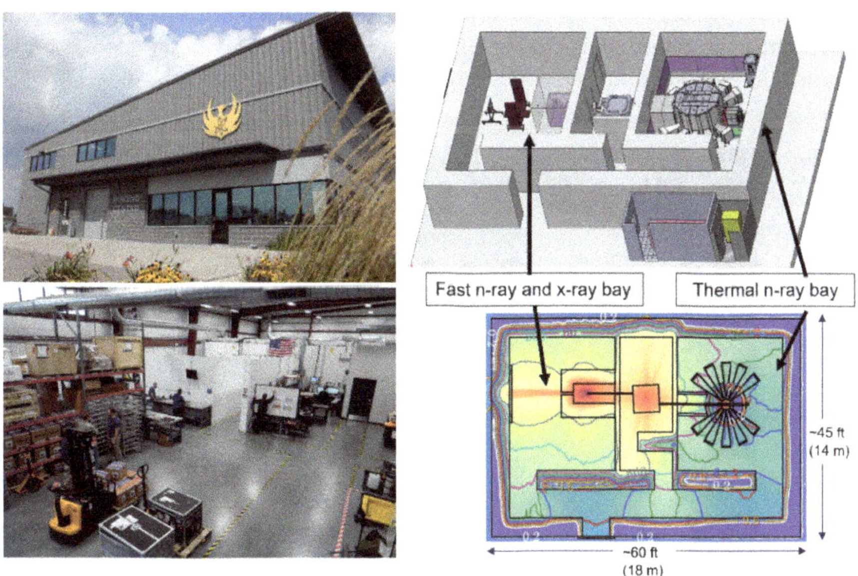

Fig. 2. Phoenix Neutron Imaging Center. Left: Photos of the facility. Right: Geometric model and radiation model of the fast and thermal neutron sources.

The imaging center operates out of a 10,000 ft^2 (929 m^2) facility located in Fitchburg, WI and maintains active compliance/permits with:

320   M. Wissink

- Bureau of Alcohol, Tobacco, Firearms and Explosives (ATF)
- Department of Defense (DoD) Ammunition & Explosives Safety and Security Manuals (4145.26 & 5100.76)
- International Traffic in Arms Regulations (ITAR)
- Aerospace & defense quality programs for process and personnel (ISO9001, AS9100, NAS410, etc.)

Since opening the facility in April 2020, Phoenix has served dozens of customers across aerospace, defense, and academia, imaging several hundred thousand components per year.

2 Current NDT Applications for Neutron Imaging

The two primary applications of neutron imaging in NDT are inspection of jet engine turbine blades to detect residual core material in cooling passages, and inspection of mission-critical aerospace pyrotechnic devices (broadly termed as "energetics").

The type of turbine blades inspected with neutron imaging are generally those used in the highest-temperature stages of a jet turbine, where the gas temperature can exceed the working range of the Ni-superalloy from which the blades are made. These blades are often fabricated as a columnar or single crystal structure on ceramic molds to create a very complex internal cooling channel geometry. After the alloy solidifies, the ceramic core material is removed using a caustic solution in a chemical leaching process [4]. Inspection for residual core material is critical to ensure that there is no blockage of the cooling channels, but as shown in Fig. 3, an x-ray image taken with sufficient energy to penetrate the Ni alloy offers no contrast to the ceramic core.

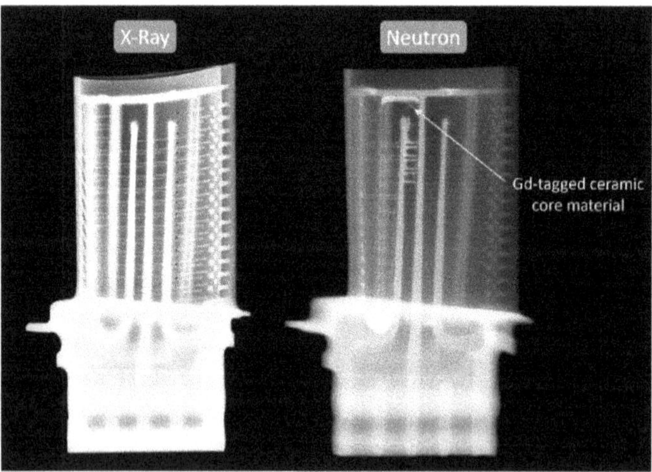

Fig. 3. Gadolinium-tagged turbine blade shows highlighted ceramic core material in neutron image that is invisible to x-ray.

With neutron imaging, the extremely high thermal neutron capture cross section of Gd is leveraged with a process known as gadolinium tagging. The turbine blades are immersed in a solution containing gadolinium(III) nitrate hexahydrate, some of which will naturally permeate into the porous ceramic core. The turbine blades are then thoroughly rinsed and dried to remove residual Gd and water from the metal surfaces. The resulting thermal neutron image, as shown in Fig. 3, offers good penetration through the Ni alloy while also highlighting the residual core material.

Energetic devices are a very diverse category as they span many applications in aerospace and defense ranging from ejection seat mechanisms to explosive bolts used to separate rocket booster stages. The types of inspection are therefore numerous and may include such items as presence of explosive charges or hydrogenous o-rings, rolled up or damaged o-rings, cracks, gaps, or discontinuities in explosive charges and trains, presence of foreign objects or debris in explosive delay columns, presence and proper installation of polyimide and ceramic insulators, proper internal orientation of safe arm devices, and many more.

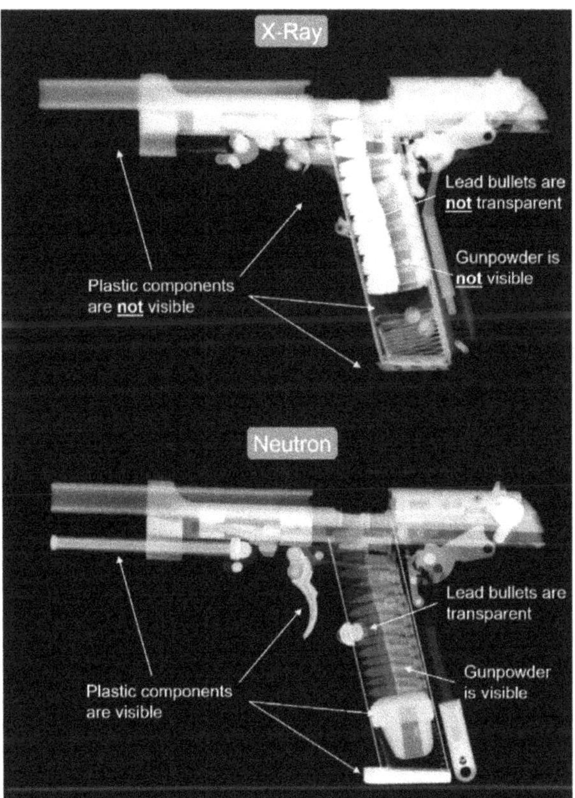

Fig. 4. Example of neutron and x-ray material contrast in an energetic assembly.

Due to ITAR restrictions, we are unable to show images of real energetic devices, but a comparative example of x-ray and neutron material contrast is provided in Fig. 4 in the form of a small caliber firearm.

Production neutron radiography of both turbine blades and energetic devices is performed on film in accordance with ASTM standards E545 [5] and E748 [6], and regularly achieves ASTM Category I image quality. Given the shift to digital imaging in both medical and industrial x-ray [7], a similar transition for neutron radiography seems inevitable, and Phoenix has been actively developing digital neutron imaging systems tailored to our neutron source in order to embrace the possibilities offered by this technology.

3 Neutron Computed Tomography for NDT

Neutron computed tomography (nCT) extends the complementary nature of neutron imaging to other NDT techniques into three dimensions, offering the ability to visualize and quantify light elements inside metallic containers even in the case of overlapping indications that would prove impossible to evaluate in a single 2D image. A case study of an aluminum pneumatic actuator is offered here. As shown in Fig. 5, the actuator was attached to a rotation stage with aluminum tape. A total of 181 projections were acquired using a golden angle scan method [8], with an exposure time of 60 s per projection. Iterative reconstruction using the SIRT method was performed using the NeuTomPy toolbox [9].

Both the neutron image and the CT slice shown in Fig. 5 clearly illustrate that the hydrogenous components such as the o-rings and lubricating grease are visible with high contrast through the aluminum body of the actuator, which is quite transparent. What is not apparent, however, in either the movie of the object rotating or in scrolling through the CT slices, is that this device has been intentionally damaged and rendered inoperable. As shown in Fig. 6, one of the o-rings was nicked using diagonal cutting pliers before the CT scan. This is faintly visible as an indentation in the o-ring when viewed as a volumetric rendering but becomes immediately obvious when the CT reconstruction is virtually unrolled and a line profile is taken along the outer radius of the o-ring. This type of inspection is virtually impossible in 2D imaging due to the presence of attenuating overlapping features but can be reduced to routine with the proper analysis methodology in 3D imaging.

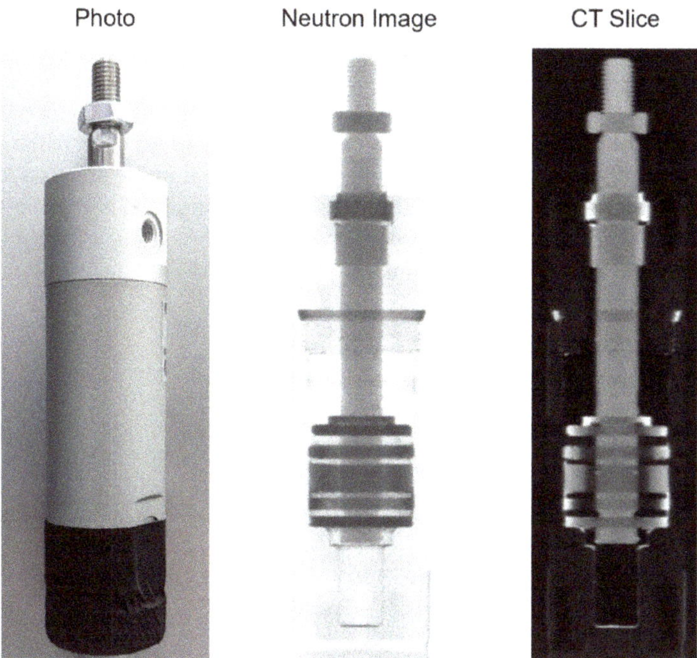

Fig. 5. Neutron CT scan of aluminum pneumatic actuator. Left: Photo of actuator taped to CT stage. Center: A single neutron image from the scan. Right: Frontal slice of the CT reconstruction.

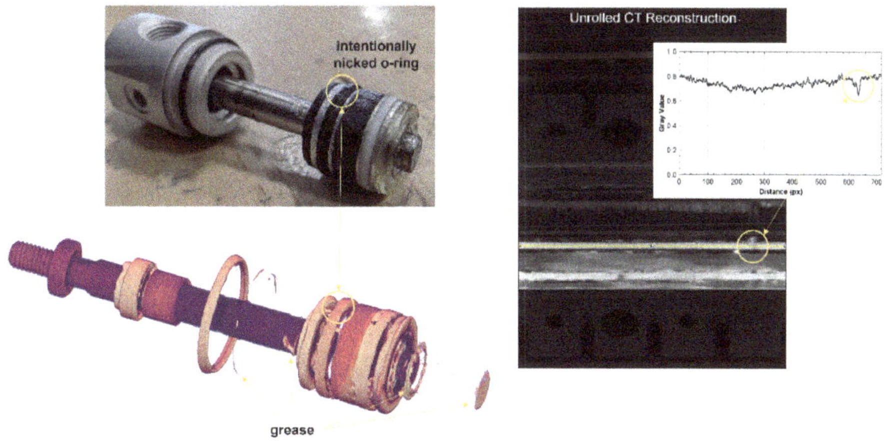

Fig. 6. Analysis of actuator nCT reveals damaged o-ring that is invisible in 2D imaging.

4 Time-Resolved Neutron Imaging

Many research applications of neutron imaging are focused on dynamic or time-resolved systems. With the ability to produce useful neutron images on timescale on the order of 60 s, the Phoenix Neutron Imaging Center may be suitable for a variety of academic and

R&D users studying slower dynamic processes. The example provided here is a time sequence of a "neutron cocktail", pictured in Fig. 7, in which H_2O ice is melting into D_2O liquid. Images were acquired at 60 s exposure time and are presented in intervals of 15 min.

Multiphase dynamics are clearly visible in this system, as we observe the ice slowly melting into the liquid below. The vertical contrast gradient inside the glass shows that the lighter and more strongly attenuating H_2O remains stratified above the D_2O even after melting. Condensation of atmospheric water onto the outside of the glass is also visible and can be seen slowly trickling down the glass and pooling at its base.

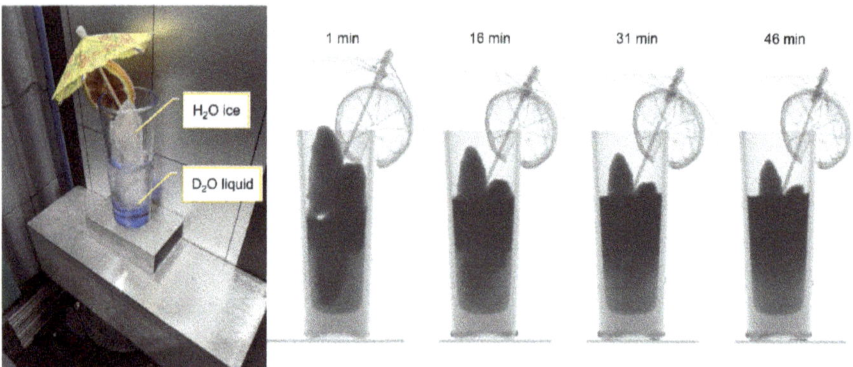

Fig. 7. Time-resolved neutron imaging of H_2O ice melting into D_2O liquid.

5 Conclusions

Phoenix has brought to market a commercial accelerator-based neutron radiography service with steadily growing operations since April 2020. This inherently scalable approach has primarily targeted aerospace and defense NDT applications but is also suitable for a variety of R&D and academic users. Development of neutron CT and time-resolved neutron imaging capabilities is poised to bring these underutilized and oversubscribed research tools to the public with short turnaround times and aerospace quality management.

References

1. Mix, P.E.: Introduction to Nondestructive Testing: A Training Guide, 2nd edn. John Wiley & Sons, New Jersey (2005)
2. IAEA Research Reactor Database. https://nucleus.iaea.org/rrdb/#/home. Accessed 30 Sept 2024
3. Neutron Imaging Facilities Survey: Jointly Prepared and Conducted by the ISNR and IAEA, https://nucleus.iaea.org/sites/neutrons/SitePages/NIDataBase.aspx?web=1. Accessed 30 Sept 2024
4. Zhu, W.J., Tian, G.Q., Lu, Y., Miao, K., Li, D.C.: Leaching improvement of ceramic cores for hollow turbine blades based on additive manufacturing. Adv. Manuf. 7, 353–363 (2019)

5. ASTM E545-19, Standard Test Method for Determining Image Quality in Direct Thermal Neutron Radiographic Examination. https://www.astm.org/e0545-19.html. Accessed 30 Sept 2024
6. ASTM E748-19, Standard Guide for Thermal Neutron Radiography of Materials. https://www.astm.org/e0748-19.html. Accessed 30 Sept 2024
7. Ou, X., et al.: Recent development in x-ray imaging technology: future and challenges. Research (2021)
8. Kohler, T.: A projection access scheme for iterative reconstruction based on the golden section. In: IEEE Symposium Conference Record Nuclear Science 2004, vol. 6, pp. 3961–3965. IEEE (2004)
9. Micieli, D., Minniti, T., Gorini, G., NeuTomPy toolbox, a Python package for tomographic data processing and reconstruction. SoftwareX **9**, 260–264 (2019). https://doi.org/10.1016/j.softx.2019.01.005

iBeatles: Bragg Edge Imaging Software for the VENUS Beamline at the Spallation Neutron Source

Jean C. Bilheux$^{(\boxtimes)}$ and Hassina Z. Bilheux

Neutron Sciences Directorate, Neutron Scattering Division, Oak Ridge National Laboratory, Oak Ridge, TN 37831, USA
bilheuxjm@ornl.gov

Abstract. The VENUS construction project ended in August 2024 and was followed by the commissioning of the beamline. The beamline is located at the Spallation Neutron Source of the Oak Ridge National Laboratory. VENUS uniquely utilizes the intrinsic properties of the neutron source to measure wavelength-dependent imaging data. Since neutron wavelength-dependent imaging is a relatively recent technique, software development is required to reduce and analyze data. Thus, we implemented the application iBeatles, a python-based software capable of fitting Bragg edges and producing strain maps in 2 dimensions. This manuscript addresses new and optimized features suggested by the scientific community during the 11th NEUtron-WAVElength dependent workshop. The manuscript also presents the new user interface.

Keywords: iBeatles · Bragg edge radiography · strain mapping

1 Introduction

Neutron radiography [1] is an imaging technique that is broadly used to non-destructively study structural properties in objects relevant to materials science, energy research, engineering, etc. Most of the research is performed at continuous neutron sources such as research reactors [2–5]. Over the past 20 years and thanks to the advances of neutron sources and detector technology [6, 7], hyperspectral or wavelength-dependent neutron radiography [8–11] has emerged as a unique tool to study crystalline properties in materials using the so-called Bragg-edge radiography technique [12, 13]. One of the advantages of using this technique is the ability to map changes in lattice spacing and hence calculate the microstrain present in the sample [14–18]. Bragg-edge radiography provides a high-resolution two-dimensional (2D) map of microstrain values averaged through the sample thickness in the beam. Efforts are ongoing to develop the mathematics to enable three-dimensional (3D) maps of the microstrain or strain tensor [14].

Obtaining microstrain information from hyperspectral radiographs requires the ability to fit the Bragg edges on each pixel of the image. Since detectors have hundreds of thousands of pixels, this analysis requires automation via software development to fit

A. E. Craft and H. Z. Bilheux (Eds.): WCNR 2024, SPPHY 348, pp. 326–336, 2026.
https://doi.org/10.1007/978-3-032-15003-5_37

Bragg edges and to back project the resulting change in lattice parameter onto the object radiograph. The lattice values are compared to a baseline lattice parameter experimentally measured with the equivalent unstrained sample. The Imaging Bragg Edge Analysis TooLs for Engineering Structures(iBeatles), follows the process described above and provides 2D strain maps via an intuitive general user interface (GUI). The software tool was developed at the Oak Ridge National Laboratory (ORNL) for the Spallation Neutron Source (SNS) VENUS hyperspectral imaging beamline which has started commissioning [19].

2 Application

The iBeatles application has been implemented using python language and is using some community librairies such as pyqtgraph (pyqtgraph.org), PyQt5 (riverbankcomputing.com/static/Docs/PyQt5), numpy (numpy.org), matplotlib (matplotlib.org), scikit-image (scikit-image.org), pandas (pandas.pydata.org), and two in-house librairies NeuNorm (github.com/ornlneutronimaging/NeuNorm) and neutron-braggedge (github.com/ornlneutronimaging/BraggEdge). The application works on Linux, Mac and Windows operating systems. The current version of the software works with the data produced by the micro-channel plate (MCP) Timepix (TPX) [20, 21], after being corrected for detector efficiency [https://github.com/ornlneutronimaging/Neutro nImagingScripts]. The program requires a stack of TIFF images as well as a time spectra file which contains the timing information of each radiograph collected in time-of-flight (TOF) mode. The MCP TPX detector produces 16-bit FITS files that are converted into 16-bit TIFF files during the correction process, which is automated at VENUS. At the beamline, each data set, stack of images corresponding to the same acquisition, is associated with a run number and moved from the local beamline computer to the SNS analysis servers.

2.1 Pre-processing

Before being loaded, data sets require some pre-processing to improve the statistics or correct the sample orientation. The application offers 3 pre-processing algorithms: sample orientation, data sets (or runs) binning, and TOF binning.

The rotation algorithm rotates a stack of radiographs (collected during the same run) to improve the coverage of the sample region with the rectangular region of interest (ROI) used by the application. This is for example the case when a long rectangular sample is measured with a 45degrees angle due to some sample environment constraints.

The runs binning algorithm combines several runs (mean or median algorithms are available) to increase the signal-to-noise ratio (SNR). The program automatically checks the compatibility of the runs by making sure the same number of images are in each TOF data set. The third pre-processing tool bins the radiographs of the same data set using linear or logarithmic bin sizes, as displayed in Fig. 1 which shows the ability to bin data based on TOF. Various binning axes (file index, TOF or lambda) are available. The manual binning mode also allows the full customization of the bins size and positions across a selected axis.

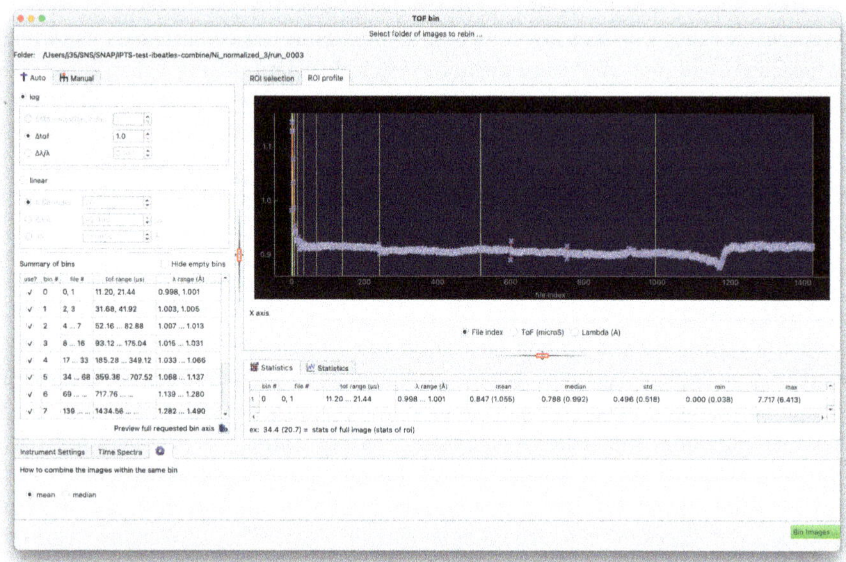

Fig. 1. Automated binning mode using TOF information where images can be binned using linear or logarithmic bins on lambda or TOF axis. The figure displays the data using the file index. Vertical yellow lines indicate the binning selection (i.e., all images comprised between two lines are combined using the average or the median of those images).

2.2 Data Import and Visualization

Two sets of data are required for normalization: the raw sample data and the open beam data or OB (i.e. the TOF radiographs without the sample in the beam). Dark current is not required when using the MCP TPX detector since the electronics noise does not significantly impact the data. The sample and OB runs are loaded in the Load Raw Data tab of the application. On this tab, one can select a set of radiographs to visualize their total counts and select a region of interest (ROI) to visualize the total counts as a function of file index, TOF or wavelength (λ). By moving the ROI in and out of the sample, one can visualize "live" the sample Bragg edges as compared to the background Bragg edges due to the presence of beamline windows throughout the beam path. A given material, from a pre-defined list of powders (unstrained material) can be selected to compare the location of the unstrained Bragg edges with the experimental measurements, as illustrated in Fig. 2. A material can also be manually defined by the user. Moreover, the instrument settings such as source-to-detector distance, detector time offset relative to the time when neutrons leave the moderator, allow the calibration of the instrument when using with a powder sample of known lattice spacing. The time spectrum axis can also be previewed using the time spectra file created with the set of radiographs. This file provides a time bin for each radiograph so that file index, TOF and λ can be used as horizontal axes.

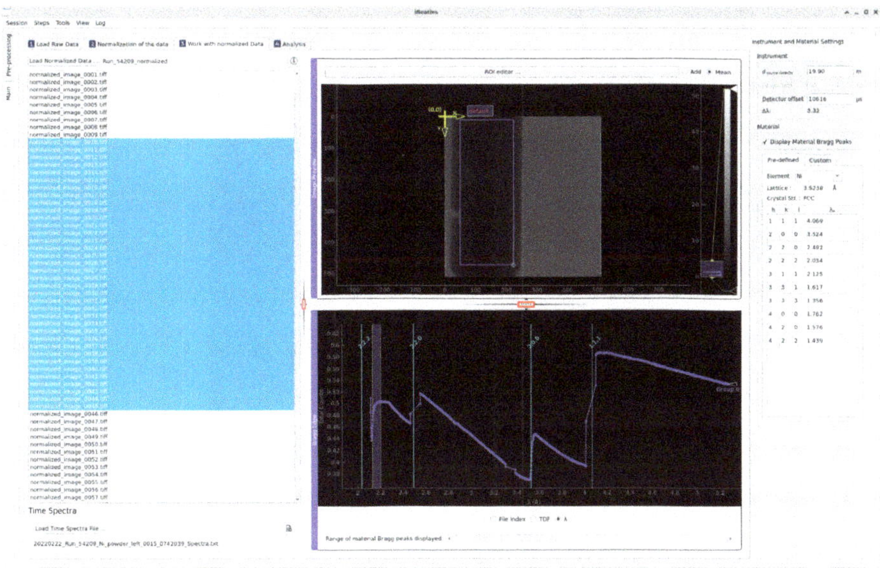

Fig. 2. Load Raw Data tab showing the list of radiographs loaded in iBeatles (left), a preview of the images selected (top center), the intensity as a function of wavelength of the selected ROI (bottom center), and the material different lattice Miller indices and corresponding lattice parameter (right).

2.3 Normalization

Normalization is performed by taking the ratio of the sample and the OB, while using a correction factor that takes the ratio of the same ROI outside the sample and in the OB. This accounts for slight variations in the incident neutron beam. Several settings and workflows are offered to normalize data such as the ability to select moving average with 2D or 3D modes, different cluster sizes, types, etc. Figure 3 shows the normalization tab with the ability to select ROIs for sample and background. Once normalized, the data are automatically saved to a folder selected by the user and then reloaded in the next tab called "Work with normalized Data". This tab is similar to the "Load Raw Data" tab and hence enables Bragg edges experimental plots comparison with predicted Bragg edges.

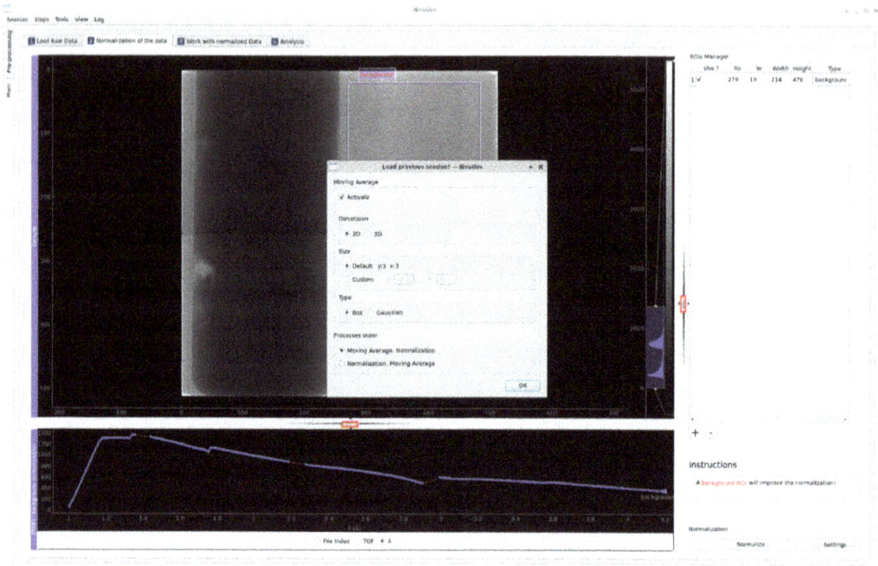

Fig. 3. Normalization tab showing the wavelength-integrated sample radiograph on the left with background and sample regions (also displayed on the right column). The pop-up window shows the normalization settings, and the different workflow options.

2.4 Binning

Fitting single pixels requires a high SNR, and thus several hours of acquisition times for highly scattering samples such as Ni superalloys. Thus, to perform measurements in a few minutes, pixels are often binned to improve the SNR ratio. This is performed in the binning GUI where one selects the total region to use to fit Bragg edges followed by the size of the bins that will divide the ROI, as illustrated in Fig. 4.

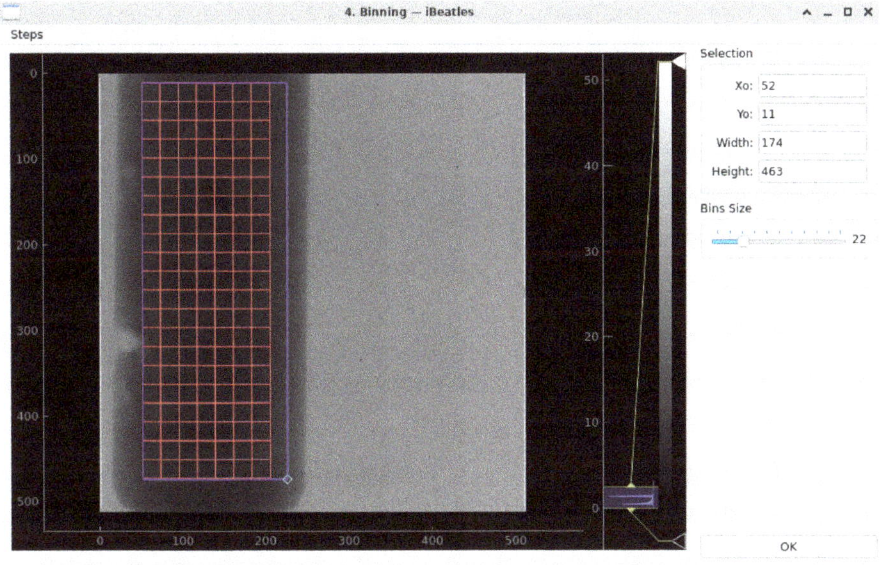

Fig. 4. Binning interface showing the ROI to fit (blue box) as well as the bins (red squares). Size and location of the bins can be modified using the input widgets on the right size.

2.5   Bragg Edge Fitting

iBeatles is designed to fit a single Bragg edge across every single bin of the selected region of interest. This is where the 'heart' of the program takes place (see Fig. 5). Once the edge is selected, the program uses the Kropff method [22] to fit each pixel. Various algorithms use the wavelength range around a Bragg edge to extract its position. These algorithms are used to split the wavelength range in 3 sets of data as required by the Kropff fitting method (github.com/neutronimaging/python_notebooks/blob/next/notebooks/__code/wave_front_dynamics/algorithms.py).

The right part (corresponding to the longer wavelengths) of the edge is fitted first to calculate the parameters (A_o and B_o which represent the transmission of that section). Then these parameters are used in the fitting algorithm of the left side of the edge to determine the value of the new parameters (A_{hkl} and B_{hkl} describe the jump in the transmission value between the low and high wavelength ranges). These latest parameters are finally used to fit the entire Bragg edge region. The fitting produces parameters such as the edge location, λ, the decay constant, τ, and the Gaussian broadening, σ. The program fits each single bin profile using various initial parameters (see Fig. 5) until it reaches a user-defined acceptable accuracy value of λ, τ, and σ. Here, a bin can mean a single pixel or a group of pixels, depending on SNR. Once the accuracy has been reached, in other words when the fitting uncertainties are below a defined value, the pixel is locked, the fitting parameters are saved, and the program moves on to the fitting of the next bin. The whole process can take up to a few minutes depending on the size of the bins and/or the overall ROI to fit.

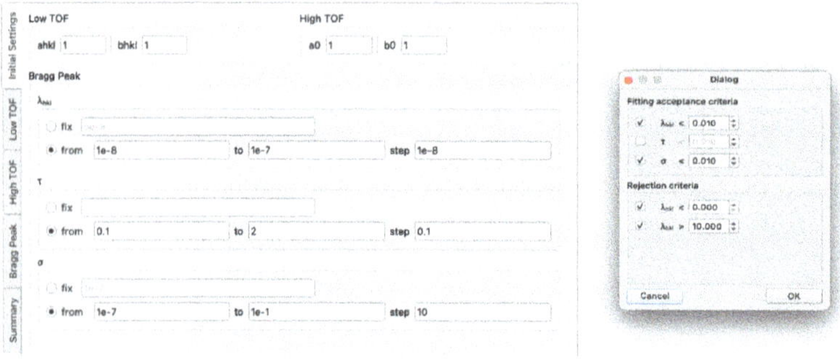

Fig. 5. Left: initial fitting parameters tab and range condition; the initial parameters can be fixed at a given value or allowed to change within a predefined range; Right: fitting convergence parameter conditions.

The status bar displays the percentage of bins that have been fitted with parameters values within the user-defined fitting wavelength range. Figure 6 displays the binned ROI pixels (top left), the selected Bragg edge (top right), the fitting parameters (bottom left), and subsequent fitted Bragg edge (bottom right).

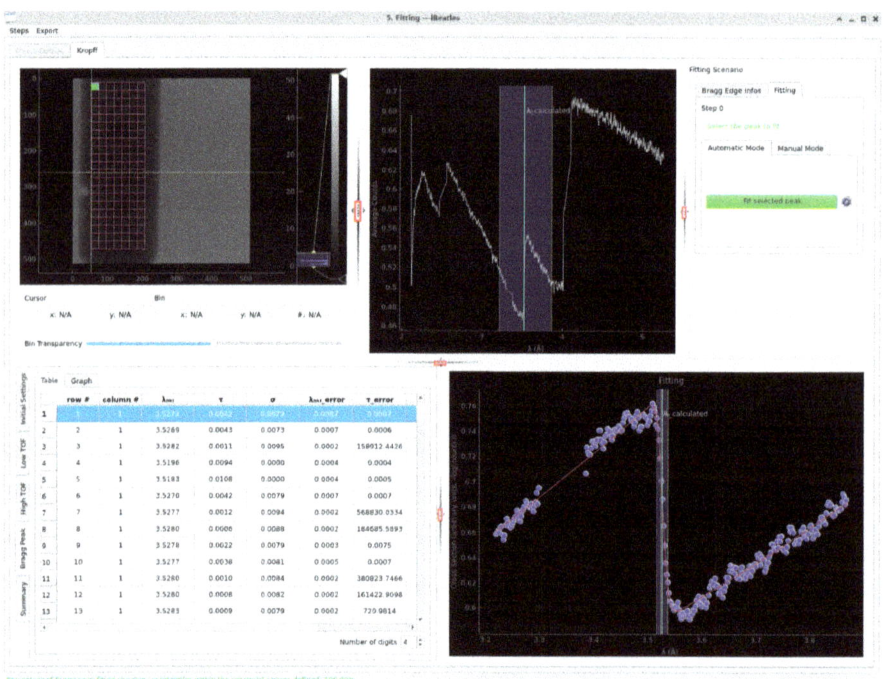

Fig. 6. Fitting tab where the Bragg edge of interest is selected, then where fitting takes place for every single binned pixel. Edge position and output of the fit are displayed (vertical green line and red fit, respectively, in bottom right plot) and can be compared to the unstrained wavelength, λ_0, position (vertical blue line).

2.6 Strain Mapping

Once all the selected binned pixels are fitted and locked (fitting within the user-defined tolerance), it is possible to display on the sample radiograph the fitted wavelength, λ, or the averaged strain, s, through the sample thickness, map as defined by (see Fig. 7):

$$s = \frac{\lambda - \lambda_0}{\lambda_0} \tag{1}$$

By default the program uses λ_0 of the selected material, with the option to modify it on this GUI.

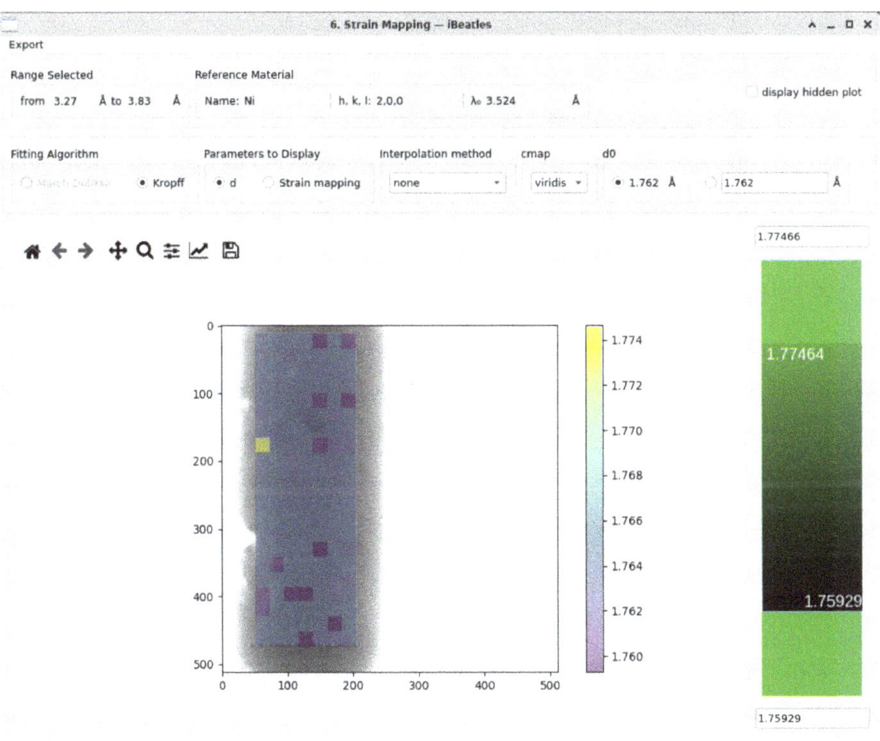

Fig. 7. Strain mapping GUI where the strain map, or lattice, *d*, are displayed over the sample radiograph. Several parameters are available, such as interpolation or colormap, to produce the desired figure.

3 Experimental Demonstration of iBeatles

Using a nickel (Ni) powder sample, we performed Bragg edge fitting with iBeatles and displayed the strain map and as expected, the strain map values are very low as the sample was made to be unstrained. This type of sample is also very useful to calibrate an instrument, since it can be used to determine with high precision the distance between the source and the detector. To do so, one must vary this distance until the fitted wavelength and the unstrained theoretical wavelength match. Figure 8 shows the fitting of the Ni powder (left part on Fig. 8) using iBeatles and the d spacing map obtained from fitting the TOF data.

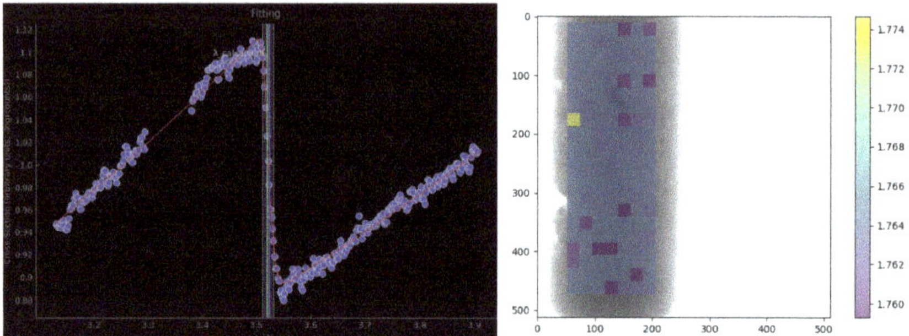

Fig. 8. Left: fitting of the < 2,0,0 > Bragg edge of Ni showing the fitting (red line), the unstrained wavelength value (vertical green light) and the lambda obtained from fitting the data with iBeatles (vertical blue line): Right: calculated lattice d of the sample over the ROI.

4 Conclusion

The implementation of the standalone iBeatles user application allows the pre-processing, normalization, fitting and display of the strain map of the sample. The current version requires a stack of TIFF radiographs and a time spectra file (as produced by the MCP TPX detector) that gives a time stamp for each image. Future work includes the ability to perform multi-edge fitting using other algorithms, the ability to read other data formats such as NeXus files for event or histogram modes and a command line version of this tool.

This research used resources at the Spallation Neutron Source, DOE Office of Science User Facilities operated by the Oak Ridge National Laboratory. The team would like to thank the user community for their many contributions in guiding the development of the software tools as well as being gracious with their time when testing the notebooks.

References

1. Anderson, I.S., McGreevy, R.L., Bilheux, H.Z.: Neutron Imaging and Applications, vol. 200, no. 2209, p. 287. Springer, Heidelberg (2009)
2. Grosse, M., Schillinger, B., Kaestner, A.: In situ neutron radiography investigations of hydrogen related processes in zirconium alloys. Appl. Sci. **11**(13), 5775 (2021)
3. Tengattini, A., et al.: NeXT-Grenoble, the Neutron and X-ray tomograph in Grenoble. Nucl. Instrum. Methods Phys. Res., Sect. A **968**, 163939 (2020)
4. Hussey, D.S., et al.: A new cold neutron imaging instrument at NIST. Phys. Procedia **69**, 48–54 (2015)
5. Bilheux, H.Z., et al.: Neutron radiography and computed tomography of biological systems at the oak ridge national laboratory's high flux isotope reactor. JoVE (J. Visual. Exp.) **171**, e61688 (2021)

6. Losko, A.S., et al., Event-Mode Neutron Imaging Using the TPX3Cam-Breaking the Boundaries of Conventional Neutron Imaging Techniques (2021)

7. Tremsin, A.S., et al.: Optimization of high count rate event counting detector with Microchannel Plates and quad Timepix readout. Nucl. Instrum. Methods Phys. Res., Sect. A **787**, 20–25 (2015)

8. Chen, J., et al.: First neutron Bragg-edge imaging experimental results at CSNS. Chin. Phys. B **30**(9), 096106 (2021)

9. Shinohara, T., et al.: The energy-resolved neutron imaging system, RADEN. Rev. Sci. Instrum. **91**(4), 043302 (2020)

10. Nelson, R., et al.: Neutron imaging at LANSCE—from cold to ultrafast. J/ Imaging **4**(2), 45 (2018)

11. Bilheux, H.Z., et al.: The construction of the VENUS imaging beamline at the Spallation Neutron Source. In: Journal of Physics: Conference Series, vol. 2605, p. 012004 (2023). https://doi.org/10.1088/1742-6596/2605/1/012004

12. Santisteban, J.R., et al.: Strain imaging by Bragg edge neutron transmission. Nucl. Instrum. Methods Phys. Res., Sect. A **481**(1–3), 765–768 (2002)

13. Steuwer, A., et al., Bragg edge determination for accurate lattice parameter and elastic strain measurement. Physica Status Solidi (A) **185**(2), 221–230 (2001)

14. Wensrich, C., et al., General reconstruction of elastic strain fields from their Longitudinal Ray Transform (2024). arXiv preprint arXiv:2408.10250

15. Van Tran, K., et al.: Phase and texture evaluation of transformation-induced plasticity effect by neutron imaging. Mater. Today Commun. **35**, 105826 (2023)

16. Ramadhan, R.S., et al.: Mechanical surface treatment studies by Bragg edge neutron imaging. Acta Mater. **239**, 118259 (2022)

17. Busi, M., et al.: Bragg edge tomography characterization of additively manufactured 316L steel. Phys. Rev. Mater. **6**(5), 053602 (2022)

18. Tremsin, A.S., et al.: Monitoring residual strain relaxation and preferred grain orientation of additively manufactured Inconel 625 by in-situ neutron imaging. Addit. Manuf. **46**, 102130 (2021)

19. Bilheux H.Z., et al.: The VENUS imaging beamline at the Spallation Neutron Source: layout, expected performance and status of the construction project. In: Journal of Physics: Conference Series, 9th International Topical Meeting on Neutron Radiography (2023)

20. Tremsin, A., Vallerga, J., Raffanti, R.: Optimization of spatial resolution and detection efficiency for photon/electron/neutron/ion counting detectors with Microchannel Plates and Quad Timepix readout. J. Instrum. **13**(11), C11005 (2018)

21. Tremsin, A., Vallerga, J.: Unique capabilities and applications of Microchannel Plate (MCP) detectors with Medipix/Timepix readout. Radiat. Meas. **130**, 106228 (2020)

22. Kropff, F., Granada, J., Mayer, R.: The bragg lineshapes in time-of-flight neutron powder spectroscopy. Nucl. Instrum. Methods Phys. Res. **198**(2–3), 515–521 (1982)

Enhancing Reconstruction of Time-of-Flight Neutron Computed Tomography Using Artificial Intelligence

Shimin Tang[1]([⊠]) , Mohammad Samin Nur Chowdhury[2] , Diyu Yang[2] ,
Singanallur V. Venkatakrishnan[3] , Kyle D. Anderson[4] , Ryan D. Ross[4] ,
Gregery T. Buzzard[5] , Charles A. Bouman[2] , Yuxuan Zhang[1] ,
and Hassina Z. Bilheux[1]

[1] Neutron Scattering Division, Oak Ridge National Laboratory, Oak Ridge, TN 37830, USA
`{tangs,bilheuxhn}@ornl.gov`
[2] Electrical and Computer Engineering, Purdue University, West Lafayette, IN 47907, USA
[3] Electrical and Engineering Infrastructure Division, Oak Ridge National Laboratory,
Oak Ridge, TN 37830, USA
[4] Department of Anatomy and Cell Biology, Rush University Medical Center, Chicago,
IL 60612, USA
[5] Department of Mathematics, Purdue University, West Lafayette, IN 47907, USA

Abstract. Neutron time-of-flight imaging can provide a unique contrast mechanism of crystalline properties. Recently, computed tomography scans using time-of-flight instruments have been used to study structural and spectral characteristics of samples in 3D. To address the challenge of long measurement times associated with hyperspectral neutron computed tomography, the Oak Ridge National Laboratory neutron imaging team has recently demonstrated an autonomous system which can significantly reduce the measurement time by enabling high quality reconstructions from a sparse set of measurements. Some of the core components of such systems are the novel tomographic reconstruction algorithms including those based on artificial intelligence methods.

In this work, a new training method is proposed to improve the performance of the artificially intelligent CT reconstruction algorithms. This training method can improve the quality of the reconstruction from very sparse time-of-flight scans. Our method helps hyperspectral tomography systems to obtain high-quality reconstructions with sparse scanning, which can potentially enable hyperspectral neutron computed tomography with reasonable acquisition times and particularly impact research projects which need to scan multiple similar sample beamlines such as the newly constructed VENUS beamline at the Spallation Neutron Source.

Keywords: time-of-flight · neutron computed tomography · artificial intelligence

© The Author(s) 2026
A. E. Craft and H. Z. Bilheux (Eds.): WCNR 2024, SPPHY 348, pp. 337–346, 2026.
https://doi.org/10.1007/978-3-032-15003-5_38

1 Introduction

Neutron hyperspectral or time-of-flight (TOF) imaging provides a unique contrast mechanism in objects by measuring their crystalline properties using Bragg edge radiography, or by detecting their elemental or isotopic content using resonance tomography. The collection of these data sets is lengthy due to the hyperspectral nature of the experiment. Recently, demonstration measurements have been performed resulting in 4 dimensional (4D) hyperspectral reconstructions, i.e. 3D structural and 1D wavelength combined information. However, due to the inherently low signal of each wavelength-dependent projection radiograph, a hyperspectral CT experiment takes longer than traditional a white beam neutron computed tomography (nCT). Therefore, the Oak Ridge National Laboratory (ORNL) neutron imaging team developed an autonomous Hyperspectral Computed Tomography (HyperCT) system to drastically reduce the overall experimental time. HyperCT reduced acquisition time by approximately a factor 3 to 5 (depending on the sample geometry and contrast) using a sample adaptive scanning angle selection method [1] combined with a convolutional neural network (CNN) streaming reconstruction quality evaluation method [2]. Generally, HyperCT can achieve fast TOF nCTs using fewer projections than traditional nCTs that use equally spaced projections and filtered-back projection (FBP) as a reconstruction method. In special cases, such as when multiple similarly shaped objects need to be scanned within a limited neutron beam time, one sample can be scanned using a large portion of the allotted time (hence becoming the ground truth training data) and HyperCT can be used to measure other samples using sparser angular hyperspectral projections.

In some instances, the object structure is very complex, thus the quality of the reconstruction from sparser projections may not provide useful details. More advanced reconstruction algorithms, such as those using deep learning approaches, may be required for sparse angular projections. Micieli *et al.* proposed the Neural Network Filtered Back-Projection (NN-FBP) method to improve reconstruction quality from fast nCTs [3]. Han *et al.* deployed a residual learning method to remove streaking artifacts caused by sparse projection views in CT images [4]. The general adversarial network (GAN) can also be used for 3D tomographic denoising. TomoGAN is an architecture capable of obtaining a high-quality output image from a low-quality input image, and it was applied to synchrotron-based x-ray CTs [5]. These methods all assumed the object was fully scanned with a fixed number of projections. However, models trained using certain samples and a fixed number of projections do not generalize well if the number of projections varies when measuring other samples. Training multiple models using the reconstruction from different numbers of projections individually could be one solution. Venkatakrishnan *et al.* trained several models (Artificial Intelligence AI-CT [6]) using the reconstructions from different numbers of projection for fast FBP and discussed the generalization of a different model [7]. However, training multiple models is a computationally intensive approach. On the other hand, training one model using reconstructions from multiple numbers of projections is a more reasonable solution. The use of deep neural network (DNN) algorithms raises another problem: how to combine data from different reconstructions during training so as to have a model that works well across different experimental scenarios? Gnanasambandam *et al.* discussed this problem for an image denoiser and used multiple noised level data to train the image denoiser for a better performance [8]. In this paper, we proposed a training method to improve the performance of a unique AI model across a range of experimental conditions. The trained model is tested on a real neutron scanning experiment of a set of rat femur. We introduced the concept of a sparsity factor to represent how sparse the projections used for reconstruction are. Then we included multiple sparsity factors into the training dataset to develop a single training model. Furthermore, we discussed the influence of different sparsity factor data distributions for the performance of the AI model. In particular, we demonstrated that our proposed method outperforms the traditional FBP method when collecting sparse projections.

2 Methodology

In this work, we are trying to shorten the neutron beam time required to perform hyperspectral nCT of multiple similar samples at the Spallation Neutron Source (SNS). In practice and due to the limited beam time, only one or two samples can be scanned using a large number of projections (of the order of the Nyquist sampling rate [9]). Hence, to complete the hyperspectral nCT of multiple samples, other samples can only be acquired using sparse scanning methods with the risk of influencing negatively the quality of the final 4D reconstructed objects. Therefore, we demonstrated a machine learning (ML) model to enhance the quality of the reconstructions using sparse projections. This model is trained to generate a high-quality reconstruction from a low-quality reconstruction. Thus, with this model, we can significantly reduce the number of projections needed for other similar samples. In this section, we will introduce details of the ML model workflow, as illustrated in Fig. 1, and its training mechanism.

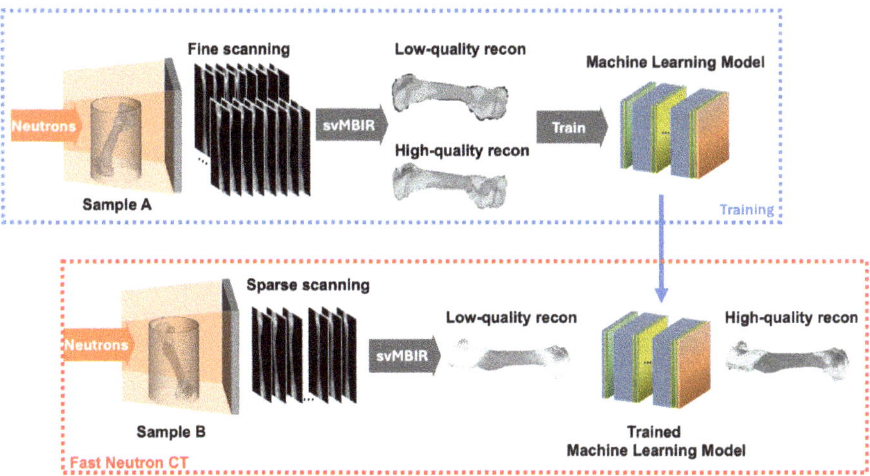

Fig. 1. The flowchart of hyperspectral nCT: the ML model is trained using the data reconstructed from a sufficiently large number of projections based on Nyquist's sampling theorem called "fine scanning" in this figure. Then the trained model is used to improve the quality of the reconstructed data from sparse projections of similar samples.

2.1 Enhancing the Image Quality of Reconstructed Data Using the AI-CT Model

We selected the 2.5D AI-CT [6] ML model to improve the quality of the reconstruction from sparse projection measurements. The structure is designed to learn the non-linear mapping between residual information and the low-quality reconstructed slices input data (from which the 4D volumes are built), where the residual information refers to the difference between high-quality and low-quality reconstructions. In this case, the low-quality reconstruction input data is generated from the sparse projections and has low SNR; we set it as X_{in}. . Moreover, there is a high-quality reconstruction (of the same object), X_{GT}, generated from sufficient projections and has a much higher SNR, which is the ground truth (GT) of X_{in}. The residual information is denoted as v, where $v = X_{GT} - X_{in}$. The non-linear mapping, $\mathcal{M}(X_{in}; \theta)$, is determined by minimizing the loss function, $\mathcal{L}(\theta)$, described as:

$$\mathcal{L}(\theta) = \frac{1}{N_T} \sum_{i=1}^{N_T} \|\mathcal{M}(X_{in}; \theta) - v\|^2 \tag{1}$$

where θ refers to the parameters used in the model and N_T is the total number of training data at each training batch. $\| \bullet \|^2$ here denotes the l^2 norm.

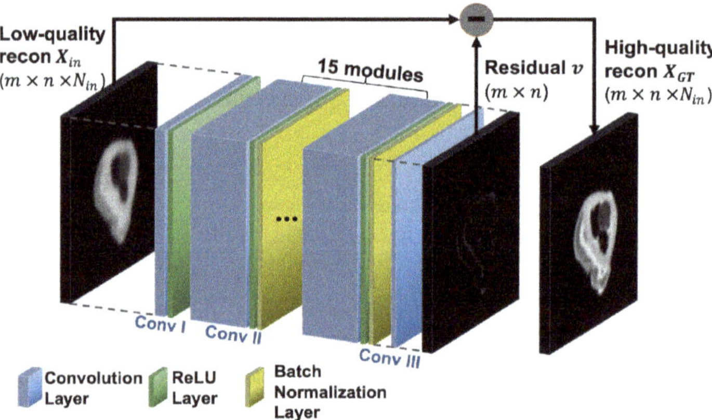

Fig. 2. The structure of the 2.5D AI-CT model which consists of 17 convolution modules defined from 3 types of convolutions: Conv I is the combination of a convolution layer and a rectifier (ReLU); Conv II is the convolution layer followed by a batch normalization and ReLU; and Conv III refers to a single convolution layer. The low-quality reconstruction patch size is $m \times n \times N_{in}$, and finally, the output residual image size is $m \times n$. Subtracting residual image from input slices can eventually improves the quality of the reconstructed slices input data.

Figure 2 shows a schematic of the 2.5D AI-CT model which consists of 17 convolution modules. There are 3 types of modules used in this model: The Conv I module refers to one convolution layer followed by a rectifier (ReLU) [10]. The Conv II module consists of one convolution layer, a batch normalization [11] and a ReLU. Finally, the Conv III module is a single convolution layer. The convolutional kernel in each convolution layer has the form $(3 \times 3) \times N_{in} \times N_{out}$, , where (3×3) is the size of convolution kernel, N_{in} and N_{out} are the input and output channel numbers, respectively. The channel refers to the neighboring slices in the input reconstruction data. The model starts with one Conv I module, then 15 Conv II modules, and ends with one Conv III module. The input data is a batch of slices (3D), and the output data from the last module is the residual 2D image. This is the reason why this model is named the 2.5D AI-CT model. The training procedure aims to tune this residual image in order to minimize the difference between the slices input data and the residual image to approach the high-quality ground truth.

2.2 Training Scheme

In order to train the AI-CT model, pairs of low-quality and high-quality reconstructions are entered into the model. In this work, we used the svMBIR algorithm [12] to generate all CT reconstructions. Our previous work [7], which also used AI-CT to improve the quality of the reconstruction, showed the model trained using the reconstructions from the same number of projections (K_1) did not significantly improve the quality of reconstructions generated from different number of projections (K_2). To solve this challenge, we adjusted the AI-CT training scheme by using reconstructions generated from different numbers of projections to train the model.

We define a sparsity factor, δ, to represent the sparsity level of projections used for each reconstruction. The number of projections used in a sparse reconstruction is $\left\lceil \frac{N_{sample}^{projection}}{\delta} \right\rceil$; where $N_{sample}^{projection}$ is fixed and is the total number of projections used for the high-quality reconstruction, and δ is an integer. When δ is large, the number of projections is sparse. When δ equals 1, all projections are used for the reconstruction. Additionally, we set the reconstruction generated using a number of projections with a sparsity factor δ_i as X^{δ_i}, where $\delta_i \in \Delta_{N_\Delta} = \{\delta_1, \ldots, \delta_{N_\Delta}\}$ and $i \leq N_\Delta$; N_Δ refers to the number of the different sparsity factors used in this work. The AI-CT model that we use is a "2.5D" network, i.e. it uses a collection of neighboring slices in order to suppress artifacts from each input slice. In our work, we set the number of neighboring slices in AI-CT to 5. For one reconstruction X^{δ_i} with a total of M cross-sections slices, we can have $N_{tot} = \frac{N_{sample}}{5}$ non-overlapping samples used to train the model. Each sample is noted as $x_j^{\delta_i}$ where $j \leq \frac{N_{sample}}{5}$, then let $\Pi_{N_{\delta_i}}^{\delta_i} = \left\{ x_0^{\delta_i} \ldots x_j^{\delta_i} \ldots x_{N_{\delta_i}}^{\delta_i} \right\}$ be the set of samples randomly selected from X^{δ_i} where N_{δ_i} is the number of samples. Therefore, the training dataset can consist of multiple sets from different X^{δ_i} and can be noted as $T = \{\Pi_{N_{\delta_1}}^{\delta_1} \ldots \Pi_{N_{\delta_{N_\Delta}}}^{\delta_{N_\Delta}}\}$. The total number of training samples, N_{tot}, is given by: $N_{tot} = \sum_{i=1}^{N_\Delta} N_{\delta_i} = \frac{N_{sample}}{5}$.

2.3 Composition of the Training Dataset

Since the training dataset contains samples from reconstructions using different sparsity factors, the number of samples (N_{δ_i}) from each X^{δ_i} is also a factor that can impact the performance of the trained model. To explore the influence of the composition of the training data set, we designed 3 training data compositions for comparison. The first training dataset, T_{single}, consists of samples from the same reconstruction $X^{\delta_{i*}}$ with a sparsity factor δ_{i*}. Let us set $N_{\delta_{i*}} = 0$, $i \neq i^*$, then $T_{single} = \left\{ \Pi_{N_{\delta_{i*}}}^{\delta_{i*}} \right\}$, where $N_{\delta_{i*}} = N_{tot}$ is the total number of training samples. The AI-CT model trained with T_{single} is called M_{single} and is considered the baseline in the performance evaluation. We composed the second training dataset, $T_{uniform}$, with the samples from reconstructions using multiple sparsity factors. In this dataset, an equal number of samples are randomly selected from each X^{δ_i}. Let $N_{\delta_i} = \frac{N_{tot}}{N_\Delta}$, and $T_{uniform} = \{\Pi_{N_{\delta_1}}^{\delta_1}, \ldots, \Pi_{N_{\delta_{N_\Delta}}}^{\delta_{N_\Delta}}\}$. The model trained by $T_{uniform}$ is denoted as $M_{uniform}$. The last training dataset T_{RoT} uses a rule-of-thumb distribution [8] to compose the training dataset. With this rule, the majority of the samples are allocated to high-quality cases and a few to low-quality cases. The model trained by T_{RoT} is denoted as M_{RoT}.

2.4 Performance Evaluation

Three models M_{single}, $M_{uniform}$, and M_{RoT} are trained using different compositions of training data, as described in Sect. 2.3. To evaluate the performance of each model, the sparse-projection reconstructions are used as input to the trained neural network, and the improved reconstructions are generated as the output. The normalized mean squared error (NRMSE) between the improved reconstructions and corresponding ground truths is used to compare the performance.

3 Implementation

Neutron computed tomography was performed on two rat femurs called Sample A and B, respectively. A 1.5 mm diameter cylindrical titanium is implanted in each femur. The samples include rat bone tissue which is primarily comprised of an inorganic calcium phosphate mineral phase and an organic phase, primarily made of type I collagen. The sample consists of a roughly cylindrical bone cross-section with an area of approximately 10 mm^2 and a wall thickness of 750 μm, containing a 1.5 mm diameter cylindrical titanium implant. Sample A is used to train the three AI-CT models with a total of 1165 reconstructed slices. Sample B is the test sample which has a total of 1250 reconstructed slices. Each sample is scanned using 694 projection angles at the High Flux Isotope Reactor (HFIR) Multimodal Advanced Radiography Station (MARS) imaging beamline. Figure 3 shows the 3D rendering of samples A and B, respectively. Sample A was used to generate 6 ($N_\Delta = 6$) reconstructions with different sparsity factors, $\delta_i \in \Delta_6 = \{2, 4, 6, 8, 16, 32\}$. Reconstructed cross-sectional slices of Sample A generated by different numbers of projection (i.e., different sparsity factors) are also displayed. As the sparsity factor increases, the quality of the reconstructed slice significantly decreases showing blurred edges and limited sample details.

The total number of cross-sectional slices is $N_A = 1165$, thus the training dataset size is $N_{tot} = \frac{N_A}{5} = 233$. There are 3 training datasets that use the different compositions $\{T_{single}, T_{unifrom}, T_{RoT}\}$. Figure 4 illustrates the dataset compositions as a function of the sparsity factors. We set $i^* = 2$, $\delta_{i*} = \delta_2 = 4$ for T_{single}, then all training samples are from X^{δ_2}, which means $N_{total} = N_{\delta_2} = 233$ and $N_{\delta_i; i \neq 2} = 0$. For $T_{uniform}$, $N_{\delta_i} = \frac{N_{tot}}{N_\Delta} =$

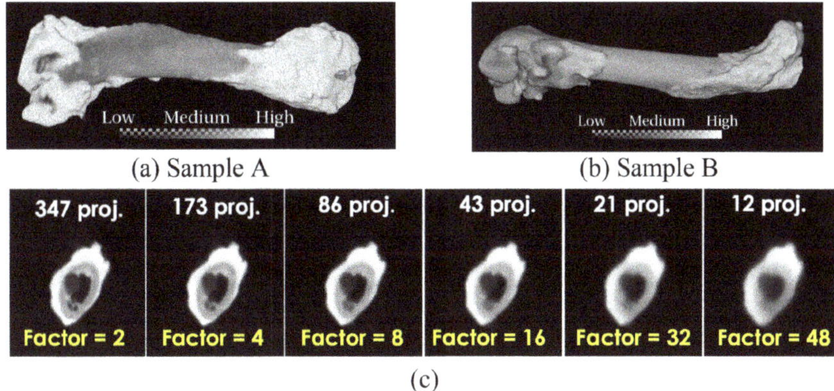

(a) Sample A   (b) Sample B

(c)

Fig. 3. 3D volume rendering of (a) Sample A used for model training; (b) Sample B used for testing models; and (c) reconstructed cross-sectional slices of Sample A using different sparsity factors (the corresponding number of projections are displayed on top of each slice). The rat femurs are approximately 40 mm long.

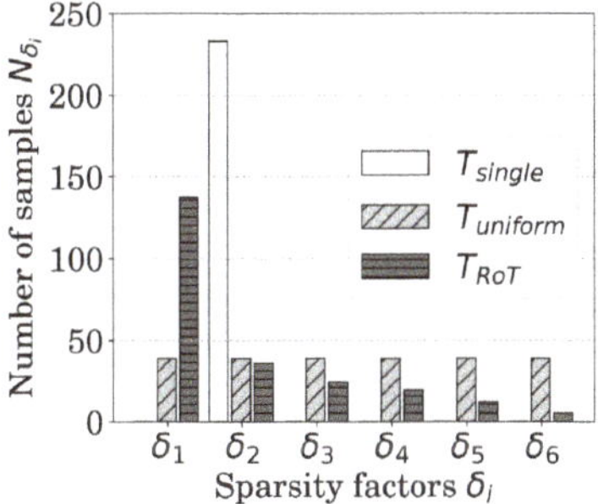

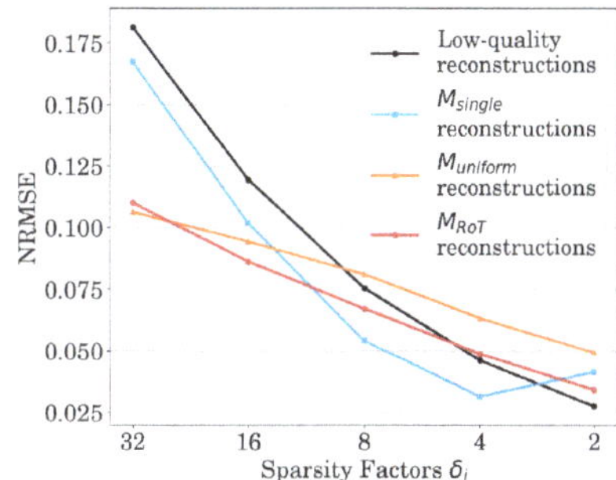

Fig. 4. The compositions of training datasets: the sample number selected from each reconstruction is plotted as a function of the sparsity factor used to generate the corresponding reconstructions.

Fig. 5. NRMSE plot as a function of the sparsity factor for sample B, the number of projections used for low-resolution reconstructions is 21, 43, 86, 173, 347.

38 which means 38 samples are randomly selected from each reconstruction $X^{\delta_i} \in \{X^{\delta_1}, \ldots, X^{\delta_6}\}$. For T_{RoT}, we set $N_{\delta_1} = 0.6N_{tot}$, $N_{\delta_2} = 0.15N_{tot}$, $N_{\delta_3} = 0.1N_{tot}$, $N_{\delta_4} = 0.08N_{tot}$, $N_{\delta_5} = 0.05N_{tot}$, and $N_{\delta_6} = 0.02N_{tot}$; then the different number of samples are selected from $X^{\delta_i} \in \{X^{\delta_1}, \ldots, X^{\delta_6}\}$ and X^{δ_1} has the most samples. The number of samples decrease as the sparsity factor increases, as expected. After training, Sample B is used to test the performance of the ML model (in this case, $N_B = 1250$). There are 5 sparsity factors ($N_\Delta = 5$) used for generating the test reconstructions $\{X^{\delta_1}, \ldots, X^{\delta_5}\}$, where $\delta_i \in \Delta_5 = \{2, 4, 6, 8, 16\}$ and $\frac{N_B}{5} = 250$ samples in each reconstruction.

4 Results and Discussion

Figure 5 illustrates the NRMSE of the reconstructed slices as a function of the sparsity factor for Sample B data using the different AI-CT models after they have been trained with Sample A. The results are compared to the original low-quality reconstructed slices (i.e. unmodified by the different ML AI-CT models) as a function of the sparsity factor. As expected, the NRMSE decreases with the decrease of the sparsity factor, i.e. as more reconstructed slices are used for both the original reconstructed slices and the ones improved with the 3 ML models. However, not all models improve the quality of the reconstructed slices. Except for $\delta = 2$, the M_{single} ML model produces an NRMSE lower than the one for the original reconstruction. However, the NRMSE is not reduced significantly when $\delta \geq 16$ indicating this training model is not appropriate for higher sparsity factors. The other two models, $M_{uniform}$ and M_{RoT}, have lower NRMSEs when $\delta > 8$. Since the information of the reconstruction with different sparsity factors is added to the training data (see Fig. 2), the NRMSE is significantly reduced for $M_{uniform}$ and M_{RoT} when $\delta > 8$. Moreover, M_{RoT} performs better across most δ values. However, when $\delta \leq 8$, $M_{uniform}$ becomes unstable and worsens the quality of the slices as compared to the original reconstruction. When the sparsity factor is not very high, for example $\delta = 2$, the reconstruction quality is sufficient and the NRMSEs for all 3 trained models are higher than the original reconstruction. M_{RoT} produces the lowest NRMSE among all three models which shows that M_{RoT} has the best generalization performance. We note that the training procedure in this work limited the size of the training dataset for fair comparison, but the training dataset size can be increased for practical cases. Thus, the performance of M_{RoT} for $\delta \leq 4$ can potentially improve further. A key insight from our work is that a model trained using the proposed "rule-of-thumb" distribution does not over-smooth the results for cases with a larger number of measurements – leading to a lower error in such cases compared to a model trained using the same number of training examples from each scenario.

Sparse projections yield low-quality reconstructed slices which may not provide detailed scientific information for researchers. Using the M_{RoT} model can boost the quality of sparse-view reconstructions. Figure 6 shows the application of a K-Means clustering based segmentation on the reconstructed data. A cross-section of the rat bone slice (for δ=32) is shown on Fig. 6(b) and the corresponding segmented result is shown on Fig. 6(d). The slice reconstructed using the M_{RoT} model is shown on Fig. 6(c) and its segmented result is showed on Fig. 6(e). The titanium implant is barely seen in the low-quality slice of Fig. 6(b) and hence the segmentation cannot provide many details. When using M_{RoT}, the details of the slice become visible as seeing in Fig. 6(c) leading to a dramatic improvement in the segmentation of the slice as evidenced by the titanium implant shown in green.

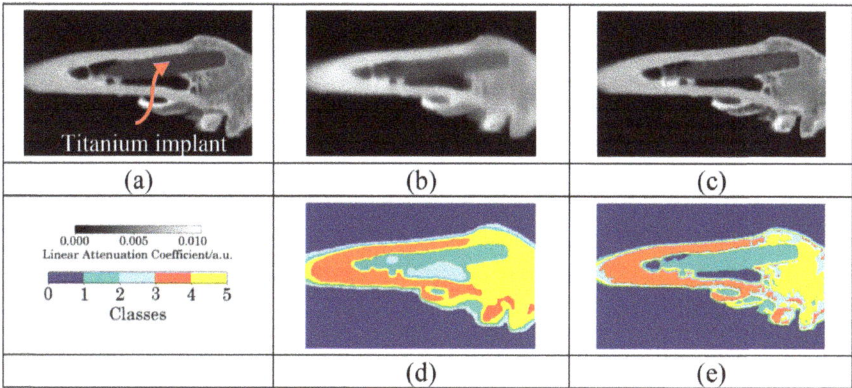

Fig. 6. An example of the application of AI-CT to the rat femurs: (a) ground truth cross-sectional slice from a high-quality scan showing the titanium implant located inside the rat femur; (b) the low-quality reconstruction using only 12 projections; (c) the reconstructed slice processed with M_{RoT}; (d) and (e) are K-Means cluster results of (b) and (c). The classes correspond to different regions in the sample and are used for visualization improvement.

5 Conclusions

In this work, we proposed a new scheme on how to train a ML model called 2.5D AI-CT to significantly improve its generalization performance across a range of experimental scenarios. With the help of the trained model, a high-quality reconstruction can be realized with sparse projections, which can potentially enable hyperspectral neutron computed tomography with reasonable acquisition times. Our research is particularly impactful when multiple similar samples need to be scanned, as required in the biomedical field for example, at beamlines such as the newly constructed VENUS instrument at the Spallation Neutron Source.

Acknowledgements. This manuscript has been authored by UT-Battelle, LLC, under contract DE-AC05-00OR22725 with the US Department of Energy (DOE). The US government retains and the publisher, by accepting the article for publication, acknowledges that the US government retains a nonexclusive, paid-up, irrevocable, worldwide license to publish or reproduce the published form of this manuscript, or allow others to do so, for US government purposes. DOE will provide public access to these results of federally sponsored research in accordance with the DOE Public Access Plan (http://energy.gov/downloads/doe-public-access-plan). The rat femurs data was measured in MARS HFIR ORNL with IPTS-24959. This material is also based upon work that was supported in part by the Orthopedic Research and Education Foundation and the Rush University Scientific Leadership Council.

References

1. Yang, D., et al.: An edge alignment-based orientation selection method for neutron tomography. In: ICASSP 2023–2023 IEEE International Conference on Acoustics, Speech and Signal Processing (ICASSP), pp. 1–5 (2023)

2. Tang, S., et al.: A machine learning decision criterion for reducing scan time for hyperspectral neutron computed tomography systems. Sci. Rep. **14**(1), 15171 (2024)
3. Micieli, D., et al.: Accelerating neutron tomography experiments through artificial neural network based reconstruction. Sci. Rep. **9**(1), 2450 (2019)
4. Han, Y.S., Yoo, J., Ye, J.C.: Deep residual learning for compressed sensing CT reconstruction via persistent homology analysis. arXiv preprint arXiv:1611.06391 (2016)
5. Liu, Z., et al.: TomoGAN: low-dose synchrotron x-ray tomography with generative adversarial networks: discussion. JOSA A **37**(3), 422–434 (2020)
6. Ziabari, A., et al.: 2.5 D deep learning for CT image reconstruction using a multi-GPU implementation. In: 2018 52nd Asilomar Conference on Signals, Systems, and Computers. IEEE (2018)
7. Venkatakrishnan, S., et al.: Convolutional neural network based non-iterative reconstruction for accelerating neutron tomography. Mach. Learn. Sci. Technol. **2**(2), 025031 (2021)
8. Gnanasambandam, A., Chan, S.: One size fits all: can we train one denoiser for all noise levels? In: Proceedings of the 37th International Conference on Machine Learning, vol. 119, pp. 3576–3586 (2020)
9. Kak, A.C., Slaney, M.: Principles of Computerized Tomographic Imaging. SIAM (2001)
10. Xu, B., et al.: Empirical evaluation of rectified activations in convolutional network. arXiv preprintarXiv:1505.00853 (2015)
11. Ioffe, S., Szegedy, C.: Batch normalization: accelerating deep network training by reducing internal covariate shift. In: International Conference on Machine Learning. PMLR (2015)
12. Team, S.D.: Super-Voxel Model Based Ierative Reconstruction (SVMBIR) (2020)

Early Commissioning Activities at the Spallation Neutron Source VENUS Imaging Beamline

Hassina Z. Bilheux[1]([envelope]) [ORCID], Harley Skorpenske[1], Kevin Yahne[1], Greg Guyotte[2], Matthew Pearson[3] [ORCID], Matthew J. Frost[2] [ORCID], Jean C. Bilheux[1] [ORCID], Shimin Tang[1] [ORCID], Chen Zhang[4] [ORCID], Yuxuan Zhang[1] [ORCID], James Torres[1] [ORCID], and Matthew Balafas[2]

[1] Neutron Scattering Division, Neutron Sciences Directorate, Oak Ridge National Laboratory, Oak Ridge, TN 37831, USA
{bilheuxhn,bilheuxhn}@ornl.gov
[2] Neutron Technologies Division, Neutron Sciences Directorate, Oak Ridge National Laboratory, Oak Ridge, TN 37831, USA
[3] Second Target Station Project, Neutron Sciences Directorate, Oak Ridge National Laboratory, Oak Ridge, TN 37831, USA
[4] Computer Science and Mathematics Division, Computing and Computational Sciences Directorate, Oak Ridge National Laboratory, Oak Ridge, TN 37831, USA

Abstract. The neutron imaging program at the Oak Ridge National Laboratory comprises capabilities at both the Spallation Neutron Source and the High Flux Isotope Reactor. Currently under commissioning, VENUS is a time-of-flight neutron imaging beamline optimized to provide unique contrast capabilities with two techniques, Bragg edge and resonance imaging. Bragg edge radiography measures transmitted neutrons through crystalline materials. Resonance imaging is based on the absorption of neutrons with energies above 0.5 eV by nuclei in the sample. VENUS is designed to allow efficient and automated change from one capability to another.

Started in October 2018 and completed in August 2024, the construction project focused on the design, procurement, testing, and installation of the beamline equipment. During the project, the beamline "cold" commissioning (without neutrons) consisted of integrating equipment controls into a user-friendly interface. The "hot" commissioning with neutrons started in September 2024 and is a continuation of the integration of equipment controls such as debugging the user interface and integrating detectors. A few experiments were performed to demonstrate the core capabilities of the instrument.

This manuscript shows a general layout of the beamline, its key capabilities and equipment, and initial commissioning experiments performed at VENUS.

Keywords: VENUS · Bragg edge imaging · resonance imaging

© The Author(s) 2026
A. E. Craft and H. Z. Bilheux (Eds.): WCNR 2024, SPPHY 348, pp. 347–354, 2026.
https://doi.org/10.1007/978-3-032-15003-5_39

1 Introduction

1.1 Overview of the VENUS Beamline

The VENUS imaging beamline [1, 2] is installed at the Spallation Neutron Source (SNS) of the Oak Ridge National Laboratory (ORNL). Installed on beamline 10, VENUS has a direct view of a decoupled and poisoned hydrogen H_2 moderator which provides sharp pulses of thermal and epithermal neutrons. VENUS is among a handful of neutron imaging facilities around the world that provide time-of-flight (TOF) or wavelength-dependent neutron imaging capabilities [3–7]. VENUS has four main modes of operation: the Bragg edge imaging mode, the resonance imaging mode, the epithermal imaging mode, and the thermal/cold neutron mode. Both Bragg edge and resonance modes rely on the intrinsic properties of the SNS, i.e., neutrons are measured based on their TOF. While Bragg edge imaging [8–10] is preferably performed using cold neutrons, there are Bragg edges in the thermal range. Resonance imaging [11, 12] utilizes neutrons with energies higher than 0.5 eV. When the Bragg condition for diffraction is met at the full backscattering angle (i.e., the neutron wavelength corresponds to twice the lattice spacing), there is an abrupt change in the transmission intensity, the so-called "Bragg edge". While the height of the edge provides a means to quantify the amount of phase in a material, its position provides information about the strain for a given < hkl > plane. Epithermal (or resonance) neutrons can be absorbed by a sample's nuclei, yielding a contrast that is specific to the elemental composition of the material. This technique does not depend on the crystallinity of the sample and thus is applicable to amorphous materials. The resonance signal shows attenuation peaks that can be used to identify elements and isotopes in materials in 3 dimensions.

The VENUS modes of operation are enabled by the selection of complex components such as choppers, collimators, an incident beam filter made of cadmium (Cd) and a suite of detectors. These components are strategically laid out (see Fig. 1) to offer maximum versatility without compromising capabilities. On an imaging beamline, one of the critical design parameters is the source-to-detector distance, i.e., where the radiograph is formed. This contrasts with the majority of time-of-flight instruments where source-to-sample distance is emphasized. For VENUS, the flightpath from source to detector is at a maximum distance of 25 m. Neutrons from the moderator pass through apertures of various sizes, selected by the user, a set of collimators, and also through tapered evacuated or helium (He) filled flight tubes equipped with beam scrapers, and finally through a sample to be detected by either a micro-channel plate (MCP) Timepix (TPX) or Timepix3 (TPX3) detector, a charge-coupled device (CCD), or a scientific complementary metal oxide semiconductor (sCMOS).

1.2 Data Acquisition Implementation at VENUS

Part of the early commissioning of the VENUS beamline includes testing the data acquisition logic as well as recording and archiving of data. Software workflow is also part of the beamline commissioning and is discussed elsewhere (see J.-C. Bilheux *et al.*, in these proceedings). The data acquisition interface is based on the Experimental Physics and Industrial Control System [13] (EPICS) and is similar to the one already deployed at the

High Flux Isotope Reactor (HFIR) Multimodal Advanced Radiography Station (MARS) imaging beamline (J. Torres *et al.*, these proceedings). EPICS is designed to allow swift (within a few minutes) and automated reconfiguration of the beamline mode, as illustrated in Fig. 2. Several modes can be queued in a scan list, along with different samples such that the instrument can run independently for hours to days reliably. This approach was developed to optimally balance the imaging team's time between contributing to the measurements and working closely on data processing and analysis with researchers who utilize VENUS. Figure 2 displays a view of the current (to an experiment) settings and experiment on the VENUS EPICS data acquisition interface (left) and a view of the beamline control (right) with the different setting options such as modes of operation, repetition rate of the choppers, collimation ratio, collimator openings, filter in or out of the beam, and the selection of detectors. There are several other windows (Fig. 2 left) on the VENUS EPICS interface. Due to the limited space on this manuscript, these windows are quickly highlighted in Fig. 2 left and will be discussed in detail in a future publication.

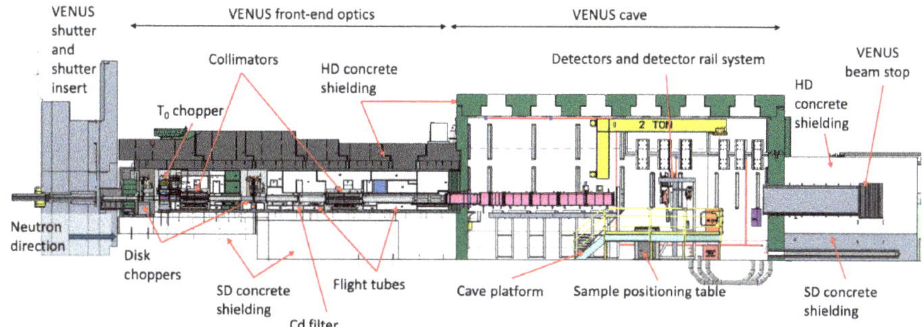

Fig. 1. Cross-sectional view of the VENUS beamline.

1.3   The VENUS Detectors

The VENUS cave is the area that is accessible when performing an experiment. It is accessible via a radiological materials area that has workbenches with tools to prepare the experimental setup, including offline testing. Detectors are mounted on a rail system and can easily be moved in the beam path, as needed for an experiment, as illustrated in Fig. 3. Three light-tight boxes are installed at VENUS and are compatible with the CCD or the sCMOS detectors, by way of a custom-made adapter. The MCP TPX and TPX3 detectors are positioned on each side of the light-tight boxes due to the required floor space for vacuum pumps. The sample elevator can be reconfigured based on the sample geometries to be measured. This elevator was designed to allow flexibility to measure several samples using different modes of operation without having to enter the VENUS cave.

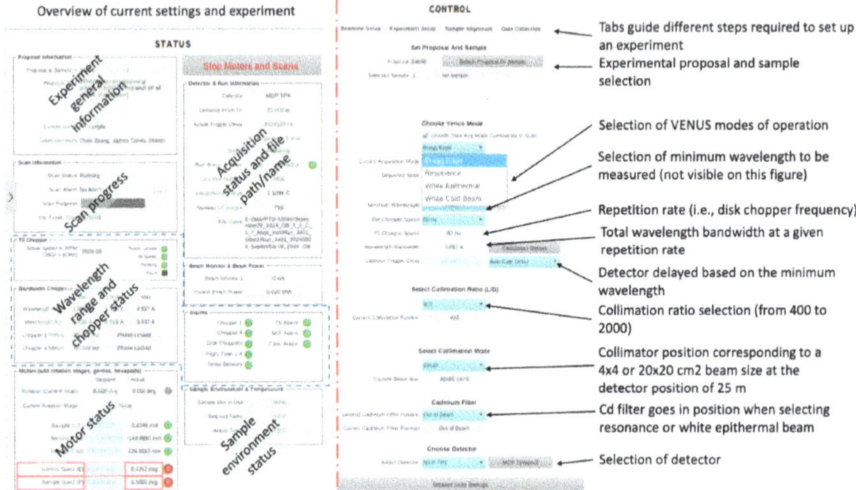

Fig. 2. EPICS data acquisition interface at VENUS: (left) overview of the current settings and experiment; (right) view of the control section of the interface showing the setting selections for the different modes of operations and beamline configuration.

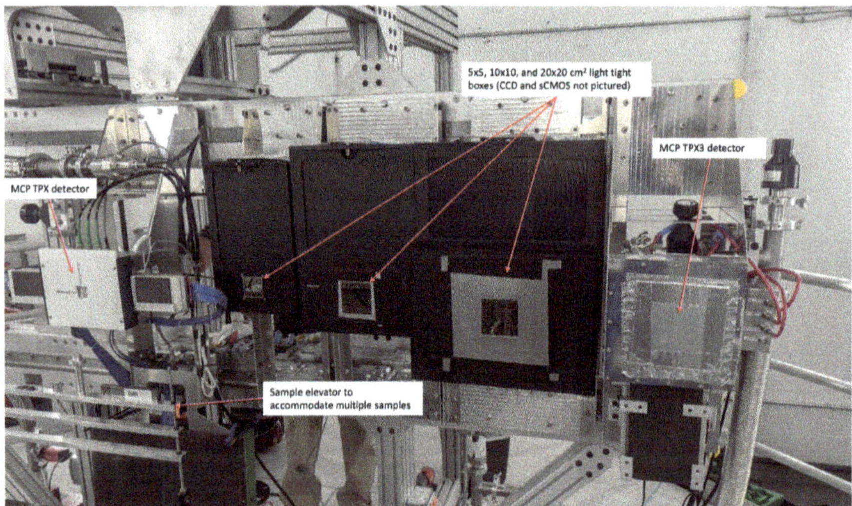

Fig. 3. Photograph of the suite of detectors installed at VENUS showing the two MCP TPX and TPX3 detectors, and the light tight boxes for the CCD and sCMOS detectors.

2 Early Measurement Activities

The focus of the early measurement activities at VENUS was the demonstration of some of the key performance criteria of the beamline. A few of these expected parameters were demonstrated during the early stages of the commissioning phase and are presented in this manuscript.

2.1 20 × 20 cm² Field-Of-View

VENUS was designed to provide a field-of-view (FOV) up to 20 × 20 cm² at 25 m. This was driven by scientific requirements to measure large objects for materials science and engineering, but also plants and geomaterials.

The FOV at VENUS was obtained using a ZWO scientific Complementary Metal Oxide Semiconductor (sCMOS), model IMX455 (SONY chip) coupled with a Nikon 50 mm f/1.2 lens and a 200 μm-thick lithium fluoride zinc sulfide (LiF/ZnS) scintillator. A radiograph of the FOV with a resolution standard made out of gadolinium (Gd) deposited on a silicon wafer is displayed in Fig. 4. The mask was manufactured by the Paul Scherrer Institute (PSI), Switzerland.

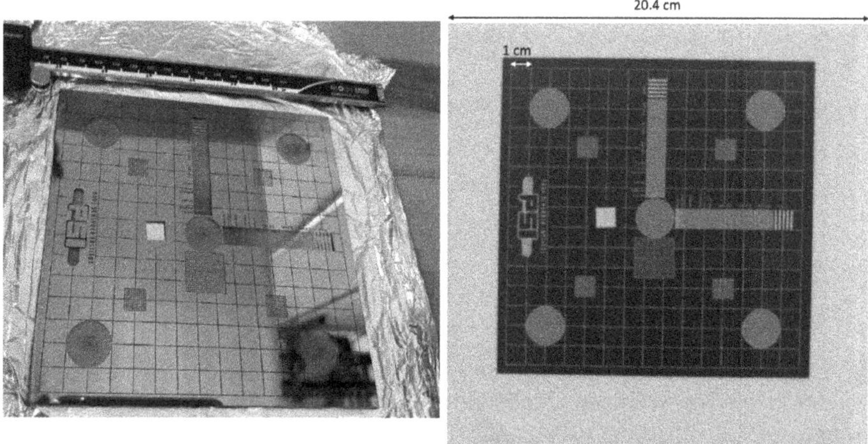

Fig. 4. 15 × 15 cm² spatial resolution mask object (left) and respective radiograph (right) demonstrating a 20 × 20 cm² field-of-view measured at the scintillator position of 25 m.

2.2 Wavelength Distribution and Measured Flux

VENUS requires the broadest range of neutron wavelengths of all SNS instruments since the beamline is optimized for Bragg edge imaging (thermal/cold neutrons) and resonance imaging (epithermal neutrons). Figure 5 displays the wavelength range accessible at VENUS when measured at 10 Hz (no frame overlap), as measured with a ORDELA beam monitor model 4560N (with an efficiency of ~ 10^{-5} at 1 Å). The integrated flux measured with the VENUS beam monitor between 1 and 3.64 Å was 2.5 × 10^6 neutrons.cm^{-2}.s at a distance of 23.7 m from the source.

2.3 Experimentally Measured Bragg Edge and Resonance Patterns

Bragg edge and resonance patterns were measured via TOF radiography using the MCP TPX detector. Figure 6 (left) shows a photograph of the experimental setup at VENUS with different calibration samples used for Bragg edge and resonance testing. Figure 6 (top right) displays Bragg edges of a 5-mm diameter nickel (Ni) powder sample in an aluminum (Al) container measured at VENUS. As seen in the figure, the Bragg edges are sharp, further indicating that the instrument wavelength resolution has been achieved. Figure 6 (bottom right) displays the resonance pattern for a 180-μm thick tantalum (Ta) foil measured at VENUS. Both measurements were performed at 24.99 m using the MCP TPX detector (also visible on the photograph). Plots were averaged across the area of the sample in the FOV.

Other samples such as copper, steel, silver, cobalt, are also planned to be measured at VENUS for calibration purposes and to test the software tools used to fit Bragg edges (see J.-C. Bilheux *et al.* manuscript in these proceedings) and resonances.

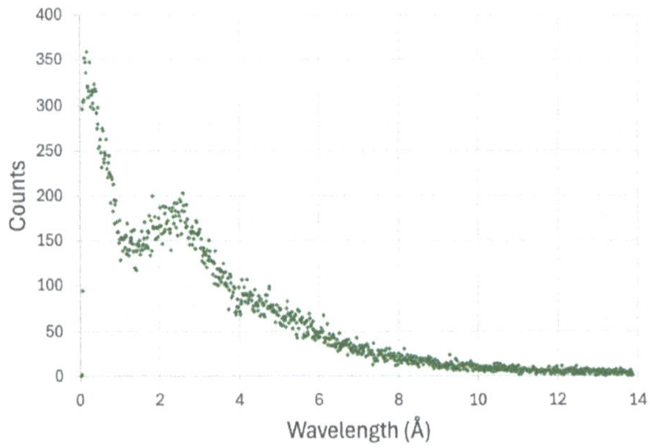

Fig. 5. Beam monitor counts as a function of neutron wavelength, measured at a distance of 23.726 m from the moderator and at 10 Hz (with all choppers parked open). The acquisition time was set to 600 s.

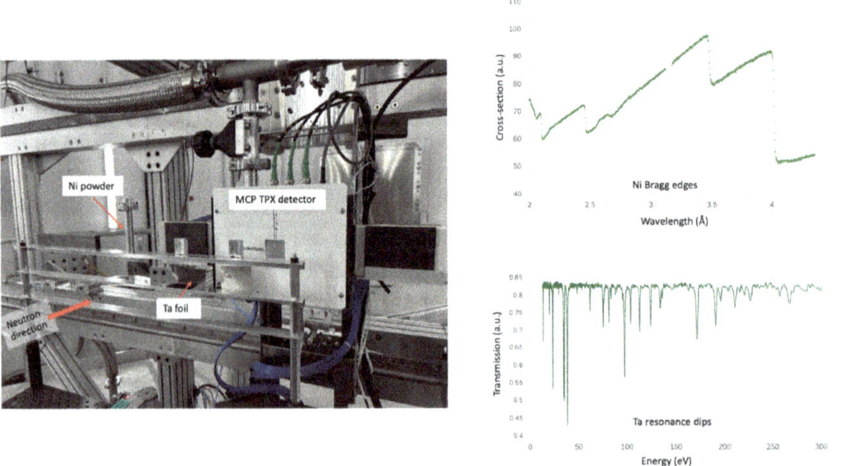

Fig. 6. 5-mm thick Ni Bragg edge pattern showing sharp Bragg edges, indicative of the appropriate wavelength resolution of the instrument at 25 m; Ta resonance transmission dips at 25 m.

3 Summary and Outlook

The VENUS imaging beamline has started commissioning. Preliminary measurements show the beamline is able to measure Bragg edge and resonance patterns using an MCP TPX detector. The beamline is designed to allow the measurements of multiple samples with different modes of operation (Bragg edge, resonance, epithermal white beam, thermal or cold white beam modes, respectively). Equipped with the ability to swiftly slide detectors in and out of the beam, this instrument is designed to minimize setup time and maximize utilization of the neutron beam, while allowing the imaging team to focus on scientific productivity of the beamline.

Further development includes software development and implementation at VENUS such as the testing of a batch mode capability of the in-house Bragg edge fitting software called iBeatles. Other software development includes the resonance fitting software. A few commercial software tools are already installed and operating at VENUS such as the visualization software VGStudio MAX [14] and the data processing and analysis software AMIRA/AVIZO [15]. Focus will also be given to the implementation of the ANDOR CCD and the QHY sCMOS detectors, and the EPICS integration of the MCP TPX3 detector.

Acknowledgments. The imaging team acknowledges the engineering team contribution in designing, fabricating and installing the VENUS beamline. This also includes but is not limited to the chopper, the detector, the data acquisition, the vacuum, and the installation teams. Special thanks to Mr. Aaron Hanks, the VENUS construction project lead engineer, Ms. Amy Byrd, the project management and schedule expert, and Ms. Molly Brewer, the engineering support specialist.

This research used resources at the Spallation Neutron Source, a DOE Office of Science User Facility operated by the Oak Ridge National Laboratory. The beam time was allocated to VENUS on proposal numbers IPTS-33479.1 and 33699.1.

References

1. Bilheux, H., et al.: The VENUS imaging beamline at the spallation neutron source: layout, expected performance and status of the construction project. J. Phys. Conf. Ser. (2023)
2. Bilheux, H.Z., et al.: The VENUS imaging beamline at the oak ridge national laboratory spallation neutron source. AM&P Tech. Articles, **181**(8), 18–23 (2023)
3. Nelson, R., et al.: Neutron imaging at LANSCE—from cold to ultrafast. J. Imaging, **4**(2) (2018)
4. Kockelmann, W., et al.: Time-of-flight neutron imaging on IMAT@ISIS: a new user facility for materials science. J. Imaging, **4**(3) (2018)
5. Shinohara, T., et al.: The energy-resolved neutron imaging system, RADEN. Rev. Sci. Instrum. **91**(4) (2020)
6. Pooley, D., et al.: Wavelength-resolved neutron imaging on IMAT. Neutron Radiograph.WCNR-11, 15: p. 29 (2020)
7. Chen, J., et al.: First neutron Bragg-edge imaging experimental results at CSNS. Chin. Phys. B **30**(9), 096106 (2021)
8. Santisteban, J.R., et al.: Engineering applications of Bragg-edge neutron transmission. Appl. Phys. A Mater. Sci. Process. **74**, s1433–s1436 (2002)

9. Santisteban, J.R., et al.: Strain imaging by Bragg edge neutron transmission. Nucl. Instrum. Methods Phys. Res., Sect. A **481**(1–3), 765–768 (2002)
10. Steuwer, A., et al.: Bragg edge determination for accurate lattice parameter and elastic strain measurement. Physica Status Solidi (a), **185**(2), 221–230 (2001)
11. Losko, A.S., Vogel, S.C.: 3D isotope density measurements by energy-resolved neutron imaging. Sci. Rep. **12**(1), 6648 (2022)
12. Tremsin, A.S., et al.: Spatially resolved remote measurement of temperature by neutron resonance absorption. Nucl. Instrum. Methods Phys. Res. Sect. a-Accelerators Spectrometers Detectors Assoc. Equipment **803**, 15–23 (2015)
13. Kozubal, A., et al.: Experimental physics and industrial control system. In: ICALEPCS89 Proceedings (1989)
14. Reinhart, C.: Industrial computer tomography–A universal inspection tool. In: 17th World Conference on Nondestructive Testing. Citeseer (2008)
15. Stalling, D., Westerhoff, M., Hege, H.-C.: Amira: a highly interactive system for visual data analysis. Vis. Handb. **38**, 749–767 (2005)

Data Workflow and Software Tools at VENUS

Jean C. Bilheux[1]([✉]) [iD], Hassina Z. Bilheux[1] [iD], Anton Khaplanov[2],
Matthew J. Bedynek[3], Su-Ann Chong[3], Cornelius Donahue Jr[3], Greg Guyotte[3],
Kazimierz J. Gofron[3], Rob Knudson IV[3], Jamie Molaison[1], Harley Skorpenske[1],
Shimin Tang[1], Zach Thurman[3], Singanallur Venkatakrishnan[4], and Chen Zhang[5]

[1] Neutron Scattering Division, Neutron Sciences Directorate, Oak Ridge National Laboratory,
Oak Ridge, TN, USA
`bilheuxjm@ornl.gov`
[2] Neutron Upgrades Project Office, Oak Ridge National Laboratory, Oak Ridge, TN, USA
[3] Neutron Sciences Directorate, Neutron Technologies Division, Oak Ridge National
Laboratory, Oak Ridge, TN, USA
[4] Electrification and Energy Infrastructures Division, Energy Science and Technology
Directorate, Oak Ridge National Laboratory, Oak Ridge, TN, USA
[5] Computer Science and Mathematics Division, Computing and Computational Sciences
Directorate, Oak Ridge National Laboratory, Oak Ridge, TN, USA

Abstract. The time-of-flight imaging beamline VENUS at the Spallation Neutron Source of the Oak Ridge National Laboratory has recently entered the commissioning phase. Based on the experience acquired at the Spallation Neutron Source SNAP and the High Flux Isotope Reactor MARS beamlines, we have designed various analysis tools and a new intuitive data acquisition system user interface. VENUS starts commissioning with the micro-channel plate Timepix detector and light sensitive cameras that include a charge-coupled device and complementary metal oxide semiconductor sensors. A micro-channel plate (MCP) Timepix3 detector is currently being tested at VENUS and is in the process of being integrated to the VENUS data acquisition system. Since this new detector is capable of measuring data in event mode, both hardware and data workflow require careful planning as described in this manuscript. A new version of the in-house Imaging Bragg Edge Analysis Tools for Engineering Structures (iBeatles) has been produced thanks to the feedback received at the 11th workshop on neutron wavelength-dependent imaging held in Japan in 2023. This paper gives an overview of the VENUS hardware and software infrastructures, the data workflow, the VENUS data acquisition interface and features available in the new version of iBeatles.

Keywords: software · python · user interface

1 Introduction

The new VErsatile NeUtron imaging instrument at the Spallation Neutron Source (VENUS) of the Oak Ridge National Laboratory (ORNL) has recently begun commissioning (see H. Bilheux et al. in these proceedings). This beam line offers unique capabilities to the suite of instruments in operation at the Spallation Neutron Source (SNS).

© The Author(s) 2026
A. E. Craft and H. Z. Bilheux (Eds.): WCNR 2024, SPPHY 348, pp. 355–363, 2026.
https://doi.org/10.1007/978-3-032-15003-5_40

Mainly, VENUS can measure contrast features in crystalline materials using Bragg edge imaging [1–3], and elemental/isotopic content in objects with resonance imaging [4–6]. However, these advanced imaging techniques come with new challenges such as large data rate when using event mode detectors, new data format, the requirement to interact with live data, streaming normalization, analysis as well as visualization, and the ability to implement machine learning tool to drive experiments and make autonomous scanning using machine learning (ML). The next section focuses on explaining the VENUS data workflow from the detector to the analysis server where users can access their data. Finally, the last section provides a brief overview of the numerous tools implemented and available to users.

2   Data Workflow

Experimental data are created by the detectors. Depending on the detector, they are produced in different formats such as histograms (tiff or fits file format) or events (binary files). In the case of the events, the files are produced when using the microchannel plate (MCP) Timepix3 (TPX3) detector. This detector produces up to a few GB of raw data per minute. The raw data is pre-processed in a streaming fashion such that the position of each incoming neutron event is determined based on a clustering algorithm (of hits), thus reducing the size of the data prior to saving them (https://zenodo.org/records/133 81303).

For all detectors, the data is streamed to allow live visualization and some basic analysis, or in-situ driven experiment workflows with ML algorithms that drive the experiment and make real-time selection of the acquisition protocol such as the next set of angles to measure for a computed tomography (CT) scan and the decision to continue or terminate it [7].

Once an MCP TPX run is completed, data and metadata (any parameters relevant to the experiment, such as motor positions, acquisition time, proton charge, etc.) are gathered and combined within a NeXus (https://zenodo.org/records/1472392) file, and simultaneously archived on the SNS analysis servers [8] and made available to users. This cluster is a set of more than 20 computing nodes, running the Red Hat 9 operating system and providing the users with many software tools (commercial or developed in-house). This cluster is accessible to any past and current SNS and/or HFIR users, bringing another challenge regarding the central processing unit (CPU) time used by these machines. Due to the high-processing time imaging data requires, the VENUS beam line has a dedicated node reserved for processing data acquired at the beamline.

Figure 1 summarizes the complexity of the VENUS data workflow showing on the right side all the hardware systems installed inside the beam line firewall, while left side displays the systems accessible to users from their home institutions.

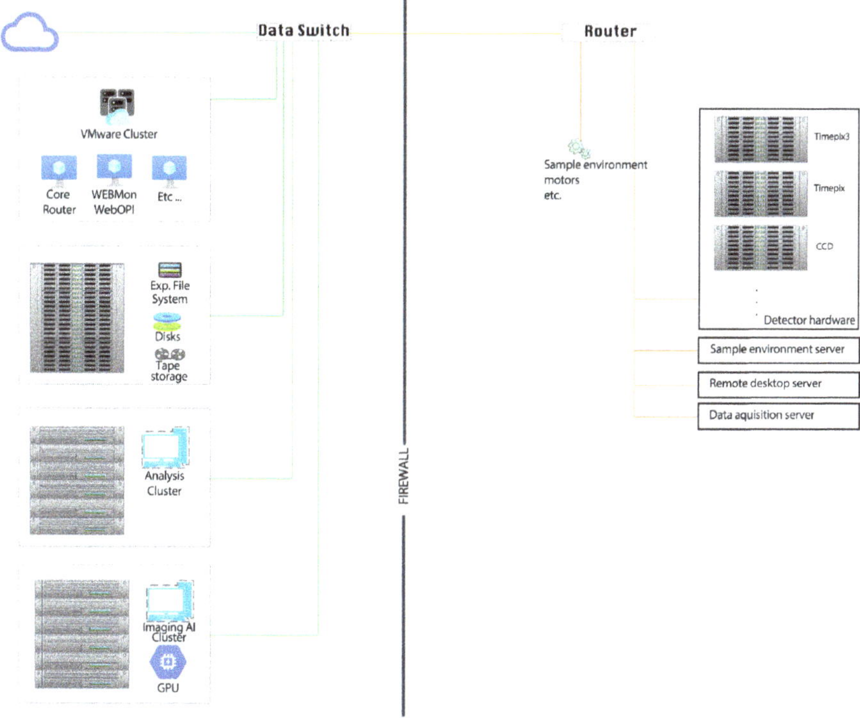

Fig. 1. VENUS data workflow showing the system living inside a firewall (right part), with the left part exposed to the outside world, where the users can access the archived data. Inside the firewall lives all the server controlling the detectors, sample enviroments and data acquisition system (for local and remote access), when all the analysis and visualization servers stay outside the firewall and can then be remotely accessed via the use of ThinLinc [https://www.cendio.com/thinlinc].

3   Software Tools

3.1   Planning Tools

In order to evaluate samples to be measured in the future, we implemented various planning tools on a publicly available URL, such as the estimation of the neutron transmission through a sample, the simulated resonance pattern, the simulated Bragg pattern, and the prediction of Bragg edges based on X-ray diffraction (XRD) input file [9–11]. This intuitive user interface called iNEUtron Imaging Toolbox (iNEUIT, pronounced "I knew it") was developed using python and DASH [https://dash.plotly.com/] and new tools are added on a regular basis. Moreover, thanks to the use of GitLab action, the deployment of new features is done automatically when the main branch of the repository is updated. At the same time, prototyping is performed on parallel branches to allow the developer to test the newest features on a private URL prior to public deployment.

3.2 Visualization

To provide 3D visualization to users, a dedicated windows machine can be accessed directly from the beam line control hutch or by using the Windows remote desktop application for off-site users. Several licenses of the AMIRA [12] software are available on this machine. For 2D visualization, ImageJ [13] and several optimized (for a specific application and task) Jupyter notebooks are available on the beamline Linux analysis machine.

3.3 Analysis

Imaging covers a broad range of neutron science. This means that users from many different scientific backgrounds and computing expertise utilized the imaging beamlines. Taking this into account, the tools we provide to the users need to satisfy the whole spectrum of user expertise, from user who are not programmers and prefer using intuitive tools with graphic user interfaces (GUIs) or notebooks, to users who want the option to modify the code provided by our team. We tackled this challenge by deploying analysis tools through Jupyter notebooks, featuring widgets that are beginner-friendly and still offer the flexibility to access and modify the underlying code. For some other more complex data analysis workflows, we implemented standalone application such as the Bragg edge fitting tool Imaging Bragg Edge Analysis TooLs for Engineering Structures (iBeatles, J.-C. Bilheux, H. Z. Bilheux, these proceedings) or grating interferometry analysis (Angel 2.0).

Jupyter Notebooks

More than 60 Jupyter notebooks are available to perform data pre-processing, management, analysis and visualization. The imaging notebooks are available on the dropdown menu of the analysis servers (analysis.sns.gov). During the launch, the notebooks are copied from the repository to the user home folder, allowing them to make changes to the notebooks without affecting the original ones which remain available to all users. This procedure ensures users benefit from the latest notebooks development without the need of applying updates. The choice of the Jupyter notebooks was made for several reasons such as the simplicity to run a notebook, the rapidity to develop and deploy a new notebook, as well as the possibility to hide most of the code by using widgets, hence reducing the possibilities to make any input errors. Some of the more complex notebooks, for example image registration, also trigger a standalone user interface displayed on top of the notebook. Moreover, every notebook has a tutorial link visible in the notebook itself. Figure 2 displays several notebooks that highlight some of their important features, such as the widgets and the possibility to launch a GUI directly from the notebook.

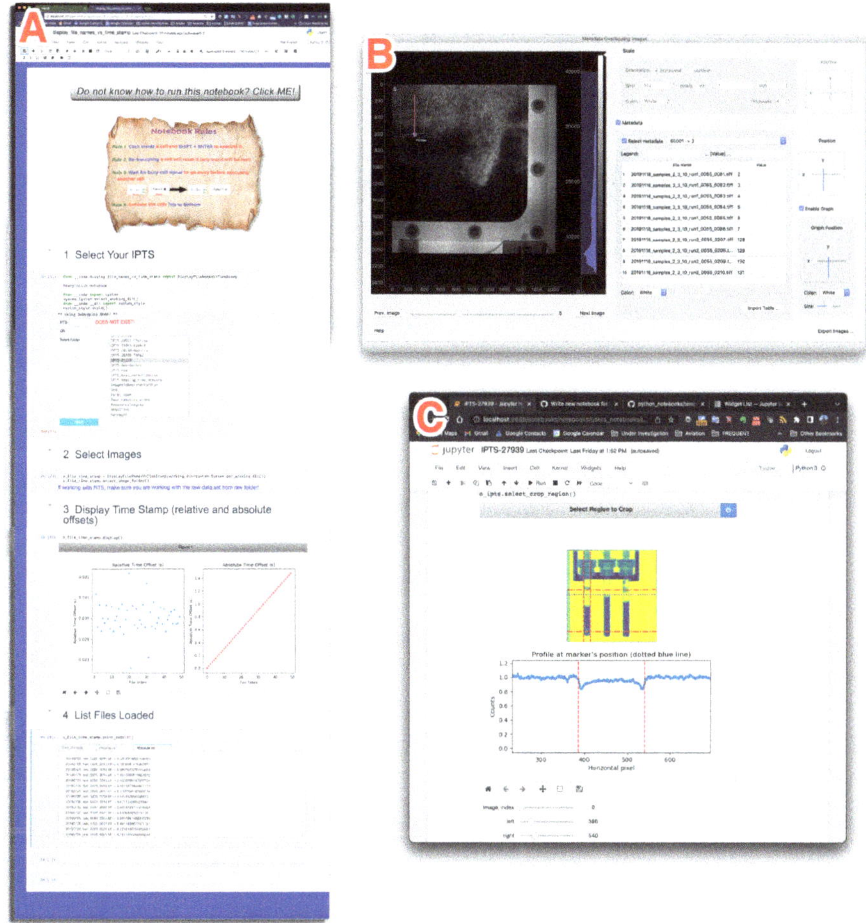

Fig. 2. Examples of available imaging notebooks. A: notebook showing interactive widgets. B: standalone user interface launched directly from a notebook. C: Interactive plots using widgets to select a region of interest, as illustrated here.

Standalone Application

Several applications have been developed to perform complex tasks such as grating interferometry analysis, strain mapping or advanced computed tomography. These standalone applications are available on the SNS analysis servers and freely available on our GitHub repository (github.com/ornlneutronimaging.ornl.gov).

iBeatles is developed to map residual stresses onto a sample radiograph by fitting Bragg edge measured by time-of-flight. The user-friendly application uses the raw TOF data, i.e., a stack of TIFF or FITS radiographs, and performs several steps such as pre-processing, normalization, pixel binning, Bragg edge selection, such that strain values can be determined for each (binned) pixel. Figure 3 corresponds to screenshots of some of the aforementioned steps (for a detailed description of the iBeatles workflow, see J.-C. Bilheux and H. Z. Bilheux, these proceedings).

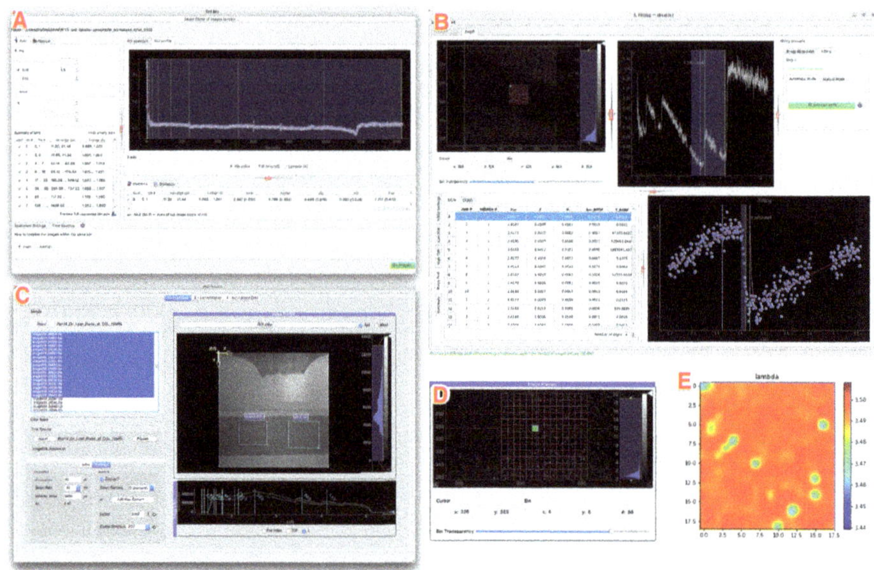

Fig. 3. The iBeatles interface: A: pre-processing step corresponding to radiograph binning in TOF. B: Selection of a single Bragg edge and its fitting. C: Normalization tab with ability to select sample and open beam radiographs, respectively. The Bragg edges contained in the TOF spectrum of a region of interest (ROI) can also be compared to any pre-selected unstrained material Bragg edges. D: Visualization of the ROI bins. E: neutron wavelength map in the selected ROI. The final step (not displayed here) corresponds to the strain map in units of micro strain.

Neutron grating interferometry (nGI) is a complex technique that leverages the wave property of the neutron to measure a phase and attenuation shift of the neutron interference pattern. Since detectors have pixels that are larger than the required resolution to measure the wave pattern, one of the gratings has to be stepped 10–12 times perpendicular, thus generating large sets of radiographs that cannot be processed and analyzed manually. Therefore, a grating interferometry data processing analysis tool was developed by FRM-II, and ORNL contributed by developing a new graphical user interface called Angel 2.0. This application allows to visualize the input data and play with different workflows in order to determine the transmitted image (TI), the differential phase contrast image (DPCI), and the dark field image (DFI). Several sets of measurements are processed as a batch job using a pre-defined excel spreadsheet directly loaded in the application. Figure 4 highlights some of the key features of Angel 2.0.

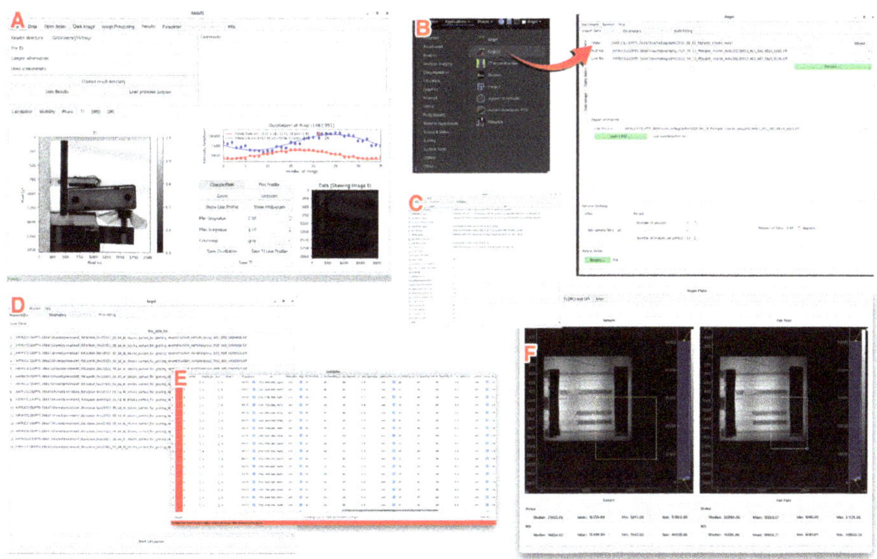

Fig. 4. A: FRM-II version of Angel adapted to work on MARS. B: The new version of Angel can easily be launched from the analysis server dropdown menu. C: Screenshot showing how the various parameters of the analysis workflow can be modified in a password-protected window. D: View of the Excel input tab where a batch of analysis can be processed. E: View of a dedicated Jupyter notebook used to simplify the editing of this Excel spreadsheet. F: One of the output views of the analysis.

Several computed tomography reconstruction tools have been developed at ORNL and new ones are under development specifically for VENUS. They all have in common a user interface that allows user to define all input parameters, calculate and visualize the center of rotation, sample tilt, apply and visualize various image filters. The reconstruction is either launched in the background or performed live prior to the reconstruction being visualized. These tools used various algorithms available in the open literature (ASTRA [astra-toolbox.com], pyMBIR [osti.gov/biblio/1550791], svMBIR [svmbir.readthedocs.io/], FBP, etc.). Figure 5 shows screenshots of the python MBIR GUI (pyMBIRui), one of the GUIs using the MBIR algorithm.

3.4 Conclusion

The new neutron imaging beamline VENUS at the Spallation Neutron Source (SNS) comes with many scientific and data processing challenges. To be a successful beamline, these challenges must provide easy solutions for novice and expert users alike without being time-consuming. Our plans are to provide intuitive tools that do not require any programming expertise. A web browser gives access to many Jupyter notebooks and standalone applications such that the data workflow is streamlined, and data are accessible during an experiment, or archived data a few minutes after the measurement is done. Because of the quantity of data VENUS will acquire, especially when working with neutron even data sets, the beamline will archive users data up to a few months

(TBD). It will be up to the users to retrieve the data using various tools we will provide and demonstrate.

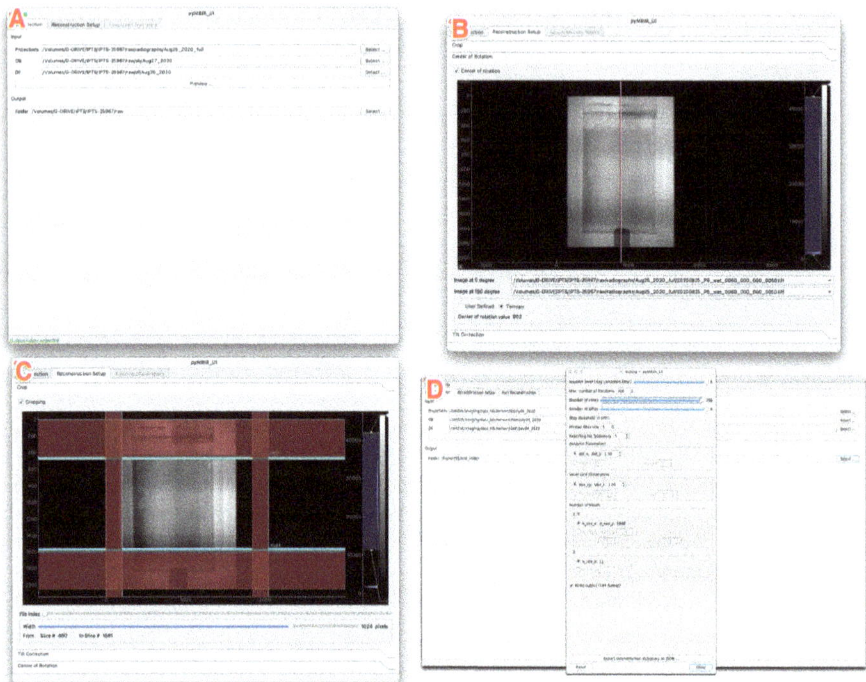

Fig. 5. A: Input window of the pyMBIRui application, where the sample, open beam and dark field images are loaded and visualized. B: Calculation of the center of rotation. C: View where the sample is cropped to narrow down the number of slices to use in the reconstruction. D: Parameters used by the reconstruction are set up in this view.

Acknowledgments. This research used resources at the Spallation Neutron Source, DOE Office of Science User Facilities operated by the Oak Ridge National Laboratory. The team would like to thank the user community for their many contributions in guiding the development of the software tools as well as being gracious with their time when testing the notebooks.

References

1. Santisteban, J.R., et al.: Strain imaging by Bragg edge neutron transmission. Nucl. Instrum. Methods Phys. Res., Sect. A **481**(1–3), 765–768 (2002)
2. Steuwer, A., et al.: Bragg edge determination for accurate lattice parameter and elastic strain measurement. Physica Status Solidi (a), **185**(2), 221–230 (2001)
3. Woracek, R., et al.: Diffraction in neutron imaging—a review. Nucl. Instrum. Methods Phys. Res. Sect. A **878**, 141–158 (2018)
4. Losko, A.S., Vogel, S.C.: 3D isotope density measurements by energy-resolved neutron imaging. Sci. Rep. **12**(1), 6648 (2022)

5. Tremsin, A.S., et al.: Spatially resolved remote measurement of temperature by neutron resonance absorption. Nucl. Instrum. Methods Phys. Res. Sect. a-Accelerators Spectrometers Detectors Assoc. Equipment **803**, 15–23 (2015)
6. Tremsin, A.S., et al.: Non-destructive studies of fuel pellets by neutron resonance absorption radiography and thermal neutron radiography. J. Nucl. Mater. **440**(1–3), 633–646 (2013)
7. Tang, S., et al.: A machine learning decision criterion for reducing scan time for hyperspectral neutron computed tomography systems. Sci. Rep. **14**(1), 15171 (2024)
8. Campbell, S.I., et al.: Accelerating Data Acquisition, Reduction, and Analysis at the Spallation Neutron Source. Oak Ridge National Lab.(ORNL), Oak Ridge, TN (United States). Oak Ridge (2014)
9. Zhang, Y., Bilheux, J.-C.: ImagingReso: a tool for neutron resonance imaging. J. Open Source Softw. **2**(19) (2017)
10. Zhang, Y., et al.: An interactive web-based tool to guide the preparation of neutron imaging experiments at oak ridge national laboratory. J. Phys. Commun. **3**(10), 103003 (2019)
11. *iNEUtron Imaging Toolbox*
12. Stalling, D., Westerhoff, M., Hege, H.-C.: Amira: a highly interactive system for visual data analysis. Vis. Handb. **38**, 749–767 (2005)
13. Schneider, C.A., Rasband, W.S., Eliceiri, K.W.: NIH Image to ImageJ: 25 years of image analysis. Nat. Methods **9**(7), 671–675 (2012)

Neutron Grating Interferometry at the High Flux Isotope Reactor

Yuxuan Zhang[1]([⊠]) [iD], Erik Stringfellow[1], Jean-Christophe Bilheux[1] [iD], Hassina Z. Bilheux[1] [iD], James Torres[1] [iD], Ian Turnbull[1], Roger Hobbs[1], Leslie G. Butler[2], and Kyungmin Ham[2]

[1] Oak Ridge National Laboratory, Oak Ridge, TN 37831, USA
zhangy6@ornl.gov
[2] Louisiana State University, Baton Rouge, LA 70803, USA

Abstract. Neutron imaging is a non-destructive probe that can spatially resolve bulk materials' internal features/structures. Neutron imaging spatial resolution is driven by detector technology, i.e., pixel size, and is currently on the order of 10–20 µm. This limitation can be alleviated by the use of neutron grating interferometry – an advanced neutron imaging technique – that enables simultaneous access to three contrast mechanisms: attenuation, differential phase, and small-angle neutron scattering. Recently, a Talbot-Lau neutron grating interferometry has been implemented at the Multimodal Advanced Radiography Station at the High Flux Isotope Reactor. In this work, the design, setup and performance of the grating apparatus are presented.

Keywords: Neutron Grating Interferometry · Neutron Instrumentation · MARS

1 Introduction

Neutron grating interferometry (nGI) is a neutron imaging technique that utilizes gratings to spatially resolve small angular deviations caused by a sample in a neutron beam [1]. Simultaneously, transmission contrast, phase contrast, and small-angle scattering (SAS) or "dark field" contrast can be obtained by analyzing the modulation in each pixel. Since the demonstration at Paul Scherrer Institute [2, 3], this technique has proven to have high potential in many research areas, including engineering, magnetism, soft matter and many more [1]. In the past decade, nGI has been implemented at a few world-class neutron imaging facilities [4–8].

Recently, a Talbot-Lau nGI system has been implemented at the Multimodal Advanced Radiography Station (MARS) at the High Flux Isotope Reactor (HFIR) [9, 10]. In this manuscript, the design, setup, and performance are described in detail. Furthermore, far-field grating interferometry has been explored at MARS in collaboration with colleagues from the National Institute of Standards and Technology (NIST). However, this is not discussed in the current manuscript.

© The Author(s) 2026
A. E. Craft and H. Z. Bilheux (Eds.): WCNR 2024, SPPHY 348, pp. 364–371, 2026.
https://doi.org/10.1007/978-3-032-15003-5_41

2 nGI at MARS

2.1 The Design

MARS is situated in the Cold Guide Hall at the High Flux Isotope Reactor. The instrument begins in a shared optics box at the end of cold guide (CG) 1 or CG1. Neutrons from the bottom section of the neutron guide go through a beam mask, diffuser, and a pinhole aperture inside the optics box. Neutrons then travel through two sections of helium filled flight tubes (with beam scrappers) to reach the sample and detector end station. Inside the optics box, a double-bounce single crystal monochromator can be selected to provide a monochromatic beam with a fixed wavelength of 2.53 Å. There is no neutron velocity selector (NVS) at MARS due to the space limitation inside the optics box. A future upgrade of the beamline does include one.

Compared to the transmission-based imaging configuration, the main difference when running nGI is the introduction of the source (G_0), phase (G_1) and analyzer gratings (G_2), respectively. These three gratings must be positioned precisely in the beam to achieve the maximum amplitude of modulation, called "visibility". In addition to visibility, a key metric, specifically for dark field imaging, is the accessible autocorrelation lengths (ACL), ξ, given by

$$\xi = \frac{\lambda L_s}{P} \tag{1}$$

where L_s is the effective sample-to-G_2 distance, λ is the neutron wavelength, and P is the period of the G_2 grating [4]. Since the monochromator at MARS is fixed at 2.53 Å, different autocorrelation lengths can only be accessed by varying either P or L_s. Given the beamline physical limitations, two nGI systems were designed to provide a broad range of ξ: one symmetric nGI with G_2 period of 56.5 μm, and one asymmetric nGI with G_2 period of 15 μm. Figure 1 shows the accessible ξ of these two systems by changing L_s between 0.5 cm and 20 cm (maximum travel allowed by the physical setup) using a monochromatic beam fixed at 2.53 Å. The symmetric setup offers a $\xi \leqslant 900$ nm while the asymmetric setup reaches higher values up to $\xi \leqslant 3400$ nm but with coarser ACL resolution (due to smaller period).

Figure 2 shows the schematic of the two nGI setups. For each setup, gratings specifications including period (p), height (h), duty cycle (dc) and distance (D) are labeled accordingly. The total length of both nGI setups is designed to be the same, so that the reconfiguration between symmetric and asymmetric setups can be easily achieved.

Both nGI setups are optimized for 2.53 Å, which is the fixed wavelength that the monochromator provides. The visibility for each grating setup are simulated as a function of neutron wavelength using the Fresnel-propagator [11] and are presented in Fig. 3. At 2.5 Å, the expected visibility is greater than 55%.

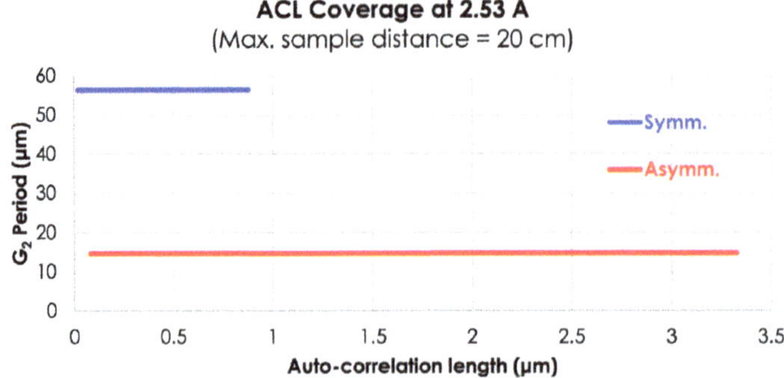

Fig. 1. Autocorrelation length coverage provided by the symmetric and asymmetric nGI setups at MARS, respectively.

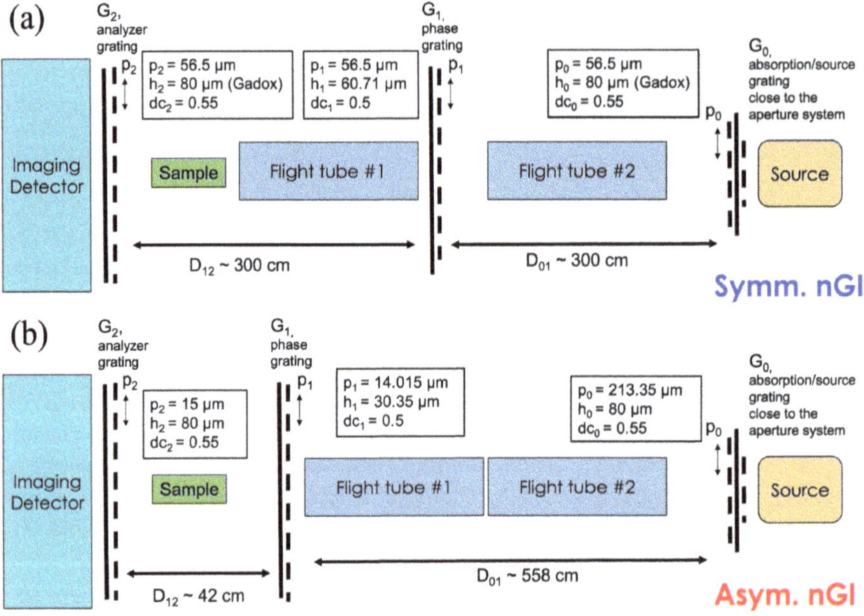

Fig. 2. Schematic of (a) symmetric and (b) asymmetric nGI setups. Neutrons are coming from the right.

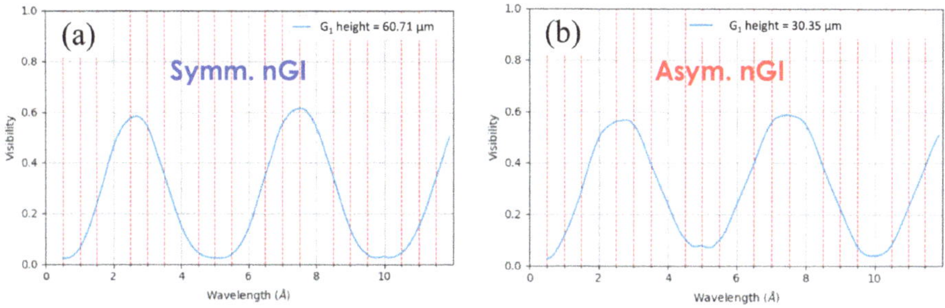

Fig. 3. Simulated visibility of (a) symmetric and (b) asymmetric nGI setups.

2.2 The Experimental Setup

Except for the location of the phase grating (G_1), the physical nGI setups are almost identical. Therefore, only an example of the symmetric nGI apparatus is shown in Fig. 4. G_0 and G_1 are mounted on a 360° rotation stage sitting on top of manual pitch/yaw stage for alignment of the gratings. G_2 is mounted directly on the face of the detector so it is perpendicular to the neutron beam. At G_1, two linear translation stages are used to fine tune the Talbot distance and to perform phase stepping, respectively.

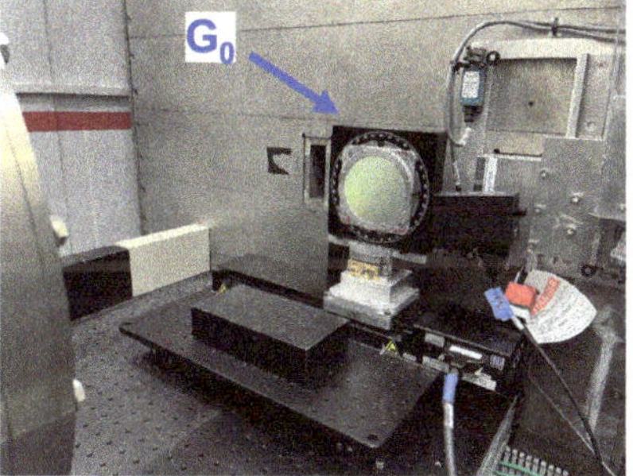

Fig. 4. Photographs of the symmetric nGI experimental apparatus at MARS.

Before aligning the gratings using neutrons, a self-leveling laser is used to roughly align the pitch, yaw and roll at all grating locations. Then a laser distance measurer is used to approximately dial in the calculated grating-to-grating distances. Subsequently, a series of motor position scans are performed to achieve the maximum visibility using a white neutron beam.

2.3 Demonstration of the Gratings' Performance

When using an nGI system, the neutron flux is significantly lower than transmission-based imaging. This is due to the beam intensity loss (~80% in total) at the absorption gratings (G_0 and G_2). Additionally, selecting a narrow wavelength band via monochromator would make the nGI measurements even slower. Since this demonstration does not require quantitative information by knowing the neutron wavelength distribution, the white beam mode was used for all the measurements presented in this work.

After aligning the gratings, an acceptable white beam visibility of ~ 25% was achieved using both symmetric and asymmetric setups. Visibility maps are shown in Fig. 5. Please note that the slightly darker stripes in these visibility maps match well with the beam structure created by the neutron guide system upstream of G_0 which produces wavelength dependent reflection angles. Therefore, the slightly lower visibility is likely due to the different neutron wavelengths in these areas.

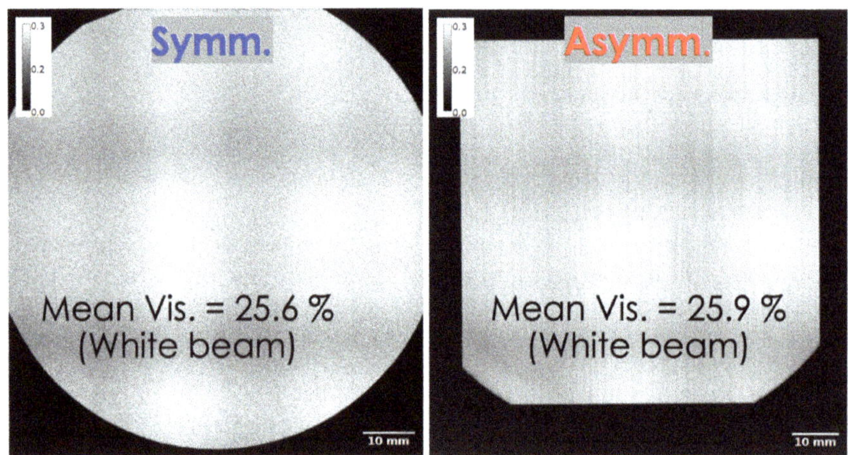

Fig. 5. Measured white beam visibility map of symmetric and asymmetric nGI setups. The effective nGI field-of-view at the detector is 90 mm diameter for the symmetric setup and 64 mm × 64 mm for the asymmetric setup.

Following the optimization of the visibility, a few metal rods (~9.5 mm in diameter) were measured in order to demonstrate the dark field contrast that can be used to differentiate metals/alloys that have very similar neutron attenuation coefficients. The result is shown in Fig. 6. A clear contrast is observed in the dark field image, as compared to the transmission image.

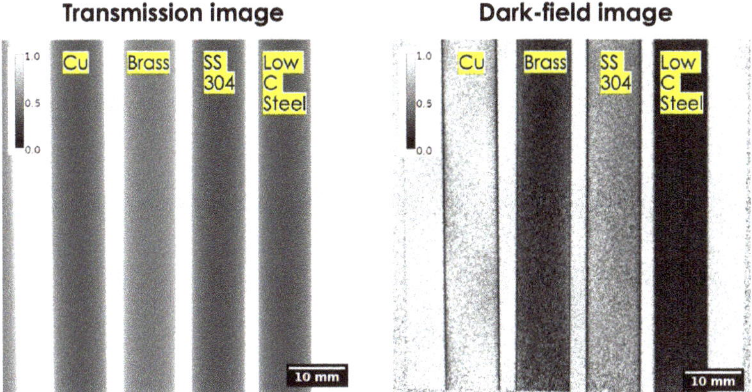

Fig. 6. Transmission and dark field images of metal rods, respectively.

To demonstrate the ability to probe nano-sized features, two suspensions that contained poly (methyl methacrylate) (PMMA) spheres with mean diameters of 100 nm and 150 nm, correspondingly, were measured using the symmetric setup. Transmission and dark field radiographs are shown in Fig. 7. Three regions-of-interest (ROIs) were identified and used to plot dark field intensity vs. ACL. The plot extracted from the blue ROI at the center of the 100 nm PMMA suspension is indicating a feature size (~100 nm) that agrees with the PMMA suspension specifications. The 150 nm PMMA suspension exhibits a gradient due to particle settling. In this case, two ROIs (red and green) were identified to assess the dark field intensity vs. ACL in the dilute and concentrated regions, respectively. Both green and red plots indicate that the feature size is identical and approximately 150 nm. The lower dark field intensity of the green plot indicates a higher number density of PMMA in the associated ROI compared to the red ROI. Please note that the ACLs presented here is calculated using the peak wavelength (~2.53 Å) at MARS. Because white beam was used for this measurement, the data is not suitable for fitting to extracting quantitative size information of the suspended PMMA particles. This demonstration aims at showing that the contrast due to nanoscale changes can be probed qualitatively.

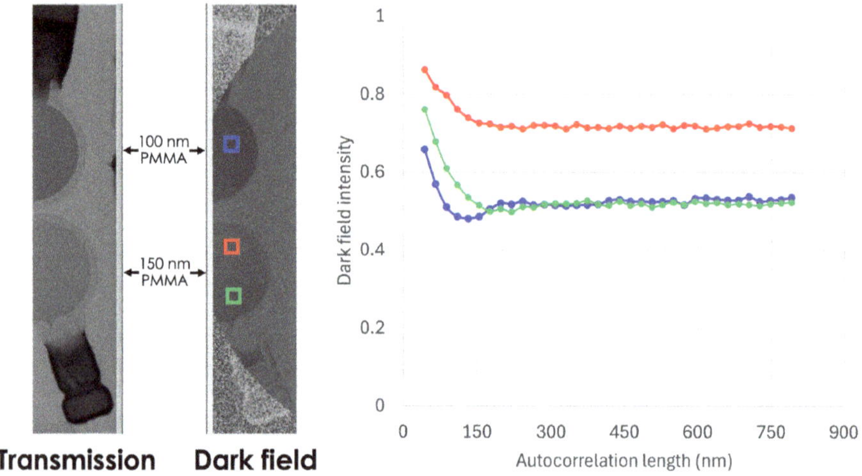

Fig. 7. nGI measurement of nano-sized PMMA spheres in suspension. Transmission and dark field images are displayed side by side. Within each region-of-interest (ROI), dark field intensity is plotted as a function of ACL.

3 Conclusion

Two nGI setups – symmetric and asymmetric Talbot-Lau interferometers – have been successfully implemented at the MARS beamline at HFIR. Without any reconfiguration at the sample position, ACLs between 40 nm and 3400 nm can be accessed by stepping the sample-to-G_2 distance up to 20 cm. Currently, white beam nGI is the primary configuration for qualitative measurements that aim to spatially resolve the location of nano-sized features. Later on, other quantitative techniques can be strategically utilized at the locations in the sample identified by nGI.

A monochromator fixed at a neutron wavelength of 2.53 Å is available if quantitative nGI measurement is desired at the cost of flux. Furthermore, a future MARS upgrade includes a neutron velocity selector that will allow time-efficient quantitative nGI measurements of such.

Acknowledgments. This research used resources at the High Flux Isotope Reactor, a DOE Office of Science User Facility operated by the Oak Ridge National Laboratory. The beam time was allocated to MARS (CG-1D) on proposal numbers IPTS-26032 and IPTS-26647. The authors thank Dr. Fankang Li and Dr. Fumiaki Funama providing the PMMA samples used in the characterization of the nGI apparatus. Y.Zhang also thank Dr. Jiao Y.Y. Lin for sharing his Fresnel-propagator codes for nGI simulation. The authors thank the Xiaosong Geng, Mike Harrington, Gary Taufer and Rob Knudson IV from DAQ group for their support in integrating all the nGI motors and stages.

References

1. Strobl, M., et al.: Wavelength-dispersive dark-field contrast: micrometre structure resolution in neutron imaging with gratings. J. Appl. Crystallogr. **49**(2), 569–573 (2016)

2. Pfeiffer, F., et al.: Neutron phase imaging and tomography. Phys. Rev. Lett. **96**(21), 215505 (2006)

3. Grunzweig, C., et al.: Design, fabrication, and characterization of diffraction gratings for neutron phase contrast imaging. Rev. Sci. Instrum. **79**(5), 053703 (2008)

4. Reimann, T., et al.: The new neutron grating interferometer at the ANTARES beamline: design, principles and applications. J. Appl. Crystallogr. **49**(5), 1488–1500 (2016)

5. Neuwirth, T., et al.: A high visibility Talbot-Lau neutron grating interferometer to investigate stress-induced magnetic degradation in electrical steel. Sci. Rep. **10**(1), 1764 (2020)

6. Sarenac, D., et al.: Three phase-grating moire neutron interferometer for large interferometer area applications. Phys. Rev. Lett. **120**(11), 113201 (2018)

7. Hidrovo, I., et al.: Neutron interferometry using a single modulated phase grating. Rev. Sci. Instrum. **94**(4) (2023)

8. Kardjilov, N., et al.: CONRAD-2: the new neutron imaging instrument at the Helmholtz-Zentrum Berlin. J. Appl. Crystallogr. **49**(1), 195–202 (2016)

9. Crow, L., et al.: The CG1 instrument development test station at the high flux isotope reactor. Nucl. Instrum. Methods Phys. Res., Sect. A **634**(1), S71–S74 (2011)

10. Santodonato, L., et al.: The CG-1D Neutron Imaging Beamline at the Oak Ridge National Laboratory High Flux Isotope Reactor. Phys. Procedia **69**, 104–108 (2015)

11. Hipp, A., et al.: Energy-resolved visibility analysis of grating interferometers operated at polychromatic X-ray sources. Opt. Express **22**(25), 30394–30409 (2014)

Conceptual Optics Design of the CUPI^{2}D Beamline at the Spallation Neutron Source Second Target Station

Hassina Z. Bilheux[1]([✉]) [iD], James Torres[1] [iD], Adrian Brügger[2] [iD], and Jiao Lin[1] [iD]

[1] Neutron Scattering Division, Neutron Sciences Directorate, Oak Ridge National Laboratory, Oak Ridge, TN 38731, USA
bilheuxhn@ornl.gov

[2] Civil Engineering and Engineering Mechanics, Columbia University, New York, NY 10027, USA

Abstract. The Oak Ridge National Laboratory's Spallation Neutron Source is planning to add new capabilities centered around high fluxes of cold neutrons as part of the Second Target Station. Among the proposed instruments is a time-of-flight imaging beamline, called CUPI^{2}D, optimized for dynamic measurements in advanced materials such as energy materials and superalloys, using two techniques: Bragg edge imaging and neutron grating interferometry. Since the Second Target Station will produce a bright pulse of cold neutrons, the beamline will capture phenomena in situ within seconds to minutes in two dimensions. It will also be capable of capturing computed tomography data using both aforementioned techniques. This manuscript explains how CUPI^{2}D complements the VENUS beamline currently under commissioning at the Spallation Neutron Source and focuses on the beamline requirements and optics.

Keywords: CUPI^{2}D · beamline optics · time-of-flight imaging

1 Introduction

1.1 State-of-the-Art Time-of-Flight Capabilities

The transmission measurement of cold neutrons through crystalline materials was first demonstrated by E. Fermi *et al.* [1]. More development followed between the 1950's and 1980's. In the past 25 + years, the measurement of crystalline properties using cold neutrons was modernized with the advances of spallation sources [2] and detector technology [3]. Time-of-flight (TOF) or hyperspectral imaging was demonstrated in the early 2000's with the measurements of strain averaged through a sample's thickness [4, 5]. Several years of development at reactor facilities (using a velocity selector or a monochromator) ensued and provided a strong justification of building beamlines at spallation sources around the world, such as RADEN at the Japanese Accelerator Research Complex (J-PARC) in Japan, the Imaging and Materials Science and Engineering (IMAT) at the Rutherford Appleton Laboratory in the UK, and the Energy Resolved Neutron Imaging

© The Author(s) 2026
A. E. Craft and H. Z. Bilheux (Eds.): WCNR 2024, SPPHY 348, pp. 372–379, 2026.
https://doi.org/10.1007/978-3-032-15003-5_42

(ERNI) beamline at Los Alamos National Laboratory (LANL) in the US. More recently, the Chinese Spallation Neutron Source (CSNS) built ERNI (unrelated to the one built at LANL) and the Spallation Neutron Source (SNS) at the Oak Ridge National Laboratory just completed the construction of the VENUS TOF imaging beamline at the First Target Station (FTS). Finally, ODIN is currently being built at the European Spallation Neutron Source (ESS).

1.2   Existing and Future SNS Imaging Capabilities

VENUS [6, 7, *H. Z. Bilheux et al., these proceedings*] and the Complex Unique and Powerful Imaging Instrument for Dynamics (CUPI^{2}D) [8] are complementary beamlines, as illustrated in Table 1. CUPI^{2}D is designed to provide simultaneous Bragg edge imaging (BEI) and neutron grating interferometry (nGI) measurements with a time resolution that enables in situ dynamic measurements in a broad range of materials such as those used for energy storage, aerospace, additive manufacturing and civil engineering. The most significant complementarity between VENUS and CUPI^{2}D is their neutron energy range. While VENUS excels at epithermal and thermal imaging, CUPI^{2}D provides the complementary cold neutrons, allowing, for example, the performance of resonance imaging in the former and the probing of large structures beyond 4 Å in the latter. Since the SNS Second Target Station (STS) will operate at a lower repetition rate, CUPI^{2}D will benefit from a larger bandwidth while accepting a lower (than VENUS) wavelength resolution driven by the moderator's intrinsic energy resolution and source-to-detector distance. Nevertheless, this wavelength resolution is sufficient for both BEI and nGI.

Table 1. VENUS and CUPI^{2}D

Parameter	CUPI^{2}D	VENUS
Source power (MW)	0.7	2
Source repetition rate (Hz)	15	60
Moderator	Cylinder [9] (cold)	Decoupled H_2 (epithermal/thermal)
Source-to-detector distance (m)	~ 22 to 35	25
Wavelength bandwidth $\Delta\lambda$ (Å)	7.89 Å (@33 m)	2.64 (@25 m)
Wavelength resolution ($\Delta\lambda/\lambda$) @3 Å	~ 0.003 (@33 m)	~ 0.0015 (@25 m)
Minimum wavelength λ_{min} (Å)	2	0.01 (with parked choppers)
Collimation ratio range	100–1000	400–2000
Max. Field-of-view (cm^2)	15 × 15	20 × 20

The collimation ratio options between the two beamlines are somewhat comparable, along with their field-of-view at the farthest respective detector position. While VENUS does not have a guide system, STS is designed such that there is a common front-end area called the bunker with optical components such as choppers (see Fig. 2). This area is inaccessible while the source is running due to shared shielding among neighboring beamlines. Hence, CUPI^{2}D necessitates the use of an elliptical guide system that fundamentally transports the source phase space to a position inside the instrument cave where components can be accessed and interchanged by imaging staff and users (i.e., a virtual source).

2 Conceptual Optics Design of the CUPI^{2}D Beamline

2.1 CUPI^{2}D Bunker

All STS beamlines share a common front-end optics area called the bunker with bandwidth (disk) choppers and a T_0 chopper, filters, collimators that are also the primary and secondary shutters, and a larger shutter when bunker access is required for maintenance activities during the outages. All these components are placed between the source and the cave shielding, which starts 13.7 m from the upstream neutron moderator (Fig. 1). The disk and double disk choppers are optimally located 6 and 9 m away from the moderator, respectively [8]. The disk chopper is necessary to tune the wavelength bandwidth, $\Delta\lambda$, to the appropriate source-to-detector distance. The T_0 chopper is located at 7.5 m and is designed to reduce the flux of epithermal and fast neutrons since CUPI^{2}D looks directly at the source. In general, the T_0 chopper location is placed upstream of the cave to minimize background in the sample/detector area, and its design is optimized to match the minimum desirable wavelength of 2 Å. Filters are judiciously positioned upstream since the beam cross-section is smaller closer to the source. Neutron transport is achieved using an elliptical guide system that "transports/projects" the moderator source parameters to the aperture located inside the cave (see next section).

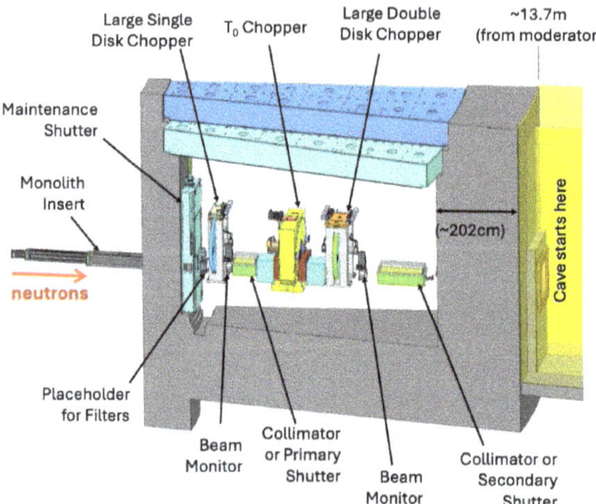

Fig. 1. Conceptual design of the CUPI^{2}D bunker components showing the maintenance shutter, disk and T_0 choppers, beam monitors, collimators/shutters, and filters.

2.2 CUPI^{2}D Conceptual Configurations

The CUPI^{2}D is carefully optimized for maximum flexibility, ensuring access to essentially three configurations defined by three L/D ratios, where L is the distance from the aperture of diameter, D, and the detector. Unlike VENUS, which has a fixed source-to-detector position and a variable aperture system (i.e., different D's) with round apertures, CUPI^{2}D has a fixed 3×3 cm^2 square aperture positioned at the entrance of the cave at a distance of 19 m from the moderator. L/D is changed by changing L, which inevitably changes the field-of-view (FOV) at the detector position, as illustrated in Table 2. The closest position corresponds to the lowest L/D of 100 and is designed for high-flux measurements such as the observations of ms dynamic events within a FOV of 4×4 cm^2. The furthest position (L/D 1000) provides the highest wavelength resolution and thus is ideal for Bragg edge measurements. Neutron interferometry data can also be measured since they require a lower wavelength resolution and thus data can be binned in TOF to alleviate the lower flux available at this position. The intermediate configuration is ideal for nGI measurements with both moderate flux and wavelength resolution. Additionally, Bragg edge imaging (i.e., the technique that requires the highest wavelength resolution as compared to nGI) can be performed at any position with the highest wavelength resolution for strain mapping at 33 m at the cost of flux.

Table 2. CUPI^{2}D conceptual configurations

Parameter	L/D 100	L/D 300	L/D 1000
Aperture-to-detector distance (m)	3	9	15
FOV (cm^2)	4×4	9×9	15×15
Source-to-detector distance (m)	21	27	33
Wavelength resolution ($\delta\lambda/\lambda$) @ 3 Å	~ 0.007	~ 0.005	~ 0.003
Simulated flux* (n.cm^{-2}.s^{-1}); $2 < \lambda < 4$ Å	5.9×10^8	6.6×10^7	6.0×10^6
Simulated flux* (n.cm^{-2}.s^{-1}); $4 < \lambda < 10$ Å	2.6×10^8	2.9×10^7	2.6×10^6
Simulated flux* (n.cm^{-2}.s^{-1}); $10 < \lambda < 14$ Å	1.3×10^7	1.5×10^6	1.4×10^5
Optimized capability	High flux	nGI	Bragg edge & nGI

(*) based on a cutoff flux of 75% of the peak flux.

A perfect elliptical guide, i.e., without gaps for the CUPI^{2}D components, was first modeled using the Monte Carlo code named McStas [10, 11]. It was optimized to reduce the guide artefacts at the detector position for each L/D (see Fig. 2(a), (b) and (c) correspond to L/D 100, 300 and 1000, respectively). Once gaps are introduced in the elliptical guide and optimized, minimal impact is observed in the corresponding FOVs (see Fig. 2(d), (e) and (f) correspond to L/D 100, 300 and 1000, respectively).

The CUPI^{2}D cave is designed to accommodate the different modes of operation of the instrument, as illustrated in Fig. 3. Every instrument at SNS occupies a "pie shape" real estate with a bisector that is shared between neighboring instruments, as illustrated in the top view of the beamline in Fig. 3(a). Figure 3(b) displays a vertical cross-section of the beamline from the bunker to the radiological materials area (RMA) with the two hutches (beamline control room called "hutch", and sample preparation laboratory) sitting behind the beam stop, similar to VENUS. The cave shielding and beam stop are modeled for the accidental scenario for which all choppers are parked open and white beam (and gammas) reach the beam stop. The cave is equipped with an elevator to allow swift changes between *L/D* configurations, nGI gratings, and potential advanced optics, while maintaining critical alignment. The farthest position (*L/D* = 1000) is outfitted with a roof hatch to allow the easy transfer of large specimen and sample environment – a load frame, for example – into the cave. Additional optimization of the cave will follow in the coming years, as the technical capabilities are further refined, and the shielding design is optimized.

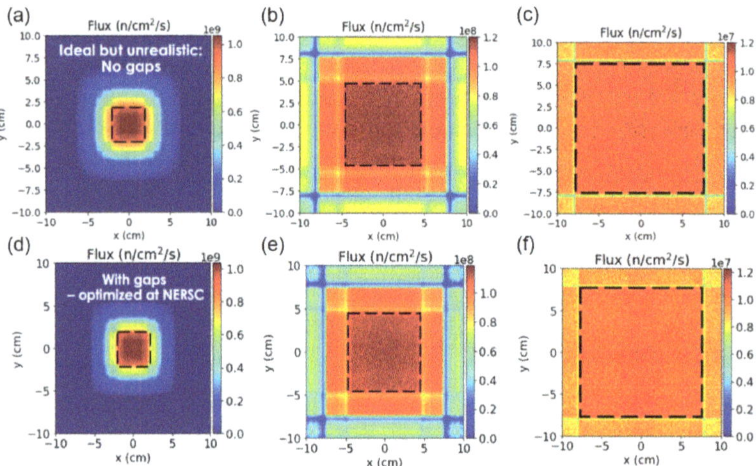

Fig. 2. Simulated radiographs and FOVs for *L/D*'s 100, 300, and 1000 (left to right); (a), (b) and (c) correspond to the ideal elliptical guide system; (d), (e), and (f) are the equivalent radiographs for an optimized elliptical guide with gaps introduced for the beamline components located in the bunker. The FOVs are indicated with a black dashed-line box.

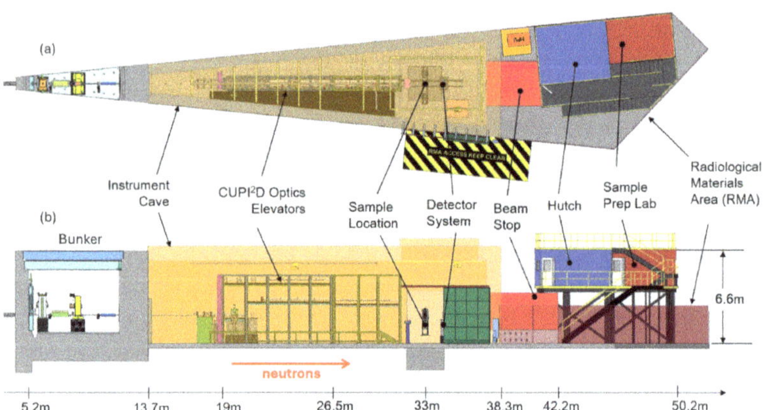

Fig. 3. Conceptual layout of CUPI²D with key positions marked relative to the moderator surface (z = 0).

An elevator system, similar to the one in operation at RADEN [12], will allow the beamline to quickly transition between the three operating modes while maintaining measurement precision, and equally importantly, it can be performed with minimal effort from beamline scientists. This versatility means that a quick turnaround can be made between experiments, especially important when the high beam brightness will result in frequent sample changes, on the order of every day or two for simple imaging/tomography measurements. Furthermore, high measurement precision will be demanded when performing Bragg edge imaging for aerospace or other failure-critical components. Such applications demand robust workflows to ensure data integrity and repeatability in the relevant beamline configuration, a need that is critically supported through this automation. Figure 4(a), (b), and (c) illustrate the $L/D = 100, 300$, and 1000 collimation positions, respectively.

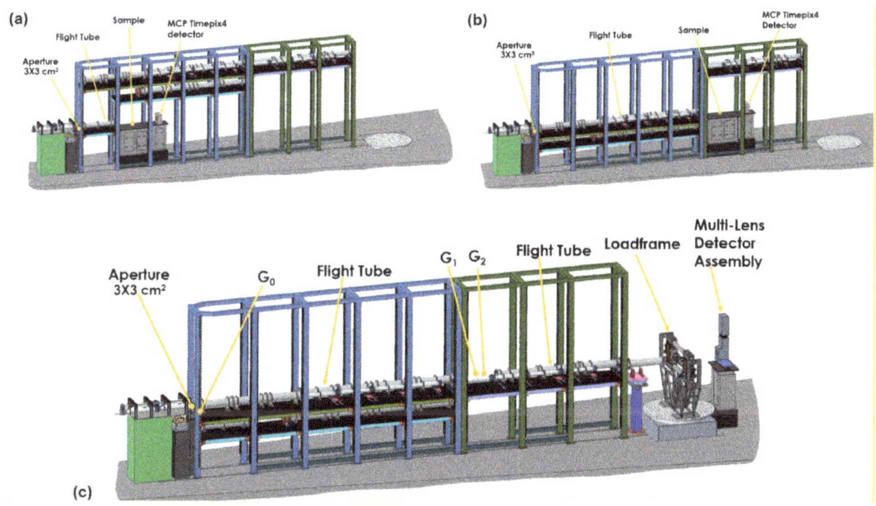

Fig. 4. Elevator system showing three main operating modes; (a) $L/D = 100$ collimation for high-flux fast measurements, (b) $L/D = 300$ intermediate position for nGI measurements at a balanced wavelength resolution, FOV, and flux, and (c) $L/D = 1000$ collimation for maximum FOV and, independently, maximum wavelength resolution required to perform BEI and nGI simultaneously. The nGI gratings are labelled G_0, G_1 and G_2, although other grating setups may be used.

The high-flux configuration in Fig. 4(a) shows the sample stage at the position nearest to the source, with only a small flight tube deployed into the beam via the elevator. The imaging detector and sample stages are each mounted on a platform that itself sits on a set of rails that allow travel along the beam direction beneath the elevator, with the elevator columns straddling the detector and sample stages. When the beamline is configured to $L/D = 300$, as shown in Fig. 4 (b), then a noble gas purged flight tube can be deployed into the beam via the elevator to bridge the additional 6 m of neutron flight path. Finally, $L/D = 1000$ can be achieved in the far field position at 33 m from the source. At the cost of flux, this position maximizes both the FOV and wavelength resolution, enabling the performance of high-fidelity BEI experiments. This location is also outfitted with a heavy-duty sample displacement/rotation table capable of carrying load frames and large samples. Figure 4(c) illustrates this position with an nGI setup as well as a large axial-torsion load frame. This grating configuration serves an illustrative purpose and is by no means the only option being considered (e.g. far field gratings, similar configuration at other *L/D*, etc.). However, it underlines the design of this beamline to perform *simultaneous* multi-modal and multiscale measurements in direct (attenuation contrast imaging) and reciprocal space (nGI, BEI).

3　Summary and Future Efforts

The conceptual design of the $CUPI^2D$ beamline is driven by the scientific requirements [8] to provide three main configurations which are optimized to provide high flux, neutron grating interferometry, and simultaneous nGI and Bragg edge imaging capabilities at three different locations in the instrument cave, as described in this manuscript. The highest wavelength resolution is on the order of 0.3% and is required for strain mapping at the most downstream detector position. While a concept has been established for $CUPI^2D$, such as the elliptical transport guide system, there remain significant research efforts in developing a detailed design concept, as well as cost and schedule for the instrument. For example, the elliptical guide system must be refined, considering the specific manufacturing technology being employed at the time of construction of the instrument. Future interactions with the scientific community will contribute to the refinement of the $CUPI^2D$ technical capabilities.

Acknowledgments. This research used resources of the Spallation Neutron Source Second Target Station Project at Oak Ridge National Laboratory (ORNL). ORNL is managed by UT-Battelle, LLC, for DOE's Office of Science, the single largest supporter of basic research in the physical sciences in the United States. This research used resources of the National Energy Research Scientific Computing Center (NERSC), a Department of Energy Office of Science User Facility The authors would like to thank Ms. Cristina Boone, Mr. Scott Dixon, and Mr. William (Bill) Turner for their invaluable contributions to the concept and engineering design of the $CUPI^2D$ beamline.

References

1. Fermi, E., Sturm, W.J., Sachs, R.G.: The transmission of slow neutrons through microcrystalline materials. Phys. Rev. **71**(9), 589–594 (1947)
2. Mason, T.E., et al.: The spallation neutron source in oak ridge: a powerful tool for materials research. Physica B **385–386**, 955–960 (2006)
3. Tremsin, A.S., et al.: Detection efficiency, spatial and timing resolution of thermal and cold neutron counting MCP detectors. Nucl. Instrum. Methods Phys. Res., Sect. A **604**(1–2), 140–143 (2009)
4. Steuwer, A., et al.: Bragg edge determination for accurate lattice parameter and elastic strain measurement. Physica Status Solidi (a), **185**(2), 221–230 (2001)
5. Santisteban, J.R., et al.: Strain imaging by Bragg edge neutron transmission. Nucl. Instrum. Methods Phys. Res., Sect. A **481**(1–3), 765–768 (2002)

6. Bilheux, H., et al.: The VENUS imaging beamline at the Spallation Neutron Source: Layout, expected performance and status of the construction project. J. Phys. Conf. Ser. (2023)
7. Bilheux, H.Z., et al.: The VENUS imaging beamline at the oak ridge national laboratory spallation neutron source. AM&P Techn. Art. **181**(8), 18–23 (2023)
8. Brügger, A., et al.: The complex, unique, and powerful imaging instrument for dynamics (CUPI2D) at the spallation neutron source. Rev. Sci. Instrum. **94**(5) (2023)
9. Ghoos, K., et al.: Optimization of the second target station cold source moderators using an automated workflow. Nucl. Instrum. Methods Phys. Res., Sect. A **1060**, 169035 (2024)
10. Willendrup, P.K., Lefmann, K.: McStas (i): introduction, use, and basic principles for ray-tracing simulations. J. Neutron Res. **22**(1), 1–16 (2020)
11. Willendrup, P.K., Lefmann, K.: McStas (ii): an overview of components, their use, and advice for user contributions. J. Neutron Res. **23**(1), 7–27 (2021)
12. Shinohara, T., et al.: The energy-resolved neutron imaging system, RADEN. Rev. Sci. Instrum. **91**(4) (2020)

Stroboscopic Bragg Edge Radiography of Twin Formation in a Magnetic Shape Memory Alloy

S. Kabra[1,2], W. Kockelmann[1(✉)], G. Burgess[1], M. J. Gutmann[1], and A. S. Tremsin[3]

[1] Science and Technology Facilities Council (STFC), Rutherford Appleton Laboratory, ISIS Facility, Harwell OX11 0QX, UK
`Winfried.kockelmann@stfc.ac.uk`
[2] Oak Ridge National Laboratory (ORNL), Oak Ridge, TN 37831, USA
[3] Space Sciences Laboratory, University of California at Berkeley, Berkeley, CA 94720, USA

Abstract. Stroboscopic neutron imaging provides opportunities to capture fast and repetitive processes in the millisecond range, even for a slow acquisition technique such as energy-resolved neutron imaging. We have implemented an asynchronous collection mode and performed stroboscopic neutron Bragg dip imaging on a pulsed source instrument to follow dynamic microstructure changes in a Ni_2MnGa magnetic shape memory alloy. The single crystalline alloy exhibited large elongations and contractions of a few percent upon application of a cyclic alternating magnetic field and an external force. We used a custom-made actuator and signal generator for the external magnetic field and a microchannel plate (MCP) detector in an event list mode for neutron detection. We achieved a spatial resolution of approximately 200 μm and a temporal resolution of a few milliseconds for observing crystal twinning. The stroboscopic method used to study twin formation and dynamics is essentially a 4D neutron imaging technique, combining radiographic imaging using a 512x512-pixel Timepix system with time-of-flight (TOF) and an actuator phase as additional dimensions.

Keywords: Stroboscopic Neutron Radiography · Bragg Edge · Bragg Dip · Magnetic Shape Memory Alloy

1 Introduction

Stroboscopic imaging is often used to showcase advantages of neutron science, as internal processes inside materials, devices and machinery can be visualized in real space. This is enabled by the high penetration depth of neutrons in materials and the exceptional in situ measurement capabilities provided on neutron beamlines. Stroboscopic neutron imaging has been demonstrated several times at steady state sources, often to visualize running motor engines [1, 2]. Stroboscopic data collection enables studies of dynamic and repetitive processes that are much faster than the typical data acquisition capability of an instrument. This acquisition mode requires a trigger from the sample process to take repeated snapshots of a frame, e.g. a fraction of a second each time, and by accumulating data until the image has sufficient statistics. One does not need a strong neutron source,

© The Author(s) 2026
A. E. Craft and H. Z. Bilheux (Eds.): WCNR 2024, SPPHY 348, pp. 380–388, 2026.
https://doi.org/10.1007/978-3-032-15003-5_43

rather one needs a sample process that is repetitive and does not change over a certain number of cycles. Over the past decades, examples of stroboscopic neutron imaging have been relatively few. Before dedicated imaging instruments were available at pulsed neutron sources, an asynchronous stroboscopic mode was proposed to capture dynamic processes [3].

We have implemented asynchronous stroboscopic imaging and, furthermore, demonstrated feasibility of a stroboscopic Bragg edge experiment on IMAT at the pulsed ISIS neutron and muon source. The sample used for this study was a Ni_2MnGa magnetic shape memory alloy which has a twinned microstructure in the martensitic phase at room temperature [4]. A single crystal of Ni_2MnGa exhibits changes in sample shape and sample size as a function of external magnetic field or external force. This is due to changing relative populations of twin variants as the external stimuli, i.e. magnetic field or stress, is applied. An external field flips a domain so that the c-axis (easy axis) of the ferromagnetic compound aligns with the field (Fig. 1). With the easy axis perpendicular to an increasing magnetic field, a crystal successively transforms to the other domain variant type and expands in the process. By application of a force perpendicular to the magnetic field, twins rotate back while the crystal contracts. Thus, the elongation is achieved magnetically and the contraction is achieved mechanically. The operating temperature for twinning to take place in Ni_2MnGa is below 60 °C, with relative elongations, i.e. strains, of up to 6% [5]. Ni_2MnGa alloys have been shown to actuate at large frequencies of up to 1 kHz and for millions of cycles [6]. These properties makes the alloy of interest for a number of applications such as actuators, circuit breakers, valves and mini-grippers, for example in robotic and medical devices.

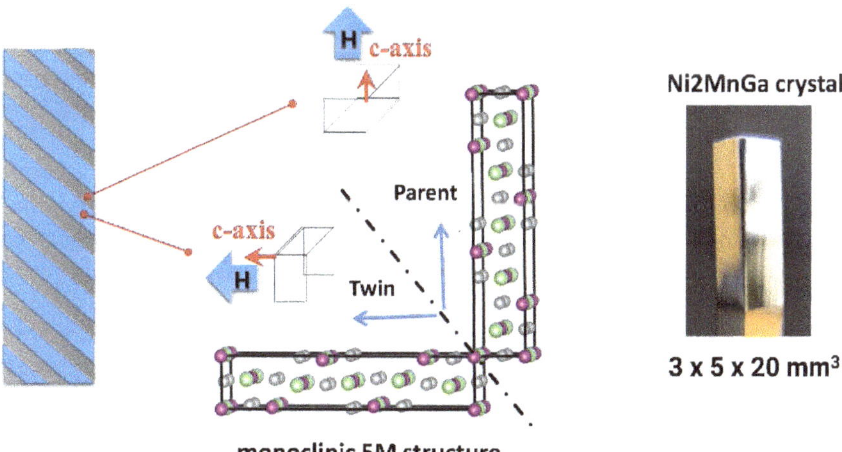

Fig. 1. Sketch of a partially twinned crystal (left) and crystal structure of ferromagnetic Ni_2MnGa (middle) exhibiting two types of domains that are related by a 90 degree rotation of the unit cell. An external magnetic field (H) aligns the c-axis of the unit cell along the field direction. Photo of a crystal with dimensions used in this study (right).

Structure and phase transitions have been investigated in numerous neutron diffraction studies [5]. Previous energy-resolved Bragg edge measurements of an ex-situ twinned Ni_2MnGa crystal [7] demonstrated that domain structures could be visualized even when white-beam neutron imaging showed no visible contrast. Here we focused on the experimental requirements for in-situ stroboscopic neutron imaging at a pulsed source, using twin formation of Ni_2MnGa as an example. Additional aims included measuring sample strain and determining twin boundary velocities and potential performance degradation over extended cycling with an alternating magnetic field and applied stress.

2 Experimental

A Ni_2MnGa crystal of dimension 3x5x20 mm^2 was obtained from ETO Magnetic GmbH, Germany. The crystal structure, examined by neutron diffraction on the SXD diffractometer [8] at the ISIS neutron source, was found (i) to be consistent with a non-stoichiometric composition around $Ni_{1.9}Mn_{1.2}Ga_{0.9}$ in agreement with [4] and (ii) to exhibit a twin structure in zero field (Fig. 1). Neutron imaging experiments were performed on IMAT [9, 10] at ISIS TS2, a source operating at 10 Hz. The neutron wavelength range was 0.7 - 6.7 Å corresponding to a TOF range from 10 to 95 ms. The experiment used an L/D of 170 and a neutron flux of about 1×10^7 n/cm^2/s [9]. A micro-channel plate (MCP)/Timepix detector [11] was used, providing 512×512 pixels for a field of view of 28×28 mm^2, with each pixel measuring 55 μm. The detector accepted two external timing signals: a 10 Hz trigger signal synchronizing the MCP to the neutron source, and a 20.01 Hz signal from the actuator device used for the experiment to apply an external field and mechanical stress. The flight path from source to detector was 56.5 m. Neutron events were collected in event list mode where a detected neutron was tagged with x, y pixels numbers, time difference of neutron detection to a 10 Hz source trigger Ts (i.e. TOF of neutron) and time difference of actuator signal Tp to the 10 Hz source trigger (i.e. sample phase start time).

The setup is shown in Fig. 2. An actuator provided by ETO Magnetic GmbH was used for the stroboscopic experiment. A Ni_2MnGa sample was installed between electromagnetic pole shoes creating a horizontal periodic magnetic field. A spring on top of the sample applied a force sufficiently high to make the crystal contract as the magnetic field dropped to zero halfway through an actuator cycle. A custom-made field generator controller was designed and built at ISIS, to create specific cyclic discharge current waveforms for in-situ measurements. The actuator control box provided magnetic field amplitudes of up to several hundred mT, as it followed voltage (up to 30 V) and current (up to 6 A) profiles. The actuator was air-cooled to keep the sample at about 40 °C. Examples of current and voltage waveforms and the corresponding elongation curve are displayed in Fig. 3. Furthermore, the controller provided a timing signal of the rising edge of the current waveform to the MCP detector which also received the ISIS synchronisation trigger.

A slight mismatch of the actuator frequency, here 20.01 Hz instead of 20 Hz, from an integer multiple to the 10 Hz source frequency is intentional. This asynchronous setup ensures that neutrons of varying TOF are recorded for a given sample displacement

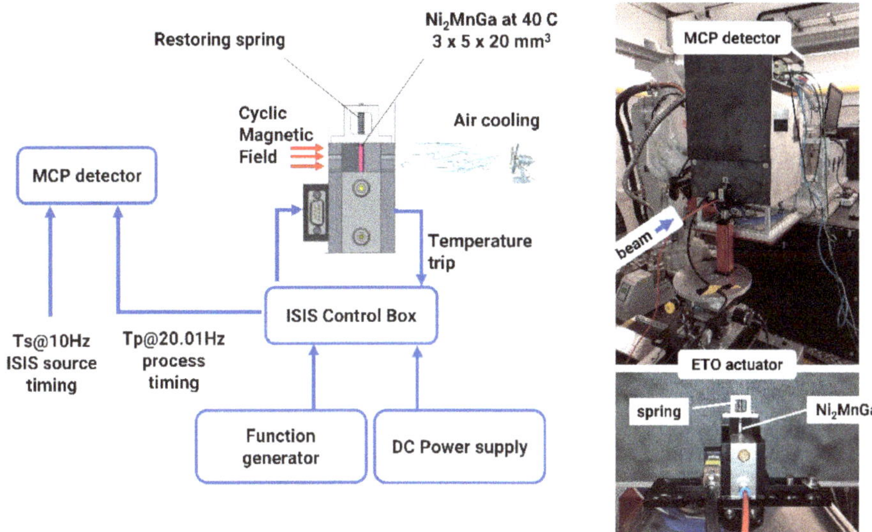

Fig. 2. (a) Schematic of ETO actuator with 20 mm long Ni_2MnGa crystal in a horizontal magnetic field. A restoring steel spring on top of the crystal reverted the effect of the magnetic field; (b) setup on IMAT with actuator in front of MCP detector.

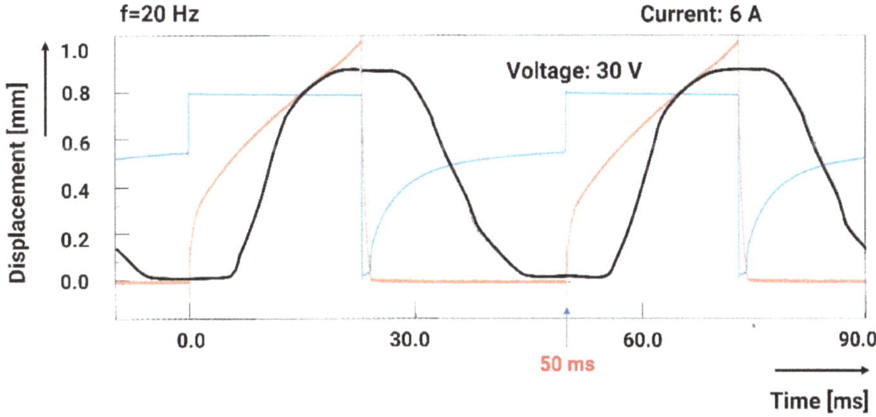

Fig. 3. Voltage (blue), current (orange) and displacement (black) curves for 20 Hz operation of the actuator as per datasheet. For 30 V and 6 A, the elongation amplitude of the crystal is about 0.9 mm.

channel [3]. Hence, with a scan over many cycles one obtains a full TOF spectrum for each displacement channel. The total collection times for the data reported here was 2 h (zero field) and 16 h (16 V), the latter corresponding to 1.15×10^6 actuator cycles. Crystal elongation and contraction closely followed the periodic, continuously varying external magnetic field and force, while at the same time the twin structure of the sample was interrogated using neutron imaging with microsecond time resolution using the event-mode setup. Event list data were processed by choosing the number of displacement phases (here 20) and by creating a TOF histogram with a sufficient number of channels (here 1000). Dividing the data into 20 displacement channels corresponds to time bins of 2.5 ms, within which any temporal information is averaged. For each displacement phase, neutron events were sorted into TOF histogram bins. In this way, the event list was converted to TOF image stacks, with one TOF stack for each displacement phase, i.e. 20 TOF stacks were obtained. Data stacks were processed using Fiji ImageJ [12] and Mantid Imaging [13] software tools. It can be noted that saving data as event lists is highly advantageous as one can carry out sorting of data into TOF and sample channels post-experiment. Furthermore, one can analyse subsets of data to study the temporal evolution of the crystal response.

For each transmission spectrum recorded by a detector pixel, a twin variant generates a distinct pattern of single-crystal Bragg dips. The positions and intensities of these dips are determined by the crystal structure and the orientation of the twin relative to the neutron beam direction [7, 14]. When neutrons pass through multiple twins, the resulting spectrum represents a superposition of all individual Bragg dip patterns. Notably, for a polycrystalline sample with randomly oriented crystallites, the superposition of numerous Bragg dip patterns leads to the formation of Bragg edges in the transmission spectrum.

3 Results

Figure 4 shows flat-fielded data from the sample in zero field. The sample outline is clearly visible between the steel boundaries of the actuator (Fig. 4a) where the white areas in-between indicate air transmission. The steel spring is not shown. Summing over all TOF channels yields a white beam image which is more or less homogenous without any diffraction contrast. Selecting a certain TOF range around 56 ms (3.9 Å) reveals a non-uniform grey value map (Fig. 4b), with transmission levels (green and blue) varying across the sample and belonging to two crystal orientations. The color scales for green and blue regions represent transmission levels in the Bragg dip spectra of Fig. 4c and Fig. 4d. In turn, such spectra can be used to identify further wavelength regions that create diffraction contrast.

With the actuator in operation, data were accumulated over repeated 50 ms cycles of crystal elongation via magnetic field and crystal contraction through restoration by the steel spring. Figure 5 shows sequences of images for such a full actuator cycle for an external voltage of 16 V. Only 10 of 20 sample channels are displayed for space reasons. Relative sample length changes $\Delta l/l$ of the crystal up to 2% were observed in the white-beam image sequence (Fig. 5a). This sequence is reminiscent of images from a running motor engine recorded stroboscopically at a steady state neutron source. By limiting data to a TOF range between 41.6–43.6 ms, corresponding to 2.9–3 Å, diffraction contrast was obtained and a twin domain structure was observed (Fig. 5b) with characteristic 45-degree stripes reminiscent of the pattern in Fig. 1. At the beginning of the sequence most of the crystal is represented by a single variant (mostly green), albeit not being in a single variant state. With increasing field during the external cycle, domains expanded to their maximum size under applied field and contracted as the field decreased and the applied spring force became dominant, over a 50ms period shown in Fig. 5 at 5 ms intervals. At the midpoint of the cycle (25 ms), a significant part of the crystal had transformed into the second twin variant, however never reaching a single variant state.

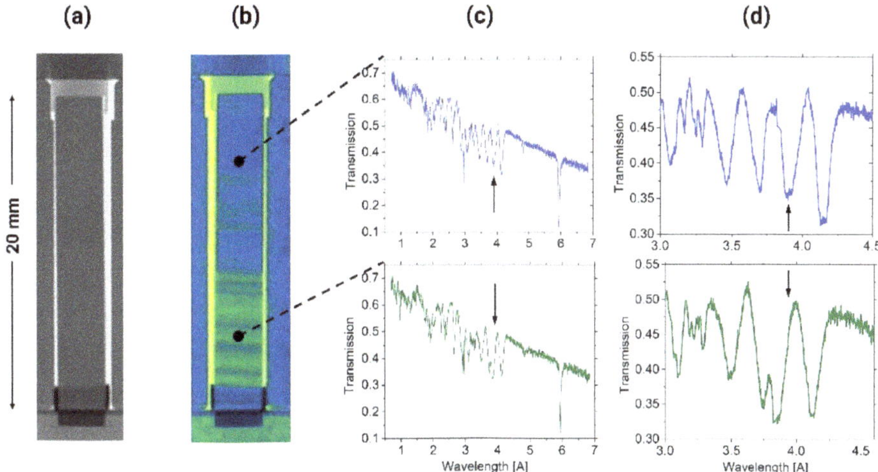

Fig. 4. Enhancing contrast by energy-selection, with sample in zero field. (a) Averaged over all TOF channels ('white beam'). (b) Radiogram for a TOF range between 56–57 ms. (c) Examples of Bragg dip spectra for higher (green) and lower (blue) transmission regions of interest. (d) zoomed regions from (c). Black arrows indicate a wavelength region of large contrast, used to plot the radiogram in (b).

Figure 5b illustrates that diffraction contrast is straightforwardly enhanced by selecting data for a narrow wavelength band. It is expected that fitting of Bragg dip spectra will help to maximize diffraction contrast even further. It is important to note that the images from the stroboscopic acquisition show the response of the sample averaged over many cycles. In other words, temporal behaviour during one cycle is averaged over, in this case, one million periods. Further analysis of the data, in terms of twin boundary velocities and evaluation of degradation of a sample will be presented elsewhere.

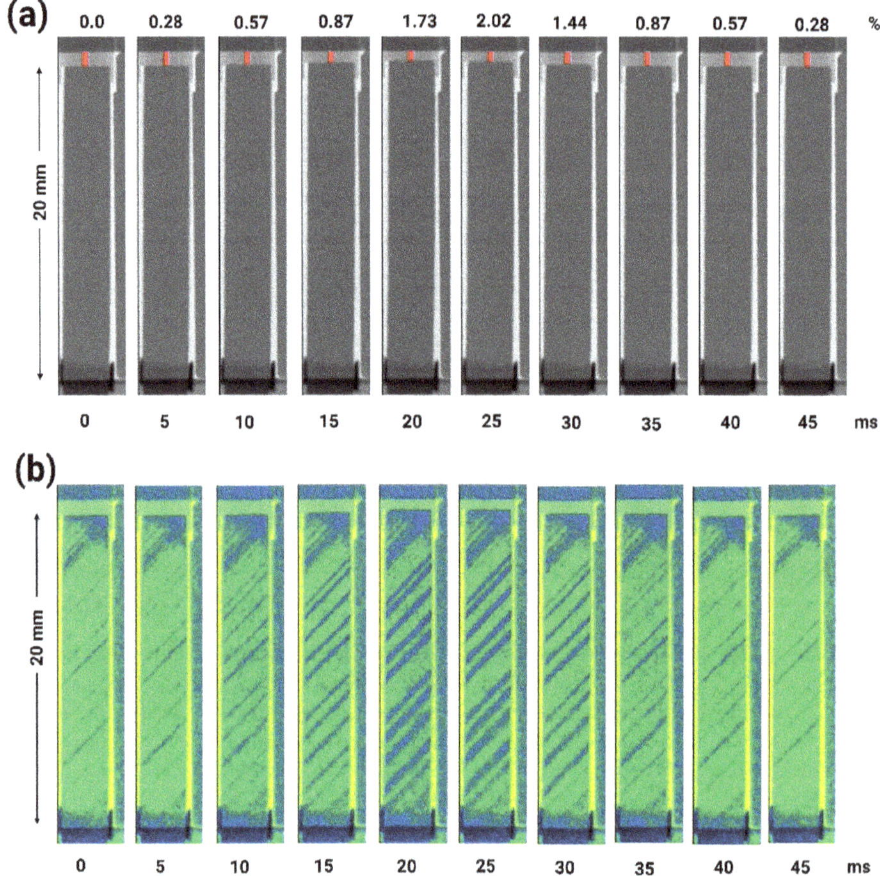

Fig. 5. Sequence of 10 radiograms in steps of 5 ms between 0 - 45 ms. (a) White beam images reveal elongation (strain) up to $\Delta l/l = 2\%$. (b) Plotted for a TOF range between 41.5 – 43.6 ms, images display evolution of two types of domains over a magnetic field cycle.

4   Conclusions

We have implemented asynchronous stroboscopic imaging on IMAT. Furthermore, we have demonstrated feasibility of stroboscopic Bragg edge imaging at a pulsed neutron source. The latter opens avenues to study microstructure transformations with millisecond time resolution. The stroboscopic image acquisition reported here was used for studying twin formation in a Ni_2MnGa magnetic shape memory alloy. Stroboscopic neutron imaging was, for the first time, applied to magnetic shape memory alloys, enabling visualization of strain and twin structures within a crystal. One can look at microstructure at short timescales while the achievable process resolution is limited by the time resolution of the pulsed source. While the MCP detector can achieve time resolutions as high as few µs, given by the accuracy to time-tag neutrons, the time resolution is limited by the widths of neutron pulses generated in the moderator of the neutron source.

On IMAT, neutron pulse widths in the Bragg dips range are approximately 0.4 ms [8]. On other beamlines with narrower pulse widths, higher temporal resolutions for stroboscopic measurements can be achieved. A source frequency of about 10 Hz is generally advantageous for asynchronous stroboscopic imaging, allowing for a time range of up to 100 ms.

The method is essentially a radiographic 4D technique (x-y pixels, TOF, magnetic field/strain phase) using a medium-flux neutron source. The technique can potentially be employed for imaging other cyclic processes, to track phase transitions in compounds and to achieve in-situ energy-resolved imaging of dynamic processes. The application range is limited to repetitive physical or chemical processes with external stimulus, however it exploits the wavelength selective capabilities on a pulsed source very well.

References

1. Schillinger, B., Brunner, J., Calzada, E.: A study of oil lubrication in a rotating engine using stroboscopic neutron imaging. Physica B **385–386**, 921–923 (2006)
2. Schillinger, B., et al.: Detection systems for short-time stroboscopic neutron imaging and measurements on a rotating engine. Nucl. Instr. Meth. A **542**, 142–147 (2005)
3. Schillinger, B.: Proposed asynchronous data collection scheme for stroboscopic neutron imaging on spallation sources for the examination of running engines. J. Neutron Res. **16**, 121–126 (2008)
4. Righi, L., Albertini, F., Pareti, L., Paoluzi, A., Calestani, G.: Commensurate and incommensurate 5M modulated crystal structures in Ni–Mn–Ga martensitic phases. Acta Mater. **55**, 5237–5245 (2007)
5. Río-López, N.R., et al.: Neutron scattering as a powerful tool to investigate magnetic shape memory alloys: a review. Metals **11**, 829 (2021)
6. Söderberg, O., et al.: Recent development of the magnetic shape memory materials research in Finland. Mater. Res. Soc. Symp. Proc. 1200-G03–0 (2010)
7. Kabra, S., Kelleher, J., Kockelmann, W., Gutmann, M.J., Tremsin, A.S.: Energy-dispersive neutron imaging and diffraction of magnetically driven twins in a Ni_2MnGa single crystal magnetic shape memory alloy. J. Phys. Conf. Ser. **746**, 012056 (2016)
8. Keen, D.A., Gutmann, M.J., Wilson, C.C.: SXD - the single-crystal diffractometer at the ISIS spallation neutron source. J. Appl. Cryst. **39**, 714–722 (2006)
9. Minniti, T., Watanabe, K., Burca, G., Pooley, D.E., Kockelmann, W.: Characterization of the new neutron imaging and materials science facility IMAT. Nucl. Instr. Meth. A **888**, 184–195 (2018)
10. Kockelmann, W., et al.: Time-of-flight neutron imaging on IMAT@ISIS: a new user facility for materials science. J. Imaging **4**, 47 (2018)
11. Tremsin, A.S., Vallerga, J.V., McPhate, J.B., Siegmund, O.H.W., Raffanti, R.: High resolution photon counting with MCP-Timepix quad parallel readout operating at > 1 KHz frame rates. IEEE Trans. Nucl. Sci. **60**, 578–585 (2013)
12. Schindelin, et al.: Fiji: an open-source platform for biological-image analysis. Nat. Methods **9**, 676–682 (2012)
13. Tygier, S., et al.: Tomographic reconstruction with Mantid Imaging. J. Phys. Conf. Ser. **2605**, 012017 (2023)
14. Sato, H., et al.: Inverse pole figure mapping of bulk crystalline grains in a polycrystalline steel plate by pulsed neutron Bragg-dip transmission imaging. J. Appl. Cryst. **50**, 1601–1610 (2017)

Preliminary Campaign of Bragg Edge Neutron Imaging Round Robin

Ranggi S. Ramadhan[1]([✉]) [iD], Winfried Kockelmann[1] [iD], Florencia Malamud[2] [iD], Matteo Busi[2] [iD], Saurabh Kabra[3] [iD], Takenao Shinohara[4] [iD], Thilo Pirling[5] [iD], and Anton Tremsin[6] [iD]

[1] STFC, Rutherford Appleton Laboratory, ISIS Facility, Harwell OX11 0QX, UK
ranggi.ramadhan@stfc.ac.uk
[2] Lab. of Neutron Scattering and Imaging, Paul Scherrer Institute, 5232 Villigen, Switzerland
[3] Oak Ridge National Laboratory, Oak Ridge, TN 37830, USA
[4] J-PARC Center, JAEA, 2-4 Shirakata, Tokai 319-1195, Ibaraki, Japan
[5] Institut Laue-Langevin, 71 Avenues Des Martyrs, 38000 Grenoble, France
[6] Space Science Laboratory Univ. of California Berkeley, Berkeley, CA 94720, USA

Abstract. This paper reports a preliminary campaign in preparation of a Bragg edge neutron imaging round robin. A series of measurements using a standard sample has been carried out at different facilities with Bragg edge imaging capabilities, with the aim of inter-laboratory comparison. The standard was also characterised at neutron diffraction beamlines to benchmark imaging results. Lessons learned from this campaign highlighted the importance of including instrument parameters in the data analysis to obtain accurate results. The need of careful interpretation of Bragg edge strain mapping result was highlighted by comparison with results of neutron diffraction experiments. Other practical lessons from the campaign are discussed to be taken into account for the future round robin activities.

Keywords: Bragg edge imaging · round robin · strain mapping

1 Introduction

In the recent years, there has been huge interest to expand the standard neutron computed tomography (CT) and to develop new kinds of tomographies, i.e., ones that visualize material properties rather than just features via greyvalues. One of the emerging techniques is Bragg edge neutron imaging, which is a form of diffraction imaging. The technique offers unique capabilities to provide 2-dimensional (2D) or 3-dimensional (3D) maps of materials information (e.g., residual strain, phase, and crystallographic information) in the interior of components, with a relatively simple setup and non-destructively. Taking advantage of these unique traits, the usefulness of Bragg edge imaging for material science and engineering applications has been shown several times for the 2D case [1–3] and a few times for the 3D case [4, 5] albeit with rather modest spatial resolution in the hundreds of microns. These studies motivate further development and establishment of Bragg edge imaging for material science at neutron facilities, for example at ISIS

© The Author(s) 2026

A. E. Craft and H. Z. Bilheux (Eds.): WCNR 2024, SPPHY 348, pp. 389–397, 2026.
https://doi.org/10.1007/978-3-032-15003-5_44

(UK), J-PARC (Japan), and ORNL (USA). A major push towards true "microstructure imaging" via Bragg edge analysis down to a few microns could be achieved on the ODIN instrument at the European Spallation Source (ESS), Sweden, which is currently under construction.

Owing to its recent development, however, many aspects pertaining to the technique need further improvement. While significant progress was achieved mostly independently at neutron facilities, and considering the small number of experimental facilities worldwide, a collaborative effort is required to take Bragg edge imaging from a proof-of-concept technique to a tool that is routinely and confidently used by academia and industries. This especially concerns the establishment and the worldwide standardisation of the method, i.e., the harmonisation of calibration and measurement protocols, data analysis algorithms, quantification of accuracy and reliability of the results, repeatability of measurements, and interchangeability between instruments.

Learning from past experiences of a neutron diffraction round robin [6], a Bragg edge imaging round robin project has been proposed to address these needs. The round robin activity refers to the measurement of a set of common/agreed benchmark samples at several neutron facilities with Bragg edge imaging capability aiming to include the major neutron imaging beamlines around the world. The objective of this paper is to report the results, i.e., comparison between instruments, and lessons learned from the preliminary campaign. The preliminary round robin campaign comprises the measurement of a round robin sample candidate at a small number of neutron imaging and diffraction facilities. Considering the wealth of information can be derived from Bragg edge data, the current scope focuses on strain analysis.

2 Materials and Methods

2.1 Materials Selection

Standard samples for Bragg edge imaging round robin need to satisfy certain requirements. The sample needs to be polycrystalline. Preferably, samples with high and low coherent cross sections and with well-separated and overlapping Bragg edges would be included. The region of interest of the sample should fit the field-of-view (FOV) of imaging detectors used on the different beamlines. The current limitations of the microchannel-plate (MCP) system [7] used at many spallation sources require a small FOV for the round. The sample is preferably homogeneous, both geometrically and microstructurally, in the beam direction, isotropic or weakly textured, and have an average grain size of smaller than 100 μm. For benchmarking strain analysis, the sample should ideally have a strain gradient exceeding 1000 $\mu\varepsilon$ within the FOV.

From the criteria above, one candidate for Bragg edge imaging round robin was selected (see Fig. 1). The specimen is an aluminium – silicon carbide metal-matrix composite (AlSiC MMC) coupons, with dimension of $35 \times 35 \times 15$ mm. Due to the water-quenching process during fabrication, the specimen exhibited parabolic through-thickness in-plane macroscopic residual stress (thus strain) variation [8]. Aluminum powder which constitutes the metal matrix composite was used as stress-free reference.

2.2 Round Robin Measurements

A total of five neutron instruments were involved in the preliminary round robin campaign: i) IMAT is a time-of-flight (TOF) imaging beamline [9], and ii) ENGIN-X is a neutron strain scanner [10], both at ISIS neutron source, UK. While primarily used as diffractometer, ENGIN-X can accommodate a TOF imaging detector, thus perform Bragg edge strain mapping experiment; iii) RADEN is a TOF imaging beamline at J-PARC, Japan [11]; iv) SALSA is a neutron strain scanner at the Institut Laue-Langevin, France [12]; v) POLDI is a neutron strain scanner at Paul Scherrer Institut, Switzerland [13]. On instruments with Bragg edge imaging capability, measurements were carried out using a microchannel plate (MCP) detector. Bragg edge imaging data were collected with schematics of the setup shown in Fig. 1(b). The matrix of the measurements are detailed in Table 1.

Table 1. Matrix of the measurements in the preliminary round robin campaign

Instrument	Technique	L/D (imaging)	GV (diffraction)	Diffraction measurement point (refer Fig. 1(c))
IMAT @ ISIS	Imaging	250	-	-
RADEN @ J-PARC	Imaging	298	-	-
ENGIN-X @ ISIS	Imaging	< 100	-	-
ENGIN-X @ ISIS	Diffraction	-	$2 \times 2 \times 10 \text{ mm}^3$	Line A3
SALSA @ ILL	Diffraction	-	$0.6 \times 0.6 \times 2 \text{ mm}^3$	Line A1-A4
POLDI @ PSI	Diffraction	-	$3.8 \times 3.8 \times 3.8 \text{ mm}^3$	Line A1-A4

2.3 Data Analysis

Figure 2(a) shows the Bragg edge and peak spectra of Al powder from IMAT, ENGIN-X and RADEN. The edge highlighted by the box is the Al(111) used for strain mapping analysis, and enlarged in Fig. 2(b). It can be seen that edge width on IMAT is larger compared to ENGIN-X and RADEN, as also indicated by resolution function in Fig. 2(c). That said, edge position determination relies more on the sharp rise of the edge rather than the decay on the tail of the edge, therefore strain information derived from the different beamlines should be comparable. The *TPX_edgefit* software carried out Bragg edge fitting using analytical function described in Eq. (1), derived from [14]. The fitted Bragg edge was averaged over a 15 × 15 macro-pixel, of which running average is applied across an image to generate maps of Bragg edge parameters. This macro-pixel averaging translates to a spatial resolution according to [2], and the effective spatial resolution for the strain maps is around 500 μm. Diffraction data from ENGIN-X was analysed using single-peak refinement consisting of the convolution of truncated exponential with a Voigt function, embedded in an *Open Genie – GSAS* based analysis suite. Diffraction data from SALSA was analysed using single-peak gaussian function on LAMP software. Diffraction data from POLDI was analysed using a single-peak refinement within *Mantid Workbench*. The Al(111). It was selected because this plane has low sensitivity to intergranular strain and produces the most intense Bragg edge among the recorded reflections. Strain was derived from the shift of Bragg edge position λ from the original "stress-free" reference position λ_0, described by Eq. (2)

$$T(\lambda) = C_1 + C_2\left[efrc\left(\frac{\lambda_t - \lambda}{\sqrt{2}\sigma}\right) - exp\left(\frac{\lambda_t - \lambda}{\tau} + \frac{\sigma^2}{2\tau^2}\right) \times efrc\left(\frac{\lambda_t - \lambda}{\sqrt{2}\sigma} + \frac{\sigma}{\sqrt{2}\tau}\right)\right]$$

(1)

$$\varepsilon = (\lambda - \lambda_0)/\lambda_0$$

(2)

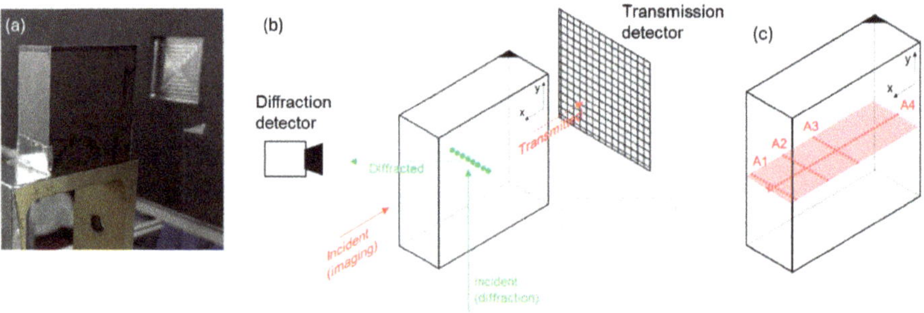

Fig. 1. (a) AlSiC metal-matrix composite; (b) Illustration of Bragg edge imaging measurements at IMAT, ENGIN-X, and RADEN.; (c) Lines of measurement points A1-A4 for neutron diffraction experiment on SALSA and POLDI.

Fig. 2.(a) Bragg edge and Bragg peak spectra of Al powder measured on IMAT, ENGIN-X and RADEN; (b) Comparison between Al(111) edge measured on the different beamlines; (c) Comparison between measured resolution function of IMAT, ENGIN-X and RADEN.

3   Results and Discussion

3.1   Importance of Instrument Resolution Function

The accuracy of strain analysis from Bragg edge imaging highly depends on the accuracy of estimating the Bragg edge position. This estimation is often made complicated by i) the low signal-to-noise ratio, especially for data averaged only from a few detector pixels, ii) the broadening of Bragg edge due to instrumental and crystallographic contributions from the sample. This necessitates fitting analytical functions, e.g., Eq. (1), to a measured Bragg edge to accurately determine its position, λ. As will be shown in this work, the fitted Bragg edge position is sensitive to other parameters including the τ and σ which represents the characteristic property of the moderator and sample broadening contribution, respectively. As the two parameters tied to physical quantities, it is important to carefully set constrains for these parameters.

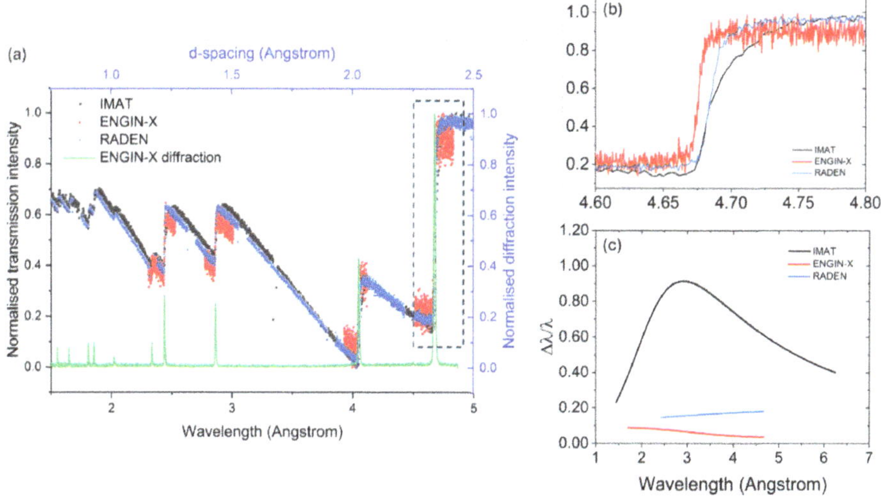

Fig. 2. (a) Bragg edge and Bragg peak spectra of Al powder measured on IMAT, ENGIN-X and RADEN; (b) Comparison between Al(111) edge measured on the different beamlines; (c) Comparison between measured resolution function of IMAT, ENGIN-X and RADEN.

Figure 3 shows the comparison of strain profiles produced on IMAT, ENGIN-X, and RADEN: a) when the fitting was performed without constraining the τ parameter, i.e., forcing best fit, while b) when the τ parameter was fixed according to the values obtained from fitting the Bragg edges of the d_0-powder. The powder has no crystallographic texture, thus the τ value should closely represent the property of instrument's moderator (and resolution function). It can be observed that fixing the τ parameter according to instrument's resolution function yields good agreement between the instruments, while not constraining it resulted in discrepancies between results although the general shape of the parabolic strain profile was still captured. The result highlights the importance of taking the instrument parameter for strain analysis using Bragg edge imaging. For instance, taking the derivative of a Bragg edge and fitting it with a simple peak-shape function (e.g., Gaussian) might not be the best method as it removes the instrument parameter from the consideration, not to mention introduce more noise.

The current campaign focused on benchmarking the instruments in terms of accuracy. In future round robin activities, experiments will be designed to allow a comparison of instruments performance, e.g., achievable spatial resolution within realistic exposure time.

3.2   Benchmarking and Interpretation of Bragg Edge Strain Map

Figure 4(a) shows the comparison between strain measured using Bragg edge imaging on IMAT, ENGIN-X, and RADEN, and using neutron diffraction on ENGIN-X, SALSA, and POLDI. Agreement between the two techniques is well within the uncertainty of the measurements. This is despite the fact results from the two techniques are averaged across different sample volumes.

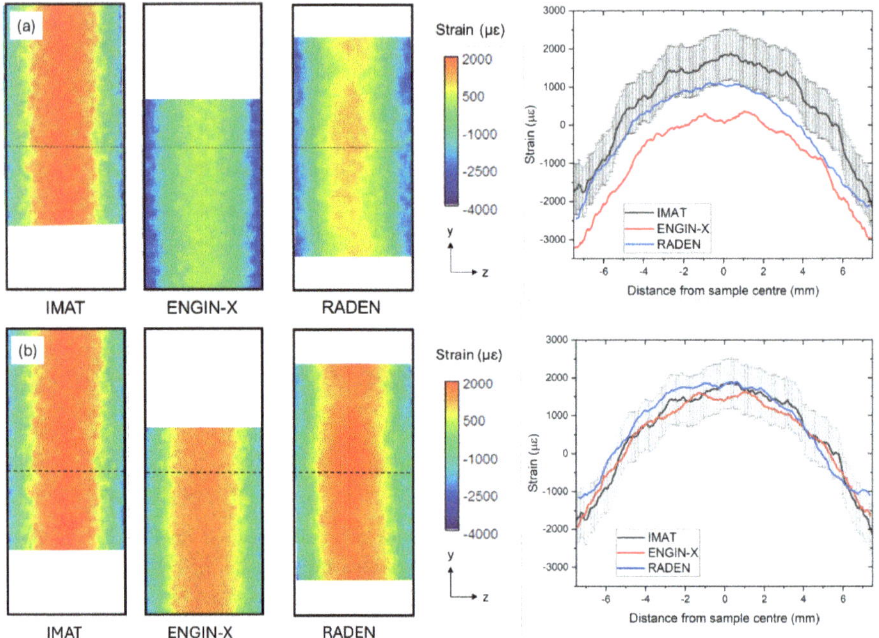

Fig. 3. Comparison of strain maps from IMAT, ENGIN-X and RADEN. The dotted line indicates the position for which strain profiles were extracted for comparison. (a) The fitting was done without constraining the τ parameter, i.e., forcing best fit. (b) The τ parameter was fixed according to the values obtained from fitting the Bragg edges of the d_0-powder. Error bars only shown for IMAT data for easier read.

Figure 4(b) shows the strains measured on SALSA and POLDI. It can be observed that the strain profile of line A1 is much flatter and closer to zero strain compared to that of lines A2 and A3. This is due to the line A1 being closer to free surface, where there is not enough material to constrain the residual stress and thus the value is close to zero. This is confirmed by the strain profiles of line A4, where strains are closer to zero near the surfaces, Fig. 4(c). This near-surface effect is not significant in many cases, and the through-thickness averaged values is often representative of the actual strain state in the bulk of the sample, as previously shown in Fig. 4(a). Nevertheless, this result serves as a reminder to carefully interpret Bragg edge strain maps as there is likely some degree of variation in the third direction that is being averaged.

4 Conclusion

The results from a preliminary campaign of Bragg edge imaging round robin has been presented. This exercise provided valuable lessons for Bragg edge analysis, especially for strain mapping. More importantly, the lessons learned serve as a springboard for the real round robin campaign. Key takeaways from this exercise are as follows:

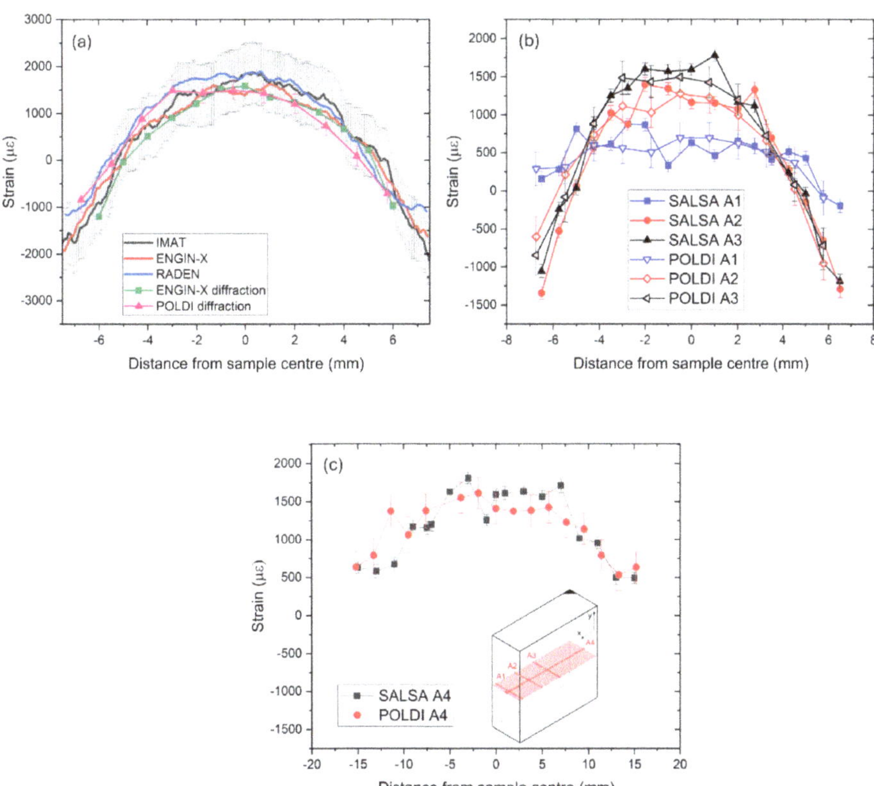

Fig. 4. (a) Comparison between strain profiles measured by Bragg edge imaging and neutron diffraction; (b) Strain profile of line A1, A2, and A3 measured at SALSA and POLDI. Strain profile is significantly lower for A1 which is close to free surface; (c) Strain profile of line A4 measured at SALSA and POLDI.

1. The importance of a round robin campaign for Bragg edge imaging was highlighted. A standard sample/ sample set is crucial to determine the accuracy of methods at the different instruments, which can be affected by a variety of factors ranging from instrument setup parameters to data analysis details. Standard samples are also useful to benchmark the performance of different instrument, e.g., practical/realistic spatial resolution of each instrument, required exposure time, etc.

2. The studied standard sample is a viable candidate; however, it has a number of limitations. It is an off-cut from a historical sample which is difficult to replicate, meaning there is a risk of losing this benchmark sample. FEA modelling is not available for the specimen. A dedicated sample for Bragg edge round robin will be designed specifically to address these issues.

3. Beyond inter-laboratory comparison, benchmarking of Bragg edge imaging results with other techniques are important. So far, comparison with a standardised neutron diffraction technique yielded excellent agreement. That said, it is important to make a comparison with techniques with different underlying principles.

Acknowledgments. The authors are grateful the provision of beam time on IMAT (RB1630006) and ENGIN-X (DOI: https://doi.org/10.5286/ISIS.E.87813178) at ISIS, on RADEN at J-PARC (2018A0115), on SALSA at ILL (DOI: https://doi.org/10.5291/ILL-DATA.1-02-315), and on POLDI at PSI.

Disclosure of Interests. The authors have no competing interests to declare that are relevant to the content of this article.

References

1. Woracek, R., et al.: Diffraction in neutron imaging—a review. Nucl. Instrum. Methods Phys. Res., Sect. A **878**, 141–158 (2018)
2. Ramadhan, R., et al.: Characterization and application of Bragg-edge transmission imaging for strain measurement and crystallographic analysis on the IMAT beamline. J. Appl. Crystallogr. **52**, 351–368 (2019)
3. Tremsin, A., Yau, T.Y., Kockelmann, W.: Non-destructive examination of loads in regular and self-Locking Spiralock® threads through energy-resolved neutron imaging. Strain **52**, 548–558 (2016)
4. Carminati, C., et al.: Bragg-edge attenuation spectra at voxel level from 4D wavelength-resolved neutron tomography. J. Appl. Crystallogr. **53**(Pt 1), 188–196 (2020)
5. Ziesche, R., et al.: 4D bragg edge tomography of directional ice templated graphite electrodes. J. Imaging **6**(12), 136 (2020)
6. Daymond, M., Johnson, M., Sivia, D.: Analysis of neutron diffraction strain measurement data from a round robin sample. J. Strain Anal. Eng. Des. **37**(1), 73–85 (2002)
7. Tremsin, A., et al.: High resolution photon counting with MCP-Timepix quad parallel readout operating at >1 KHz frame rates. IEEE Trans. Nucl. Sci. **60** (2013)
8. Fitzpatrick, M., Hutchings, M., Withers, P.: Separation of macroscopic, elastic mismatch and thermal expansion misfit stresses in metal matrix composite quenched plates from neutron diffraction measurements. Acta Materiallia **45**, 4867–4876 (1997)
9. Kockelmann, W., et al.: Wavelength-resolved neutron imaging on IMAT. Mater. Res. Proc. **15**, 29–34 (2020)
10. Santisteban, J.R., et al.: ENGIN-X: a third-generation neutron strain scanner. J. Appl. Crystallogr. **39**(6), 812–825 (2006)
11. Shinohara, T., et al.: The energy-resolved neutron imaging system. RADEN, Rev Sci Instrum. **91**(4), 043302 (2020)
12. Pirling, T., Bruno, G., Withers, P.: SALSA: advances in residual stress measurement at ILL. Mater. Res. Forum **524–525**, 217–222 (2006)
13. Stuhr, U., et al.: Time-of-flight diffraction with multiple frame overlap Part II: the strain scanner POLDI at PSI. Nucl. Instrum. Methods Phys. Res., Sect. A **545**, 330–338 (2005)
14. Tremsin, A., et al.: Energy-resolving neutron transmission radiography at the ISIS pulsed spallation source with a high-resolution neutron counting detector. IEEE Trans. Nucl. Sci. **56**, 2931–2937 (2009)

Author Index

A

Alvarado Alvarez, M.   55
Alvarez, Miguel A. Vicente   84
Aminzadeh, Alaleh   93
An, Hyeonjun   235
Anastasi, Nicholas   276
Anderson, Kyle D.   337
Armitage, Douglas P.   308
Asano, Hitoshi   180

B

Balafas, Matthew   347
Batha, S. H.   55
Bedynek, Matthew J.   355
Benard, P.   1
Benkechkache, M. A.   64
Bevitt, Joseph J.   93, 259
Bhattacharya, Pijush   276
Bilheux, Hassina Z.   206, 290, 308, 326, 337, 347, 355, 364, 372
Bilheux, Jean C.   326, 347, 355
Bilheux, Jean-Christophe   290, 308, 364
Bily, Tomas   109
Boillat, Pierre   40
Bouman, Charles A.   206, 337
Brambilla, Laura   140
Broughton, D. P.   55
Brown, Jeremy M. C.   93
Brügger, Adrian   372
Burgess, G.   380
Busi, Matteo   389
Butler, Leslie G.   364
Butler, Leslie G   299
Buzzard, Gregery T.   206, 337

C

Capps, Nathan   32
Carminati, A.   1
Cheon, Jiyong   235
Chiang, Cheng-i   93
Chong, Su-Ann   355

Chowdhury, Mohammad Samin Nur   206, 337
Christian, J.   64
Chuirazzi, William   168
Chulapakorn, Thawatchart   149
Číhal, Petr   40
Cocen, Ocson   140
Colldeweih, Aaron W.   32
Cool, Steven   168
Craft, Aaron   32, 119, 168
Crow, Lowell   290

D

Daemen, Luke L.   308
Di Bert, S.   1
Ding, Huan   299
Donahue Jr, Cornelius   355
Drezet, Jean-Marie   140
Durďáková, Tereza-Markéta   40
Dzramado, Eric   180

E

Erdem, O. F.   55

F

Finney, Charles E. A.   308
Freudenberg, Violet   290
Frost, Matthew J.   347
Fullagar, Wilfred K.   93

G

Gallagher, J.   64
Garbe, Ulf   93
Gofron, Kazimierz J.   355
Golduber, R.   64
Granget, Elodie   140
Griesche, Axel   252
Grosse, Mirco   158
Guo, Shengmin   299
Gutmann, M. J.   380
Guyotte, Greg   347, 355

© Idaho National Laboratory 2026
A. E. Craft and H. Z. Bilheux (Eds.): WCNR 2024, SPPHY 348, pp. 399–401, 2026.
https://doi.org/10.1007/978-3-032-15003-5

H

Hall, Stephen A.   149
Ham, Kyungmin   299, 364
Harayama, Isao   244, 268
Harrington, Mike   290
Heitkam, Sascha   47
Heyda, Jan   40
Hobbs, Roger   290, 364
Hoegberg, E.   64
Hong, H.   64
Huang, C.-K.   55
Huegle, Thomas   290

I

Iikura, Hiroshi   244
Imam, S.   64
Inukai, Junji   180
Ito, Daisuke   180, 216, 222, 244, 268
Iverson, Erik   290

J

Jamdee, Tawan   276
Jeong, Joonwoo   235
Jia, R.   1
Junghans, A.   55

K

Kabra, S.   380
Kabra, Saurabh   389
Kaestner, A. P.   1
Kaestner, A.   10
Kaestner, Anders   76, 140, 158
Kai, Tetsuya   244
Kamerman, David W.   32
Kanatzidis, M.   64
Kardjilov, Nikolay   101, 119, 149, 252
Kargar, A.   64
Khaplanov, Anton   355
Kingston, Andrew M.   93
Kis, Zoltan   168
Kiyanagi, Yoshiaki   216, 222
Knudson IV, Rob   355
Knuepfer, Leon   47
Kockelmann, W.   380

Kockelmann, Winfried   389
Koo, Youngtak   235
Krieg, M.   10
Kurita, Keisuke   244, 268
Kwon, Saerom   222

L

Lappan, Tobias   47
Lee, Jongmin   40
Lehmann, E.   10, 188
Lehmann, Eberhard H.   200
Lehmann, Eberhard   19, 25
Lin, Jiao   372
Lukosi, E.   64

M

Malamud, Florencia   84, 389
Manke, Ingo   149
Mannes, D.   10
Mannes, David   25, 76, 140
Markötter, Henning   149
Massey, Caleb   32
Matouskova, Jana   101
Matoušková, Jana   109
Matsubayashi, Masahito   244
Melčák, Martin   40
Mente, Tobias   252
Meyer, Michael   32
Miller, Stuart   276
Mills, Rebecca A.   308
Molaison, Jamie   355
Morooka, Satoshi   244
Murakawa, Hideki   180

N

Nagarkar, Vivek V.   276
Nakagawa, Katsufumi   180
Needham, Andrew W.   206
Nelson, Malachi   32
Nemati, Saber   299

O

Obayashi, Yuto   180

Ochiai, Kentaro   222
Odaira, Naoya   180, 216, 222, 244, 268
Ossler, Frederik   308

P
Paganin, David M.   93
Pandian, Lakshmi Soundara   276
Pearson, Matthew   347
Pirling, Thilo   389
Pont, Stefano Dal   252

Q
Qian, E.   64

R
Ramadhan, Ranggi S.   389
Reinovsky, R. E.   55
Ridley, Mackenzie   32
Roessger, Conrado   158
Rose, Paul   276
Ross, Ryan D.   337

S
Saheb, K.   64
Saito, Yasushi   180, 216, 222, 244, 268
Salvemini, Filomena   93
Santisteban, Javier R.   84
Santodonato, Louis J.   308
Sato, Satoshi   222
Schillinger, Burkhard   101, 109, 119, 129,
     168, 200
Schmidt, T. R.   55
Schreiber, Christopher   180
Šercl, Jonatan   40
Shakoorioskooie, M.   10
Shakoorioskooie, Mahdieh   19, 76, 140
Shinohara, Takenao   216, 222, 244, 389
Shirase, Yuto   180
Singh, Bipin   276
Sklenka, Lubomir   101
Skorpenske, Harley   347, 355
Sosa, Charles   276
Sponar, Stephan   129
Srivastava, Rohit   290
Steinbrueck, Martin   158
Stella, Valentina   84
Stringfellow, Erik   290, 364
Strobl, M.   10

Strobl, Markus   84
Sugimoto, Katsumi   180
Szentmiklosi, Laszlo   168

T
Tang, Shimin   206, 337, 347, 355
Tengattini, Alessandro   129, 252
Thurman, Zach   355
Torres, James   290, 299, 347, 364, 372
Tremsin, A. S.   380
Tremsin, Anton   299, 389
Trtik, P.   10
Trtik, Pavel   32, 40, 47, 235
Trunner, Clemens   129
Tsuchikawa, Yusuke   222, 244
Turnbull, Ian   290, 364
Tvrdý, Štěpán   40

V
van Loef, Edgar   276
Van Tran, Khanh   149
Velasco, Guillermo   276
Venkatakrishnan, Singanallur V.   206, 337
Venkatakrishnan, Singanallur   355
Vopička, Ondřej   40

W
Wakolo, Solomon   180
Walfort, B.   188
Walzel, Friedrich   47
Wang, Z.   55
Weick, Sarah   158
Wissink, Martin   317
Wolfe, B. T.   55
Wong, C.-S.   55
Woracek, Robin   149

Y
Yahne, Kevin   347
Yang, Diyu   206, 337
Yetik, Okan   32

Z
Zhan, Q.   10
Zhan, Qianru   76
Zhang, Chen   290, 347, 355
Zhang, Yuxuan   290, 308, 337, 347, 364
Ziauddin, Muhammad   47

GPSR Compliance
The European Union's (EU) General Product Safety Regulation (GPSR) is a set
of rules that requires consumer products to be safe and our obligations to
ensure this.

If you have any concerns about our products, you can contact us on

ProductSafety@springernature.com

In case Publisher is established outside the EU, the EU authorized
representative is:

Springer Nature Customer Service Center GmbH
Europaplatz 3
69115 Heidelberg, Germany